Lecture Notes in Computer Science

T0250685

Commenced Publication in 1973
Founding and Former Series Editors:
Gerhard Goos, Juris Hartmanis, and Jan van Leeuwen

Patrick Healy   Nikola S. Nikolov (Eds.)

Graph Drawing

13th International Symposium, GD 2005
Limerick, Ireland, September 12-14, 2005
Revised Papers

 Springer

Volume Editors

Patrick Healy
Nikola S. Nikolov
University of Limerick
CSIS Department
National Technological Park
Limerick, P.O. Box , Ireland
E-mail:{patrick.healy,nikola.nikolov}@ul.ie

Library of Congress Control Number: 2005938800

CR Subject Classification (1998): G.2, F.2, I.3, E.1

LNCS Sublibrary: SL 1 – Theoretical Computer Science and General Issues

ISSN   0302-9743
ISBN-10   3-540-31425-3 Springer Berlin Heidelberg New York
ISBN-13   978-3-540-31425-7 Springer Berlin Heidelberg New York

Springer is a part of Springer Science+Business Media

springer.com

© Springer-Verlag Berlin Heidelberg 2006
Printed in Germany

Typesetting: Camera-ready by author, data conversion by Scientific Publishing Services, Chennai, India
Printed on acid-free paper   SPIN: 11618058   06/3142   5 4 3 2 1 0

In Memoriam

Dedicated to the memory of Ondrej Sýkora.

Preface

The 13th International Symposium on Graph Drawing (GD 2005) was held in Limerick, Ireland, September 12-14, 2005. One hundred and fifteen participants from 19 countries attended GD 2005.

In response to the call for papers the Program Committee received 101 submissions, each detailing original research or a system demonstration. Each submission was reviewed by at least three Program Committee members; each referee's comments were returned to the authors. Following extensive discussions, the committee accepted 38 long papers, 3 short papers and 3 long system demos, each of which were presented during one of the conference's 12 sessions. Eight posters were also accepted and were on display throughout the conference.

Two invited speakers, Kurt Mehlhorn and George Robertson, gave fascinating talks during the conference. Prof. Mehlhorn spoke on the use of minimum cycle bases for reconstructing surfaces, while Dr. Robertson gave a perspective, past and present, on the visualization of hierarchies.

As is now traditional, a graph drawing contest was held during the conference. The accompanying report, written by Stephen Kobourov, details this year's contest. This year a day-long workshop, organized by Seok-Hee Hong and Dorothea Wagner, was held in conjunction with the conference. A report on the "Workshop on Network Analysis and Visualization," written by Seok-Hee Hong, is included in the proceedings.

We are indebted to many people for the success of the conference. The Program Committee and external referees worked diligently to select only the best of the submitted papers. The Organizing Committee under the co-chairmanship of Nikola Nikolov worked tirelessly in the months leading up to the conference. In particular, a big debt is owed to Aaron Quigley for his Herculean fund-raising efforts, to Alex Tarassov for his system maintenance, to Karol Lynch for his web page development, and to Gemma Swift and Nuala Kitson for their administrative support and constant good humor. Thanks are also due to Vincent Cunnane, who opened the conference. Last, but not least, we thank Peter Eades, who provided valuable direction and kept a steady head throughout.

The conference received assistance from Science Foundation Ireland (Benefactor); Intel Corp., Microsoft Corp. and Tom Sawyer Software (Gold Sponsors); National ICT Australia, Enterprise Ireland, Fáilte Ireland, ILOG Inc., AbsInt Angewandte Informatik GmbH (Silver Sponsors); Lucent Technologies, Jameson Irish Whiskey and Dell Inc.

The 14th International Symposium on Graph Drawing (GD 2006) will be held September 18-20, 2006 in Karlsruhe, Germany, co-chaired by Michael Kaufmann and Dorothea Wagner.

October 2005

Patrick Healy
Nikola S. Nikolov
Limerick

Organization

Steering Committee

Franz-J. Brandenburg	Universität Passau
Giuseppe Di Battista	Università degli Studi Roma
Peter Eades	National ICT Australia Ltd., Univ. of Sydney
Hubert de Fraysseix	Centre d'Analyse et de Mathematique Sociale
Patrick Healy	University of Limerick
Michael Kaufmann	University of Tübingen
Takao Nishizeki	Tohoku University
Janos Pach	City College and Courant Institute, New York
Pierre Rosenstiehl	Centre National de la Recherche Scientifique
Roberto Tamassia	Brown University
Ioannis (Yanni) G. Tollis	University of Crete
Dorothea Wagner	Universität Karlsruhe
Sue Whitesides	McGill University

Program Committee

Ulrik Brandes	Universität Konstanz
Giuseppe Di Battista	Università degli Studi Roma
Peter Eades	NICTA, University of Sydney (*Co-chair*)
Jean-Daniel Fekete	INRIA, Paris
Emden Gansner	AT&T Labs
Patrick Healy	University of Limerick (*Co-chair*)
Seok-Hee Hong	NICTA, University of Sydney
Michael Kaufmann	Universität Tübingen
Jan Kratochvil	Charles University
Giuseppe Liotta	Università degli Studi di Perugia
Kim Marriott	Monash University
Patrice de Mendez	Centre National de la Recherche Scientifique
Petra Mutzel	Universität Dortmund
János Pach	City College and Courant Institute
Helen Purchase	University of Glasgow
Md. Saidur Rahman	BUET
Ben Shneiderman	University of Maryland
Ondrej Sýkora (R.I.P.)	Loughborough University
Sue Whitesides	McGill University
Steve Wismath	University of Lethbridge
David Wood	Universitat Politècnica de Catalunya

Organizing Committee

Patrick Healy   University of Limerick (*Co-chair*)
Stephen Kobourov   University of Arizona
Karol Lynch   University of Limerick
Joseph Manning   University College Cork
Nikola S. Nikolov   University of Limerick (*Co-chair*)
Aaron Quigley   University College Dublin (*Treasury Chair*)
Gemma Swift   University of Limerick
Alexandre Tarassov   University of Limerick

Contest Committee

Christian Duncan   University of Miami
Stephen Kobourov   University of Arizona (*Chair*)
Dorothea Wagner   Universität Karlsruhe

External Referees

Greg Aloupis   Markus Geyer   Christian Pich
Radoslav Andreev   Carsten Gutwenger   Emmanuel Pietriga
Christian Bachmaier   Stefan Hachul   Maurizio Pizzonia
Therese Biedl   Martin Harrigan   Catherine Plaisant
Manuel Bodirsky   Hongmei He   Rados Radoicic
Nicolas Bonichon   Nathaline Henry   Aimal Tariq Rextin
Prosenjit Bose   Petr Hliněný   Bruce Richter
Christoph Buchheim   Martin Hoefer   Adrian Rusu
Markus Chimani   David Kirkpatrick   Georg Sander
Robert Cimikowski   Karsten Klein   Thomas Schank
Pier Francesco Cortese   Yehuda Koren   Barbara Schlieper
Jurek Czyzowicz   Katharina Lehmann   Karl-Heinz Schmitt
Walter Didimo   Jürgen Lerner   Martin Siebenhaller
Emilio Di Giacomo   Karol Lynch   Matthew Suderman
Vida Dujmović   Jiří Matoušek   Laszlo Szekely
Adrian Dumitrescu   Sascha Meinert   Gabor Tardos
Zdeněk Dvořák   Bernd Meyer   Geza Toth
Tim Dwyer   Kazuyuki Miura   Imrich Vrťo
Daniel Fleischer   Pat Morin   Michael Wybrow
Michael Forster   Maurizio Patrignani
Hubert de Fraysseix   Merijam Percan

Sponsoring Institutions

Table of Contents

Papers

Software Demonstrations

Posters

Workshop on Network Analysis and Visualisation

Graph Drawing Contest

Invited Talks

Crossings and Permutations[*]

Therese Biedl[1], Franz J. Brandenburg[2], and Xiaotie Deng[3]

[1] School of Computer Science, University of Waterloo, ON N2L3G1, Canada
biedl@uwaterloo.ca
[2] Lehrstuhl für Informatik, Universität Passau, 94030 Passau, Germany
brandenb@informatik.uni-passau.de
[3] Department of Computer Science, City University of Hong Kong,
83 Tat Chee Avenue, Kowloon Tong, Hong Kong, SAR, China
csdeng@cityu.edu.hk

Abstract. We investigate crossing minimization problems for a set of permutations, where a crossing expresses a disarrangement between elements. The goal is a common permutation π^* which minimizes the number of crossings. This is known as the Kemeny optimal aggregation problem minimizing the Kendall-τ distance. Recent interest into this problem comes from application to meta-search and spam reduction on the Web.

This rank aggregation problem can be phrased as a one-sided two-layer crossing minimization problem for an edge coloured bipartite graph, where crossings are counted only for monochromatic edges.

Here we introduce the max version of the crossing minimization problem, which attempts to minimize the discrimination against any permutation. We show the NP-hardness of the common and the max version for $k \geq 4$ permutations (and k even), and establish a 2-2/k and a 2-approximation, respectively. For two permutations crossing minimization is solved by inspecting the drawings, whereas it remains open for three permutations.

1 Introduction

One-sided crossing minimization is a major component in the Sugiyama algorithm. The one-sided crossing minimization problem has gained much interest and is one of the most intensively studied problems in graph drawing [8, 15]. For general graphs the crossing minimization problem is known to be NP-hard [13]. The NP-hardness also holds for bipartite graphs where the upper layer is fixed, and the graphs are dense with about $n_1 n_2 / 3$ crossings [10], or alternatively, the graphs are sparse with degree at least four on the free layer [17]. The special case with degree 2 vertices on the free layer is solvable in linear time, whereas the degree 3 case is open.

The rank aggregation problem finds a consensus ranking on a set of alternatives, based on preferences of individual voters. The roots for a mathematical

---

[*] The work of the first author was supported by NSERC, and done while the author was visiting Universität Passau. The work of the second and third authors was partially supported by a grant from the German Academic Exchange Service (Project D/0506978) and from the Research Grant Council of the Hong Kong Joint Research Scheme (Project No. G_HK008/04).

P. Healy and N.S. Nikolov (Eds.): GD 2005, LNCS 3843, pp. 1–12, 2005.
© Springer-Verlag Berlin Heidelberg 2005

investigation of the problem lie in voting theory and go back to Borda (1781) and Condorcet (1785). Rank aggregations occur in many contexts, including sport, voting, business, and most recently, the Internet. *"Who is the winner?"* In gymnastics, figure skating or dancing this is decided by averaging or ranking the points of the judges. In Formula 1 racing and similarly at the annual European Song Contest the winner is who has the most points. Is this scheme fair? Why not deciding the winner by the majority of first places?

Also, the organizers of GD2005 are confronted with our crossing minimization problem. They have to make many decisions. For example, which beer (wine, food) shall be served at the GD conference dinner? What is the best choice for the individual taste of the participants? Or, more specific: *which beer is the best?*

In their seminal paper from the WWW10 conference, Dwork et al. [9] have used rank aggregation methods for web searching and spam reduction. A search engine is called *good* if it behaves close to the aggregate ranking of several search engines. Besides experimental results they have investigated the theoretical foundations of the rank aggregation problem. One of the main results is the NP-hardness of computing a so-called Kemeny optimal permutation of just four permutations, here called PCM-4. However, the given proof has some flaws, and is repaired here. In addition, we show a relationship to the feedback arc set problem and establish a $2-2/k$ approximation, which is achieved by the best input permutation.

The common rank aggregation methods take the sum of all disagreements over all permutations. Here we introduce the *maximum version*, PCM_{max}-k, which expresses a fair aggregation and attempts to avoid a too severe discrimination of any participant or permutation. With the optimal solution, nobody should be totally unhappy. We show the NP-hardness of PCM_{max}-k for all $k \geq 4$ and establish a 2-approximation, which is achieved by any input permutation. This parallels similar results for the Kemeny aggregation problem [1, 9] and for the Coherence aggregation problem [5]. The case PCM_{max}-2 with two permutations is efficiently solvable, whereas the case $k = 3$ remains open.

Besides the specific results, this work aims to bridge the gap between the combinatorics of rank aggregations and crossing minimizations in graph drawing, with a mutual exchange of notions, insights, and results.

In Section 2 we introduce the basic notions from graph drawing and rank aggregations, and show how to draw rank aggregations. In Section 3 we state the NP-hardness of the crossing minimization problems for just four permutations, and prove the approximation results, and in Section 4 we investigate the special cases with two and three permutations.

2   Preliminaries

Given a set of alternatives U, a *ranking* π with respect to U is an ordering of a subset S of U such that $\pi = (x_1, x_2, \ldots, x_r)$ with $x_i > x_{i+1}$, if x_i is ranked higher than x_{i+1} for some total order $>$ on U.

For convenience, we assign unique integers to the items of U and let $U = \{1, \ldots, n\}$. We call π a *(full) permutation*, if $S = U$, and a *partial permutation*, if $S \subseteq U$. A permutation is represented by an ordered list of items, where the rank of an item is given by its position in the ordered list, with the highest, most significant, or best item in first place.

The rank aggregation or the *crossings of permutations problem* is to combine several rankings $\pi_1, \ldots, \pi_k$ on U, in order to obtain a common ranking π^*, which can be regarded as the compromise between the rankings. The goal is the best possible common ranking, where the notion of 'better' depends on the objective. It is formally expressed as a cost measure or a penalty between the π_i and π^*; the *common* version takes the sum of the penalties, the *max* version is introduced here. Several of these criteria have a correspondence in graph drawing.

A prominent and frequently studied criterion is the Kendall-τ distance [3, 5, 9, 16]. The *Kendall-τ distance* of two permutations over $U = \{1, \ldots, n\}$ measures the number of pairwise disagreements or inversions, $K(\pi, \tau) = |\{(u, v) \,|\, \pi(u) < \pi(v) \text{ and } \tau(u) > \tau(v)\}|$. This value is invariant under renaming, or the application of a permutation σ on both π and τ, and such that τ becomes the identity. For a set of permutations $P = \{\pi_1, \ldots, \pi_k\}$ this generalizes by collecting all disagreements, $K(P, \pi^*) = \sum_{i=1}^{k} K(\pi_i, \pi^*)$.

The value $K(P, \pi^*)$ can be expressed in various ways. For every pair of distinct items (u, v), the *agreement* $A_P(u, v)$ is the number of permutations from P which rank u higher than v, and the *disagreement* is $D_P(u, v) = k - A_P(u, v)$. Clearly, the agreement on (u, v) equals the disagreement on the reverse ordering (v, u). For every (unordered) pair of items, let $\Delta(u, v) = |k - 2A_P(u, v)|$ express the difference between the agreement and the disagreement of u and v.

There is an established lower bound for the number of unavoidable crossings for the permutations of P, which is the sum over the least of the agreements and disagreements,

$$LB(P) = \sum_{u < v} \min\{A_P(u, v), D_P(u, v)\}.$$

Then the disagreement against a common permutation π^* is

$$K(P, \pi^*) = LB(P) + \sum_{\pi^*(u) < \pi^*(v) \text{ and } D_P(u,v) > A_P(u,v)} \Delta(u, v).$$

Thus $\Delta(u, v)$ is added as a penalty if π^* disagrees with the majority of the permutations. If there is a tie for the ranking of u and v in P, then just the term from the lower bound is taken into account.

Recall that for the crossing minimization problem of two layered graphs the agreement and disagreement of two free vertices u and v is the crossing number of the edges incident with u and v and placing u left of v, or vice versa. The so obtained lower bound is often 'good' and close to the optimum value [14].

Another popular measure for the distance between permutations is the *Spearman footrule distance*, which accumulates the linear arrangement or the length between two permutations over $\{1, \ldots, n\}$ by $f(\pi, \tau) = \sum_i |\pi(i) - \tau(i)|$. Again this extends to a set P of permutations by summation $f(P, \pi^*) = \sum_{j=1}^{k} f(\pi_i, \pi^*)$.

These measures can be scaled by individual weights, and they can be extended to partial permutations $\pi_1, \ldots, \pi_k$, where each permutation operates on its subset of the universe, see [9].

Given a set of (full or partial) permutations $P = \{\pi_1, \ldots, \pi_k\}$ on a universe $U = \{1, \ldots, n\}$, the *crossing number* of P is the number of crossings against the best permutation π^* with respect to the Kendall-τ-distance, i.e., $CR(P) = \min_{\pi^*} K(P, \pi^*)$. The crossing minimization problem is finding such a permutation π^*. We will refer to the crossing minimization problem of k permutations as the PCM-k problem.

A new cost measure is the *max crossing number*, which attempts to minimize the number of crossings for any permutation. For a set of k permutations P and a target permutation π^* let $K_{max}(P, \pi^*) = \max\{K(\pi_i, \pi^*) \mid \pi_i \in P\}$ and define the max crossing number of P by $CR_{max}(P) = \min_{\pi^*} K_{max}(P, \pi^*)$. The permutation π^* giving the value $CR_{max}(P)$ is a solution to the max crossing minimization problem. This problem is referred to as the PCM_{max}-k problem. One could similarly consider a maximum version for the Spearman footrule distance; we have not investigated the latter further.

The following fact is readily seen.

Lemma 1. *For a set of k permutations $P = \{\pi_1, \ldots, \pi_k\}$,*

$$CR_{max}(P) \le CR(P) \le k \cdot CR_{max}(P).$$

The crossing number represents an aggregation, which is the best compromise for the given lists of preferences and minimizes the number of disagreements. The minimal number of crossings does not necessarily distribute them uniformly among the given permutations; one can construct examples where $CR_{max}(P) \ge \lceil CR(P)/2 \rceil$ and not $CR_{max}(P) = \lceil CR(P)/k \rceil$ as one would hope. The latter equation holds for $k = 2$. The objective behind the max crossing number is an aggregation, which is fair and treats every permutation equally well and minimizes the discrimination of each participant. Clearly, both objectives can be combined to the best possible permutation π^* which minimizes the sum of crossings and then balances their distribution.

2.1   Drawing Permutations

We now translate rank aggregations to graph drawing. Two permutations π and τ on a universe $U = \{1, \ldots, n\}$ are drawn as a two-layer bipartite graph with the vertices $1, \ldots, n$ on each layer in the order given by π and τ and a straight-line edge between the two occurrences of each item v on the two layers.

A set of k permutations $\pi_1, \ldots, \pi_k$ and a common permutation π^* are represented by a sequence of pairs of permutations, where the lower layer is fixed in all drawings. For convenience, we let the lower layer be the identity with $\pi^*(i) = i$. We can merge the permutations into the *coloured permutation graph* G, which is a bipartite graph with k edge colours, such that there are vertices $1, \ldots, n$ on each layer. There is an edge in the i-th colour between u on the upper layer and j on the lower layer if and only if $\pi_i(u) = j$. See also Fig. 1.

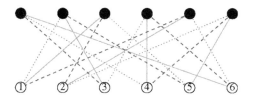

Fig. 1. Coloured permutation graph for $\pi_1 = (6,3,1,4,2,5)$ (green and solid), $\pi_2 = (3,5,2,6,1,4)$ (blue and dashed), and $\pi_3 = (4,1,5,3,6,2)$ (red and dotted)

Obviously, for two full or partial permutations π and τ, the Kendall-τ distance $K(\pi, \pi^*)$ is the number of edge crossings in a straight-line drawing of their bipartite graph. It ranges between 0 and $n(n-1)/2$ and can be efficiently computed either by accumulating for every i the number of items, which are greater than i and occur to the left of i in π, provided π^* is the identity, or by techniques from counting crossings in two-layer graphs in [21].

Lemma 2. *The Kendall-τ distance $K(\pi, \pi^*)$ of two permutations over $U = \{1, \ldots, n\}$ can be computed in $O(n \log n)$ time.*

2.2   Penalty Graphs

There is a direct relationship between the crossing minimization problem and the *feedback arc set problem*, which has been established at several places. Recall that the feedback arc set problem is finding the least number of arcs F in a directed graph $G = (V, E)$, such that every directed cycle contains at least one arc from F, i.e., the graph $G' = (V, E - F)$ is acyclic. In the more general weighted case, the objective is a set of arcs with least weight. In the two-layer crossing minimization problem, the *penalty graph* has arcs with weights corresponding to the difference between the number of crossings among the edges incident with two vertices u and v, if u is placed left of v, or vice versa.

In their seminal paper, Sugiyama et al. [20] have introduced the penalty digraph for the two-layer crossing minimization problem, and in [2] it is used for voting tournaments. Demetrescu and Finocchi [6] have used this approach for the two-sided crossing minimization problem and have tested several heuristics. Recently, Ailon et al. [1] have established improved randomized approximations for aggregation and feedback arc set problems.

For the crossing minimization problem for permutations, the penalty graph can be applied in the same spirit, but we use the difference in the majority counts $\Delta(u, v)$ as edge weights. Thus, for a set of permutations P over $\{1, \ldots, n\}$ the *penalty digraph* of P is a weighted directed graph $H = (V, A, w)$ with a vertex for each item u and an arc (u, v) with weight $\Delta(u, v)$ if and only if a strict majority of permutations rank u higher than v, i.e., if $(u - v) \cdot (D_P(u, v) - A_P(u, v)) < 0$. Let $w(FAS(P))$ denote the *weight* of the optimum feedback arc set in the penalty digraph.

First, we establish the connection between the crossing number and the feedback arc set of the penalty graph. For the two-layer crossing minimization problem it was first observed by Sugiyama [20], and used in various places, [1, 6, 10, 17]. As a consequence, the crossing minimization problem can be reduced to a feedback arc set problem.

Theorem 1. *Let $P = \{\pi_1, \ldots, \pi_k\}$ be a set of permutations. Then the crossing number of P equals the lower bound plus the weight of the feedback arc set*

$$CR(P) = LB(P) + w(FAS(P)).$$

Proof. For any permutation π there are $LB(P)$ unavoidable inversions or crossings and $K(P, \pi) = LB(P) + \sum_{\pi(u)<\pi(v) \text{ and } D_P(u,v)>A_P(u,v)} \Delta(u, v)$. Now, the deletion of all arcs (u, v) with $u < v$ and $\pi(u) > \pi(v)$ from the penalty digraph of P leaves an acyclic digraph, since there are no cycles in a single permutation π. If π is such that $K(P, \pi)$ is minimal, then the set of arcs removed from the penalty graph is a feedback arc set.

Conversely, consider the penalty graph of P and remove any set of arcs F to make the remainder acyclic. Consider any permutation π which is in conformity with a topological ordering. Then $K(P, \pi) \leq LB(P) + \sum_{f \in F} \Delta(f)$, and if F is such that its weight is $w(FAS(P))$, then π is such that $K(P, \pi)$ is minimal.

3   Complexity of Optimal Permutations

In this section we study the complexity of finding an optimal permutation for the common and the max crossing numbers. There are strong similarities to the one-sided crossing minimization problem, which go through to the number of permutations and the degrees of the free vertices.

Crossing minimization in graphs is NP-hard. This holds true for general graphs [13], and even for two-layer graphs with the upper layer fixed. These graphs may be dense [10] or sparse with degree $k = 4$ for the vertices on the free layer [17]. The case of degree 3-graphs for the free layer is still open.

Correspondingly, there are NP-hardness results for permutations. For many partial permutations with just two elements the crossing minimization problem is in one-to-one correspondence with the feedback arc set problem, where every two element permutation represents an arc, and thus is NP-hard [11, 12]. By a different reduction from the feedback arc set problem, Bartholdi et al. [3] have proved the NP-hardness of Kemeny optimal permutations for many permutations. In [2] the first NP-hardness proof is credited to Orlin (1981, unpublished manuscript).

A major strengthening has been claimed by Dwork [9] with a reduction from the feedback arc set problem to just four permutations. However, the construction in [9] has some flaws and needs some minor corrections.

Theorem 2. *The (common) crossing minimization problem PCM-k is NP-hard for k full permutations, where $k \geq 4$ and k even.*

Proof. (Sketch). We follow the construction in [9] and reduce from the feedback arc set problem. We only explain the case $k = 4$ here; for $k \geq 6$ use the technique of [9]. Let $G = (V, E)$ be a directed graph with $|V| = \{v_1, \ldots, v_n\}$ and $|E| = m$ in which we want to find the smallest feedback arc set. For every vertex v let $out(v)$ be the sequence of outgoing edges in any order, and let $in(v)$ denote the sequence of incoming edges. Finally, for a sequence x let x^r denote its reversal, reading the elements right-to-left. Now, construct two pairs of permutations from the vertices and edges of G.

$$\pi_1 = v_1, out(v_1), v_2, out(v_2), \ldots, v_n, out(v_n),$$
$$\pi_2 = v_n, out(v_n)^r, \ldots, v_2, out(v_2)^r, v_1, out(v_1)^r,$$
$$\pi_3 = in(v_1), v_1, in(v_2), v_2, \ldots, in(v_n), v_n, \text{ and}$$
$$\pi_4 = in(v_n)^r, v_n, \ldots, in(v_2)^r, v_2, in(v_1)^r, v_1.$$

In [9] the incoming edges are listed to the right of their vertices in π_3 and π_4, but then the construction does not work.

Let $K' = 2\binom{n}{2} + 2\binom{m}{2} + 2m(n-1)$. The claim is now that G has a feedback set of size at most f iff $CR(P) \leq K = K' + 2f$. Dwork et al. [9] use a different value for K. We omit the (straightforward) proof of this claim for space reasons.

For the common crossing minimization problem we sum the number of crossings of monochrome edges. In the max problem we wish to minimize the maximal number of such crossings, i.e., we wish to treat every arrangement as fair as possible.

Theorem 3. *The max crossing minimization problem* $\mathrm{PCM_{max}}$-k *is NP-hard for any $k \geq 4$ (full or partial) permutations.*

Proof. (Sketch) Consider the permutations $\pi_1, \ldots, \pi_4$ from Theorem 2, and construct four new permutations over four copies of pairwise disjoint elements, namely

$$\sigma_1 = \pi_1 \cdot \pi_2 \cdot \pi_3 \cdot \pi_4, \quad \sigma_2 = \pi_2 \cdot \pi_3 \cdot \pi_4 \cdot \pi_1, \quad \sigma_3 = \pi_3 \cdot \pi_4 \cdot \pi_1 \cdot \pi_2, \quad \sigma_4 = \pi_4 \cdot \pi_1 \cdot \pi_2 \cdot \pi_3.$$

One can show that the permutation that minimizes the maximal number of crossings to $\sigma_1, \ldots, \sigma_4$ solves again the feedback arc problem.

3.1   Approximation Algorithms

Since the crossing minimization problems are NP-hard for any (even) $k \geq 4$, we cannot hope to find the best solution in polynomial time, and hence study other ways to attack the problem. One easy way is to use integer programming; the problem can be formulated, in a relative straightforward way (we omit details) as a 0/1 program with $O(n^4 + k)$ variables and constraints. Another way is to consider approximation algorithms, which we study next.

There is a close connection between the number of crossings, i.e., the Kendall-τ distance and the Spearman-footrule distance, as established in [7]. For a pair of permutations, every move induces a disarrangement and each crossings implies that at most two elements must move each by one position. Hence, $K(\pi, \tau) \leq f(\pi, \tau) \leq 2K(\pi, \tau)$ for full permutations π and τ. The optimal permutation for the Spearman-footrule distance can be computed by solving a weighted perfect bipartite matching problem, as explained in [9].

An alternative 2-approximation is obtained by choosing the best among the given permutations, see [1], and there is a simple 2-approximation for the coherence complexity [5]. We now show that the technique of choosing the best among the given permutations in fact gives an even better approximation, in particular for small values of k.

Theorem 4. *There is a $(2 - \frac{2}{k})$-approximation for the (common) crossing minimization problem* PCM-k.

Proof. Let $P = \pi_1, \ldots, \pi_k$ be the input permutations. For $a > d$ and $a + d = k$, let $E_{a,d}$ be those arcs $u \rightarrow v$ for which $A_P(u, v) = a$ and $D_P(u, v) = d$, i.e., u comes before v in a permutations, and after v in d permutations. Denote $m_{a,d} = |E_{a,d}|$.

Consider the k vertex orderings defined by the k permutations, and count the number of arcs that are reversed in them. For $a > d$, each arc in $E_{a,d}$ must be reversed in exactly d of the permutations, hence the total number of reversed arcs is

$$L = m_{k-1,1} + 2m_{k-2,2} + \cdots + jm_{k-j,j} + \ldots^1 = \sum_{a>d, a+d=k} dm_{a,d}. \qquad (1)$$

By the pigeon hole principle, therefore in at least one of the permutations (say in π_1), the number of reversed arcs is at most $1/k$th of Equation 1. Denote by $r_{a,d}$ the number of arcs in $E_{a,d}$ that are reversed in π_1, then we therefore have

$$r_{k-1,1} + r_{k-2,2} + \cdots + r_{k-j,j} + \cdots \leq \frac{1}{k}(m_{k-1,1} + 2m_{k-2,2} + \cdots + jm_{k-j,j} + \ldots)$$

Each arc in $E_{a,d}$ has weight $a - d$ in the feedback arc set problem, so the weight of the feedback arc set solution defined by π_1 is

$$\begin{aligned} w(FAS) &= (k-2)r_{k-1,1} + (k-4)r_{k-2,2} + \cdots + (k-2j)r_{k-j,j} + \cdots \\ &\leq (k-2)r_{k-1,1} + (k-2)r_{k-2,2} + \cdots + (k-2)r_{k-j,j} + \cdots \\ &\leq (k-2)\frac{1}{k}(m_{k-1,1} + 2m_{k-2,2} + \cdots + jm_{k-j,j} + \ldots) = \frac{k-2}{k}L \end{aligned}$$

Now note that L of Equation 1 also exactly equals the lower bound $LB(P)$, since we only consider edges in $E_{a,d}$ with $a > d$. Therefore, the number of crossings obtained with π_1 is

$$LB(P) + w(FAS) \leq L + \frac{k-2}{k}L = (2 - \frac{2}{k})L \leq (2 - \frac{2}{k})OPT,$$

where OPT is the number of crossings in the optimal solution.

---

[1] The series ends for $j = \lceil (k-1)/2 \rceil$, but in order not to clutter the equations, we will not write this explicitly here.

We note here that if the target permutation is taken from the given set of permutations, the $(2 - \frac{2}{k})$-approximation is best possible for PCM-k. Namely, let $\sigma_1, \ldots, \sigma_k$ be k permutations (over distinct elements) of length $N = n/k$, and consider the following k permutations:

$$\pi_1 = \sigma_1^r \cdot \sigma_2 \cdot \sigma_3 \cdots \cdots \sigma_k$$
$$\pi_2 = \sigma_1 \cdot \sigma_2^r \cdot \sigma_3 \cdots \cdots \sigma_k$$
$$\pi_3 = \sigma_1 \cdot \sigma_2 \cdot \sigma_3^r \cdots \cdots \sigma_k$$
$$\vdots \quad \vdots$$
$$\pi_k = \sigma_1 \cdot \sigma_2 \cdot \sigma_3 \cdots \cdots \sigma_k^r$$

Then $\pi^* = \sigma_1 \cdot \sigma_2 \cdots \cdots \sigma_k$ achieves $k\binom{N}{2}$ crossings. However, any π_i disagrees with any π_j on the directions of both σ_i and σ_j, and hence creates $2(k-1)\binom{N}{2}$ crossings, which is $\frac{2k-2}{k} = 2 - \frac{2}{k}$ times the optimum.

Now we turn to approximation algorithms for the max version of the problem. Here, choosing any of the input permutations yields a 2-approximation, and again, this cannot be improved.

Theorem 5. *There is a 2-approximation for the max crossing minimization problem* PCM$_{\mathrm{max}}$-k.

Proof. Let $\pi_1, \ldots, \pi_k$ be a given set of permutations. We claim that any of these permutations is a 2-approximation, and prove this for π_1.

Let π^* be the optimal permutation for the PCM$_{\mathrm{max}}$-k problem, and let j^* be the index of the permutation where the maximum is achieved in the optimal solution, i.e.,

$$K(\pi_{j*}, \pi^*) \geq K(\pi_i, \pi^*) \text{ for all } i.$$

Note that the optimal value OPT equals therefore $K(\pi_{j*}, \pi^*)$. Now for any permutation π_i, we have

$$K(\pi_i, \pi_1) \leq K(\pi_i, \pi^*) + K(\pi^*, \pi_1) \leq K(\pi_{j*}, \pi^*) + K(\pi^*, \pi_{j*}) = 2\mathrm{OPT},$$

so $\max_i K(\pi_i, \pi_1) \leq 2\mathrm{OPT}$, and therefore π_1 is a 2-approximation for the max crossing number problem.

Clearly, if the target permutation is taken from the given set of permutations, the 2-approximation is best possible for PCM$_{\mathrm{max}}$-k. To see this use any permutation π and its reversal π^r. Then $CR(\pi, \pi^r) = n(n-1)/2$ and $CR_{max}(\pi, \pi^r) = \lceil n(n-1)/4 \rceil$.

It remains open whether the approximation bound could be improved by choosing some other permutations. Note that for the one-sided two-layer crossing minimization, the best approximation bound long stood at 2 as well [22], but was recently improved to 1.4664 [19]. Some randomized approximations have been established in [1].

4   The Small Cases

We now consider PCM-k and PCM$_{max}$-k for small values of k. Clearly, for $k = 1$, a single user will take his preferences for the optimal arrangement, and then there are no crossings.

Consider the case $k = 2$. For bipartite graphs with vertices of degree 2 on the lower layer the one-sided crossing minimization problem is solvable in linear time by the barycenter heuristic, and due to the nesting structure of the neighbours on the upper layer determines the left-right positions in an optimal layout, see [17]. The main ingredient here is that the penalty digraph is acyclic.

Similarly, the permutation crossing number can be found easily for two permutations π_1 and π_2; π_1 itself is optimal with value $c = K(\pi_1, \pi_2)$. Many optimal permutations can be found from a straight-line drawing of π_1 and π_2, see also Figure 2. Consider an arbitrary poly-line from left to right that crosses each straight line (v, v) for $v = 1, \dots, n$ exactly once (we call such a line a *pseudo-line*.) This yields a permutation π^* by listing the elements in the order in which they were crossed. Any permutation obtained in such a way is optimal for PCM-2.

For example, for $\pi_1 = (6, 3, 1, 4, 2, 5)$ and $\pi_2 = (3, 5, 2, 6, 1, 4)$, π_1 and π_2 themselves and also $(6, 3, 5, 2, 1, 4)$ are optimal, see Fig. 2.

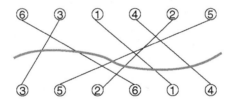

Fig. 2. Crossings for 2 permutations

Using these "intermediate" permutations, the max crossing problem can be solved in polynomial time by a sweep-line technique. Since the sum of the number of crossings c is determined, the max crossing minimization problem is solved by distributing these crossings uniformly to either side such that $CR_{max}(\pi_1, \pi_2) = \lceil c/2 \rceil$. An optimal permutation—which is best possible both for the sum and for the maximum—can be computed in $O(n + r) \log n$ time by a standard sweep-line technique, where r is the number of crossings, by searching among all pseudo-lines.

Now we address the case $k = 3$. Here, the complexity is open, both for permutations and for one-sided two-layered graphs with degree k on the free layer [17].

There is a 3-D drawing of the crossing minimization problem, where the permutations are represented on three piles in parallel to the Z-axis, and for every item i there is a triangle between the three occurrences of i. Whether such a drawing can be used to find the optimal solution (or even a good approximation), similar as for $k = 2$, remains open.

For the crossings of permutations problem the case with odd numbers is special. For every pair of items u and v there is a clear winner. There are no ties and the penalty graph is a complete tournament, i.e., there is exactly one directed arc (u, v) or (v, u) between each pair of vertices. Then every cycle c has a subcycle of length three [18]. There are simple permutations including a cycle, e.g. $(1, 2, 3), (2, 3, 1)$ and $(3, 2, 1)$. The feedback arc set problem in tournaments has been discussed at several places, see e.g. [1, 4]. It is NP-hard in the weighted version, and still open in the unweighted case.

5   Conclusion

In this paper, we investigated the problem of rank aggregation, which corresponds to find a permutation that minimizes the number of crossings with a given set of permutations. We introduced a variant that instead considers the maximum number of crossings among those permutations. We investigated complexity results and approximation algorithms.

This problem is a one-sided two-layer crossing minimization problem in an edge-coloured bipartite graph, where only crossings between equally coloured edges are counted. As such, it is not surprising that the complexity results for our problem mirror the ones for one-sided two-layer crossing minimization. We end by mentioning some of the numerous open problems that remain in this field:

1. How do the common techniques from one-sided two-layer crossing minimization, such as barycenter and median heuristics, sifting, or ILP approaches perform for the crossing minimization of permutations?
2. How can the Spearman footrule distance be used for the one-sided two-layer crossing minimization problem? How does it relate to sorting the barycenters?
3. Investigate the *max* versions, e.g., max Spearman footrule distance and the maximum number of crossings for any edge in the one-sided two-layer crossing minimization problem.
4. Improve the approximations and establish bounds for partial permutations.
5. The case $k = 3$ remains wide open. Is it NP-hard or polynomial?

Acknowledgments

The authors would like to thank Wolfgang Brunner, Christof König and Marcus Raitner for inspiring discussions.

References

1. N. Ailon, M. Charikar, and A. Newman. Aggregating inconsistent information: ranking and clustering. STOC (2005), 684–693.
2. J.P. Barthelemy, A. Guenoche, and O. Hudry. Median linear orders: heuristics and a branch and bound algorithm. Europ. J. Oper. Res. 42, (1989), 313–325.
3. J. Bartholdi III, C.A. Tovey, and M.A. Trick. Voting schemes for which it can be difficult to tell who won the election. Soc. Choice Welfare 6, (1989), 157–165.

4. I. Charon, A. Guenoche, O. Hudry, and F. Woirgard. New results on the computation of median orders. Discrete Math. 165/166 (1997), 139–153.

5. F.Y.L. Chin, X. Deng, Q. Feng, and S. Zhu. Approximate and dynamic rank aggregation. Theoret. Comput. Sci. 325, (2004), 409–424.

6. C. Demetrescu and I. Finochi. Breaking cycles for minimizing crossings. Electronic J. Algorithm Engineering 6, No. 2, (2001).

7. P. Diaconis and R. Graham. Spearman's footrule as a measure for disarray. Journal of the Royal Statistical Society, Series B, 39, (1977), 262–268.

8. G. Di Battista, P. Eades, R. Tamassia, and I.G. Tollis. Graph Drawing: Algorithms for the Visualization of Graphs. Prentice Hall, (1999).

9. C. Dwork, R. Kumar, M. Noar, and D. Sivakumar. Rank aggregation methods for the Web. Proc. WWW10 (2001), 613–622.

10. P. Eades and N.C. Wormald. Edge crossings in drawings of bipartite graphs. Algorithmica 11, (1994), 379–403.

11. G. Even, J. Naor, B. Schieber, and M. Sudan. Approximating minimum feedback sets and multicuts in directed graphs. Algorithmica 20, (1998), 151–174.

12. M.R. Garey and D.S. Johnson. Computers and Intractability: A Guide to the Theory of NP-Completeness. W.H. Freeman, San Francisco, (1979).

13. M.R. Garey and D.S. Johnson. Crossing number is NP-complete. SIAM J. Alg. Disc. Meth. 4, (1983), 312–316.

14. M. Jünger and P. Mutzel. 2-layer straightline crossing minimization: performance of exact and heuristic algorithms. J. Graph Alg. Appl. 1, (1997), 1–25.

15. M. Kaufmann and D. Wagner (Eds.). Drawing Graphs: Methods and Models, LNCS 2025, (2001).

16. J. G. Kemeny. Mathematics without numbers. Daedalus 88, (1959), 577–591.

17. X. Munos, W. Unger, and I. Vrto. One sided crossing minimization is NP-hard for sparse graphs. Proc. GD 2001, LNCS 2265, (2002), 115–123.

18. J.W. Moon. Topics on Tournaments. Holt, New York (1968).

19. H. Nagamochi. An Improved approximation to the One-Sided Bilayer Drawing. Discr. Comp. Geometry 33(4), (2005), 569–591.

20. K. Sugiyama, S. Tagawa, and M. Toda. Methods for visual understanding of hierarchical systems structures. IEEE Trans. SMC 11, (1981), 109–125.

21. V. Waddle and A. Malhotra An $E \log E$ line crossing algorithm for leveled graphs. Proc. GD 99, LNCS 1731 (2000), 59–70.

22. A. Yamaguchi and A. Sugimoto. An approximation algorithm for the two-layered graph drawing problem. Discrete Comput. Geom. 33, (2005), 565–591.

Morphing Planar Graphs While Preserving Edge Directions

Therese Biedl, Anna Lubiw, and Michael J. Spriggs

School of Computer Science, University of Waterloo, Waterloo, Ontario,
Canada, N2L 3G1
{biedl, alubiw, mjspriggs}@uwaterloo.ca

Abstract. Two straight-line drawings P, Q of a graph (V, E) are called
parallel if, for every edge $(u, v) \in E$, the vector from u to v has the same
direction in both P and Q. We study problems of the form: given simple, parallel drawings P, Q does there exist a continuous transformation
between them such that intermediate drawings of the transformation
remain simple and parallel with P (and Q)? We prove that a transformation can always be found in the case of orthogonal drawings; however,
when edges are allowed to be in one of three or more slopes the problem
becomes NP-hard.

1 Introduction

The process of drawing a graph is rarely a one-time task devoid of prior geometric
information. In many situations we already have a drawing of a graph, and the
graph may change or the requirements on the drawing may change. *Dynamic
graph drawing* [6] deals with the situation where the graph changes incrementally.
The goals—to avoid recomputing the drawing from scratch, and to preserve the
user's mental map [22]—are accomplished by altering the drawing as little as
possible, which makes it straightforward to animate the changes.

There are situations however, where the graph changes more dramatically or
the requirements on the drawing change, and the best approach is to compute
a new drawing. Preserving the user's mental map is still desirable, but it is no
longer straightforward to animate a continuous transformation from the original
drawing to the new drawing [14, 15].

Transforming one geometric object to another in a continuous way is called
morphing, and is well-studied in graphics [16], where it is often accomplished
in *image space* by transforming each pixel. More appropriate for graph drawing
applications are *object space* morphs, which operate on geometric objects.

In addition to the visualization applications just mentioned, morphing graph
drawings also finds application in the medical imaging problem of creating a
3-dimensional model from 2-dimensional slices obtained e.g. by X-rays [2].

Morphing without maintaining geometric structure is easy but usually unhelpful. The *linear morph*, for example, moves every vertex in a straight line
from its position in the source to its position in the target. It has the desirable
property of making minimal changes to vertex positions, but has the undesirable

P. Healy and N.S. Nikolov (Eds.): GD 2005, LNCS 3843, pp. 13–24, 2005.

property of producing intersections between disjoint objects—for example, you and your dance partner would change places by moving *through* each other.

Besides avoiding intersections, some other criteria for quality morphs are that a vertex should not stray too far from the line between its initial and final positions, and the length and direction of an edge should not deviate radically from the initial and final values. Criteria for evaluating interactive graph drawings also apply—see Bridgeman and Tamassia [7] for the case of orthogonal drawings.

Our aim in this paper is not to develop heuristics to address the many (conflicting) criteria. Rather, we concentrate on morphs that exactly preserve two properties: planarity (i.e. simplicity) and edge directions—we call these *parallel morphs*. The source and target drawings are simple straight-line drawings that represent the same graph embedded the same way, and such that each edge in the source drawing is parallel to its counterpart in the target drawing.

Our main result is an algorithm to find a parallel morph for the case of orthogonal graph drawings. The morphs produced by our algorithm are composed of $O(n)$ linear morphs where n is the size of the graphs. The user's mental model should be well preserved by these morphs. We briefly address the issue of how edge lengths change during the morph. One application of this result arises when VLSI compaction techniques [20] (which preserve edge directions) are used to reduce the area of an orthogonal drawing—our morph provides a continuous motion from the original drawing to the compacted one.

Recently, Lubiw, Petrick and Spriggs [21] devised an algorithm for morphing between two orthogonal drawings of a graph, where in these drawings vertices are points and edges are orthogonal paths. Morphs produced by the algorithm maintain both planarity and orthogonality. The algorithm employs—as a subroutine—the parallel-morphing algorithm described in the present paper.

On the negative side, we show that it is NP-hard to decide whether a parallel morph exists for the case of general planar graph drawings—in fact, in a typical 2-3 dichotomy, the problem is hard for 3 edge directions, and easy for 2.

1.1 Background

There is a broad, rich body of work on transforming one object to another while maintaining some geometric structure. Included are problems of morphing, animation, motion planning, folding, linkage reconfiguration, rigidity theory, etc. We will mention some of the most relevant background.

Preserving the Mental Map. Friedrich et al. [14, 15] considered the problem of "animating" the transformation from one graph drawing (not necessarily planar) to another. They do not insist on any geometric structure being strictly maintained, but their goal is to produce an animation that preserves the users mental map, and the criteria they formulate to accomplish this include minimizing temporary edge crossings and maintaining some minimal distance between nodes. Their method uses a combination of rigid motions and linear morphs, with the addition of clustering techniques in the second paper.

Preserving Simplicity. In 1944, long before the word "morph" was coined, Cairns [9] showed that there is a non-intersecting morph from any planar triangulation to any isomorphic one with the same fixed triangle as a boundary. Thomassen [23] strengthened this in two ways: First, he generalized to convex subdivisions and morphs preserving convexity. Secondly, he generalized to straight-line drawings of planar graphs, using the technique of "compatible triangulation" (discovered independently by Aronov et al. [1]) to augment both drawings to isomorphic triangulations, thus reducing to Cairns' result. These results are constructive, but algorithmic issues are not explored. Although only one vertex moves at a time, the graph is contracted down to a triangle which does nothing for the user's mental map.

Independently, Floater and Gotsman [13] proved Thomassen's convex morphing result using an entirely different approach based on Tutte's method of embedding graphs using barycentric coordinates. Their morph moves all vertices at once, and computes snapshots of the graph at intermediate time points. Combining this result with compatible triangulation [1] gives a different non-intersecting morph for straight line drawings [17]. These morphs can be visually pleasing, but there are no analytical results on the complexity of the vertex trajectories, or the number of time steps required to give the appearance of continuous motion. Erten, Kobourov, and Pitta [11, 12] have implemented the Floater-Gotsman method, with a preliminary phase that attempts to align the two drawings using rigid planar transformations.

Preserving Edge Directions. In addition to preserving simplicity and convexity, Thomassen [23] considered the problem of preserving edge directions. He showed that between any two simple orthogonal cycles with corresponding edges parallel, there is a parallel morph. Thomassen's morphs shrink edges to infinitesimal lengths. Our main result in this paper generalizes Thomassen's result to orthogonal graphs, rather than just cycles, and we do not shrink edges to infinitesimal lengths.

Thomassen's result was extended in a different direction to the case of simple non-orthogonal cycles by Guibas et al. [19], and independently by Grenander et al. [18]. In related work we show that there exists a parallel morph between any two trees in any dimension, but not for orthogonal cycles in 3D even if they represent the trivial knot [3], and not for edge graphs of genus-0 orthogonal polyhedra in 3D [5].

We have also explored the possibility of parallel morphs that change edge lengths monotonically—the most stringent condition for nice edge-length behavior. We show [4] that such morphs are possible for convex and orthogonally convex polygons, but that the decision problem becomes NP-hard for orthogonal polygons.

Preserving Edge Lengths: Linkage Reconfiguration and Rigidity. When a morph must preserve simplicity and edge *lengths* (rather than directions) we arrive at linkage reconfiguration problems, a topic of considerable recent interest—see [10] and references therein. For connections with rigidity theory and parallel redrawings, please see [3].

2   Preliminaries

Let (V, E) be an undirected graph with vertex set V and edge set E, and let $p : V \to \mathbb{R}^2$. The triple $P = (V, E, p)$ uniquely determines a bend-free straight-line *drawing* of graph (V, E) in the plane. Each edge $(u, v) \in E$ is represented in this drawing by the line segment between $p(u)$ and $p(v)$. We will use $p(u, v)$ to refer to this edge, and $|p(u, v)|$ to denote its length. A drawing $P = (V, E, p)$ is called *simple* if each vertex lies at unique coordinates and each pair of (non-equal) edges may intersect each other only at a common vertex. A drawing is *orthogonal* if each edge of the drawing is parallel with one of the axes.

Two drawings $P = (V, E, p)$ and $Q = (V, E, q)$ of the same graph are called *parallel* if for each edge $(u, v) \in E$, there exists some $\lambda > 0$ such that $p(u) - p(v) = \lambda(q(u) - q(v))$. When this expression holds for a particular edge (u, v), we say that (u, v) has the same *direction* in both P and Q.

Given two simple, parallel drawings P, Q of a graph (V, E) a *parallel morph* from P to Q is a continuous motion of the vertices that takes us from P to Q such that at all times the positions of the vertices determine a drawing of (V, E) that is both simple and parallel with P and Q. Formally, a parallel morph from P to Q is a continuously changing family of drawings R such that $R(0) = P$, $R(1) = Q$, and for every $t \in [0, 1]$, $R(t) = (V, E, r_t)$ where $r_t : V \to \mathbb{R}^2$ determines a simple drawing $R(t)$ that is parallel with P and Q.

Given drawings $P = (V, E, p)$ and $Q = (V, E, q)$, the *linear morph* between them is the morph in which each vertex $v \in V$ moves continuously from $p(v)$ to $q(v)$ at constant velocity—i.e. using the notation above, $r_t(v) = tq(v) + (1-t)p(v)$ for each vertex $v \in V$. Notice, by this definition $R(0) = P$ and $R(1) = Q$. One can show easily that a linear morph between two parallel simple drawings keeps each edge parallel with its realization in $R(0)$ and $R(1)$, and changes edge-lengths monotonically. However, it may destroy simplicity. At the heart of our algorithm is the result that a linear morph does maintain simplicity in some situations: when the ordering of the coordinates of the vertices is the same in P and Q; and more generally, when P and Q are *rectangular drawings* as defined in the next section. These results are proved in [5].

3   Morphing Orthogonal Drawings

This section contains our main result—an algorithm to find a parallel morph between any two simple parallel orthogonal graph drawings that are "bend-free"—i.e. in which each edge is a single line segment.

Traditionally, an orthogonal graph drawing represents each edge as a path with bends. We find it more convenient to deal with edges that are single line segments (e.g. for defining "parallel"). Morphing of traditional orthogonal graph drawings can be achieved via our method if each edge has the same number and direction of bends in the source and target drawings—we simply replace each bend by a vertex. Henceforth, "orthogonal drawing" will mean "bend-free orthogonal drawing".

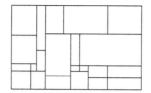

Fig. 1. A rectangular drawing

Theorem 1. *Any two simple parallel orthogonal drawings P, Q of a connected graph (V, E) admit a parallel morph that is composed of $O(|V|)$ linear morphs.*

3.1 Overview of the Morphing Algorithm

A *rectangular drawing* is a drawing in which the boundary of every face—including the outer face—is a rectangle (Fig. 1); the side of a rectangular face may be subdivided by any number of vertices. (A rectangular drawing is a type of *turn-regular* drawing as defined by Bridgeman et al. [8], i.e. no face has "kitty corners".) One can show that for a pair of parallel rectangular drawings, the linear morph is a parallel morph, i.e. it maintains both simplicity and edge directions.

So given two parallel orthogonal drawings P and Q, if they are rectangular drawings, we can morph them by applying a linear morph. Otherwise, our approach is to augment the drawings (by adding vertices, subdividing edges, and/or adding edges) to turn them into parallel rectangular drawings. Clearly, if we can morph two parallel augmented drawings, then we can also morph the original drawings by using the induced morph.

Our algorithm has three stages. The first stage ensures that the boundary of the exterior face of each drawing is a rectangle. Adding a new bounding rectangle around each drawing is easy; the only complication is maintaining connectedness of the graph and keeping the drawings parallel. In the target drawing Q, add a non-intersecting vertical edge between some vertex v and a new vertex u placed along the upper edge of the boundary rectangle. See Fig. 2 (a) and (b). We want the source drawing P to be parallel with the new target. In the source drawing, we can subdivide the upper edge of the bounding rectangle by vertex u, and position it above v, but the line segment (u, v) may cross parts of the drawing. We fix this by performing a parallel morph of the source so that (u, v) can be added, while maintaining simplicity. The fact that such a morph can always be performed on an orthogonal drawing is the key idea underlying our algorithm. Details are given in Sect. 3.2.

This completes the first stage of the algorithm. At this point, we have a new source and new target drawing. The drawings are parallel, the underlying graph is connected, and the the exterior face is bounded by a rectangle.

The second stage of the algorithm further modifies the drawings obtained in the first stage so that the boundary of each interior face is a rectangle. Until

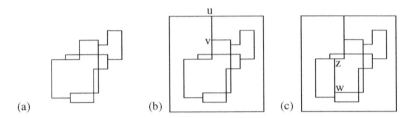

Fig. 2. Modifying the target drawing: (a) The original target. (b) The target following the first stage. (c) The target after adding an edge in the second stage.

every face of the target is a rectangle, iterate as follows. Pick a face f that is not a rectangle, and add a vertical edge from one reflex vertex w of f to the nearest edge e of f, which we subdivide by a new vertex z; see Fig. 2(c). To maintain parallel drawings, subdivide edge e by vertex z in the source drawing. Then, morph the source so that the vertical edge (w, z) can be added to it while maintaining the simplicity of the drawing; refer again to Sect. 3.2 for details.

The third stage of the algorithm is a linear morph between the rectangular source and rectangular target drawings. With that, the morph is complete.

3.2   Morphing to Add a New Edge

The first two stages of our morphing algorithm depend on the ability to morph the source drawing to a parallel drawing that admits a non-intersecting vertical edge between two given vertices. The idea is to draw a non-intersecting orthogonal path between the two vertices, and then morph the drawing (including the path) in order to straighten the path until it has no bends—at which point it forms the desired edge.

Not every orthogonal path can be straightened. Let Φ be a simple orthogonal drawing of a path. Φ is *balanced* if we encounter an equal number of left and right turns as we follow the path from one end to the other. In the remainder of this section we show that a balanced path can be straightened, and that in the above situation we can always find a balanced path between the two vertices we wish to join by an edge. Together with an analysis of the number of morphing steps, this will complete the proof of Theorem 1.

Straightening a Balanced Path. In this section we show that a balanced path of m bends can be straightened using $O(m)$ linear morphs.

Suppose that P and Φ are drawings. We define $P \cup \Phi$ in the natural way, noting that any vertex common to P and Φ must be in the same location in both drawings.

Lemma 1. *Let $P = (V, E, p)$ and $\Phi = (V_\Phi, E_\Phi, \phi)$ be simple orthogonal drawings with $v_\alpha, v_\beta \in V \cap V_\Phi$ such that Φ is a balanced drawing of a path with end-vertices v_α and v_β, and $P \cup \Phi$ is simple. There exists a parallel morph from P to a drawing $P' = (V, E, p')$ such that $p'(v_\alpha)$ and $p'(v_\beta)$ can be connected by a horizontal or vertical line segment whose interior does not intersect P'. Further,*

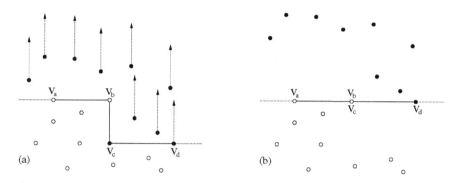

Fig. 3. The arrangement of the path vertices

the morph is composed of a sequence of $O(m)$ linear morphs, where m is the number of vertices in Φ.

Proof. As we follow Φ from v_α to v_β, we pass an equal number of left and right turns. We prove the lemma by induction on the number of left turns in Φ. If Φ contains no left turns, then it contains no right turns either and must be a line segment, and we are done. So assume that Φ contains $k > 0$ left turns. Since Φ is balanced, somewhere a left turn must be followed by a right turn or vice versa; so assume that $v_a, v_b, v_c, v_d \in V_\Phi$ is a sub-path with a right turn at v_b followed by a left turn at v_c. We will show below how to remove these two turns with a linear morph; this proves the lemma by induction.

Assume w.l.o.g. that the arrangement of $\phi(v_a)$, $\phi(v_b)$, $\phi(v_c)$, $\phi(v_d)$ is as shown in Fig. 3(a). Let $\mathcal{V} \subset V \cup V_\Phi$ be those vertices that lie either:

1. Strictly above the ray originating at $\phi(v_b)$ and going leftward; or
2. On or above the ray originating at $\phi(v_c)$ and going rightward.

The vertices in $\mathcal{V}$ are shown black in Fig. 3, while the others are drawn white.

Let R be the linear morph from $P \cup \Phi$ in which each $v \in \mathcal{V}$ moves upward at a uniform rate a distance of $|\phi(v_b, v_c)|$ while other vertices remain fixed; see Fig. 3(b). Let $P' = (V, E, p')$ denote the drawing of graph (V, E) following this linear morph. Notice, R reduces the distance between v_b and v_c to zero. Simplify the path graph (V_Φ, E_Φ) by removing v_b and v_c and adding the horizontal edge (v_a, v_d). The resulting path Φ' has one fewer left turn and one fewer right turn than Φ, so it is a balanced path between v_α and v_β with fewer than k left turns.

To complete the proof we must show that R keeps edges parallel, and—excepting vertices v_α and v_β—maintains simplicity. This is proved easily (we omit details) by observing the following properties of our morph: (1) Vertices move only vertically and upward. (2) If a vertex moves, then any vertex vertically above it (with the exception of v_b) moves by exactly the same amount. (3) By simplicity of $\Phi \cup P$, no horizontal edge of P has one vertex in $\mathcal{V}$ and the other vertex outside $\mathcal{V}$. □

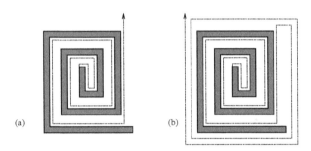

Fig. 4. A path from a vertex in P to infinity: In (a) the path has an excess of eight left turns, and in (b) the path is balanced

Finding a Balanced Path in the First Stage. Recall that for the first stage we want a balanced path in the source drawing between vertex v lying on the original outer face and a vertex u on the upper edge of the bounding rectangle. It suffices to show that we can build a simple balanced path from v that ends in an upward-directed vertical ray.

Lemma 2. *Let $P = (V, E, p)$ be an orthogonal drawing of a connected graph and let $v \in V$ be a vertex that has no incident vertical segment above it, such that the face immediately above $p(v)$ is the outer face. Drawing P admits a balanced simple path Φ of complexity $O(|V|)$ that starts at $p(v)$, goes upward, and ends with an upward-directed vertical ray.*

Proof. We construct a path that goes upward some small distance ϵ from $p(v)$ and then walks around the boundary of the outer face until we reach a point where an upward-directed ray does not intersect P. If this path is balanced we are done. Otherwise, add the appropriate number of turns of opposite direction, as illustrated in Fig. 4(b). □

Lemma 1 and Lemma 2 together prove that the first stage of the algorithm runs correctly, and is composed of $O(n)$ morphing steps.

Finding a Balanced Path in the Second Stage. We augment the target drawing by $\Theta(n)$ edges to produce a rectangular drawing, and, for each such edge, find a corresponding orthogonal path in the source which we then straighten by morphing. If we were to add the target edges in arbitrary order, each of the $\Theta(n)$ paths in the source might have $\Theta(n)$ bends to straighten, for a total of $\Theta(n^2)$ morphing steps in this stage. We can avoid this by choosing the new target edges carefully. We use only vertical edges. Each new vertical edge cuts a face in two. We choose an edge s.t. one of the new faces is a rectangle. In the source drawing, we find a balanced path with $O(1)$ turns by walking just inside the perimeter of this rectangular face. Straightening this balanced path takes $O(1)$ morphing steps, for a total of $O(n)$ morphing steps in the second stage.

This finishes the proof of Theorem 1. We note here that while only $O(n)$ linear morphs are needed, each of them might require $\Omega(n)$ time for updating the coordinates of vertices (which are needed for computing later morphs correctly). Hence the total time to perform all morphs is $O(n^2)$.

4   Edge Lengths in Morphing Orthogonal Drawings

In this section we explore how edge lengths change during parallel morphs between orthogonal drawings. There seems to be a trade-off between the number of times an edge increases and decreases in length, and the amount by which an edge deviates from its lengths in the source and target. Morphs produced by the algorithm of Sect. 3 are well-behaved with respect to the first measure, but not the second. In these morphs, each edge is non-decreasing in length until the third stage when a linear morph to the target is performed. If, prior to the final linear morph, we scale up the drawing so that every edge is longer than its target length, we obtain a two-phase morph where edges are non-decreasing in the first phase, and non-increasing in the second phase. Call this a $(+,-)$-morph. We can prove any $(+,-)$-morph will, in some cases, dramatically alter edge lengths.

For a parallel morph $R(t) = (V, E, r_t)$, define the *stretch factor* $\Delta(R)$ as:

$$\Delta(R) = \max_{(u,v)\in E} \left\{ \frac{\max_{t\in[0,1]}\{|r_t(u,v)|\}}{\max\{|r_0(u,v)|,|r_1(u,v)|\}}, \frac{\min\{|r_0(u,v)|,|r_1(u,v)|\}}{\min_{t\in[0,1]}\{|r_t(u,v)|\}} \right\} \quad (1)$$

The stretch factor is the largest factor by which some edge of the graph deviates from the range delimited by its lengths in the source and target drawings.

Theorem 2. *For any positive integer n there exists a pair of parallel orthogonal drawings with n vertices such that for any $(+,-)$-morph R between the drawings, $\Delta(R) \geq 2^{\Omega(n)}/n$.*

Due to space limitations we omit the proof, but an example of the construction is given in Fig. 5. Curiously, we have been unable to construct situations where $(-,+)$-morphs have such bad stretch factors. If we allow more fluctuations in edge lengths we can do much better.

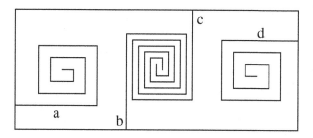

Fig. 5. The source drawing for Theorem 2. The target drawing is similar, except that spirals b and c are "disentwined" while spirals a and d are "entwined."

Theorem 3. *There exists a parallel morph R between any two simple orthogonal drawings P, Q of a graph (V, E) such that $\Delta(R) \leq n - 1$, where $n = |V|$.*

By Theorem 1 there exists a parallel morph R' from P to Q. The idea—for proving Theorem 3—is to decompose R' by a sequence of breakpoints such that between breakpoints no two vertices change order in w.r.t. a coordinate axis. The drawings at the breakpoints can be realized on nice-sized grids, and a new parallel morph R can be generated by a sequence of linear morphs between successive drawings on these nice-sized grids. Edge lengths are well-behaved on the grid, and linear morphs change edge-lengths monotonically. R has a linear stretch factor and each edge will alternately expand and shrink $O(n^3)$ times.

5   Non-orthogonal Morphing Is NP-Hard

Previous sections deal with orthogonal drawings. We now consider general drawings, and prove that it is NP-hard to decide whether parallel non-orthogonal drawings of a graph admit a parallel morph—even if there are only three possible edge directions. We note that the algorithm of Sect. 3 together with a shear can be used to morph any parallel graphs drawn using two edge directions.

Our NP-hardness reduction is from a closely related problem called *Parallel Morphing with Static Edges* (PM-STATIC):

– Given parallel orthogonal polygons $P = (V, E, p)$ and $Q = (V, E, q)$ and a subset $\mathcal{E} \subset E$ such that for each edge $(u, v) \in \mathcal{E}$, $|p(u, v)| = |q(u, v)|$,
– does P, Q admit a parallel morph such that all edges in $\mathcal{E}$ remain of fixed length throughout the morph?

We call the edges in $\mathcal{E}$, *static edges*, and the remaining edges of E are called *elastic edges*. The proof that PM-STATIC is NP-hard appears in [5]; we use a similar reduction to prove it NP-hard to decide whether two parallel orthogonal polygons admit a monotone morph [4].

Theorem 4. *Given two parallel drawings of a graph, it is NP-hard to decide whether there exists a parallel morph between them—even in the case where edges can only be horizontal, vertical, or of slope 1.*

Proof. We reduce from PM-STATIC. Let $P = (V, E, p)$ and $Q = (V, E, q)$ be a pair of parallel orthogonal polygons and let $\mathcal{E} \subseteq E$ be a set of static edges, whose lengths in P and Q are equal. Assume w.l.o.g. that both P and Q are embedded on a unit grid, i.e., all vertices are located at integer coordinates; one can show (details omitted) that this can be done with coordinates polynomial in $n = |V|$.

Construct a drawing P' from P as follows. Fix a value $\epsilon = (4n)^{-1}$. For each vertex $v \in V$, include a drawing of an $\epsilon \times \epsilon$-square in P', centered at $p(v)$, with a diagonal edge between the lower-left and upper-right corners. Observe that such a square permits only translation and scaling during a parallel morph.

For each edge $(u, v) \in E$, in P' connect the diagonalized squares corresponding to u and v as follows. An elastic edge of P is encoded in P' by two parallel axis-aligned edges, and a static edge is encoded by a series of diagonalized squares; see Fig. 6. The encoding of a static edge in P' permits only translation and scaling in a parallel morph, while the encoding of an elastic edge also permits changes

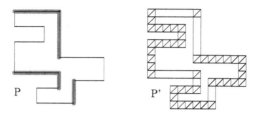

Fig. 6. An orthogonal polygon P and corresponding drawing P'

to the length of the two parallel edges. We construct Q' from Q in the same way. One can easily verify that P' and Q' are simple, and also show (details are omitted here) that they admit a parallel morph if and only if P and Q admit a parallel morph that does not change the length any static edge.   □

6   Conclusion

This paper addressed the problem of morphing one planar graph drawing to another when corresponding edges have the same direction and the morph should maintain this property. We showed how to morph orthogonal graph drawings; our morphs are computationally and visually well-behaved. However, as soon we allow edges to have one of three slopes the problem becomes NP-hard. We conclude with some open problems.

The morphing algorithm of Sect. 3 works for orthogonal *point*-drawings (vertices are points), but not necessarily for orthogonal *box*-drawings (vertices are disjoint boxes that must remain of the same dimensions throughout the morph). What is the complexity of this problem? In more practical situations, corresponding edges will not be parallel in the source and target drawings. A morph should not change edge directions more than necessary. Is it possible to design morphs that minimize changes to edge directions, or to angles? Even the following is open: given two polygons, is there a non-intersecting morph between them that preserves convexity/non-convexity of angles?

References

1. B. Aronov, R. Seidel, and D. Souvaine. On compatible triangulations of simple polygons. *Computational Geometry: Theory and Applications*, 3:27–35, 1992.
2. G. Barequet and M. Sharir. Piecewise-linear interpolation between polygonal slices. *Comput. Vis. Image Underst.*, 63(2):251–272, 1996.
3. T. Biedl, A. Lubiw, and M. Spriggs. Parallel morphing of trees and cycles. In *15th Canadian Conference on Computational Geometry (CCCG)*, pages 29–32, 2003.
4. T. Biedl, A. Lubiw, and M. J. Spriggs. Angles and lengths in reconfigurations of polygons and polyhedra. *Proc. Mathematical Foundations of Computer Science (MFCS'04)*, pages 748–759, 2004.
5. T. Biedl, A. Lubiw, and Michael J. Spriggs. Angles and lengths in reconfigurations of polygons and polyhedra (long version), 2005. See *http://arxiv.org/*.

6. J. Branke. Dynamic graph drawing. In D. Kaufmann and D. Wagner, editors, *Drawing Graphs – Methods and Models, Lecture Notes in Computer Sciecne 2025*, pages 228–246. Springer, 2001.

7. S. Bridgeman and R. Tamassia. Difference metrics for interactive orthogonal graph drawing algorithms. *Journal of Graph Algorithms and Applications*, 4 (3):47–74, 2000.

8. S. S. Bridgeman, G. Di Battista, W. Didimo, G. Liotta, R. Tamassia, and L. Vismara. Turn-regularity and optimal area drawings of orthogonal representations. *Computational Geometry*, 16(1):53–93, 2000.

9. S.S. Cairns. Deformations of plane rectilinear complexes. *American Math. Monthly*, 51:247–252, 1944.

10. J. Cantarella, E.D. Demaine, H. Iben, and J. O'Brien. An energy-driven approach to linkage unfolding. In *20th Annual ACM Symposium on Computational Geometry*, pages 134–143, 2004.

11. C. Erten, S. Kobourov, and C. Pitta. Intersection-free morphing of planar graphs. In G. Liotta, editor, *Graph Drawing 2003, Lecture Notes in Computer Science 2912*, pages 320–331. Springer, 2004.

12. C. Erten, S. Kobourov, and C. Pitta. Morphing planar graphs. In *20th Annual ACM Symposium on Computational Geometry*, pages 451–452, 2004.

13. M. Floater and C. Gotsman. How to morph tilings injectively. *Journal of Computational and Applied Mathematics*, 101:117–129, 1999.

14. C. Friedrich and P. Eades. Graph drawing in motion. *Journal of Graph Algorithms and Applications*, 6 (3):353–370, 2002.

15. C. Friedrich and M.E. Houle. Graph drawing in motion II. In P. Mutzel, M. Jünger, and S. Leipert, editors, *Graph Drawing 2001, Lecture Notes in Computer Science 2265*, pages 220–231. Springer, 2002.

16. J. Gomes, L. Darsa, B. Costa, and L. Velho. *Warping and Morphing of Graphical Objects*. Morgan Kaufmann, 1999.

17. C. Gotsman and V. Surazhsky. Guaranteed intersection-free polygon morphing. *Computers and Graphics*, 25:67–75, 2001.

18. U. Grenander, Y. Chow, and D. M. Keenan. *Hands: A Pattern Theoretic Study of Biological Shapes (Appendix D)*. Springer-Verlag, 1991.

19. L. Guibas, J. Hershberger, and S. Suri. Morphing simple polygons. *Discrete and Computational Geometry*, 24(1):1–34, 2000.

20. T. Lengauer. *Combinatorial Algorithms for Integrated Circuit Layout*. Wiley, 1990.

21. A. Lubiw, M. Petrick, and Michael J. Spriggs. Morphing orthogonal planar graph drawings. In *Proceedings ACM-SIAM Symposium on Discrete Algorithms*, 2006. To appear.

22. K. Misue, P. Eades, W. Lai, and K. Sugiyama. Layout adjustment and the mental map. *J. Visual Lang. Comput.*, 6 (2):183–210, 1995.

23. C. Thomassen. Deformations of plane graphs. *Journal of Combinatorial Theory*, Series B: 34:244–257, 1983.

Dynamic Spectral Layout of Small Worlds

Ulrik Brandes, Daniel Fleischer, and Thomas Puppe

Department of Computer & Information Science, University of Konstanz, Germany

Abstract. Spectral methods are naturally suited for dynamic graph layout, because moderate changes of a graph yield moderate changes of the layout under weak assumptions. We discuss some general principles for dynamic graph layout and derive a dynamic spectral layout approach for the animation of small-world models.

1 Introduction

The main problem in dynamic graph layout is the balance of layout quality and mental-map preservation [17]. Typically, the problem is addressed by adapting a static layout method such that it produces similar layouts for successive graphs. While these adaptations are typically ad-hoc [8], others [2, 1] are based on the formally derived method [3] of integrating difference metrics [5] into the static method. See [4] for an overview of the dynamic graph drawing problem.

Spectral layout denotes the use of eigenvectors of graph-related matrices such as the adjacency or Laplacian matrix as coordinate vectors. See, e.g., [15] for an introduction. We argue that spectral methods are particularly suited for dynamic graph layout both from a theoretical and practical point of view, because moderate changes in the graph naturally translate into moderate changes of the layout, and updates can be computed efficiently.

This paper is organized as follows. In Sect. 2, we define some basic notation and recall the principles of spectral graph layout. The dynamic graph layout problem is reviewed briefly in Sect. 3, and methods for updates between layouts of consecutive graphs are treated in more detail in Sect. 4. In Sect. 5, our approach for small worlds is introduced, and we conclude with a brief discussion in Sect. 6.

2 Preliminaries

For ease of exposition we consider only two-dimensional straight-line representations of simple, undirected graphs $G = (V, E)$ with positive edge weights $\omega : E \to \mathbb{R}^+$, although most techniques and results in this paper easily carry over to other classes of graphs.

In straight-line representations, a two-dimensional *layout* is determined by a vector $(p_v)_{v \in V}$ of *positions* $p_v = (x_v, y_v)$. Most of the time we will reason about one-dimensional layouts x that represent the projection of p onto one component.

P. Healy and N.S. Nikolov (Eds.): GD 2005, LNCS 3843, pp. 25–36, 2005.

For any graph-related matrix $M(G)$, a *spectral layout* of G is defined by two eigenvectors x and y of $M(G)$. For simplicity, we will only consider layouts derived from the *Laplacian matrix* $L(G)$ of G, which is defined by elements

$$\ell_{v,w} = \begin{cases} \sum_{u \in V} \omega(u, v) & , v = w , \\ -\omega(v, w) & , v \neq w , \end{cases}$$

The rows of $L(G)$ add up to 0, thus, the vector $\mathbf{1} = (1, \ldots, 1)^{\mathrm{T}}$ is a trivial eigenvector for eigenvalue 0. Since $L(G)$ is symmetric all eigenvalues are real, and the theorem of Gershgorin [13] yields, that the spectrum is bounded to the interval $[0, g]$, for an upper bound $g \geq 0$. Hence, the spectrum can be written as $0 = \lambda_1 \leq \lambda_2 \leq \ldots \leq \lambda_n \leq g$ with corresponding unit eigenvectors $1/\sqrt{n} = v_1, \ldots, v_n$.

Based on the Laplacian, a spectral layout is defined as $p = (v_2, v_3)$, where v_2 and v_3 are unit eigenvectors to the second and third smallest eigenvalues of the corresponding Laplacian matrix $L(G)$. This has already been used for graph drawing in 1970 by Hall [14].

For sparse graphs of moderate size, a practical method to determine the corresponding eigenvectors is *power iteration*. For an initial vector x the matrix multiplication $L(G)x/\|L(G)x\|$ is iterated until it converges to a unit eigenvector associated with the largest eigenvalue. Since we are not interested in v_n, we use matrix $\hat{L} = g \cdot I - L(G)$, which has the same eigenvectors with the order of their eigenvalues $g = g - \lambda_1 \geq g - \lambda_2 \geq \ldots \geq g - \lambda_n$ reversed. To obtain v_2 and v_3, respectively, x is orthogonalized with v_1 (and in the case of v_3 also with v_2) after each iteration step, i.e., the mean value $\sum_{i=1}^{n} x_i/n$ is subtracted from every element of x to ensure $x \perp \mathbf{1}$. Spectral layouts of larger graphs can be computed efficiently using multiscale methods [16].

3   Dynamic Layout

In our setting, a *dynamic graph* is a sequence $G^{(1)}, \ldots, G^{(r)}$ of graphs with, in general, small edit distance, i.e. $G^{(t)}$ is obtained from $G^{(t-1)}$, $1 < t \leq r$, by adding, changing, and deleting only a few vertices and edges.

There are two main scenarios for the animation of a dynamic graph, depending on whether the individual graphs are presented to the layout algorithm one at a time, or the entire sequence is known in advance. Layout approaches for the *offline* scenario (e.g., [7]) are frequently based on a layout of the union of all graphs in the sequence. A variant are 2.5D representations in which all graphs are shown at once (e.g., [9]). In the *online* scenario, the typical approach is to consider only the previous layout (e.g., [8]). A variant in which provisions for likely future changes are made is presented in [6].

Since, typically, spectral layouts of similar graphs do not differ much anyway, it is reasonable to ignore the fact that a graph is but one graph in a sequence altogether and compute static layouts for each of them. We rather concentrate on the update step between consecutive layouts.

4 Updates

Assume we are given a sequence of layouts $p_1, \ldots, p_r$ for a dynamic graph $G^{(1)}, \ldots, G^{(r)}$. The step from p_t to p_{t+1} is called *logical update*, whereas the actual animation of the transition is referred to as the *physical update*.

While simple, say, *linear interpolation* of two layouts is most frequently used in graph editors, more sophisticated techniques for morphing are available (see, e.g., [11, 12, 10]. General morphing strategies do not take into account the method by which origin and target layout are generated.

For dynamic spectral layout, at least two additional strategies are reasonable.

4.1 Iteration

If the target layout x_{t+1} is a spectral layout, the iteration for its own computation can and should be initialized with x_t, that will usually be close to the target layout. The power iteration then produces intermediate layouts which can be used for the physical update. A way to enhance the smoothness of morphing is needed because of the observation, that the first steps of the iteration yield greater movement of the vertices when compared to later steps. Let $\hat{L} = g \cdot I - L(G^{(t+1)})$. An iteration step then consists of computing the new layout $\hat{L}x/\|\hat{L}x\|$ from a given layout $x \perp \mathbf{1}$. Let $g = \lambda_1 \geq \lambda_2 \geq \ldots \geq \lambda_n$ and $1/\sqrt{n} = v_1, v_2, \ldots, v_n$ be the eigenvalues and unit eigenvectors of $\hat{L}$, respectively. Then if $\lambda_1 > \lambda_2 > \lambda_3$ (otherwise just proper eigenvectors and eigenvalues would have to be chosen in what follows) and $x = \sum_{i=2}^{n} a_i v_i, a_2 \neq 0$ we have

$$\frac{\hat{L}^k x}{\|\hat{L}^k x\|} \longrightarrow v_2 \quad \text{and}$$

$$\left\| v_2 - \frac{\hat{L}^k x}{\|\hat{L}^k x\|} \right\| = \left\| v_2 - \frac{\sum_{i=2}^{n} \lambda_i^k a_i v_i}{\|\hat{L}^k x\|} \right\|$$

$$\leq 1 - \frac{\lambda_2^k a_2}{\|\hat{L}^k x\|} + \frac{\sum_{i=3}^{n} \lambda_i^k a_i}{\|\hat{L}^k x\|}$$

$$= \mathcal{O}\left((\lambda_3/\lambda_2)^k\right).$$

One way to handle this non-linear decay is to use layouts after appropriately spaced numbers of steps, or to use layouts only if the difference to the last used layout exceeds some threshold c in some metric, e.g., if $\|x - x'\|_2 > c$. Both ways will enhance the smoothness of morphing by avoiding the drawing of many small movements at the end of the iteration process.

4.2 Interpolation

If both origin and target layout x_t and x_{t+1} are spectral layouts, intermediate layouts can also be obtained by computing eigenvectors of some intermediate matrices from $L(G^{(t)})$ to $L(G^{(t+1)})$. We interpolate linearly by

$$\alpha L(G^{(t)}) + (1 - \alpha)L(G^{(t+1)}), \quad 1 \geq \alpha \geq 0.$$

Layouts are computed for a sequence of breakpoints $1 \geq \alpha_1 > \alpha_2 > \ldots > \alpha_k \geq 0$ ($\alpha_{j+1} - \alpha_j$ constant, or proportional to $\sin(\pi j/k)$, depending on what kind of morphing seems to be appropriate, the latter one slowing down at the beginning and end). For every breakpoint α_j the iteration is initialized with the layout of α_{j-1}, which allows fast convergence and small movements between two succeeding breakpoints. Deletion and insertion of vertices have to be handled in a different manner, since the matrix dimension changes. See Sect. 5 for details.

Figs. 2 and 4 show smooth animations of this method. Theoretical justification for smoothness comes along with a theorem by Rellich [18] applied to the finite dimensional case. Matrix $\alpha L(G^{(t)}) + (1-\alpha)L(G^{(t+1)})$ can be seen as a perturbed self-adjoint operator $L(\varepsilon) = L(G^{(t)}) + \varepsilon(L(G^{(t+1)}) - L(G^{(t)}))$ with corresponding eigenvalues $\lambda_i(\varepsilon)$ and eigenvectors $v_i(\varepsilon)$, that are holomorphic with respect to ε, i.e.

$$L(\varepsilon)v_i(\varepsilon) = \lambda_i(\varepsilon)v_i(\varepsilon), \tag{1}$$

where $v_i(0)$ are eigenvectors at time t and $v_i(1)$ can be permuted by a permutation π, such that $v_{\pi(i)}(1)$ are (ordered) eigenvectors at time $t+1$. Note that two eigenvectors may only have to be exchanged if its corresponding eigenvalues intersect during the time from t to $t+1$. And even then the power iteration exchanges these eigenvectors sufficiently smooth for pleasing animations, because the corresponding eigenvalues remain within the same range for some time due to smooth functions $\lambda_i(\varepsilon)$. Consider λ_2 and v_2 of the following small-world example with $n = 100$ vertices and $k = 7$, where starting from a circle each vertex is connected to its $2k$ nearest neighbors. Both $\lambda_2(\varepsilon)$ and $v_2(\varepsilon)$ can locally be written as power series

$$\begin{aligned} \lambda_2(\varepsilon) &= \mu_0 + \varepsilon\mu_1 + \varepsilon^2\mu_2 + \ldots , \\ v_2(\varepsilon) &= w_0 + \varepsilon w_1 + \varepsilon^2 w_2 + \ldots . \end{aligned} \tag{2}$$

We show that $\|w_i\| \leq 1$ and $|\mu_i| \leq 2/\sqrt{n}$ for $i > 0$, hence $\lambda_2(\varepsilon)$ and $v_2(\varepsilon)$ will be smooth functions, say, within $[0, 1/2]$ (and by the same construction within the remaining interval, too). We write $L(\varepsilon) = L + \varepsilon P$, where P is the insertion of an edge between two non-adjacent vertices (with indices 1 and r), and denote by I the identity matrix. From (1) and (2) we get

$$Lw_0 = \mu_0 w_0 ,$$
$$(L - \mu_0 I)w_j = \sum_{i=0}^{j-1} \mu_{j-i} w_i - Pw_{j-1} , (j > 0) , \tag{3}$$

where we can recursively choose w_j, $(j > 0)$ such that $w_j \perp w_0$. Since the right hand sides of (3) need to be orthogonal to w_0 for $j > 0$ this yields

$$\mu_j = \langle Pw_{j-1}, w_0 \rangle , (j > 0) .$$

Now we can recursively compute upper bounds for $|\mu_j|$ and $\|w_j\|$. Note that λ_2 has multiplicity 2, and the right hand sides of (3) are also orthogonal to v_1, such that $1/(\lambda_4 - \lambda_2)$ is the least upper bound of the inverse mapping of $L - \mu_0 I$ applied to the right hand side.

$$|\mu_1| = (w_{0,1} - w_{0,r})^2 \leq 2/n \leq 2/\sqrt{n} ,$$

$$\|w_1\| \leq \frac{2|w_{0,1} - w_{0,r}|}{\lambda_4 - \lambda_2} \leq \frac{2\sqrt{2/n}}{1.568} =: c \leq 1/4.35 ,$$

$$|\mu_j| \leq 2\kappa_{j-1}c^{j-1}/\sqrt{n} \leq 2/\sqrt{n} ,$$

$$\|w_j\| \leq \kappa_j c^j \leq 1 ,$$

where κ_j is defined by $\kappa_0 = 1$ and $\kappa_j = \sum_{i=0}^{j-1} \kappa_i \kappa_{j-i-1}$ for $j > 0$.

Lemma 1. $\kappa_j \leq 4.35^j$.

Proof. We show $\kappa_j \leq 4.35^j/(j+1)^2$, which holds for all $j < 144$ (by evaluating). For $j \geq 144$

$$\kappa_j \leq \sum_{i=0}^{j-1} \frac{4.35^i}{(i+1)^2} \cdot \frac{4.35^{j-i-1}}{(j-i)^2} \leq 2 \cdot 4.35^{j-1} \sum_{i=0}^{\lfloor j/2 \rfloor} \frac{1}{(i+1)^2} \cdot \frac{1}{(j-i)^2}$$

$$\leq 2 \cdot 4.35^{j-1} \left(\frac{1}{(j-9)^2} \sum_{i=1}^{10} \frac{1}{i^2} + \frac{4}{j^2} \left(\zeta(2) - \sum_{i=1}^{10} \frac{1}{i^2} \right) \right) \leq \frac{4.35^{j-1}}{(j+1)^2} \cdot 4.348 .$$

$\square$

Note that $\|w_j\| \leq 1$ could also be shown for much weaker assumptions than $c \leq 1/4.35$, which was sufficient for our example. Lemma 1 is only very close to optimal, the least upper bound of $1/(\lambda_4 - \lambda_2)$ is in general not achieved, and $|\mu_j| = |\langle Pw_{j-1}, w_0 \rangle|$ can in general be better bounded than by $\|w_{j-1}\|\sqrt{4/n}$.

5 Application to Small Worlds

Spectral layout methods are naturally suited for smooth dynamic layout, because the influence of vertices and edges that are subject to change can be increased or decreased gradually. Moreover, each can be determined by iterative computations that benefit from good initialization, so that moderate changes leads to moderate and efficient updates.

Watts and Strogatz [19] introduced a random graph model that captures some often-observed features of empirical graphs simultaneously: sparseness, local clustering, and small average distances. This is achieved by starting from a cycle and connecting each node with its $2k$ nearest neighbors for some small, fixed k. The resulting graph is sparse and has a high clustering coefficient (average density of vertex neighborhoods), but also high (linear) average distance.

The average distance drops quickly when only a few random edges are rewired randomly. If each edge is rewired independently with some probability p, there is a large interval of p in which the average distance is already logarithmic while the clustering coefficient is still reasonably high.

5.1   Dynamic Laplacian Layout

Interestingly, spectral layouts highlight the construction underlying the above model and thus point to the artificiality of generated graphs. This is due to the

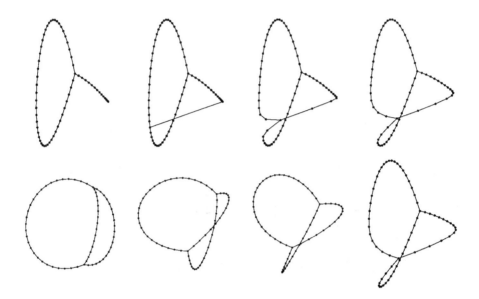

Fig. 1. Update by iteration (read top left to top right to bottom right to bottom left). Note the spread of change along the graph structure.

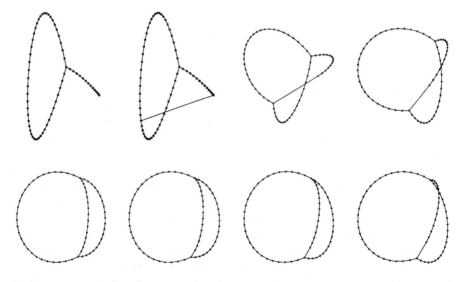

Fig. 2. Update by interpolation. Layout anomalies are restricted to modified part of graph.

fact that spectral layouts of regular structures display their symmetry very well, and are only moderately disturbed by small perturbations in the graph (mirroring the argument for their use in dynamic layout). The initial ring structure of the small world in Fig. 5 is therefore still apparent, even though a significant number of chords have been introduced by random rewiring. In fact, the layout conveys very well which parts of the ring have been brought together by short-cut edges.

Figs. 1 and 2 point out differences between the two approaches using intermediate layouts obtained from the power iteration and from matrix interpolation.

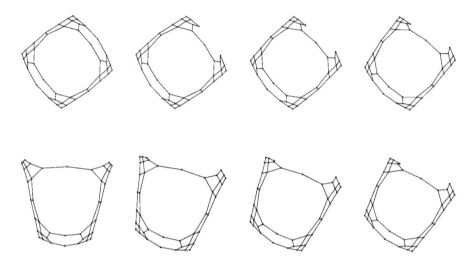

Fig. 3. Update by simple linear interpolation. Intermediate layouts are less symmetric.

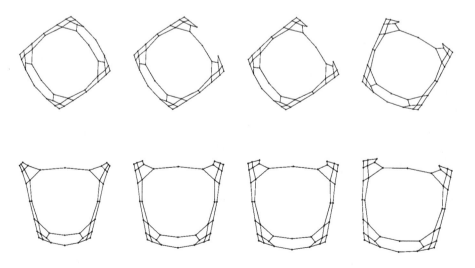

Fig. 4. Interpolation updates maintain symmetry

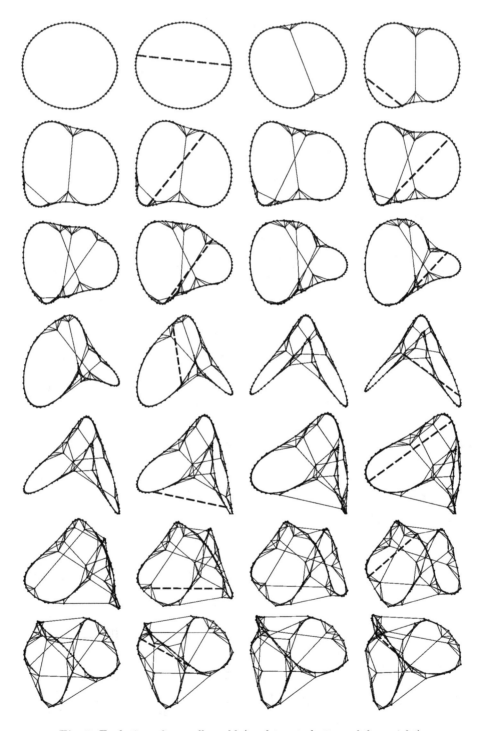

Fig. 5. Evolution of a small world (read top to bottom, left to right)

It can be seen that the power iteration first acts locally around the changes. This stems from the fact that in the first multiplication only the neighborhood of the change, i.e., the two incident vertices of an edge with changed weight or the neighbors of a deleted or inserted vertex, is affected. The next step also affects vertices at distance 2, and so on. Hence, the change spreads like a wavefront. The matrix interpolation approach acts globally at every step. Interpolating the Laplacian matrices corresponds to gradually changing edge weights. The animation therefore is much more smooth.

Figs. 3 and 4 show differences between simple linear interpolation of the positions and matrix interpolation. In Fig. 3 can be seen, that the symmetry of the graph to its vertical axis is not preserved during the animation, whereas in Fig. 4 each intermediate layout preserves this symmetry.

Fig. 5 finally shows some snapshots of a small world evolving from a torus. The layouts were obtained by using matrix interpolation (one intermediate step per change shown). Note that deletion and insertion of vertices requires some extra efforts, in particular, if the deletion of a vertex disconnects the graph.

5.2   Deletion and Insertion of Vertices

Consider deletion of a single vertex v, that does not disconnect the graph. Matrix $L(G^{(t+1)})$ is then expanded by one row and column of zeros corresponding to vertex v, such that $L(G^{(t)})$ and $L(G^{(t+1)})$ have the same dimension. This derived matrix has a double eigenvalue 0. A new corresponding eigenvector is, e.g., $(0, \ldots, 0, 1, 0, \ldots, 0)^{\mathrm{T}}$, where the 1 is at position corresponding to v. This eigenvector will cause vertex v to drift away during power iteration, and thus all other vertices stick together. This can be prevented by defining $\ell_{v,v} = g$ in matrix $L(G^{(t+1)})$, leading to a movement of v towards 0. But in practice, the following method proved to be successful. After every matrix multiplication reset the position of v to the barycenter of its neighbors. This either prevents a drifting away or an absorbing to 0, which would otherwise be hard to manage. Apart from using matrix $\alpha L(G^{(t)}) + (1 - \alpha)L(G^{(t+1)})$ for the power iteration, orthogonalization and normalization also have to be adapted. For time $t + 1$ we only need $x_{t+1} \perp (1, \ldots, 1, 0, 1, \ldots, 1)$, instead of $x_{t+1} \perp \mathbf{1}$, and only the restriction to the elements not corresponding to v have to be normalized. Both can be done by linear interpolation of these operations.

Insertion of a vertex v is treated analogously. Expand matrix $L(G^{(t)})$ by one row and column of zeros as above. Orthogonalization and normalization again have to be adapted.

5.3   Disconnected Graphs

The deletion of a *cut vertex* (or a *bridge*) disconnects the graph $G^{(t+1)}$ into $k \geq 2$ components $G_1, \ldots, G_k$. Each component is drawn separately by spectral methods and afterwards these layouts are merged to a layout for $G^{(t+1)}$. Basically, there are three parameters for each component, that have to be determined after a layout x_j for each G_j was computed. The first one determines the size of each component, i.e., find a constant s_j, that scales x_j to $s_j x_j$. The second one determines where the barycenter of each component is set to. The rotation angle of each component could also be considered, but we concentrate on the first two parameters only.

The removal of a cut vertex (or a bridge) yields a matrix $L(G^{(t+1)})$, that, after rearranging, consists of k blocks $L_1, \ldots, L_k$, which are Laplacian matrices of lower dimensions

$$L(G^{(t+1)}) = \begin{pmatrix} L_1 & & & 0 \\ & L_2 & & \\ & & \ddots & \\ 0 & & & L_k \end{pmatrix}.$$

Each of the components is now drawn separately, simply by the common power iteration of the whole matrix $L(G^{(t+1)})$, where only normalization and orthogonalization have to be modified appropriately. The barycenter c_j of each component thus is 0, which we now reset to a new position. For notational purposes identify the 2-dimensional plane with complex numbers. Let the current barycenters c_j be sorted increasingly by their angle to the positive real axis and reset them to

$$c_j := \frac{\eta}{2\sqrt{2}} \exp\left(\frac{2\pi i}{\eta}\left(-\frac{\eta_j}{2} + \sum_{\ell=1}^{j} \eta_\ell\right)\right), \quad \eta_j = \sqrt{\frac{|G_j|}{|G^{(t+1)}|}}, \quad \eta = \sum_{j=1}^{k} \eta_j .$$

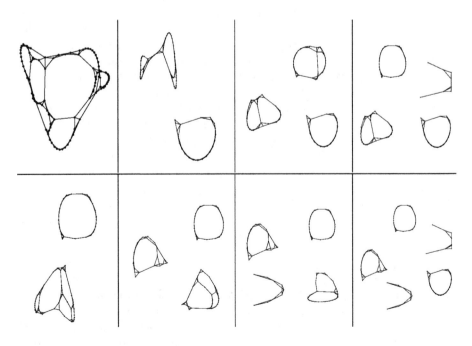

Fig. 6. Drawing connected components (top: left to right, bottom: right to left)

Together with a normalization $s_j = \eta_j$, this has the effect, that the components are distributed on a circle with radius $\eta/(2\sqrt{2})$, each on an area proportional to the size of the component, and none of them overlap. Depending on the shape of the components the radius can also be decreased. Note that the normalization is also well suited, because $\sum_{j=1}^{k} \eta_j^2 = 1$ – analogous to $||x|| = 1$.

Altogether, when removing a cut vertex, new barycenters are computed, power iteration with modified orthogonalization/normalization is applied, and meanwhile each component moves to its new barycenter linearly to the chosen breakpoints.

Further splitting and merging of connected components are handled analogously, see Fig. 6 for an example.

6   Discussion

We have proposed a dynamization scheme for spectral layout and applied it to changing small-world graphs. While there is no need to make special provisions for logical updates, it turns out that matrix interpolation is the method of choice for the physical update. Despite its simplicity, the scheme achieves both static layout quality and mental-map preservation, because it utilizes stability inherent in spectral layout methods.

Much of the dynamization scheme directly applies to force-directed methods as well, and is in fact driven by common practices [8].

For both spectral and force-directed layout update computations are rather efficient, since the preceding layouts are usually very good initializations for iterative methods. For large graphs, it will be interesting to generalize the approach to multilevel methods, possibly by maintaining (at least part of) the coarsening hierarchy and reusing level layouts for initialization.

In general, spectral layouts are not suitable for graphs with low connectivity, even in the static case. However, our dynamic approach is likely to work with any improved methods for static spectral layout as well.

References

1. U. Brandes, M. Eiglsperger, M. Kaufmann, and D. Wagner. Sktech-driven orthogonal graph drawing. In *Proc. GD 2002*, LNCS 2528, pages 1–11. Springer, 2002.
2. U. Brandes, V. Kääb, A. Löh, D. Wagner, and T. Willhalm. Dynamic WWW structures in 3D. *Journal of Graph Algorithms and Applications*, 4(3):103–114, 2000.
3. U. Brandes and D. Wagner. A Bayesian paradigm for dynamic graph layout. In *Proc. GD 1997*, LNCS 1353, pages 236–247. Springer, 1997.
4. J. Branke. Dynamic graph drawing. In M. Kaufmann and D. Wagner, editors, *Drawing Graphs*, LNCS 2025, pages 228–246. Springer, 2001.
5. S. Bridgeman and R. Tamassia. Difference metrics for interactive orthogonal graph drawing algorithms. *Journal of Graph Algorithms and Applications*, 4(3):47–74, 2000.

6. C. Demestrescu, G. Di Battista, I. Finocchi, G. Liotta, M. Patrignani, and M. Pizzonia. Infinite trees and the future. In *Proc. GD 1999*, LNCS 1731, pages 379–391. Springer, 1999.

7. S. Diehl, C. Görg, and A. Kerren. Preserving the mental map using forsighted layout. In *Proc. VisSym 2001*. Springer, 2001.

8. P. Eades, R. F. Cohen, and M. Huang. Online animated graph drawing for web navigation. In *Proc. GD 1997*, LNCS 1353, pages 330–335. Springer, 1997.

9. C. Erten, P. J. Harding, S. G. Kobourov, K. Wampler, and G. Yee. GraphAEL: Graph animations with evolving layouts. In *Proc. GD 2003*, LNCS 2912, pages 98–110. Springer, 2003.

10. C. Erten, S. G. Kobourov, and C. Pitta. Intersection-free morphing of planar graphs. In *Proc. GD 2003*, LNCS 2912, pages 320–331. Springer, 2003.

11. C. Friedrich and P. Eades. Graph drawing in motion. *Journal of Graph Algorithms and Applications*, 6(3):353–370, 2002.

12. C. Friedrich and M. E. Houle. Graph drawing in motion II. In *Proc. GD 2001*, LNCS 2265, pages 220–231. Springer, 2001.

13. G. H. Golub and C. F. van Loan. *Matrix Computations*. John Hopkins University Press, 1983.

14. K. M. Hall. An r-dimensional quadratic placement algorithm. *Management Science*, 17(3):219–229, 1970.

15. Y. Koren. Drawing graphs by eigenvectors: Theory and practice. *Computers and Mathematics with Applications*, 2005. To appear.

16. Y. Koren, L. Carmel, and D. Harel. Drawing huge graphs by algebraic multigrid optimization. *Multiscale Modeling and Simulation*, 1(4):645–673, 2003.

17. K. Misue, P. Eades, W. Lai, and K. Sugiyama. Layout adjustment and the mental map. *Journal of Visual Languages and Computing*, 6(2):183–210, 1995.

18. F. Rellich. *Perturbation Theory of Eigenvalue Problems*. Gordon and Breach Science Publishers, 1969.

19. D. J. Watts and S. H. Strogatz. Collective dynamics of "small-world" networks. *Nature*, 393:440–442, 1998.

Exact Crossing Minimization

Christoph Buchheim[1], Dietmar Ebner[2], Michael Jünger[1], Gunnar W. Klau[3], Petra Mutzel[4], and René Weiskircher[5]

[1] Department of Computer Science, University of Cologne
{buchheim, mjuenger}@informatik.uni-koeln.de
[2] Institute of Computer Graphics and Algorithms, Vienna University of Technology
ebner@ads.tuwien.ac.at
[3] Department of Mathematics and Computer Science, FU Berlin
gunnar@math.fu-berlin.de
[4] Department of Computer Science, University of Dortmund
petra.mutzel@uni-dortmund.de
[5] CSIRO Mathematical and Information Sciences
Rene.Weiskircher@csiro.au

Abstract. The crossing number of a graph is the minimum number of edge crossings in any drawing of the graph into the plane. This very basic property has been studied extensively in the literature from a theoretic point of view and many bounds exist for a variety of graph classes. In this paper, we present the first algorithm able to compute the crossing number of general sparse graphs of moderate size and present computational results on a popular benchmark set of graphs. The approach uses a new integer linear programming formulation of the problem combined with strong heuristics and problem reduction techniques. This enables us to compute the crossing number for 91 percent of all graphs on up to 40 nodes in the benchmark set within a time limit of five minutes per graph.

1 Introduction

Crossing minimization is among the oldest and most fundamental problems arising in the areas of automatic graph drawing and *VLSI* design. At the same time, it is very easy to formulate: "*Given a graph $G = (V, E)$, draw it in the plane with a minimum number of edge crossings*". A drawing of G is a mapping of each vertex $v \in V$ to a distinct point and each edge $e = (v, w) \in E$ to a curve connecting the incident vertices v and w without passing through any other vertex. Common points of two edges that are not incident vertices are called *crossings*. The minimum number of crossings among all drawings of G is denoted by $\operatorname{cr}(G)$.

The main goal in automatic graph drawing is to obtain a layout that is easy to read and understand. Although the definition of layout quality often depends on the particular application and is hard to measure, the number of edge crossings is among the most important criteria [18]. Figure 1 shows a comparison of different drawings for the same graph preferring different aesthetic criteria.

In fact, the crossing minimization problem is even older than the area of automatic graph drawing. It goes back to P. Turán, who proposed the problem

P. Healy and N.S. Nikolov (Eds.): GD 2005, LNCS 3843, pp. 37–48, 2005.

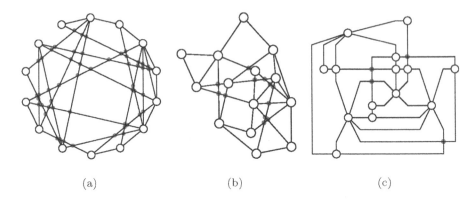

(a)   (b)   (c)

Fig. 1. Three drawings of the same graph with 51 (a), twelve (b), and four crossings (c). Most aesthetic criteria like few edge bends, uniform edge lengths, or a small drawing area favor the first two drawings, while the last drawing is preferable with respect to the number of edge crossings.

in his "Notes of Welcome" in the first issue of the *Journal of Graph Theory* [19]. While working in a labor camp during the Second World War, he noted that crossings of the rails between kilns and storage yards caused the trucks to jump the rails. Minimizing these crossings corresponds to the crossing minimization problem for a complete bipartite graph $K_{m,n}$.

In 1953, K. Zarankiewicz and K. Urbaník independently claimed a solution for this problem by providing a drawing rule for complete bipartite graphs $K_{m,n}$ with $\lfloor \frac{m}{2} \rfloor \lfloor \frac{m-1}{2} \rfloor \lfloor \frac{n}{2} \rfloor \lfloor \frac{n-1}{2} \rfloor$ crossings. About ten years later, their proof of optimality was shown to be wrong and it is still unknown whether the conjecture holds. The situation for complete graphs K_n is similar. Their crossing number is conjectured to be $\frac{1}{4} \lfloor \frac{n}{2} \rfloor \lfloor \frac{n-1}{2} \rfloor \lfloor \frac{n-2}{2} \rfloor \lfloor \frac{n-3}{2} \rfloor$, which has been verified for graphs of up to ten nodes by Guy [10]. However, both conjectures are based on a drawing rule and therefore serve as an upper bound for $cr(G)$.

It is well known that the general crossing minimization problem is *NP*-hard [7]. More precisely, it is shown that the *crossing number problem, i.e., "given a graph G and a non-negative integer K, decide whether there is a drawing of G with at most K edge crossings"*, is *NP*-complete. However, for fixed K, we can obtain a polynomial time algorithm by examining all possible configurations with up to K crossings. Clearly, this algorithm is not appropriate in practical applications for larger values of K. Recently, Grohe could show that this problem can be solved in time $O(|V|^2)$ [8]. Even though the exponent is independent of K, the constant factor of his algorithm grows doubly exponentially in K. Therefore, this method is also of little relevance in practice.

The search for approximation algorithms did not lead to significant results either. While there is no known polynomial time approximation algorithm with any type of quality guarantee for the general problem, Bhatt and Leighton could derive an algorithm for graphs with *bounded degree* that approximates the number of crossings *plus the number of nodes* in polynomial time [2]. Due to the complexity of the crossing minimization problem, many restricted versions have been considered in the literature. However, in most cases, *e.g.*, for bipartite, linear, and circular drawings, the problem remains *NP*-hard [6, 16, 15].

The most prominent and practically successful approach for solving the crossing minimization problem heuristically is the *planarization approach* [1], which addresses the problem by a two step strategy. The idea is to remove a preferably small number of edges in order to obtain a planar subgraph and reinsert them into a planar drawing with as few crossings as possible. For each step, various algorithms can be applied. Pre- and post-processing procedures have been developed to improve the solution quality. A computational study on state-of-the-art heuristics can be found in [9].

Contribution and Structure. In this paper, we present the first algorithm able to compute the crossing number of general sparse graphs of moderate size. We state computational results on a popular benchmark set of graphs, the so-called Rome library [5]. The approach uses a new integer linear programming formulation of the problem combined with strong heuristics and problem reduction techniques. This enables us to compute the crossing number for 91 percent of all graphs on up to 40 nodes in the Rome library within a time limit of five minutes per graph. In Sect. 2, we show how to reduce the problem to the easier problem of computing crossing-minimal drawings where each edge is involved in at most one crossing. We give an integer linear programming formulation for the simpler problem and a branch-and-cut algorithm to compute provably optimal solutions for this formulation in Sect. 3. Section 4 summarizes the computational results obtained with our new approach for the simple as well as the general crossing number problem for Rome library graphs. We present conclusions and further work in Section 5.

2   Reduction to Simple Drawings

The area of crossing minimization is closely related to the field of planarity testing, which aims to decide whether a given graph G can be drawn in the plane without any edge crossings. This task can be performed surprisingly fast, more precisely in linear time [11, 4]. Beyond doubt, one of the ground-breaking results in this research area was *Kuratowski's theorem*, which provides a full characterization of planar graphs based on the complete graph K_5 and the complete bipartite graph $K_{3,3}$.

Theorem 1 (Kuratowski's theorem). *A finite graph is planar if and only if it contains no subgraph that is a subdivision of K_5 or $K_{3,3}$.*

We can obtain a subdivision S of a graph G by repeatedly replacing its edges by a path of length two.

As a consequence of Theorem 1, at least two edges in every Kuratowski subdivision, *i.e.*, a subdivision of K_5 or $K_{3,3}$, have to cross in every planar drawing of a graph G. As we describe in Section 3, we can obtain inequalities from this observation that fully characterize the set of *realizable* crossing configurations (corresponding to drawings in the plane).

Unfortunately, even deciding whether there is a drawing for a given set of edge crossings is NP-complete [13]. This problem is known as the *realizability problem* and can be stated as follows: *"Given a set of edge pairs D, does there exist a drawing of G such that two edges $e, f \in E$ cross each other if and only if $\{e, f\} \in D$?*

In order to efficiently answer this question, we also need to know the *order* of the edge crossings for a particular edge e. With this additional information, it is easy to solve the problem by placing dummy vertices on all chosen crossings and testing the resulting graph for planarity.

One way to work around the realizability problem is the reduction to *simple drawings*. A drawing is called simple if each edge crosses at most one other edge. Not surprisingly, there are graphs that do not admit any simple drawing. Pach and Tóth [17] showed the following more general theorem:

Theorem 2. *Let $G = (V, E)$ be a simple graph drawn in the plane so that every edge is crossed by at most k others. If $0 \le k \le 4$, then we have*

$$|E| \le (k + 3)(|V| - 2) . \tag{1}$$

They could further prove that this bound cannot be improved for $0 \le k \le 2$ and that for *any* $k \ge 1$ the following inequality holds:

$$|E| \le \sqrt{16.875\,k}\,|V| \approx 4.108\sqrt{k}\,|V| \tag{2}$$

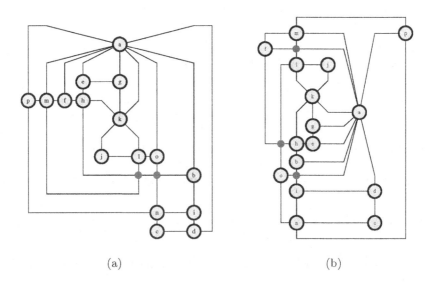

(a)   (b)

Fig. 2. Optimal drawing of a graph with two crossings (a) and an optimum simple drawing of the same graph with three crossings (b). Both drawings were produced with our exact algorithm presented in this paper.

Furthermore, Bodlaender and Grigoriev proved that it is NP-complete to determine whether there is a simple drawing for a given graph G [3]. If there is such a drawing, we denote the minimum number of crossings among all simple drawings of G by $\mathrm{crs}(G)$.

Even if there is a simple drawing for G, its crossing number $\mathrm{crs}(G)$ does not necessarily coincide with $\mathrm{cr}(G)$. Consider the sample graph in Figure 2. The left drawing shows an optimum drawing with two crossings while the right drawing shows an optimum drawing among all simple drawings.

However, given a graph $G = (V, E)$ we can create a graph $G^* = (V^*, E^*)$ by replacing every edge $e \in E$ with a path of length $|E|$. It is easy to show that for any non-negative number K the graph G can be drawn with K crossings if and only if there is a *simple drawing* of G^* with K crossings. Therefore, it is "sufficient" to solve the crossing minimization problem restricted to simple drawings in order to solve the general crossing minimization problem, clearly at significant computational expense. Since the transformation obviously can be done in polynomial time, the NP-completeness of the corresponding decision problem for simple drawings follows immediately from the NP-completeness for the general crossing number problem [7].

It is well-known that every graph G admits a *good drawing* with a minimum number of crossings, *i.e.*, a drawing that satisfies the following conditions:

1. no edge crosses itself
2. adjacent edges do not cross each other
3. non-adjacent edges cross each other at most once

Therefore it is sufficient to replace every edge $e = (v, w) \in E$ with a path of length $|E| - |\delta(v)| - |\delta(w)| - 1$. We can further lower the number of required dummy edges by using any upper bound for $\mathrm{cr}(G)$, since no edge can cross more than $\mathrm{cr}(G)$ other edges in any optimal solution.

3   An Integer Linear Program for Simple Drawings

Mathematical programming is a powerful tool to address NP-hard combinatorial optimization problems. Starting from an integer linear program (ILP) modeling the problem under consideration, *i.e.*, a linear program with integer variables, sophisticated techniques like branch-and-cut can be applied. In the following, we present an integer linear programming formulation for the crossing minimization problem restricted to simple drawings. It is described in Sect. 2 how this method can be used to solve the general crossing minimization problem.

Let $G = (V, E)$ be a graph and let D be a set of unordered pairs of edges of G. We call D *simple* if for every $e \in E$ there is at most one $f \in E$ such that $(e, f) \in D$. Furthermore, D is called *realizable* if there is a drawing of G such that there is a crossing between edges e and f if and only if $(e, f) \in D$.

For every graph G and every simple D, we denote with G_D the graph that is obtained by introducing a dummy node $d_{e,f}$ for each pair of edges $(e, f) \in D$. More precisely, we introduce dummy nodes on both e and f and identify them.

Note that G_D is only well-defined if D is simple, as otherwise it would not be clear where to place the dummy nodes. For both edges e_1 and e_2 resulting from splitting e, we set $\hat{e}_1 = \hat{e}_2 = e$, analogously for f.

Corollary 1. *Let D be simple. Then D is realizable if and only if G_D is planar.*

Using a linear time planarity testing and embedding algorithm, we can thus test in time $O(|V| + |D|)$ whether D is realizable, and compute a realizing drawing in the affirmative case.

Definition 1. *For a set of pairs of edges $D \subseteq E^2$ we define*

$$x_{e,f}^D = \begin{cases} 1 & \text{if } (e,f) \in D \\ 0 & \text{otherwise} . \end{cases}$$

Next, for every subgraph $H = (V', E')$ of G_D, let $\hat{H} = \{\hat{e} \mid e \in E'\} \subseteq E$. Less formally, $\hat{H}$ contains all edges of G involved in the subgraph H of G_D.

Proposition 1. *Let D be simple and realizable. For an arbitrary simple set of pairs of edges $D' \subseteq E^2$ of $G = (V, E)$ and any subdivision H of K_5 or $K_{3,3}$ in $G_{D'}$, the following inequality holds:*

$$C_{D',H} : \sum_{(e,f) \in \hat{H}^2 \setminus D'} x_{ef}^D \geq 1 - \sum_{(e,f) \in \hat{H}^2 \cap D'} (1 - x_{ef}^D) \tag{3}$$

Proof. Suppose (3) is violated. Since $x_{e,f}^D \in \{0,1\}$ for all $e, f \in E$, the left hand side of (3) must be zero and the right hand side must be one, which means that

$$x_{e,f}^D = 0 \quad \text{for all } (e,f) \in \hat{H}^2 \setminus D', \text{ and}$$
$$x_{e,f}^D = 1 \quad \text{for all } (e,f) \in \hat{H}^2 \cap D' .$$

It follows from the definition of x^D that $\hat{H}^2 \cap D' = \hat{H}^2 \cap D$, in other words, that G_D corresponds to $G_{D'}$ on the subgraph induced by $\hat{H}$, so that H is also a forbidden subgraph in G_D, i.e., a subdivision of K_5 or $K_{3,3}$. It follows from Kuratowski's Theorem that G_D is not planar. This contradicts the realizability of D by Corollary 1.   □

Theorem 3. *Let $G = (V, E)$ be a simple graph. A set of pairs of edges $D \subseteq E^2$ is simple and realizable if and only if the following conditions hold:*

$$x_{e,f}^D \in \{0,1\} \qquad\qquad\qquad \forall\, e, f \in E, \ e \neq f$$

$$\sum_{f \in E} x_{e,f}^D \leq 1 \qquad\qquad\qquad \forall\, e \in E$$

$$C_{D',H} \qquad\qquad\qquad \begin{array}{l} \text{for every simple } D' \subseteq E^2 \text{ and every} \\ \text{forbidden subgraph } H \text{ in } G_{D'} \end{array}$$

Proof. It is easy to see that the constraints from the second row are satisfied if and only if D is simple. It remains to show that a simple D is realizable if and only if the conditions $C_{D',H}$ from the last row hold. For a realizable D every $C_{D',H}$ is satisfied according to Proposition 1.

We have to show that any non-realizable set D violates at least one of the constraints $C_{D',H}$. It follows from Corollary 1 that G_D is not planar if D is not realizable and we know from Theorem 1 that there exists a subdivision H of K_5 or $K_{3,3}$ in G_D. Let $D' = D$ and consider the constraint $C_{D,H}$:

$$C_{D,H} : \sum_{(e,f)\in\hat{H}^2\setminus D} x_{ef}^D \geq 1 - \sum_{(e,f)\in\hat{H}^2\cap D} (1 - x_{ef}^D) \qquad (4)$$

It follows from the definition of x^D that every $x_{e,f}^D \in \hat{H}^2 \setminus D$ is zero, hence the left hand side of (4) is also zero. Since $\hat{H}^2 \cap D \subseteq D$ we also know that

$$\sum_{(e,f)\in\hat{H}^2\cap D} (1 - x_{ef}^D) = 0 \,,$$

so that the right hand side of $C_{D,H}$ is one. Thus $C_{D,H}$ is violated. $\qquad\square$

For every simple and realizable set $D \subseteq E^2$, we can compute a corresponding drawing in polynomial time. Thus we can reformulate the crossing minimization problem for simple drawings as *"Given a graph $G = (V, E)$, find a simple and realizable subset $D \subseteq E^2$ of minimum cardinality"*. This immediately leads to the following ILP-formulation, where we use $x(F)$ as an abbreviation for the term $\sum_{(e,f)\in F} x_{e,f}$:

$$\min\ x(E^2)$$

s.t. $\sum_{f\in E} x_{e,f} \leq 1 \qquad\qquad\qquad\qquad\qquad\qquad \forall e \in E$

$\qquad x(\hat{H}^2 \setminus D') - x(\hat{H}^2 \cap D') \geq 1 - |\hat{H}^2 \cap D'| \qquad$ for every simple D'
$\qquad\qquad\qquad\qquad\qquad\qquad\qquad\qquad$ and every forbidden subgraph H in $G_{D'}$

$\qquad x_{e,f} \in \{0,1\} \qquad\qquad\qquad\qquad\qquad\qquad \forall\ e,f \in E$

It is clearly impractical to generate all constraints $C_{D,H}$ in advance and solve the *ILP* in a single step. Instead, we embed the given formulation into a branch-and-cut framework, separating violated inequalities dynamically during runtime according to the proof of Theorem 3.

A crucial factor in this approach is the *separation problem*: "Given a class of valid inequalities and a vector $y \in \mathbb{R}^n$, either prove that y satisfies all inequalities in the class, or find an inequality which is violated by y.". Although we can easily separate violated inequalities for integral solution vectors according to the proof of Theorem 3, the problem is more complex within the branch-and-cut framework since we have to deal with fractional values. A heuristic for separating the inequalities is to round variables to either zero or one, and then check for violated inequalities. The problem is that the inequalities produced by this heuristic might not be violated by the current fractional solution. In this case we select a branching variable and split the current problem into two subproblems by setting the branching variable to zero, respectively one. The same is done if no inequalities at all are produced by the separation heuristic.

In some cases, we can omit variables from our ILP. For instance, we can split the graph into its *blocks* (two-connected components) and solve these blocks independently—the crossing number of a graph is equal to the sum of the crossing numbers of its blocks. Furthermore, it is easy to show that adjacent edges do not cross in an optimal drawing and no edge crosses itself, *i.e.*, we can restrict ourselves to good drawings.

4   Computational Results

We have implemented the presented algorithm in C^{++} using the class library *LEDA* and solve the linear programs arising during the optimization process with the commercial optimization library *CPLEX* (version 8.1). We have integrated our new algorithm into *AGD*, a powerful library of Algorithms for Graph Drawing. This enables us to use any of the existing planar layout algorithms to produce a drawing for G with $\mathrm{cr}(G)$ crossings.

In order to decrease the computational expense of many input graphs, we applied a number of correctness preserving pre-processing procedures. These remove edges temporarily that do not influence the crossing number of the resulting graph. More precisely we repeatedly remove nodes of degree one and merge paths such that each vertex except the start and the end vertex has degree two to a single edge. The latter approach cannot be applied if we intend to determine $\mathrm{crs}(G)$, since we may exclude the optimal solutions from consideration.

In order to obtain good upper bounds, we apply additionally to known bounds based on the number of vertices and edges the well known planarization approach to the input graph [1]. Furthermore, we make use of an exact algorithm proposed by Jünger and Mutzel [12] that computes the *skewness* $\mathrm{sk}(G)$ of G. It is defined as the minimum number of edges that must be removed from G in order to obtain a planar subgraph. It is well-known that the crossing number of a graph cannot be smaller than its skewness. Hence we have that $\mathrm{cr}(G) \geq \mathrm{sk}(G)$. Computing the skewness is equivalent to the *maximum planar subgraph problem*, which was shown to be *NP*-hard by Liu and Geldmacher [14]. However, medium sized instances can be solved to optimality in reasonable computation time.

To test the performance of our new algorithm and to compare its solution quality to heuristic approaches, we used a benchmark set of graphs of the University of Rome III, introduced in [5]. The set contains 11, 389 graphs that consist of 10 to 100 vertices and 9 to 158 edges. These graphs were generated from a core set of 112 "real life" graphs used in database design and software engineering applications. Most of the graphs are sparse, which is a common property in most application areas of automatic graph drawing. The average ratio between the number of edges and the number of nodes of the graphs from the benchmark set is about 1.35.

Due to the complexity of the crossing minimization problem we only consider graphs of up to 40 nodes. We need to round the current fractional solution to integer values in order to separate violated inequalities. Therefore we experimented with different strategies and compared their performance against each other.

- *R1* We round every value that is greater than $1 - \varepsilon$ to one. All other variables are mapped to zero. In our implementation, we used $\varepsilon = 10^{-10}$.
- *R05* Every variable with a value greater or equal than 0.5 is rounded to one.
- *R0208* If the value of a variable is less than 0.2 or greater than 0.8 it is mapped to zero or one, respectively. In the interval [0.2, 0.8] a coin flip decides if we round to zero or one.

It turns out that *R1* performs best on average and is therefore presented in the following figures. While this strategy often leads to constraints that are already satisfied by the current fractional solution, the generated cuts are usually stronger.

Figure 3 shows the percentage of graphs that could be solved within a time limit of five minutes on an Intel Pentium 4 with 2.4 GHz and 1 GB of main memory. As expected, the difference between our implementation for $cr(G)$ (crossing number) and $crs(G)$ (simple crossing number) grows with the size of the graphs. The smaller number of variables needed for the computation of $crs(G)$ (where we do not need edge decomposition) leads to a significantly higher number of instances that could be solved within the time limit. While the percentage of

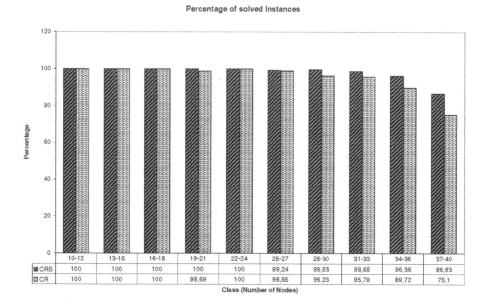

	10-12	13-15	16-18	19-21	22-24	25-27	28-30	31-33	34-36	37-40
CRS	100	100	100	100	100	99,24	99,53	98,68	96,36	86,63
CR	100	100	100	98,69	100	98,85	96,23	95,78	89,72	75,1

Class (Number of Nodes)

Fig. 3. Percentage of graphs solved by our exact algorithm for graphs on up to 40 nodes with (*CR*) and without (*CRS*) supporting multiple crossings per edge

Table 1. The computation time strongly depends on the number of crossings. The values K, $\bar{t}_{crs}$, and $\bar{t}_{cr}$ denote the number of crossings and the average computation time to compute the (simple) crossing-minimal drawings.

K	$\bar{t}_{cr}$	$\bar{t}_{crs}$
1	0.16s	0.16s
2	6.40s	0.79s
3	52.90s	8.20s
4	155.05s	31.57s

graphs with 40 nodes where we can compute $cr(G)$ within the time limit goes down to about 65%, we can still compute $crs(G)$ for about 80% of these instances.

Table 1 shows the average computation time for instances that could be solved within five minutes by all of the considered rounding strategies. The required time to solve a particular instance strongly depends on its crossing number, as the table illustrates.

Clearly we are interested in the quality of our results in comparison to heuristic approaches. For the computation of heuristic values we used the planarization approach. Gutwenger and Mutzel presented an extensive computational study of crossing minimization heuristics [9]. The authors investigate the effects of various methods for the computation of a maximal planar subgraph and different edge re-insertion strategies for the planarization approach. Furthermore, they study the impact of post-processing heuristics.

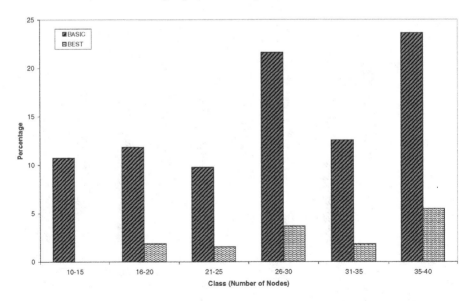

Fig. 4. Comparison between heuristic results and the crossing numbers computed with our exact algorithm

Figure 4 shows the average improvement in percent that could be achieved in comparison to the basic planarization approach (*BASIC*) and the improved version (*BEST*), for different classes of graphs. To highlight the improvements, we only considered graphs that could be solved to optimality within the time limit. We can clearly improve the heuristic results for the basic approach, even for the relatively small instances considered in our computational study. Even compared to the best known heuristic methods we achieve a notable improvement for some larger instances. The average improvement over the whole considered benchmark set is about 19.6% for the basic heuristic and 4.1% for the best known strategy.

5   Conclusion and Future Work

In this paper, we have presented the first algorithm that is able to compute the crossing number for sparse graphs of moderate size. We achieved this by combining a new integer linear programming formulation for the problem with sophisticated problem reduction techniques and the best known heuristics for the problem. Our implementation of the algorithm is able to compute the crossing number for 91 percent of all graphs on up to 40 vertices in a popular benchmark set of graphs within five minutes.

One way of improving the performance of the approach would be to use column generation for subdividing edges. As our computational results have shown, the problem is much easier to solve for simple drawings than for non-simple drawings. If we found a way of testing efficiently if subdividing a certain edge, and thus allowing more crossings on it, would decrease the overall number of crossings, we could expect to be able to solve much larger problem instances.

Another way forward is studying the polyhedron defined by the set of realizable crossing vectors. By adding new constraints that exclude fractional solutions that are not excluded by our current constraints we should be able to compute the crossing number for larger graphs with more crossings.

References

1. C. Batini, M. Talamo, and R. Tamassia.   Computer aided layout of entity-relationship diagrams. *Journal of Systems and Software*, 4:163–173, 1984.
2. S. N. Bhatt and F. T. Leighton. A framework for solving VLSI graph layout problems. *Journal of Computer and System Sciences*, 28:300–343, 1984.
3. H. Bodlaender and A. Grigoriev.   Algorithms for graphs embeddable with few crossings per edge. Research Memoranda 036, Maastricht : METEOR, Maastricht Research School of Economics of Technology and Organization, 2004.
4. K. S. Booth and G. S. Lueker. Testing for the consecutive ones property, interval graphs, and graph planarity using PQ-tree algorithms. *Journal of Computer and System Sciences*, 13(3):335–379, 1976.
5. G. Di Battista, A. Garg, G. Liotta, R. Tamassia, E. Tassinari, and F. Vargiu. An experimental comparison of four graph drawing algorithms. *Computational Geometry: Theory and Applications*, 7(5-6):303–325, 1997.

6. P. Eades and N. C. Wormald. Edge crossings in drawings of bipartite graphs. *Algorithmica*, 11(4):379–403, 1994.

7. M. R. Garey and D. S. Johnson. Crossing number is NP-complete. *SIAM Journal on Algebraic and Discrete Methods*, 4:312–316, 1983.

8. M. Grohe. Computing crossing numbers in quadratic time. In *STOC 2001: Proceedings of the 33rd Annual ACM Symposium on Theory of Computing*, 2001.

9. C. Gutwenger and P. Mutzel. An experimental study of crossing minimization heuristics. In G. Liotta, editor, *GD 2003: Proceedings of the 11th International Symposium on Graph Drawing*, volume 2912 of *LNCS*, pages 13–24. Springer-Verlag, 2004.

10. R. K. Guy. Crossing numbers of graphs. In *Graph Theory and Applications (Proceedings)*, Lecture Notes in Mathematics, pages 111–124. Springer-Verlag, 1972.

11. J. Hopcroft and R. Tarjan. Efficient planarity testing. *Journal of the ACM*, 21(4):549–568, 1974.

12. M. Jünger and P. Mutzel. Maximum planar subgraphs and nice embeddings: Practical layout tools. *Algorithmica*, 16(1):33–59, 1996.

13. J. Kratochvíl. String graphs. II.: Recognizing string graphs is NP-hard. *Journal of Combinatorial Theory, Series B*, 52(1):67–78, 1991.

14. P. Liu and R. Geldmacher. On the deletion of nonplanar edges of a graph. In *Proceedings of the 10th Southeastern Conference on Combinatorics, Graph Theory, and Computing*, pages 727–738, Boca Raton, FL, 1977.

15. S. Masuda, T. Kashiwabara, K. Nakajima, and T. Fujisawa. On the NP-completeness of a computer network layout problem. In *Proceedings of the IEEE International Symposium on Circuits and Systems*, pages 292–295, 1987.

16. S. Masuda, K. Nakajima, T. Kashiwabara, and T. Fujisawa. Crossing minimization in linear embeddings of graphs. *IEEE Transactions on Computers*, 39(1):124–127, 1990.

17. J. Pach and G. Tóth. Graphs drawn with few crossings per edge. *Combinatorica*, 17(3):427–439, 1997.

18. H.C. Purchase. Which aesthetic has the greatest effect on human understanding? In Giuseppe Di Battista, editor, *GD '97: Proceedings of the 5th International Symposium on Graph Drawing*, volume 1353 of *LNCS*, pages 248–261. Springer-Verlag, 1997.

19. P. Turán. A note of welcome. *Journal of Graph Theory*, 1:7–9, 1977.

On Embedding a Cycle in a Plane Graph[*]
(Extended Abstract)

Pier Francesco Cortese, Giuseppe Di Battista,
Maurizio Patrignani, and Maurizio Pizzonia

Università Roma Tre
{cortese, gdb, patrigna, pizzonia}@dia.uniroma3.it

Abstract. Consider a planar drawing Γ of a planar graph G such that the vertices are drawn as small circles and the edges are drawn as thin strips. Consider a cycle c of G. Is it possible to draw c as a non-intersecting closed curve inside Γ, following the circles that correspond in Γ to the vertices of c and the strips that connect them? We show that this test can be done in polynomial time and study this problem in the framework of clustered planarity for highly non-connected clustered graphs.

1 Introduction

Let Γ be a planar drawing of a planar graph G and c be a cycle composed of vertices and edges of G. We deal with the problem of testing if c can be drawn on Γ without crossings.

Of course, if the vertices of G are drawn as points, the edges as simple curves, and the drawing of c must coincide with the drawing of its vertices and edges, then the problem is trivial. In this case c can be drawn without crossings if and only if it is simple.

We consider the problem from a different point of view. Namely, we suppose that the vertices of G are drawn in Γ as "small circles" and the edges as "thin strips". Hence, c can pass several times through a vertex or through an edge without crossing itself. In this case even a non-simple cycle can have a chance to be drawn without crossings.

The problem, in our opinion, is interesting in itself. However, we study it because of its meaning in the field of *clustered planarity* [11, 10].

Clustered planarity is a classical Graph Drawing topic (see [4] for a survey). A *cluster* of a graph is a non empty subset of its vertices. A *clustered graph* $C(G, T)$ is a graph G plus a rooted tree T such that the leaves of T are the vertices of G. Each node ν of T corresponds to the cluster $V(\nu)$ of G whose

---

[*] Work partially supported by European Commission - Fet Open project DELIS - Dynamically Evolving Large Scale Information Systems - Contract no 001907, by "Project ALGO-NEXT: Algorithms for the Next Generation Internet and Web: Methodologies, Design, and Experiments", MIUR Programmi di Ricerca Scientifica di Rilevante Interesse Nazionale, and by "The Multichannel Adaptive Information Systems (MAIS) Project", MIUR–FIRB.

P. Healy and N.S. Nikolov (Eds.): GD 2005, LNCS 3843, pp. 49–60, 2005.

vertices are the leaves of the subtree rooted at ν. The subgraph of G induced by $V(\nu)$ is denoted as $G(\nu)$. An edge e between a vertex of $V(\nu)$ and a vertex of $V - V(\nu)$ is *incident* to ν. Graph G and tree T are called *underlying graph* and *inclusion tree*, respectively. A clustered graph is *connected* if for each node ν of T we have that $G(\nu)$ is connected.

In a *drawing* of a clustered graph vertices and edges of G are drawn as points and curves as usual [8], and each node ν of T is a simple closed region $R(\nu)$ such that: (i) $R(\nu)$ contains the drawing of $G(\nu)$; (ii) $R(\nu)$ contains a region $R(\mu)$ if and only if μ is a descendant of ν in T; and (iii) any two regions $R(\nu_1)$ and $R(\nu_2)$ do not intersect if ν_1 is not a descendant or an ancestor of ν_2. Consider an edge e and a node ν of T. If e is incident on ν and e crosses the boundary of $R(\nu)$ more than once, we say that edge e and region $R(\nu)$ have an *edge-region crossing*. Also, edge e and region $R(\nu)$ have an *edge-region crossing* if e is not incident on ν and e crosses the boundary of $R(\nu)$. A drawing of a clustered graph is *c-planar* if it does not have edge crossings and edge-region crossings. A clustered graph is *c-planar* if it has a c-planar drawing. C-planarity testing algorithms for connected clustered graphs are shown in [13, 11, 6]. A planarization algorithm for connected clustered graph is shown in [7].

However, the complexity of the c-planarity testing for a non connected clustered graph is still unknown. A contribution on this topic has been given by Gutwenger et al. who presented a polynomial time algorithm for c-planarity testing for *almost connected* clustered graphs [12].

Another contribution studying the interplay between c-planarity and connectivity has been presented in [3] by Cornelsen and Wagner. They show that a *completely connected* clustered graph is c-planar if and only if its underlying graph is planar. A completely connected clustered graph is so that not only each cluster is connected but also its complement is connected.

A clustered graph $C(G, T)$ is *flat* if all the leaves of T have distance two from the root. This implies that all the non-root clusters have depth 1 in T. Hence, in a flat clustered graph $C(G, T)$ a *graph of the clusters* $G^1(C)$ can be identified. Vertices of $G^1(C)$ are the children of the root of T and an edge (μ, ν) exists if and only if an edge of G exists incident to both μ and ν.

Flat clustered graphs offer a way to deepen our insight into the properties of non-connected c-planar clustered graphs. In fact, by changing the families of the graphs G and $G^1(C)$, c-planarity problems of increasing complexity can be identified. The works in [2, 1] by Biedl, Kaufmann, and Mutzel can be interpreted as a linear time c-planarity test for non connected flat clustered graphs with exactly two clusters.

A *clustered cycle* is a flat clustered graph whose underlying graph is a cycle. In [5] it is shown that for a clustered cycle $C(G, T)$ where $G^1(C)$ is also a cycle, the c-planarity testing and embedding problem can be solved in linear time.

A *rigid clustered cycle* is a clustered cycle C in which $G^1(C)$ has a prescribed planar embedding. In this paper we tackle the c-planarity testing and embedding problem for rigid clustered cycles. Namely, consider again the problem stated at the beginning of this section according to the above definitions. The cycle is the underlying graph of a flat clustered graph and the nodes of the graph are the clusters. If you are able to find a drawing of the cycle without intersections you are also able to find a c-planar embedding for the rigid clustered cycle and vice versa.

In this paper we present the following results. We develop a new theory for dealing with rigid clustered cycles, based on operations that preserve their c-planarity (Section 3). We show that the c-planarity of a rigid clustered cycle can be tested in polynomial time (Section 4). As a side effect we also solve in polynomial time the cycle drawing problem stated at the beginning of the section. If the rigid clustered cycle is c-planar we also show a simple method for computing a planar embedding of it (Section 5). Section 2 contains basic definitions, while conclusions and open problems are in Section 6.

2   Basic Definitions

We assume familiarity with connectivity and planarity of graphs [9, 8].

In the following we need a slightly wider definition of clustered cycle in which $G^1(C)$ is allowed to have multiple edges between two nodes. We define a *clustered cycle* $C(G, G^1, \Phi_V, \Phi_E)$, where G^1 is a graph, possibly with multiple edges, G is a cycle, Φ_V maps each vertex of G to a vertex of G^1, and Φ_E maps each edge of G between vertices $v_1 \in \mu_1$ and $v_2 \in \mu_2$, where $\mu_1 \neq \mu_2$, to an edge of G^1 between vertices μ_1 and μ_2.

In the following, to avoid ambiguities, we denote G^1 as $G^1(C)$, its edges will be called *pipes* while its vertices will be called *nodes* or *clusters*.

Given a cluster $\mu \in G^1(C)$, we denote by $deg(\mu)$ the number of pipes that are adjacent to μ in $G^1(C)$, where multiple pipes count for their multiplicity. The *size* of a pipe of $G^1(C)$ is the number of edges of G it contains.

It is easy to see that a path in G whose vertices belong to the same cluster can be collapsed into a single vertex without affecting the c-planarity property of the clustered cycle. Hence, in the following we consider only clustered cycles where consecutive vertices belong to distinct clusters. We call *cusp* a vertex v of G whose incident edges e_1 and e_2 are such that $\Phi_E(e_1) = \Phi_E(e_2)$.

Given a rigid clustered cycle C the *embedding* Λ of C is the specification, for each pipe a in $G^1(C)$ and for each end node μ of a, of the total ordering $\lambda_\mu(a)$ of the edges contained in a when turning around μ clockwise. An embedding of a clustered cycle is c-planar if there exists a planar drawing of C that respects such embedding. If an embedding is c-planar, for each pipe $a = (\mu, \nu)$, we have that $\lambda_\mu(a) = \overline{\lambda}_\nu(a)$, where $\overline{\lambda}_\nu(a)$ denotes the reverse of $\lambda_\nu(a)$.

3   Fountain Clusters

Consider a clustered cycle C and one of its clusters $\mu = \{v_1, \ldots, v_q\}$. For each v_i let w_i and z_i be its neighbors. Cluster μ is a *fountain cluster* if there exists a cluster ν different from μ such that for each v_i we have that $w_i \in \nu$ or $z_i \in \nu$ (see Fig. 1 for an example). We call *base* of μ the pipe of $G^1(C)$ between μ and ν.

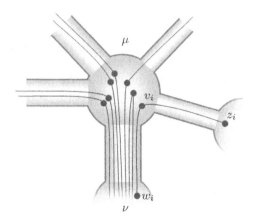

Fig. 1. A fountain cluster

A *fountain clustered cycle* is a clustered cycle in which each cluster is a fountain cluster.

Let μ be a fountain cluster and let b be a base of μ. The following properties hold:

Property 1. Cluster μ has a second base $b' \neq b$ if and only if $deg(\mu) = 2$ and no cusps belongs to μ. Otherwise μ has a single base.

Property 2. The edges incident to a cusp v of μ belong to b.

Property 3. Let a be a pipe incident to μ. If a is also a base for μ then $size(a) = size(b)$, otherwise $size(a) < size(b)$.

3.1   Cluster Expansion

Given a cluster μ of C, we call *cluster expansion* of μ the following operation (see Fig. 2), that produces the clustered cycle C'.

Let $a_1, \ldots, a_k$ be the pipes incident to μ, where $k = deg(\mu)$. Let v a vertex belonging to μ, and let e_i and e_j be the edges incident to v, where $e_i \in a_i$ and $e_j \in a_j$, respectively. Note that if v is a cusp, then $a_i = a_j$.

Cluster μ is replaced in C' with k new clusters $\mu_1, \ldots \mu_k$, each one incident to pipes $a_1, \ldots, a_k$, respectively. All the other clusters of C are unchanged in C'. Each non-cusp vertex v in μ having edges $e_i \in a_i$ and $e_j \in a_j$ is represented in C' by two new vertices v' and v'', with $\Phi_V(v') = \mu_i$ and $\Phi_V(v'') = \mu_j$. A new pipe (μ_i, μ_j) is inserted (if not already present) and a new edge (v', v'') is added such that $\Phi_E(v', v'') = (\mu_i, \mu_j)$. Each cusp vertex v having its edges in pipe a_i stays unchanged in C', and belongs to cluster μ_i.

Note that a cluster μ_i produced by the cluster expansion is a fountain cluster with base a_i. Hence, after one expansion the number of non-fountain clusters of C' is not greater than the number of non-fountain clusters of C. Also, before applying the cluster expansion, μ could be the end node of multiple pipes. After the cluster expansion these multiple pipes are eliminated.

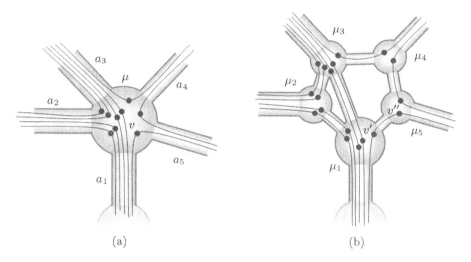

Fig. 2. An example of cluster expansion: (a) A non-fountain cluster μ. (b) The result of the cluster expansion.

Up to now, the expansion operation has been defined whithout considering the embedding of C and C'. If C is embedded (rigid) it is easy to extend the definition of cluster expansion considering also embedding issues. Namely, we embed the new pipes around the new nodes with the same order the old edges had in C. Note that, even if the starting embedding is planar, the resulting embedding may be not planar due to the new pipes inserted among the clusters $\mu_1, \ldots \mu_k$.

Given a rigid clustered cycle C, a cluster expansion of one of its clusters μ is *feasible* if the embedding induced on $G^1(C')$ is planar, that is, if C' is a rigid clustered cycle.

Lemma 1. *Given a rigid clustered cycle C, if a cluster expansion of one of its clusters μ is not feasible, then C is not c-planar.*

Proof. If the cluster expansion of μ is not feasible, then the induced embedding on $G^1(C')$ contains a crossing, that is, it contains two pipes (μ_i, μ_h) and (μ_j, μ_l), with $i < j < h < l$. This implies that there exist two paths of G, one traversing clusters ν_i, μ, ν_h and the other traversing ν_j, μ, ν_l. Since the embedding of μ is fixed, this two paths cannot be drawn without intersections. □

A *cluster expansion* operation on a clustered cycle C is done performing a cluster expansion for each non-fountain cluster of C. A cluster expansion of a rigid clustered cycle is *feasible* if all the required cluster expansions are feasible, that is if the result is a rigid clustered cycle.

Property 4. The cluster expansion of a clustered cycle produces a fountain clustered cycle.

Lemma 2. *Let C be a rigid clustered cycle and let μ be a cluster of C. Let C' be the result of a feasible cluster expansion applied to μ. C is c-planar iff C' is c-planar.*

Proof sketch. Suppose that C is c-planar, and let Γ be a c-planar embedding of C. A c-planar embedding Γ' of C' can be computed as follows. For each pipe that is present both in C and in C', including pipes $a_1, \ldots, a_k$ incident to μ, we assume that the order of edges in Γ' is the same as in Γ. The order of the edges inside the pipes added among nodes $\mu_1, \ldots, \mu_k$ is determined by the their order in the bases $a_1, \ldots, a_k$. Hence, the c-planarity of Γ' follows from the c-planarity of Γ.

Suppose now that C' is c-planar, and let Γ' be a c-planar embedding of C'. A c-planar embedding Γ of C can directly obtained from Γ'. Since all pipes of C are also present in C', the order of their edges can be assumed to be the same as in Γ'. Consider edge e of pipe (μ_i, μ_j) in Γ'. The path e_i, e, e_j of Γ', where $e_i \in a_i$ and $e_j \in a_j$ corresponds to path e_i, e_j in Γ. Hence, the c-planarity of Γ' implies the c-planarity of Γ. □

By repeatedly applying Lemma 2 we have:

Lemma 3. *Let C be a rigid clustered cycle and let C' be a feasible cluster expansion of C. C is c-planar iff C' is c-planar.*

3.2   Pipe Contraction

We call a pipe b between two fountain clusters μ and ν *contractible* if (i) b is the only pipe between μ and ν, (ii) b is a base for both μ and ν, and (iii) b is the only base for one of them.

We define the *pipe contraction* operation on a contractible pipe b as follows. The pipe contraction produces a clustered cycle C' starting from a clustered cycle C by replacing μ, ν, and b, with a new cluster μ', which is adjacent to all the clusters which μ and ν were adjacent to.

If μ and ν were adjacent to the same cluster ρ, μ' is doubly adjacent to ρ; that is, the pipe contraction may introduce multiple pipes incident to μ'.

Each edge e_{in} entering μ or ν belongs to a path $p_C = e_{in}, v, e_1, v_1, \ldots, e_k, v_k, e_{out}$, where e_{out} is the first edge exiting μ or ν and $\Phi_E(e_i) = b$, $i = 1, \ldots, k$. Since b is a base for both μ and ν, $k \geq 1$. Path p_C is replaced by $p_{C'} = e_{in}, v_{\mu'}, e_{out}$, with $\Phi_V(v_{\mu'}) = \mu'$.

An example of pipe contraction is shown in Fig. 3. Note that the new cluster μ' is, in general, not a fountain cluster. If C has a prescribed embedding we assume that the result has also a prescribed embedding in which the circular order of the pipes around μ' is the same as the circular order they have in C around the subgraph composed of μ, ν, and b.

Lemma 4. *Let C be a fountain clustered cycle and C' be obtained from C by applying a pipe contraction operation. C is c-planar iff C' is c-planar.*

Proof sketch. Suppose that C is c-planar, let Γ be a c-planar drawing of C, we show how to build a c-planar drawing Γ' of C' by slightly modifying Γ. Namely,

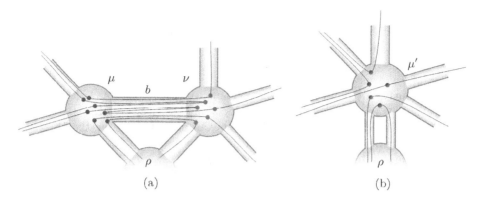

Fig. 3. An example of pipe contraction: (a) pipe b before contraction; (b) The result of the contraction of b

region $R(\mu')$ is the union of $R(\mu)$, $R(\nu)$, and the stripe corresponding to b. (Observe that $R(\mu')$ is connected.) Each path $p_C = e_{in}, v, e_1, v_1, \ldots, e_k, v_k, e_{out}$ of C, with $\Phi_E(e_i) = b$, is replaced by $p_{C'} = e_{in}, v_{\mu'}, e_{out}$, where $v_{\mu'}$ replaces v, and all vertices v_i, with $i = 1, \ldots, k$, are removed joining their incident edges. It is easy to see that the obtained drawing is a c-planar drawing of C'.

Suppose now that C' is c-planar, and let Γ' be a c-planar drawing of C'. We provide a c-planar drawing Γ of C by suitably modifying Γ'. We take region $R(\mu) = R(\mu')$. Observe that in Γ' all the pipes that were incident to ν are consecutively attached to the border of $R(\mu')$. Hence, it is possible to add two arbitrarily thin stripes, corresponding to b and $R(\nu)$, respectively, along the border of $R(\mu')$ in such a way to intersect those pipes only (see Fig 4.b).

Now, consider the edges entering $R(\mu')$ that were incident to μ before contraction in counterclockwise order. Let e_{in} be the current edge and $p_{C'} = e_{in}, v_{\mu'}, e_{out}$

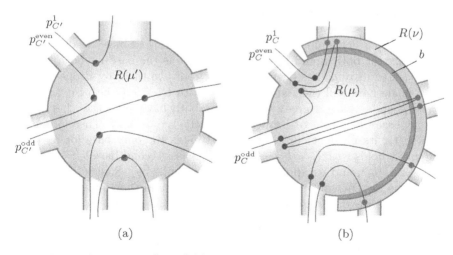

Fig. 4. A drawing Γ' of C' (a) and the corresponding drawing Γ of C (b)

be the path of C' that replaced $p_C = e_{in}, v, e_1, v_1, \ldots, e_k, v_k, e_{out}$. (Remember that $k \geq 1$.) If $k = 1$, it is easy to obtain a drawing of $p_C = e_{in}, v, e_1, v_1, e_{out}$ starting from the drawing of $p_{C'} = e_{in}, v_{\mu'}, e_{out}$ by replacing $v_{\mu'}$ with v and splitting e_{out} with a vertex v_1 in such a way that v_1 is into $R(\mu)$ (see paths $p_{C'}^1$ and p_C^1 of Fig 4 for an example).

Analogously, if k is odd (e_{out} was incident to ν) it is possible to draw $p_C = e_{in}, v, e_1, v_1, \ldots, e_k, v_k, e_{out}$ in a thin stripe along the drawing of $p_{C'} = e_{in}, v_{\mu'}$, e_{out} (see paths $p_{C'}^{\text{odd}}$ and p_C^{odd} of Fig 4 for an example). If k is even, then both e_{in} and e_{out} were incident to μ in C. In this case the drawing of $p_{C'} = e_{in}, v_{\mu'}, e_{out}$ does not immidiately provide a drawing of $p_C = e_{in}, v, e_1, v_1, \ldots, e_k, v_k, e_{out}$, which can be built as follows. Vertex v is placed into $R(\mu)$ as edge e_{in} crosses the border of $R(\mu)$. Edge e_1 follows clockwise the border of $R(\mu)$ till the previous edge e'_{in} entering $R(\mu)$ is found (or $R(\nu)$ is reached). Since edges e_{in} are considered in counterclockwise order and since b was a base for both μ and ν, path p_C', starting with edge e'_{in}, always has vertex v' into $R(\mu)$ and v'_1 into $R(\nu)$. Therefore, edge e_1 can be drawn arbitrarily near to path p_C' and can be terminated with v_1 placed into $R(\nu)$. Edges e_i, with $i = 2, \ldots, k$, can be drawn in an arbitrarily thin stripe adjacent to e_1, positioning v_i alternately into $R(\mu)$ and $R(\nu)$. Finally, edge e_{out} can follow path $p_{C'}$ to exit $R(\nu)$ (see paths $p_{C'}^{\text{even}}$ and p_C^{even} of Fig 4 for an example). □

4   C-Planarity Testing of Clustered Cycles

In this section we describe a c-planarity testing algorithm for rigid clustered cycles. The following lemmas state properties of clustered cycles which are needed to prove the correctness of the algorithm.

Lemma 5. *Let C be a fountain clustered cycle such that $G^1(C)$ is not a simple cycle and has not multiple pipes. There exists at least one contractible pipe b^* in $G^1(C)$.*

Proof sketch. Consider a pipe $b = (\mu, \nu)$ of maximum size. Since b is the pipe of maximum size for both μ and ν, by Property 3, b is the base for both. If one between μ and ν (say μ) has degree different from two then, by Property 1, μ admits a single base and the statement holds with $b^* = b$. Otherwise, suppose that both μ and ν have degree two and that both have two bases. Let b_1 be the second base of μ. Due to Property 3, $size(b_1) = size(b)$. Therefore b_1 is also a base for its incident cluster $\mu_1 \neq \mu$. If b_1 is the only base for μ_1 then the statement holds with $b^* = b_1$, otherwise μ_1 has a second base $b_2 \neq b_1$, with $size(b_2) = size(b_1)$, and we apply the same argument to b_2. Since $G^1(C)$ is not a simple cycle the current pipe b_i is different from b and there exists at least a j for which b_j is the only base for μ_j. □

We introduce a quantity that will be used to analyze the algorithm both in terms of correctness and in terms of time complexity. Intuitively, it is an indicator of the structural complexity of $G^1(C)$. We denote by $\mathcal{E}(C)$ the following quantity:

$$\mathcal{E}(C) = \sum_{a \in \{\text{pipes of } G^1(C)\}} (size(a))^2.$$

We now concentrate on a pair of consecutive contraction-expansion operations and show how $\mathcal{E}$ changes.

Lemma 6. *Let C be a fountain clustered cycle and let $b = (\mu, \nu)$ be a contractible pipe which is the only base for μ. Let C^* be the clustered cycle obtained by applying a pipe contraction to b followed by a cluster expansion of the obtained cluster μ'. We have that $\mathcal{E}(C^*) < \mathcal{E}(C)$.*

Proof. Let C' be the clustered cycle generated by the pipe contraction applied to b. C' contains all the pipes of C with the exception of b, then $\mathcal{E}(C') = \mathcal{E}(C) - (size(b))^2$. Clustered cycle C^* has the same pipes of C' plus a set of new pipes $a_1, \ldots, a_k$. If $k = 0$ then $\mathcal{E}(C^*) = \mathcal{E}(C') < \mathcal{E}(C)$. If $k = 1$ then $deg(\mu') = deg(\mu) = deg(\nu) = 2$. Since b is the only base for μ by Properties 1 and 2, b contains edges incident to cusps which are not present in a_1. Therefore $\mathcal{E}(C^*) < \mathcal{E}(C)$. Suppose $k > 2$. We have that $\mathcal{E}(C^*) = \mathcal{E}(C') + \sum_{j=1}^{k}(size(a_j))^2 = \mathcal{E}(C) - (size(b))^2 + \sum_{j=1}^{k}(size(a_j))^2$. Observe that each edge contained in the pipes $a_1, \ldots, a_k$ is generated by the split of a vertex in μ', and that the number of vertices in μ' is at most $size(b)$. Then, $\sum_{j=1}^{k} size(a_j) \le size(b)$. Hence, $\sum_{j=1}^{k}(size(a_k))^2 < (size(b))^2$, and the statement follows. □

Lemma 7. *A clustered cycle C whose graph of the clusters $G^1(C)$ is a path is c-planar.*

Proof sketch. Let $\mu_1, \ldots, \mu_m$ be the nodes of $G^1(C)$ in the order in which they appear in the path. A planar embedding of C can be built as follows. Traverse the cycle G starting from a vertex in μ_1. Each edge e belonging to pipe $a = (\mu_i, \mu_j)$ is inserted at the last position of $\lambda_{\mu_i}(a)$ and at the first position of $\lambda_{\mu_j}(a)$. When the path comes back to μ_1 for the last time it can be connected to the starting point preserving c-planarity. □

We are now ready to introduce the c-planarity testing algorithm for a rigid clustered cycle C. First, the algorithm performs a cluster expansion for each non-fountain cluster. If one of such expansions is not feasible, then, according to Lemma 1, C is not c-planar. If all the expansions are feasible, according to Property 4, we obtain a fountain clustered cycle C^f, which is c-planar iff C is c-planar. If the clusters of C^f form a cycle, then the c-planarity can be easily tested using the results described in [5]. If $G^1(C^f)$ is a path, then Lemma 7 states that C^f is c-planar. If the clusters of C^f form neither a cycle nor a path, then Lemma 5 ensures that there exists a contractible pipe $b^* = (\mu, \nu)$. Perform a contraction operation on b^*. Perform a cluster expansion on the resulting cluster. These last two steps are performed until the clusters of the clustered cycle form a cycle, or a path, or a cluster expansion fails. Note that a pipe contraction may temporarily generate multiple pipes; however, the subsequent cluster expansion produces a new clustered cycle which has no multiple pipes. The algorithm, called *ClusteredCyclePlanarityTesting*, is formally described below.

Algorithm *ClusteredCyclePlanarityTesting*

input A rigid clustered cycle C
output True if C is c-planar, false otherwise

  for all non-fountain clusters μ in C **do**
    perform a cluster expansion of μ
    if the cluster expansion of μ is not feasible **then**
      return false
    end if
  end for
  {at this point C is a fountain clustered cycle}
  while C is not a cycle or a path **do**
    let b be a contractible pipe of C
    apply a pipe contraction to b, obtaining cluster μ'.
    perform a cluster expansion of μ'
    if the cluster expansion of μ' is not feasible **then**
      return false
    end if
  end while
  {at this point C is a cycle or a path}
  if C is a cycle **then**
    return the result of the c-planarity testing on C
  else
    return true
  end if

Theorem 1. *There exists a polynomial time algorithm to test if a rigid clustered cycle is c-planar.*

Proof. First, we prove that algorithm ClusteredCyclePlanarityTesting can be always executed in a polynomial number of steps. Let C be a rigid clustered cycle whose underlying cycle is G and be n the number of vertices of G. In the first phase of the algorithm a cluster expansion is performed for all the non-fountain clusters. Each cluster expansion can be performed in polynomial time. At the end of this phase the number of vertices is at most $2n$. Suppose that $\overline{E}$ is the value of $\mathcal{E}(C)$ at the end of this phase. We have that $\overline{E} = O(n^2)$.

By Lemma 6 each pair of pipe contraction and cluster expansion decreases $\mathcal{E}(C)$ of at least one unit. Hence, the body of the **while** cycle is executed at most $\overline{E}$ times. Also, a contractible pipe always exists (see Lemma 5) and can be determined in constant time using a suitable data structure that contains the candidate bases and that is updated after each operation. This proves that algorithm ClusteredCyclePlanarityTesting terminates in polynomial time.

Second, we prove that algorithm ClusteredCyclePlanarityTesting gives the correct result. Lemmas 2, 3, and 4 guarantee that the cluster expansion and pipe contraction operations can be applied without modifying the c-planarity property of the graph, while if a cluster expansion is not feasible the graph is not c-planar. If none of the cluster expansions fails, either the algorithm produces a k-cluster cycle and applies the c-planarity testing algorithm shown in [5], or produces a clustered path, which by Lemma 7 is always c-planar. Also, (see the above discussion) the algorithm always terminates.   □

5   Computing C-Planar Embeddings of Clustered Cycles

In this section we show how to build an embedding for a c-planar rigid clustered cycle. We assume that Algorithm ClusteredCyclePlanarityTesting, described in Section 4, has been applied, and that each step of the algorithm has been recorded. The clustered cycle C_{end} obtained at the last step of the execution of that algorithm is such that $G^1(C_{end})$ is a cycle or a path. A c-planar embedding of C_{end} can be easily computed by using the results described in [5], if $G^1(C_{end})$ is a cycle, or by using the technique introduced in the proof of Lemma 7, if $G^1(C_{end})$ is a path.

The embedding of the input clustered cycle can be obtained by going through the transformations operated by Algorithm ClusteredCyclePlanarityTesting in reverse order starting from a c-planar embedding of C_{end}. Algorithm Clustered-CyclePlanarityTesting performs two kind of operations: pipe contraction and cluster expansion.

For each cluster expansion on a clustered cycle C, which produces a cluster cycle C', the embedding of C is directly obtained from the embedding of C' as described in the proof of Lemma 2 since all pipes in C' are also in C and their embedding do not change.

For each pipe contraction on a clustered cycle C, which produces a cluster cycle C', only part of the embedding of C can be directly obtained from the embedding of C' since C has one more pipe (the contracted one) with respect to C'. The proof of Lemma 4 describes how to compute a c-planar embedding of C starting from a c-planar embedding of C'.

From the above discussion and from the fact that ClusteredCyclePlanarityTesting has a polynomial time complexity we can state the following result.

Theorem 2. *Given a c-planar rigid clustered cycle, a c-planar embedding of it can be computed in polynomial time.*

6   Conclusions

In this paper we addressed the problem of drawing, without crossings, a cycle in a planar embedded graph and have shown that the problem can be solved in polynomial time.

If we interpret the problem and the result from the clustered planarity perspective it turns out that we have identified a new family of flat clustered graphs that are highly non-connected and whose c-planarity can be tested in polynomial time. This might be useful for deepening the insight into the general problem of testing the c-planarity of non-connected clustered graphs, whose computational complexity is still unknown.

However, we point out that a trivial generalization of the result to flat clustered graphs whose underlying graph is a general graph fails. In fact, it is easily to find clustered graphs which are not c-planar while all cycles of their underlying graphs are separately c-planar.

References

1. T. C. Biedl. Drawing planar partitions III: Two constrained embedding problems. Tech. Report RRR 13-98, RUTCOR Rutgen University, 1998.
2. T. C. Biedl, M. Kaufmann, and P. Mutzel. Drawing planar partitions II: HH-Drawings. In *Workshop on Graph-Theoretic Concepts in Computer Science (WG'98)*, volume 1517, pages 124–136. Springer-Verlag, 1998.
3. S. Cornelsen and D. Wagner. Completely connected clustered graphs. In *Proc. 29th Intl. Workshop on Graph-Theoretic Concepts in Computer Science (WG 2003)*, volume 2880 of *LNCS*, pages 168–179. Springer-Verlag, 2003.
4. P. F. Cortese and G. Di Battista. Clustered planarity. In *SCG '05: Proceedings of the twenty-first annual symposium on Computational geometry*, pages 32–34, New York, NY, USA, 2005. ACM Press.
5. P. F. Cortese, G. Di Battista, M. Patrignani, and M. Pizzonia. Clustering cycles into cycles of clusters. In János Pach, editor, *Proc. Graph Drawing 2004 (GD'04)*, volume 3383 of *LNCS*, pages 100–110. Springer-Verlag, 2004.
6. E. Dahlhaus. Linear time algorithm to recognize clustered planar graphs and its parallelization. In C.L. Lucchesi, editor, *LATIN 98, 3rd Latin American symposium on theoretical informatics, Campinas, Brazil, April 20–24, 1998*, volume 1380 of *LNCS*, pages 239–248, 1998.
7. G. Di Battista, W. Didimo, and A. Marcandalli. Planarization of clustered graphs. In *Proc. Graph Drawing 2001 (GD'01)*, LNCS, pages 60–74. Springer-Verlag, 2001.
8. G. Di Battista, P. Eades, R. Tamassia, and I. G. Tollis. *Graph Drawing*. Prentice Hall, Upper Saddle River, NJ, 1999.
9. S. Even. *Graph Algorithms*. Computer Science Press, Potomac, Maryland, 1979.
10. Q. W. Feng, R. F. Cohen, and P. Eades. How to draw a planar clustered graph. In Ding-Zhu Du and Ming Li, editors, *Proc. COCOON'95*, volume 959 of *LNCS*, pages 21–30. Springer-Verlag, 1995.
11. Q. W. Feng, R. F. Cohen, and P. Eades. Planarity for clustered graphs. In P. Spirakis, editor, *Symposium on Algorithms (Proc. ESA '95)*, volume 979 of *LNCS*, pages 213–226. Springer-Verlag, 1995.
12. C. Gutwenger, M. Jünger, S. Leipert, P. Mutzel, M. Percan, and René Weiskircher. Advances in *C*-planarity testing of clustered graphs. In Stephen G. Kobourov and Michael T. Goodrich, editors, *Proc. Graph Drawing 2002 (GD'02)*, volume 2528 of *LNCS*, pages 220–235. Springer-Verlag, 2002.
13. T. Lengauer. Hierarchical planarity testing algorithms. *J. ACM*, 36(3):474–509, 1989.

On Rectilinear Duals for
Vertex-Weighted Plane Graphs

Mark de Berg*, Elena Mumford, and Bettina Speckmann

Department of Mathematics & Computer Science, TU Eindhoven, The Netherlands
{mdberg, speckman}@win.tue.nl, e.mumford@tue.nl

Abstract. Let $\mathcal{G} = (V, E)$ be a plane triangulated graph where each vertex is assigned a positive weight. A rectilinear dual of $\mathcal{G}$ is a partition of a rectangle into $|V|$ simple rectilinear regions, one for each vertex, such that two regions are adjacent if and only if the corresponding vertices are connected by an edge in E. A rectilinear dual is called a cartogram if the area of each region is equal to the weight of the corresponding vertex. We show that every vertex-weighted plane triangulated graph $\mathcal{G}$ admits a cartogram of constant complexity, that is, a cartogram where the number of vertices of each region is constant.

1   Introduction

Motivation. Cartographers have developed many different techniques to visualize statistical data about a set of regions like countries, states or counties. *Cartograms* are among the most well known and widely used of these techniques. The regions of a cartogram are deformed such that the area of a region corresponds to a particular geographic variable [4]. The most common variable is population: In a population cartogram, the areas of the regions are proportional to their population. There are several types of cartograms. Of particular relevance for this paper are the *rectangular cartograms* introduced by Raisz in 1934 [12], where each region is represented by a rectangle. This has the advantage that the areas (and thereby the associated values) of the regions can be easily estimated by visual inspection.

Whether a cartogram is good is determined by several factors. In this paper we focus on two important criteria, namely the correct adjacencies of the regions of the cartogram and the *cartographic error* [5]. The first criterion requires that the dual graph of the cartogram is the same as the dual graph of the original map. Here the *dual graph* of a map—also referred to as *adjacency graph*—is the graph that has one node per region and connects two regions if they are adjacent, where two regions are considered to be adjacent if they share a 1-dimensional part of their boundaries (see Fig. 1). The second criterion, the cartographic error, is defined for each region as $|A_c - A_s|/A_s$, where A_c is the area of the region in the cartogram and A_s is the specified area of that region, given by the geographic variable to be shown.

* Supported by the Netherlands' Organisation for Scientific Research (NWO) under project no. 639.023.301.

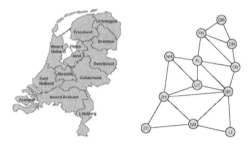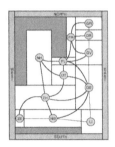

Fig. 1. The provinces of the Netherlands, their adjacency graph, a population carto-gram—here additional "sea rectangles" were added to preserve the outer shape

From a graph-theoretic point of view constructing rectangular cartograms with correct adjacencies and zero cartographic error translates to the following problem. We are given a plane graph $\mathcal{G} = (V, E)$ (the dual graph of the original map) and a positive weight for each vertex (the required area of the region for that vertex). Then we want to construct a partition of a rectangle into rect-angular regions whose dual graph is $\mathcal{G}$—such a partition is called a *rectangular dual* of $\mathcal{G}$—and where the area of each region is the weight of the corresponding vertex. As usual, we assume the input graph $\mathcal{G}$ is plane and triangulated, except possibly the outer face; this means that the original map did not have four or more countries whose boundaries share a common point and that $\mathcal{G}$ does not have degree-2 nodes.[1]

Unfortunately not every vertex-weighted plane triangulated graph admits a rectangular cartogram, even if we ignore the vertex weights and concentrate only on the correct adjacencies. There are several possibilities to address this problem. One is to relax the strict requirements on the adjacencies and areas. For example, Van Kreveld and Speckmann [14] gave an algorithm that constructs rectangular cartograms that in practice have only a small cartographic error and mild disturbances of the adjacencies. Heilmann et al. [6] gave an algorithm that always produces regions with the correct areas; unfortunately the adjacencies can be disturbed badly. The other extreme is to ignore the area constraints and focus only on getting the correct adjacencies—that is, to focus on rectangular duals rather than cartograms. This setting is relevant for computing floor plans in VLSI design. As mentioned above, ignoring the area constraints still does not guarantee that a solution exists. But, if the input graph is a triangulated plane graph without separating triangles—a separating triangle is a 3-cycle with vertices both inside and outside the cycle—then a rectangular dual always exists [1,8] and can be computed in linear time [7].

Another option is to use different shapes for the regions. We restrict our attention to so-called *rectilinear cartograms*, which use rectilinear polygons as regions—see [10, 4] for some examples from the cartography community. If we

---

[1] Degree-2 nodes can easily be handled using suitable pre- and postprocessing steps [14].

now ignore the area requirement then things become much better: Any plane triangulated graph admits a rectilinear dual. In fact, Liao et al. [9] recently showed
that any plane triangulated graph admits a rectilinear dual with regions of small
complexity, namely rectangles, L-shapes, and T-shapes. The main questions now
are: Does any plane triangulated vertex-weighted graph admit a rectilinear cartogram with zero cartographic error and correct adjacencies? And if so, can it
always be done with a constant number of vertices per region?

This problem was studied by Rahman et al. [11] for a very special class of
graphs, namely a certain subclass of graphs that admit a sliceable dual—see
below. They showed that such graphs admit a rectilinear cartogram where every
region has at most 8 vertices. Biedl and Genc [2] showed that it is NP-hard to
decide if a rectilinear cartogram that uses regions with at most 8 vertices exists
for a given graph. Furthermore, a rectangular layout can be interpreted as a
plane, cubic graph. Thomassen showed [13] that any such graph can be drawn
with straight (but not necessarily horizontal or vertical) edges such that every
bounded face has any prescribed area. These results leave the two questions
stated above still unanswered. Our paper answers them: We prove that any
plane triangulated vertex-weighted graph admits a rectilinear cartogram all of
whose regions have constant complexity. Before we describe our results in more
detail we first define the terminology we use more precisely.

Terminology. A *layout* $\mathcal{L}$ is a partition of a rectangle R into a finite set of
interior-disjoint regions. We consider only *rectilinear layouts*, where every region
is a simple rectilinear polygon whose sides are parallel to the edges of R. We
define the complexity of a rectilinear polygon as the total number of its vertices
and the complexity of a rectilinear layout as the maximum complexity of any
of its regions. A rectilinear layout is called *rectangular* if all its regions are
rectangles. Thus, a rectangular layout is a rectilinear layout of complexity 4.
Finally, a rectangular layout is called *sliceable* if it can be obtained by recursively
slicing a rectangle by horizontal and vertical lines, which we call *slice lines*. (In
computational geometry, such a recursive subdivision is called a (rectilinear)
binary space partition, or *BSP* for short.)

We denote the dual graph (also called connectivity graph) of a layout $\mathcal{L}$ by
$\mathcal{G}(\mathcal{L})$. Given a graph $\mathcal{G}$, a layout $\mathcal{L}$ such that $\mathcal{G} = \mathcal{G}(\mathcal{L})$ is called a *dual layout* (or
simply a *dual*) for $\mathcal{G}$. The dual $\mathcal{G}(\mathcal{L})$ is unique for any layout $\mathcal{L}$. Note that not
every graph $\mathcal{G}$ has a dual layout. If it does, then the dual layout is not necessarily
unique.

Fig. 2. A graph $\mathcal{G}$ with a rectangular, rectilinear, and sliceable dual

Every vertex v of a vertex-weighted graph $\mathcal{G}$ has a positive weight $w(v)$ associated with it. Given a vertex-weighted plane graph $\mathcal{G}$ that admits a dual $\mathcal{L}$, we say that $\mathcal{L}$ is a *cartogram* if the area of each region of $\mathcal{L}$ is equal to the weight of the corresponding vertex of $\mathcal{G}$. The cartogram is called *rectangular (rectilinear, sliceable)* if the corresponding layout is rectangular (rectilinear, sliceable).

Results. In Section 2 we show how to construct a cartogram of complexity 12 for any vertex-weighted plane triangulated graph that has a sliceable dual. We extend our results in Section 3 to general vertex-weighted plane triangulated graphs $\mathcal{G}$. Specifically, if $\mathcal{G}$ admits a rectangular dual then we can construct a cartogram of complexity at most 20, otherwise we can construct a cartogram of complexity at most 60. In Section 4 we conclude with several open problems.

2   Graphs That Admit a Sliceable Dual

Let $\mathcal{G} = (V, E)$ be a vertex-weighted plane triangulated graph with n vertices that admits a sliceable dual. The exact characterization of such graphs is still unknown, but Yeap and Sarrafzadeh [15] proved that every triangulated plane graph without separating triangles and without separating 4-cycles has a sliceable dual. W.l.o.g. we assume that the vertex weights of $\mathcal{G}$ sum to 1, and that the rectangle R that we want to partition is the unit square.

Let $\mathcal{L}_1$ be a sliceable dual for $\mathcal{G}$. We scale and stretch $\mathcal{L}_1$ such that it becomes a partition of the unit square R. We will transform $\mathcal{L}_1$ into a cartogram for $\mathcal{G}$ in three steps. In the first step we transform $\mathcal{L}_1$ into a layout $\mathcal{L}_2$ where every region has the correct area. In doing so, however, we may loose some of the adjacencies, that is, $\mathcal{L}_2$ may no longer be a dual layout for $\mathcal{G}$. This is remedied in the second step, where we transform $\mathcal{L}_2$ into a layout $\mathcal{L}_3$ whose dual is $\mathcal{G}$. In this step we re-introduce some errors in the areas. But these errors are small, and we can remove them in the third step, which produces the final cartogram, $\mathcal{L}_4$. Below we describe each of these steps in more detail.

Step 1: Setting the Areas Right
The first step is relatively easy. Recall that a sliceable layout is a recursive partition of R into rectangles by vertical and horizontal slice lines. This recursive partition can be modelled as a BSP tree $\mathcal{T}$. Each node ν of $\mathcal{T}$ corresponds to a rectangle $R(\nu) \subseteq R$ and the interior nodes store a slice line $\ell(\nu)$. The rectangles $R(\nu)$ are defined recursively, as follows. We have $R(\text{root}(\mathcal{T})) = R$. Furthermore, $R(\text{leftchild}(\nu)) = R(\nu) \cap \ell^-(\nu)$ and $R(\text{rightchild}(\nu)) = R(\nu) \cap \ell^+(\nu)$, where $\ell^-(\nu)$ and $\ell^+(\nu)$ denote the half-space to the left and right of $\ell(\nu)$ (or, if $\ell(\nu)$ is horizontal, above and below $\ell(\nu)$). The rectangles $R(\nu)$ corresponding to the leaves are precisely the regions of the sliceable layout. See for example Figure 3—the shaded rectangle corresponds to the shaded node. The BSP tree for a sliceable layout is not necessarily unique, because different recursive partition processes may lead to the same layout.

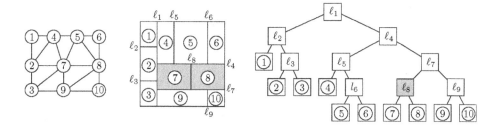

Fig. 3. A graph $\mathcal{G}$, the layout $\mathcal{L}_1$, and the BSP tree $\mathcal{T}$

The point where two or maximally three slice lines meet is called a *junction (point)*. We distinguish between T- and X-junctions. A T-junction involves two slice lines while an X-junction involves three slice lines, two of which are aligned.

Now, let $\mathcal{T}$ be a BSP tree that models the sliceable layout $\mathcal{L}_1$. We will transform $\mathcal{L}_1$ into $\mathcal{L}_2$ by changing the coordinates of the slice lines used by $\mathcal{T}$ in a top-down manner. We maintain the following invariant: When we arrive at a node ν in $\mathcal{T}$, the area of $R(\nu)$ is equal to the sum of the required areas of the regions represented by the leaves below ν. Clearly this is true when we start the procedure at the root of $\mathcal{T}$. Now assume that we arrive at a node ν which stores a slice line $\ell(\nu)$. We simply sum up all the required areas in the left subtree of ν and adjust the position of the $\ell(\nu)$ in the unique way that assigns the correct areas to $R(\text{leftchild}(\nu))$ and $R(\text{rightchild}(\nu))$. When we reach a leaf there is nothing to do; the rectangle it represents now has the required area.

Step 2: Setting the Adjacencies Right
The movement of the slice lines in Step 1 may have changed the adjacencies between the regions. To remedy this, we will use the BSP tree $\mathcal{T}$ again.

Before we start, we define two strips for each slice line $\ell(\nu)$. These strips are centered around $\ell(\nu)$ and are called the *tail strip* and the *shift strip*. The width of the tail strip is $2\varepsilon_\nu$ and the width of the shift strip is $2\delta_\nu$, where $\varepsilon_\nu < \delta_\nu$ and ε_ν and δ_ν are sufficiently small. The exact values of ε_ν and δ_ν will be specified in Step 3. At this point it is relevant only that we can choose them in such a way that the shift strips of two slice lines are disjoint except when two slice lines meet.

We will make sure that the changes to the layout in Step 2 all occur within the tail strips and that the changes in Step 3 all occur within the shift strips. Due to the choice of the δ_ν's all the junction points within the shift strip will lie on the slice line $\ell(\nu)$.

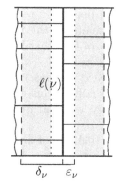

Fig. 4. The shift and tail strips for ℓ_i

To restore the correct adjacencies, we traverse the BSP tree bottom-up. We maintain the invariant that after handling a node ν, all adjacencies between regions inside $R(\nu)$ have been restored. Now suppose that we reach a node ν. The invariant tells us that all adjacencies inside $R(\text{leftchild}(\nu))$ and $R(\text{rightchild}(\nu))$ have been restored. It remains to restore the correct adjacencies between regions on different sides of the slice line $\ell(\nu)$. We will describe how to restore the adjacencies for the case where $\ell(\nu)$ is vertical; horizontal slice lines are handled in a similar fashion, with the roles of the x- and y-coordinates exchanged.

Let $A_1, A_2, \ldots, A_k$ be the set of regions inside $R(\nu)$ bordering $\ell(\nu)$ from the left, and let $B_1, B_2, \ldots, B_m$ be the set of regions inside $R(\nu)$ bordering $\ell(\nu)$ from the right. Both the A_i's and the B_j's are numbered from top to bottom—see Figure 5. We write $A_i \prec A_j$ to indicate that A_i is above A_j; thus $A_i \prec A_j$ if and only if $i < j$. The same notation is used for the B_j's. Now consider the tail strip centered around $\ell(\nu)$. All slice lines ending on $\ell(\nu)$ are straight lines within the tail strip (and, in fact, even within the shift strip). This is true before Step 2, but as we argue later, it is still true when we start to process $\ell(\nu)$.

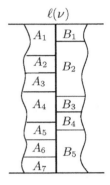

Fig. 5. Left and right neighbors

In Step 1 (and when Step 2 was applied to $R(\text{leftchild}(\nu))$ and $R(\text{rightchild}(\nu))$), the slice lines separating the A_i's from each other and the slice lines separating the B_j's from each other may have shifted, thus disturbing the adjacencies between the A_i's and B_j's. For each A_i, we define $\text{top}(A_i) := B_k$ if B_k is the highest region (among the B_j's) adjacent to A_i in the original layout $\mathcal{L}_1$. Similarly, $\text{bottom}(A_i)$ is the lowest such region. This means that in $\mathcal{L}_1$, the region A_i was adjacent to all B_j with $\text{top}(A_i) \preceq B_j \preceq \text{bottom}(A_i)$. We restore these adjacencies for A_i by adding at most two so-called *tails* to A_i, as described below. This is done from top to bottom: We first handle A_1, then A_2, and so on. During this process the slice line $\ell(\nu)$ will be deformed—it will no longer be a straight line, but it will become a rectilinear poly-line. However, the part of $\ell(\nu)$ bordering regions we still have to handle will be straight. More precisely, we maintain the following invariant: When we start to handle a region A_i, the part of $\ell(\nu)$ that lies below the bottom edge of $\text{top}(A_i)$ is straight and the right borders of all $A_j \succeq A_i$ are collinear with that part of ℓ.

Next we describe how A_i is handled. There are two cases, which are not mutually exclusive: Zero, one, or both of them may apply. When both cases apply, we treat first (a) and then (b).

(a) If A_i is not adjacent to $\text{top}(A_i)$ and $\text{top}(A_i)$ is higher than A_i, then we add a tail from A_i to $\text{top}(A_i)$. (If A_i is not adjacent to $\text{top}(A_i)$ and $\text{top}(A_i)$ is lower than A_i, then case (b) will automatically connect A_i to $\text{top}(A_i)$.) More precisely, we add a rectangle to the right of A_i whose bottom edge is collinear with the bottom edge of A_i and whose top edge is contained in the bottom edge of $\text{top}(A_i)$. The width of this rectangle is $\frac{\varepsilon_\nu}{n}$. Moreover, we shift the part of the slice line below $\text{top}(A_i)$ by $\frac{\varepsilon_\nu}{n}$ to the right. Observe that this will make all the B_j below $\text{top}(A_i)$ smaller and all A_j below A_i larger.

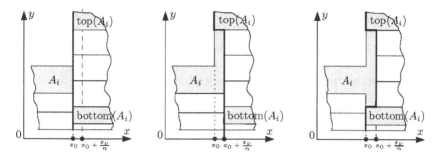

Fig. 6. Both case (a) and case (b) apply

(b) If A_i is not adjacent to bottom(A_i) and bottom(A_i) is lower than A_i, then
we also add a tail, as follows. (If A_i was not adjacent to bottom(A_i) and
bottom(A_i) was higher than A_i, then necessarily case (a) has already been
treated and in fact A_i is now adjacent to bottom(A_i).) First, we shift the
part of the slice line below the top edge of bottom(A_i) by $\frac{\varepsilon_\nu}{n}$ to the left.
Observe that this will enlarge bottom(A_i) and all the B_j below it, and make
all $A_j \succ A_i$ smaller. Next, we add a rectangle of width $\frac{\varepsilon_\nu}{n}$ to A_i, which connects A_i to bottom(A_i). Its top edge is contained in the bottom edge of A_i,
its right edge is collinear to A_i's right edge, and its bottom edge is contained
in the top edge of bottom(A_i).

Note that every tail "ends" on some B_j, that is, no tail extends all the way to
the slice lines on which $\ell(\nu)$ ends. This implies that

- no bends are introduced inside the shift strips of the two slice lines on which
  $\ell(\nu)$ ends (as we already claimed earlier).
- the bordering sequence (the sets of countries along each side of a slice line
  and their order) of any other slice line remains unchanged.
- the bottom end of $\ell(\nu)$ shifts only within the tail strip of $\ell(\nu)$.

Lemma 1. *The layout $\mathcal{L}_3$ obtained after Step 2 has the following properties:*

(i) If two regions are adjacent in $\mathcal{L}_1$, then they are also adjacent in $\mathcal{L}_3$.
*(ii) The tails that are added when handling a slice line ℓ all lie within the tail
strip of ℓ.*
(iii) Each region gets at most three tails.

Proof.

(i) It follows from the construction that each region A_i along a slice line $\ell(\nu)$ has
the required adjacencies after $\ell(\nu)$ has been handled. Hence, the construction
maintains the invariant that all adjacencies within $R(\nu)$ are restored after
$\ell(\nu)$ has been handled. Therefore, after the slice line that is stored at the
root of $\mathcal{T}$ is handled, all adjacencies have been restored.
(ii) A tail inside a tail strip of width $2\varepsilon_\nu$ has width $\frac{\varepsilon_\nu}{n}$ and is always adjacent
to the current slice line. A slice line is shifted at most $n - 2$ times by $\frac{\varepsilon_\nu}{n}$.
Hence, the tails lie within the tail strip, as claimed.

(iii) A region can get tails only when the slice line ℓ_r on its right and the slice line ℓ_t along its top are handled. Since a region must be either the topmost region along ℓ_r or the rightmost region along ℓ_t it can only get a double tail along one of these slice lines. Thus each region receives at most 3 tails. Note that since the tails along the same slice line are aligned, a region does not get more than three concave vertices.   □

Note that if $\mathcal{G}$ is triangulated then Lemma 1 (i) implies that two regions in $\mathcal{L}_3$ are adjacent if and only if they are adjacent in $\mathcal{L}_1$: All required adjacencies are present and in a plane triangulated graph there is no room for additional adjacencies.

Step 3: Repairing the Areas

When we repaired the adjacencies in Step 2, we re-introduced some small errors in the areas of the regions. We now set out to remedy this. In Step 2, the slice lines actually became rectilinear poly-lines. These poly-lines, which we will keep on calling slice lines for convenience, are monotone: A horizontal (resp. vertical) line intersects any vertical (resp. horizontal) slice line in a single point, a segment, or not at all. We will repair the areas by moving the slice lines in a top-down manner, similar to Step 1. But because we do not want to loose any adjacencies again, we have to be more careful in how we exactly move a slice line. This is described next.

Assume that we wish to move a horizontal slice line $\ell = \ell_\nu$; vertical slice lines are treated in a similar manner. Let ℓ_1 and ℓ_2 be the slice lines to the left and to the right of ℓ, that is, the slice lines on which ℓ ends. We define a so-called *container* for ℓ, denoted by $C(\ell)$. The container $C(\ell)$ is a rectangle containing most of ℓ, as well as parts of the other slice lines ending on ℓ. Instead of moving the slice line ℓ we will move the container $C(\ell)$ and its complete contents.

We first define the container $C(\ell)$ more precisely. The top and bottom edge of $C(\ell)$ are contained in the boundary of the tail strip of ℓ. The position of the right edge of ℓ is determined by what happened at the junction between ℓ and ℓ_2 when ℓ_2 was processed during Step 2. Let A_i and A_{i+1} be the regions above and below ℓ and bordering ℓ_2.

(i) A_i *did not get a downward tail and A_{i+1} did not get an upward tail.*
   In this case either there is no other junction on ℓ_2 within ℓ's shift strip, or there is exactly one and it lies within ℓ's tail strip (see Fig. 7(a)). If there is a junction on ℓ_2 within ℓ's tail strip in the direction in which $C(\ell)$ should be moved, then we set the right edge of the container $C(\ell)$ at distance ε_ν/n from ℓ_2 (see Fig. 7(b)). Otherwise, the right edge of the container is collinear with the part of ℓ_2 lying within ℓ's shift strip (see Fig. 7(c)).

(ii) A_i *got a downward tail or A_{i+1} got an upward tail.*
   Note that in this case more tails may have entered the tail strip of ℓ. For example, if A_{i+1} got an upward tail then some other regions below A_{i+1} possibly got an upward tail as well. In this case the right edge of $C(\ell)$ will go through the leftmost such tail edge—see Fig. 8. Figures 9 and 10 illustrate

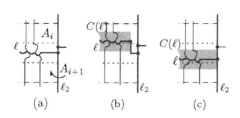

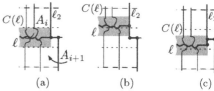

Fig. 7. (a) ℓ and ℓ_2 form a T-junction; (b) $C(\ell)$ is moved up and there is a junction in its way; (c) $C(\ell)$ is moved down and there is no junction in its way

Fig. 8. (a) A_{i+1} has an upward tail; (b) moving $C(\ell)$ up; (c) moving $C(\ell)$ down

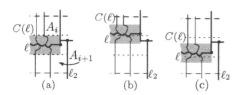

Fig. 9. (a) A_{i+1} got an upward tail with its end inside ℓ's tail strip; (b) moving $C(\ell)$ up; (c) moving $C(\ell)$ down

Fig. 10. (a) A_{i+1} got an upward tail of length 0; (b) moving $C(\ell)$ up; (c) moving $C(\ell)$ down

the case, when ℓ and ℓ_2 were involved in an X-junction in $\mathcal{L}_1$—hence A_{i+1} could have a tail within ℓ's tail strip.

The position of the left edge of $C(\ell)$ is determined in a similar fashion, the details can be found in the full paper. Note that no matter what was going on on the other sides of ℓ_1 and ℓ_2, the adjacencies are preserved when $C(\ell)$ is moved.

Recall that we are repairing the areas in a top-down manner. When we get to slice line ℓ, we need to make sure that the total area above ℓ—or rather the total area of the regions corresponding to the left subtree of the node corresponding to ℓ in the BSP tree—is correct. We do this by moving the container $C(\ell)$. We will show below that the error we have to repair is so small that it can be repaired by moving $C(\ell)$ within the shift strip of ℓ. The parts of the slice lines ending on ℓ that are inside the shift strip and outside the tail strip are all straight segments; this follows from Lemma 1 (ii). Hence, when we move $C(\ell)$ we can simply shrink or stretch these segments, and the topology does not change. We first analyze what happens to the complexity of the regions when we move the containers.

Lemma 2. *After Step 3 a region gets at most 4 concave vertices in total.*

Proof. We might only "bend" a slice line ℓ, ending on slice lines ℓ_1 and ℓ_2, when moving its container $C(\ell)$. Thus we can introduce concave vertices to two regions adjacent to ℓ and ℓ_1 (ℓ_2), denoted above as B_j and B_{j+1} (A_i and A_{i+1}). It is easy to verify—see Figures 7–10—that a region can only get an extra concave vertex at the junction of ℓ and ℓ_2 when the corner of the region did not yet get a tail in Step 2. The same is true for the junction of ℓ and ℓ_1. Hence the total number of concave vertices is bounded by four—at most one for each corner of the region in $\mathcal{L}_1$.   □

It remains to prove that we can choose the widths of the tail strip and shift strip appropriately. The two properties that we require are as follows.

Requirement 1. *The shift strips of slice lines do not intersect if the slice lines do not intersect after Step 1.*

Requirement 2. *The shift strip of each slice line ℓ is wide enough so that, when handling ℓ in Step 3, moving the container $C(\ell)$ can repair the areas while staying within the shift strip.*

For the first requirement it is sufficient to take the width of the shift strip to be smaller than $\Delta/2$, where $\Delta := \min(\Delta_x, \Delta_y)$ and Δ_x (Δ_y) is the minimum difference between any two distinct x-coordinates (y-coordinates) of the vertical (horizontal) slice lines after Step 1.

As for the second requirement, we provide a very rough estimate of the values for the width of the shift and tails strips, just to show that suitable values exist. Number the slice lines $\ell_1, \ldots \ell_{n-1}$ in the same order in which we handle them. (For example, the slice line at the root of the BSP tree will be ℓ_1.)

Lemma 3. *If the width of the shift strip of slice line ℓ_k is set to $\delta_k := \Delta/4 \cdot ((\Delta(1 - \Delta))/10)^{n-k-1}$ and the width of the tail strip is set to $\varepsilon_k := \delta_k \cdot \Delta/2$, for $1 \leqslant k \leqslant n - 1$, then Requirements 1 and 2 are fulfilled.*

The proof of Lemma 3 can be found in the full version of the paper. We conclude this section with the following theorem:

Theorem 1. *Let $\mathcal{G}$ be a vertex-weighted plane triangulated graph that admits a sliceable dual. Then $\mathcal{G}$ admits a cartogram of complexity at most 12.*

3   General Graphs

In the previous section we described an algorithm to construct cartograms for graphs that admit a sliceable dual. Next we consider more general graphs, namely graphs that admit a rectangular dual and arbitrary triangulated plane graphs. These more general classes of graphs are handled by adding an extra step before the three steps described in the previous section.

We begin with graphs that admit a rectangular dual, that is, plane triangulated graphs without separating triangles. Such a rectangular dual can be constructed, for example, by the algorithm of Kant and He [7]. Let now $\mathcal{G}$ be a plane triangulated graph without separating triangles and $\mathcal{L}_0$ a rectangular dual of $\mathcal{G}$. We construct a rectilinear BSP on $\mathcal{L}_0$, that is, we recursively partition $\mathcal{L}_0$ using horizontal or vertical splitting lines until each cell in the partitioning intersects a single rectangle from $\mathcal{L}_0$. This can be done in such a way that each rectangle in $\mathcal{L}_0$ is cut into at most four subrectangles [3]. The resulting layout of these subrectangles, $\mathcal{L}_1$, is sliceable by construction.

We then assign weights to the subrectangles. If a rectangle in $\mathcal{L}_0$ representing a vertex v of $\mathcal{G}$ was cut into k subrectangles in $\mathcal{L}_1$ then each subrectangle is assigned weight $w(v)/k$. (In practice it may be better to make the weight of each subrectangle proportional to its area.) Next, we perform Step 1–3 of the previous section on the layout $\mathcal{L}_1$ with these weights. Each rectilinear region in the layout $\mathcal{L}_4$ obtained after Step 3 corresponds to a subrectangle in $\mathcal{L}_1$. Finally, we merge the regions corresponding to subrectangles coming from the same rectangle in $\mathcal{L}_0$—and, hence, from the same vertex of $\mathcal{G}$—thus obtaining a layout $\mathcal{L}_5$ with one region per vertex of $\mathcal{G}$. The next lemma guarantees the correctness of our approach, its proof can be found in the full paper.

Lemma 4. *The algorithm described above produces a layout where each region has the correct area and adjacencies.*

It remains to analyze the complexity of the regions in the final layout. Of course we can just multiply the bound from the previous section by four, since each vertex in $\mathcal{G}$ is represented by four rectangles in $\mathcal{L}_1$. This results in a bound of 48. The next lemma shows that things are not quite that bad, its proof can be found in the full paper.

Lemma 5. *The algorithm described above produces regions of complexity at most 20.*

The next theorem summarizes our result for graphs that admit a rectangular dual.

Theorem 2. *Let $\mathcal{G}$ be a vertex-weighted plane triangulated graph that admits a rectangular dual, i.e., $\mathcal{G}$ has no separating triangles. Then $\mathcal{G}$ admits a cartogram of complexity at most 20.*

We now turn our attention to general plane triangulated graphs. As mentioned earlier, Liao et al. [9] showed that any plane triangulated graph has a rectilinear dual that uses L- and T-shapes—that is, regions of maximal complexity 8—in addition to rectangles. We cut each region into at most three subrectangles and then proceed as in the previous case: We cut the collection of subrectangle with a BSP to obtain a sliceable layout $\mathcal{L}_1$, we assign weights to the rectangles in $\mathcal{L}_1$, run Step 1–3, and merge regions belonging to the same vertex in $\mathcal{G}$. This immediately gives the following corollary.

Corollary 1. *Any vertex-weighted plane triangulated graph $\mathcal{G}$ admits a cartogram of complexity at most 60.*

4   Conclusions

We proved that every plane triangulated vertex-weighted graph admits a rectilinear cartogram of constant complexity. Currently, however, our method is not practical. First of all, although the complexity of the cartogram is bounded by a constant, it is rather high. So interesting open problems are to give an algorithm that produces cartograms of smaller complexity and to give lower bounds on the minimum complexity required to guarantee the existence of a cartogram. It would also be useful to give an exact characterization of the graphs that admit a sliceable dual, since the bound we obtain for such graphs is much better. A second problem with our algorithm from a practical point of view is that the tails we add to get the correct adjacencies can be quite thin. It would be nice to see if it is possible to do with wider tails.

References

1. J. Bhasker and S. Sahni. A linear algorithm to check for the existence of a rectangular dual of a planar triangulated graph. *Networks*, 7:307–317, 1987.
2. T. Biedl and B. Genc. Complexity of octagonal and rectangular cartograms. In *Proceedings of the 17th Canadian Conference on Computational Geometry*, pages 117–120, 2005.
3. F. d'Amore and P. G. Franciosa. On the optimal binary plane partition for sets of isothetic rectangles. *Information Processing Letters*, 44(5):255–259, 1992.
4. B. Dent. *Cartography - thematic map design*. McGraw-Hill, 5th edition, 1999.
5. J. A. Dougenik, N. R. Chrisman, and D. R. Niemeyer. An algorithm to construct continous area cartograms. *Professional Geographer*, 37:75–81, 1985.
6. R. Heilmann, D. A. Keim, C. Panse, and M. Sips. Recmap: Rectangular map approximations. In *Proceedings of the IEEE Symposium on Information Visualization (INFOVIS)*, pages 33–40, 2004.
7. G. Kant and X. He. Regular edge labeling of 4-connected plane graphs and its applications in graph drawing problems. *Theoretical Computer Science*, 172:175–193, 1997.
8. K. Koźmiński and E. Kinnen. Rectangular dual of planar graphs. *Networks*, 5:145–157, 1985.
9. C.-C. Liao, H.-I. Lu, and H.-C. Yen. Floor-planning using orderly spanning trees. *Journal of Algorithms*, 48:441–451, 2003.
10. NCGIA / USGS. Cartogram Central, 2002. `http://www.ncgia.ucsb.edu/projects/Cartogram_Central/index.html`.
11. M. S. Rahman, K. Miura, and T. Nishizeki. Octagonal drawings of plane graphs with prescribed face areas. In *Proceedings of the 30th International Workshop on Graph-Theoretic Concepts in Computer Science (WG)*, number 3353 in LNCS, pages 320–331. Springer, 2004.
12. E. Raisz. The rectangular statistical cartogram. *Geographical Review*, 24:292–296, 1934.
13. C. Thomassen. Plane cubic graphs with prescribed face areas. *Combinatorics, Probability and Computing*, 1:371–381, 1992.
14. M. van Kreveld and B. Speckmann. On rectangular cartograms. *Computational Geometry: Theory and Applications*, 2005. To appear.
15. G. K. Yeap and M. Sarrafzadeh. Sliceable floorplanning by graph dualization. *SIAM Journal of Discrete Mathematics*, 8(2):258–280, 1995.

Bar k-Visibility Graphs: Bounds on the Number of Edges, Chromatic Number, and Thickness

Alice M. Dean[1], William Evans[2], Ellen Gethner[3], Joshua D. Laison[4], Mohammad Ali Safari[5], and William T. Trotter[6]

[1] Department of Mathematics and Computer Science, Skidmore College
adean@skidmore.edu
[2] Department of Computer Science, University of British Columbia
will@cs.ubc.ca
[3] Department of Computer Science and Engineering,
University of Colorado at Denver
ellen.gethner@cudenver.edu
[4] Department of Mathematics and Computer Science, Colorado College
jlaison@coloradocollege.edu
[5] Department of Computer Science, University of British Columbia
safari@cs.ubc.ca
[6] Department of Mathematics, Georgia Institute of Technology
trotter@math.gatech.edu

Abstract. Let S be a set of horizontal line segments, or bars, in the plane. We say that G is a bar visibility graph, and S its bar visibility representation, if there exists a one-to-one correspondence between vertices of G and bars in S, such that there is an edge between two vertices in G if and only if there exists an unobstructed vertical line of sight between their corresponding bars. If bars are allowed to see through each other, the graphs representable in this way are precisely the interval graphs. We consider representations in which bars are allowed to see through at most k other bars. Since all bar visibility graphs are planar, we seek measurements of closeness to planarity for bar k-visibility graphs. We obtain an upper bound on the number of edges in a bar k-visibility graph. As a consequence, we obtain an upper bound of 12 on the chromatic number of bar 1-visibility graphs, and a tight upper bound of 8 on the size of the largest complete bar 1-visibility graph. We conjecture that bar 1-visibility graphs have thickness at most 2.

1 Introduction

Recent attention has been drawn to a variety of generalizations of bar visibility graphs [2, 3, 6, 5, 7, 8, 11, 12, 14, 15]. In this note, we report on a new generalization of bar visibility graphs called *bar k-visibility graphs*, and discuss some of their properties; complete details can be found in [4]. In what follows, we use the standard graph theory terminology found in [9, 17].

Let S be a set of disjoint horizontal line segments, or *bars*, in the plane. We say that a graph G is a *bar visibility graph*, and S a *bar visibility representation*

P. Healy and N.S. Nikolov (Eds.): GD 2005, LNCS 3843, pp. 73–82, 2005.
© Springer-Verlag Berlin Heidelberg 2005

Fig. 1. The bar visibility representation shown is an ε-visibility representation of G and a strong visibility representation of H

of G, if there exists a one-to-one correspondence between vertices of G and bars in S, such that there is an edge between two vertices x and y in G if and only if there exists a vertical line segment L, called a *line of sight*, whose endpoints are contained in X and Y, respectively, and which does not intersect any other bar in S. [1, 12, 13, 18].

If each line of sight is required to be a rectangle of positive width, then S is an ε-*visibility representation* of G, and when each line of sight is a line segment, then S is a *strong visibility representation* of G [16]. In general, these definitions are not equivalent; $K_{2,3}$ admits an ε-visibility representation but not a strong visibility representation, as shown in Figure 1.

Given a set of bars S in the plane, suppose that an endpoint of a bar B and an endpoint of a bar C in S have the same x-coordinate. We elongate one of these two bars so that their endpoints have distinct x-coordinates. If S is a strong visibility representation of a graph G, then we may perform this elongation so that S is still a strong visibility representation of G. If S is an ε-visibility representation of G, then we may perform this elongation so that S is an ε-visibility representation of a new graph H with $G \subseteq H$. Since we are interested in the maximum number of edges obtainable in a representation, we may consider the graph H instead of the graph G. Repeating this process yields a set of bars with pairwise distinct endpoint x-coordinates. For the remainder of this paper, we assume that all bar visibility representations are of this form.

If a set of bars S has all endpoint x-coordinates distinct, the graphs G and H that have S as a strong bar visibility representation and an ε-visibility representation, respectively, are isomorphic. Hence without loss of generality, for the remainder of the paper, all bar visibility representations are strong bar visibility representations.

By contrast, suppose that S is a set of closed intervals on the real line. The graph G is called an *interval graph* and S an *interval representation* of G if there exists a one-to-one correspondence between vertices of G and intervals in S, such that x and y are adjacent in G if and only if their corresponding intervals intersect. Suppose we call a set S of horizontal bars in the plane an *x-ray-visibility representation* if we allow sight lines to intersect arbitrarily many bars in S. Then we can easily transform an x-ray-visibility representation into an interval representation by vertically translating the bars in S, and vice-versa. Therefore G is an x-ray-visibility graph if and only if G is an interval graph.

Motivated by this correspondence, we define a *bar k-visibility graph* to be a graph with a bar visibility representation in which a sight line between bars

X and Y intersects at most k additional bars. As a first step on the road to a characterization of bar k-visibility graphs, since all bar visibility graphs are planar, we seek measurements of closeness to planarity for bar k-visibility graphs.

2   An Edge Bound for Bar 1-Visibility Graphs

Suppose G is a graph with n vertices, and S is a bar 1-visibility representation of G. Since we consider S to be a strong visibility representation of G, without loss of generality, we may assume that all endpoints of all bars in S have distinct x-coordinates, and all bars in S have distinct y-coordinates.

It will be convenient to use four different labeling systems for the bars in S. Label the bars 1_l, 2_l, ..., n_l in increasing order of the x-coordinate of their left endpoint. Label them 1_r, 2_r, ..., n_r in decreasing order of the x-coordinate of their right endpoint. Label them 1_b, 2_b, ..., n_b in increasing order of their y-coordinate. Finally, label them 1_t, 2_t, ..., n_t in decreasing order of their y-coordinate. So the bar 1_l has leftmost left endpoint, the bar 1_r has rightmost right endpoint, the bar $1_b = n_t$ is bottommost in the representation, and the bar $1_t = n_b$ is topmost in the representation. We use this notation for the remainder of the paper.

Remark 1. Suppose S is a bar k-visibility representation of a graph G with n vertices. We elongate the top and bottom bars of S to obtain a new bar k-visibility representation S' of a new graph G', with the additional property that $1_t = 1_r = 1_l$ and $1_b = 2_r = 2_l$ in S'. The graph G' has n vertices and contains G as a subgraph. We may therefore assume that every edge-maximal bar k-visibility graph has such a bar k-visibility representation.

Lemma 1. *If G is a bar 1-visibility graph with $n \geq 4$ vertices, then G has at most $6n - 17$ edges.*

Proof. Suppose G is a graph with n vertices, and S is a bar 1-visibility representation of G. We define the following correspondence between bars in S and edges of G. Let U be the bar in S associated with vertex u. For every edge $\{u, v\}$ in G, let $\ell(\{u, v\})$ be the vertical line segment from a point in U to a point in V whose x-coordinate is the infimum of x coordinates of lines of sight between U and V. An edge $\{u, v\}$ is called a *left edge* of U (respectively V) if $\ell(\{u, v\})$ contains the left endpoint of U (respectively V). If $\ell(\{u, v\})$ contains neither U nor V's left endpoint then it must contain the right endpoint of some bar B (that blocks the 1-visibility of U from V from that point on). In this case, we call $\{u, v\}$ a *right edge of B*. Note that the right edges of B are not incident to the vertex b of G corresponding to the bar B. Each bar B can have at most 4 left edges (two to bars above B in S and two to bars below B in S) and at most 2 right edges, as shown in Figure 2.

Counting both left and right edges, each bar in S is associated with at most 6 edges. So there are at most $6n$ edges in G. However, the bars 1_l, 2_l, 3_l, and 4_l have at most 0, 1, 2, and 3 left edges, respectively. Similarly, the bars 1_r, 2_r, 3_r,

Fig. 2. The two right edges associated to bar B

and 4_r have at most 0, 0, 0, and 1 right edges, respectively. Therefore there are at most $4n - 10$ left edges and at most $2n - 7$ right edges, for a total of at most $6n - 17$ edges in G. □

Theorem 1. *If G is a bar 1-visibility graph with $n \geq 5$ vertices, then G has at most $6n - 20$ edges.*

Proof. We improve the bound given in Lemma 1 by using a slightly more sophisticated technique. We follow the notation of Lemma 1.

By Remark 1, the edge $\{1_t, 1_b\}$ will always be a left edge. Since the edge associated with the right endpoint of the bar 4_r can only be this edge, the bar 4_r must have 0 right edges. So there are at most $2n - 8$ right edges in G, and $6n - 18$ edges in total. If G has exactly $6n - 18$ edges, then bars 1_l, 2_l, 3_l, and 4_l must have at least 0, 1, 2, and 2 left edges, respectively.

Suppose that bar 4_l has only two left edges. Then it does not have a line of sight to bar 3_l, which can happen only if 3_l ends before 4_l begins. Then $3_l = n_r$, and 3_l has 0 right edges. Therefore G has at most $6n - 20$ edges. The only remaining possibility is that bar 4_l has exactly three left edges.

If S had at most $4n - 12$ left edges, then S would have at most $6n - 20$ edges in total. The remaining possibilities are that S has either $4n - 11$ or $4n - 10$ left edges. Since 1_l, 2_l, 3_l, and 4_l have exactly 0, 1, 2, and 3 left edges, respectively, all other bars in S must have exactly four left edges, except perhaps for one bar i_l, which may have three left edges. By the same argument, since 1_r, 2_r, 3_r, and 4_r have no right edges, every additional bar must have exactly two right edges, except one additional bar, which may have only one.

Consider the four edges $e_1 = \{1_t, 1_b\}$, $e_2 = \{1_t, 2_b\}$, $e_3 = \{2_t, 1_b\}$, and $e_4 = \{2_t, 2_b\}$. If $i_l = 2_b$, then the edges e_1 and e_3 are left edges, but the edges e_2 and e_4 may not be. If $i_l = 2_t$, then the edges e_1 and e_2 are left edges, but the edges e_3 and e_4 may not be. If i_l is neither of these bars, then all four of these edges are left edges.

Since the bars 2_t and 2_b have at most one right edge each, one of them must be bar 3_r or bar 4_r. Without loss of generality, assume that bar 2_t is either bar 3_r or bar 4_r. So in the order of the bars 1_r through 5_r given by increasing y-coordinate, the bar 5_r must appear either second or third. Figure 3 shows the four possibilities that may occur.

In each of the four cases shown, and for each of the three possibilities for the bar i_l, one can check that 5_r has at most one right edge. So the remaining bars must all have exactly two right edges. Therefore the bars 2_t and 2_b must be two

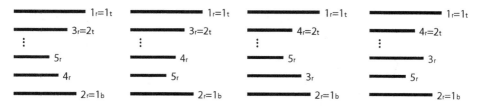

Fig. 3. The four possible arrangements of bars 1_r, 2_r, 3_r, 4_l, and 5_r

of the three bars 3_r, 4_r, and 5_r. But this implies that any right edge associated to 5_r must be between a pair of the bars 1_t, 2_t, 1_b, and 2_b. Therefore 5_r must have no right edges, and G has at most $6n - 20$ edges in total.   □

Corollary 2. *The graph K_9 is not a bar 1-visibility graph.*

Proof. Any bar 1-visibility graph with 9 vertices has at most 34 edges, whereas K_9 has 36 edges.   □

Corollary 3. *If G is a bar 1-visibility graph, then $\chi(G) \leq 12$.*

Proof. We proceed by induction. Assume that all bar 1-visibility graphs with $n - 1$ vertices have $\chi \leq 12$, and suppose that G is a bar 1-visibility graph with n vertices. By Theorem 1, $\sum_{v \in V(G)} \deg(v) < 12n$, so the average degree of a vertex in G is strictly less than 12. Then there must exist a vertex v in G of degree at most 11. We consider the graph $G - v$. Although this graph may not be a bar 1-visibility graph, it is a subgraph of the graph G' with bar 1-visibility representation obtained from a representation of G by deleting the bar corresponding to v. Therefore the edge bound in Theorem 1 still applies to H. By the induction hypothesis, we may color the vertices of H with 12 colors, replace v, and color v with a color not used on its neighbors.   □

Corollary 4. *There are thickness-2 graphs with n vertices that are not bar 1-visibility graphs for all $n \geq 15$.*

Proof. Note that there are no thickness-2 graphs with n vertices and more than $6n - 12$ edges, since if G has thickness 2 then G is the union of two planar graphs, each of which have at most $3n - 6$ edges. Consider the graph $G = C_3 \boxtimes C_5$ formed by replacing each vertex in C_5 with C_3 and taking the join of neighboring C_3's. G has 15 vertices and $6 \cdot 15 - 12 = 78$ edges. Since G is the union of the two planar graphs shown in Figure 4, G has thickness 2.

Let $G_{15} = G$ and suppose L_1 and L_2 are the two plane layers of G_{15}. Let $\{a, b, c\}$ be a face in L_1 and $\{d, e, f\}$ be a face in L_2 such that $\{a, b, c\} \cap \{d, e, f\} = \varnothing$. Add a new vertex v to G_{15} adjacent to $\{a, b, c\}$ in L_1 and $\{d, e, f\}$ in L_2; define the new graph to be G_{16}. The graph G_{16} has 16 vertices and $6 \cdot 16 - 12$ edges, and thickness 2. Following the same procedure, inductively we construct an infinite family of graphs G_n such that for all $n \geq 15$, G_n has n vertices and $6n - 12$ edges, and thickness 2. Therefore none of these graphs can be a bar 1-visibility graph by Theorem 1.   □

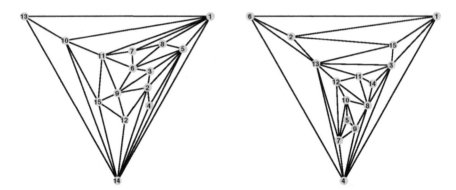

Fig. 4. Two planar graphs whose union is not a bar 1-visibility graph

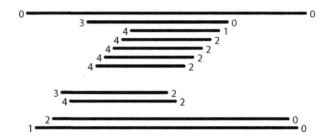

Fig. 5. A bar 1-visibility representation with $6n - 20$ edges

Note that the graphs $\{G_n\}$ given in the proof of Corollary 4 have the largest possible number of edges while having thickness 2.

Theorem 5. *There exist bar 1-visibility graphs with $6n - 20$ edges, $n \geq 5$.*

Proof. The graph with representation shown in Figure 5 is a bar 1-visibility graph with $6n - 20$ edges. For ease of counting, the left and right endpoints of bars in this representation are labeled with the number of left and right edges associated to each bar. Note that this representation has $4n - 11$ left edges and $2n - 9$ right edges. Although $n = 11$ in this representation, more bars can easily be deleted to create a representation with as few as 5 bars, or added to create a representation with arbitrarily many bars. For the values $n = 5$ through 8, this representation yields a complete graph. □

Corollary 6. *The graph K_8 is a bar 1-visibility graph.*

Proof. Take only eight bars in the representation shown in Figure 5. □

By Corollary 6, if G is a bar 1-visibility graph, then $\chi(G)$ may be 8. No bar 1-visibility graph is known with chromatic number 9. The standard example of a graph with chromatic number 9 but clique number smaller than 9 is the Sulanke graph $K_6 \vee C_5$ [17], which is not a bar 1-visibility graph since it has 11 vertices and 50 edges.

3   Edge Bounds on Bar k-Visibility Graphs

The following theorem generalizes Lemma 1 for $k > 1$. The proof is entirely analogous to the proof of Lemma 1, and can be found in [4].

Theorem 7. *If G is a bar k-visibility graph with $n \geq 2k + 2$ vertices, then G has at most $(k + 1)(3n - \frac{7}{2}k - 5)$ edges.*

Theorem 8. *There exist bar k-visibility graphs with n vertices and $(k+1)(3n - 4k - 6)$ edges for $k \geq 0$ and $n \geq 3k + 3$.*

Proof. Figure 6 shows a bar k-visibility representation of a graph with n vertices and $(k + 1)(3n - 4k - 6)$ edges. As in Figure 5, the left and right endpoints of bars in this representation are labeled with the number of left and right edges associated to each bar. Although $n = 4k + 4$ in this representation, more bars can easily be deleted to create a representation with as few as $3k + 3$ bars, or added to create a representation with arbitrarily many bars. □

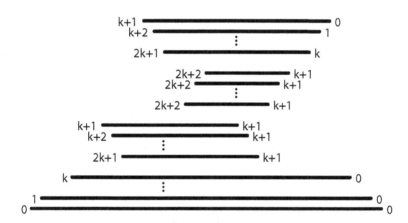

Fig. 6. A bar k-visibility graph with n vertices and $(k + 1)(3n - 4k - 6)$ edges

Note that Theorem 8 gives the largest number of edges in a bar k-visibility graph for $k = 0, 1$. We believe that this is the case for larger k as well. We state this as a conjecture.

Conjecture 1. If G is a bar k-visibility graph, then G has at most $(k + 1)(3n - 4k - 6)$ edges.

The following theorem is a corollary of Theorem 7.

Theorem 9. *K_{5k+5} is not a bar k-visibility graph.*

Proof. By way of contradiction, suppose that G is a graph with $n = 5k + 5$ vertices. Then by Theorem 7, G has at most $(k + 1)(3(5k + 5) - \frac{7}{2}k - 5) = \frac{23}{2}k^2 + \frac{43}{2}k + 10$ edges. However, K_{5k+5} has $\binom{5k+5}{2} = \frac{25}{2}k^2 + \frac{45}{2}k + 10$ edges. □

Note that if Conjecture 1 is true, we immediately obtain the following conjecture as a corollary.

Conjecture 2. K_{4k+4} is the largest complete bar k-visibility graph.

Proof (Assuming Conjecture 1). Figure 6 shows a bar k-visibility representation of K_{4k+4}. Conversely, suppose that G is a graph with $n = 4k + 5$ vertices. Then by Conjecture 1, G has at most $(k + 1)(3(4k + 5) - 4k - 6) = 8k^2 + 17k + 9$ edges. However, K_{4k+5} has $\binom{4k+5}{2} = 8k^2 + 18k + 10$ edges.   □

Conjecture 1 is not required to prove Conjecture 2 when $k = 0$ or 1; we have already proved these cases in the previous section. Note also that the graph K_{4k+4} exactly achieves the bound given by Conjecture 1. So if this conjecture is correct, the family of complete graphs K_{4k+4} is an example of a family of edge-maximal bar k-visibility graphs.

4   Thickness of Bar k-Visibility Graphs

By Corollary 6, K_8 is a bar 1-visibility graph, and thus there are non-planar bar 1-visibility graphs. Motivated by the fact that all bar 0-visibility graphs are planar [10], we are interested in measuring the closeness to planarity of bar 1-visibility graphs. The *thickness* $\Theta(G)$ of a graph G is the minimum number of planar graphs whose union is G. K_8 has thickness 2 [12], so there exist bar 1-visibility graphs with thickness 2. Conversely, the following theorem from [4] gives an upper bound for the thickness of a bar 1-visibility graph.

Suppose G is a bar 1-visibility graph, and S is a bar 1-visibility representation of G. We define the *underlying bar visibility graph* G_0 of S to be the graph with bar visibility representation S. The following theorem relates the thickness of G to the chromatic number of G_0.

Theorem 10. *If G is a bar 1-visibility graph and G_0 an underlying bar visibility graph of G, then $\Theta(G) \leq \chi(G_0)$. In particular, the thickness of any bar 1-visibility graph is at most four.*

We conjecture that bar 1-visibility graphs have thickness no greater than 2. More generally, we know that the thickness of a bar k-visibility graph is bounded by some function of k [4]. The smallest such function of k is still open.

5   Future Work

We close with a list of open problems inspired by the results of this note.

1. What is the largest number of edges in a bar 2-visibility graph with n vertices?
2. What is the largest number of edges in a bar k-visibility graph with n vertices?

3. Are there bar 1-visibility graphs with thickness 3?
4. More generally, what is the largest thickness of a bar k-visibility graph? Is it $k + 1$?
5. Are there bar 1-visibility graphs with chromatic number 9?
6. More generally, what is the largest chromatic number of a bar k-visibility graph?
7. What is the largest crossing number of a bar k-visibility graph?
8. What is the largest genus of a bar k-visibility graph?
9. What is a complete characterization of bar k-visibility graphs?
10. Is there an efficient recognition algorithm for bar k-visibility graphs?
11. *Rectangle visibility graphs* are defined in [7, 8, 15]. Generalize the results of this note to rectangle visibility graphs.
12. *Arc-* and *circle-visibility graphs* are defined in [11]. Generalize the results of this note to arc- and circle-visibility graphs.

References

1. T. Andreae. Some results on visibility graphs. *Discrete Appl. Math.*, 40(1):5–17, 1992. Combinatorial methods in VLSI.
2. P. Bose, A. Dean, J. Hutchinson, and T. Shermer. On rectangle visibility graphs. In *Lecture Notes in Computer Science 1190: Graph Drawing*, pages 25–44. Springer-Verlag, 1997.
3. G. Chen, J. P. Hutchinson, K. Keating, and J. Shen. Characterizations of $1, k$-bar visibility trees. In preparation, 2005.
4. A. Dean, W. Evans, E. Gethner, J. D. Laison, M. A. Safari, and W. T. Trotter. Bar k-visibilty graphs. Submitted, 2005.
5. A. M. Dean, E. Gethner, and J. P. Hutchinson. A characterization of triangulated polygons that are unit bar-visibility graphs. In preparation, 2005.
6. A. M. Dean, E. Gethner, and J. P. Hutchinson. Unit bar-visibility layouts of trian-gulated polygons: Extended abstract. In J. Pach, editor, *Lecture Notes in Computer Science 3383: Graph Drawing 2004*, pages 111–121, Berlin, 2005. Springer-Verlag.
7. A. M. Dean and J. P. Hutchinson. Rectangle-visibility representations of bipartite graphs. *Discrete Appl. Math.*, 75(1):9–25, 1997.
8. A. M. Dean and J. P. Hutchinson. Rectangle-visibility layouts of unions and prod-ucts of trees. *J. Graph Algorithms Appl.*, 2:no. 8, 21 pp. (electronic), 1998.
9. G. Di Battista, P. Eades, R. Tamassia, and I. G. Tollis. *Graph Drawing*. Prentice Hall Inc., Upper Saddle River, NJ, 1999.
10. M. R. Garey, D. S. Johnson, and H. C. So. An application of graph coloring to printed circuit testing. *IEEE Trans. Circuits and Systems*, CAS-23(10):591–599, 1976.
11. J. P. Hutchinson. Arc- and circle-visibility graphs. *Australas. J. Combin.*, 25:241–262, 2002.
12. J. P. Hutchinson, T. Shermer, and A. Vince. On representations of some thickness-two graphs. *Computational Geometry*, 13:161–171, 1999.
13. P. Rosenstiehl and R. E. Tarjan. Rectilinear planar layouts and bipolar orientations of planar graphs. *Discrete Comput. Geom.*, 1(4):343–353, 1986.
14. T. Shermer. On rectangle visibility graphs III. External visibility and complexity. In *Proc. 8th Canad. Conf. on Comp. Geom.*, pages 234–239, 1996.

15. I. Streinu and S. Whitesides. Rectangle visibility graphs: characterization, construction, and compaction. In *STACS 2003*, volume 2607 of *Lecture Notes in Computer Science*, pages 26–37. Springer, Berlin, 2003.
16. R. Tamassia and I. G. Tollis. A unified approach to visibility representations of planar graphs. *Discrete Comput. Geom.*, 1(4):321–341, 1986.
17. D. B. West. *Introduction to Graph Theory, 2E.* Prentice Hall Inc., Upper Saddle River, NJ, 2001.
18. S. K. Wismath. Characterizing bar line-of-sight graphs. In *Proceedings of the First Symposium of Computational Geometry*, pages 147–152. ACM, 1985.

Drawing K_n in Three Dimensions with One Bend Per Edge[*]

Olivier Devillers[1], Hazel Everett[2], Sylvain Lazard[2],
Maria Pentcheva[2], and Stephen K. Wismath[3]

[1] INRIA, Sophia-Antipolis, France
Olivier.Devillers@inria.fr
[2] INRIA, Université Nancy 2, LORIA, Nancy, France
Firstname.Name@loria.fr
[3] Dept. of Math and Comp Sci, U. of Lethbridge, Canada
wismath@cs.uleth.ca

Abstract. We give a drawing of K_n in 3D in which vertices are placed at integer grid points and edges are drawn crossing-free with at most one bend per edge in a volume bounded by $O(n^{2.5})$.

1 Introduction

Drawing graphs in three dimensions has been considered by several authors in the graph-drawing field under a variety of models. One natural model is to draw vertices as points at integer-valued grid points in a 3D Cartesian coordinate system and represent edges as straight line segments between adjacent vertices with no pair of edges intersecting. The *volume* of such a drawing is typically defined in terms of a smallest bounding box containing the drawing and with sides orthogonal to one of the coordinate axes. If such a box B has width w, length l and height h, then we refer to the *dimensions* of B as $(w + 1) \times (l + 1) \times (h + 1)$ and define the volume of B as $(w + 1) \cdot (l + 1) \cdot (h + 1)$.

It was shown by Cohen *et al.* [3] that it is possible to draw *any* graph in this model, and indeed the complete graph K_n is drawable within a bounding box of volume $\Theta(n^3)$. Restricted classes of graphs may however be drawn in smaller asymptotic volume. For example, Calamonieri and Sterbini [2] showed that 2-, 3-, and 4-colourable graphs can be drawn in $O(n^2)$ volume. Pach *et al.* [11] showed a volume bound of $\Theta(n^2)$ for r-colourable graphs (r a constant). Dujmović *et al.* [4] investigated the connection of bounded tree-width to 3D layouts. Felsner *et al.* [8] showed that outerplanar graphs can be drawn in $O(n)$ volume. Establishing tight volume bounds for planar graphs remains an open problem. Dujmović and Wood [5] showed an upper bound of $O(n^{1.5})$ on the volume of planar graphs at Graph Drawing 2003.

In 2-dimensional graph drawing, the effect of allowing bends in edges has been well studied. For example, Kaufmann and Wiese [9] showed that all planar graphs can be drawn with only 2 bends per edge and all vertices located on a straight line.

---

[*] Supported in part by the NSERC Canada.

P. Healy and N.S. Nikolov (Eds.): GD 2005, LNCS 3843, pp. 83–88, 2005.

The consequences of allowing bends in 3D has received less attention. Note that bend points must also occur at integer grid points. Bose *et al.* [1] showed that the number of edges in a graph provides an asymptotic lower bound on the volume regardless of the number of bends permitted, thus establishing $\Omega(n^2)$ as the lower bound on the volume for K_n. This lower bound was explicitly achieved by Dyck *et al.* [7] who presented a construction with at most 2 bends per edge. The upper bound is also a consequence of a more general result of Dujmović and Wood [6]. In [10], Morin and Wood presented a one-bend drawing of K_n that achieves $O(n^3/\log^2 n)$ volume. It is the gap between this result and the $\Omega(n^2)$ lower bound that motivates this paper; we improve the Morin and Wood result to achieve a one-bend drawing with volume $O(n^{2.5})$.

2   Preliminaries

We call the axes of our 3D Cartesian coordinate system respectively X, Y and Z.

The one-bend construction of K_n by Morin and Wood [10] considers $O(\log n)$ packets of $O(\frac{n}{\log n})$ collinear vertices. All the vertices lie in the XY-plane and edges joining vertices of different packets lie above this plane. Edges joining vertices within a packet lie below this plane and the volume of these (complete) subgraphs is a consequence of the following lemma.

Lemma 1 ([10]). *For all $q \geqslant 1$, K_m has a one-bend drawing in an axis-parallel box of size $q \times m \times \left\lceil \frac{\pi^2}{3} \frac{m^2}{q} \right\rceil$ with all the vertices on the Y-axis.*

Indeed, $\Omega(n^3)$ volume is required for a collinear one-bend drawing of K_n as shown by Morin and Wood. We present here a brief description of the construction behind Lemma 1 because we will use it in our construction. The edges are divided into $\Theta(m^2)$ chains of edges (i.e., sequences of edges). A chain connects all vertices with index equal to i modulo j such that the vertices on the chain are ordered with increasing indices. In each chain, the bends are placed on a line parallel to the Y-axis through a point of integer coordinates (x, z) in the XZ-plane. The chains thus lie in planes that contain the Y-axis (where the vertices lie). In the XZ-plane, the points (x, z) are chosen so that they are all strictly visible from the origin. The well-known fact that there are $\Theta(m^2)$ such choices in a rectangle of size $q \times \frac{m^2}{q}$ ensures that all the $\Theta(m^2)$ chains can be placed in distinct planes, and thus that the edges do not cross.

3   The Construction

Our construction is roughly as follows. We split the n vertices into k packets of $\frac{n}{k}$ vertices, where all vertices in one packet have the same X and Z coordinates. All edges of the complete graph contain a bend. All edges joining two vertices of one packet are placed below and right (positive X direction) of the packet, and all edges joining two vertices of different packets are placed "above". We present our construction for an arbitrary k and show later that the volume of the

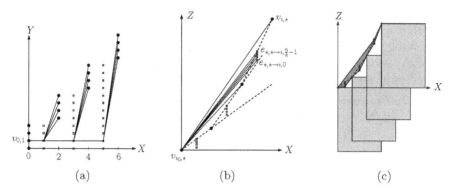

(a)   (b)   (c)

Fig. 1. (a) Projection on the XY-plane of the vertices (dots), the bends (squares), and the edges leaving vertex $v_{0,1}$. Here, $n = 16$ and $k = 4$. (b) Construction of the Z-coordinates. (c) XZ-projection of the bounding boxes of the interpacket edges.

bounding box of the drawing is minimized for $k = n^{\frac{1}{4}}$. We assume for simplicity that $\frac{n}{k}$ is an integer.

X and Y coordinates of the vertices. We first describe the X and Y-coordinates of the vertices. Refer to Figure 1(a). All vertices have different Y-coordinates, ranging between 0 and $n - 1$. We divide the set of n vertices into k packets, denoted $V_0, \ldots, V_{k-1}$, of $\frac{n}{k}$ vertices. All vertices in the same packet V_i have the same X-coordinate $2i$, the same Z-coordinate z_i (defined later), and consecutive Y-coordinates. Precisely, the j-th vertex in the i-th packet, denoted $v_{i,j}$, has coordinates $(2i, i\frac{n}{k} + j, z_i)$, with $0 \leqslant j \leqslant \frac{n}{k} - 1$ and $0 \leqslant i \leqslant k - 1$.

Edges joining vertices of one packet. Since all the vertices of one packet are collinear, we can draw the complete graph on these vertices using the $q \times m \times O(m^2/q)$ volume construction of [10] described in Section 2. In that construction there are m collinear vertices; here we have $m = \frac{n}{k}$ vertices. We choose $q = k$ and draw the edges so that the bounding box of this complete subgraph is below (negative Z) and to the right (positive X) of the vertices. Notice that we have chosen $q = k$ so that these complete subgraphs do not asymptotically increase the width of the final drawing. *In the sequel of the construction, we only consider edges that join vertices of distinct packets, and their bends.*

X and Y coordinates of the bends. Refer to Figure 1(a). The bend of an edge joining vertex v_{i_1,j_1} to vertex v_{i_2,j_2}, with $i_1 < i_2$, is denoted $e_{i_1,j_1 \to i_2,j_2}$. It separates the edge into two distinct segments, the *outgoing* segment which starts at v_{i_1,j_1} and ends at $e_{i_1,j_1 \to i_2,j_2}$, and the *incoming* segment which starts with $e_{i_1,j_1 \to i_2,j_2}$ and ends at v_{i_2,j_2}.

A bend $e_{i_1,j_1 \to i_2,j_2}$ has coordinates $(2i_2 - 1, i_1\frac{n}{k} + j_1, z_{i_2,j_2})$, that is, its X-coordinate is one less than the X-coordinate of v_{i_2,j_2}, its Y-coordinate is the same as for v_{i_1,j_1}, and its Z-coordinate, which only depends on v_{i_2,j_2}, will be defined later.

Z-coordinates of the vertices and bends. We will assign values to z_i and $z_{i,j}$ so that edges do not cross. In fact, our construction is designed to verify the following lemma. In the following we consider the projection on the XZ-plane of the vertices $v_{i,\star}$ and bends $e_{\star,\star \to i,j}$ where $\star$ can take any value since the projected points are identical.

Lemma 2. *Projected onto the XZ-plane, the polar ordering $\prec_{i_0}$ viewed from a vertex $v_{i_0,\star}$ of the vertices $v_{i,\star}$ and bends $e_{\star,\star \to i,j}$ with $i_0 < i < k$ and $0 \leqslant j < \frac{n}{k}$ satisfies $v_{i-1,\star} \prec_{i_0} e_{\star,\star \to i,0} \prec_{i_0} \cdots \prec_{i_0} e_{\star,\star \to i,j} \prec_{i_0} \cdots \prec_{i_0} e_{\star,\star \to i,\frac{n}{k}-1} \preccurlyeq_{i_0} v_{i,\star}$.*

Our construction is as follows. First, let $z_0 = 0$ and $z_{1,j} = j + 1$, then z_1 is chosen such that $v_{1,\star}$ is at the same polar angle about $v_{0,\star}$ as $e_{0,\star \to 1,\frac{n}{k}-1}$, which gives $z_1 = 2z_{1,\frac{n}{k}-1} = 2\frac{n}{k}$ (see Figure 1(b)).

Assume now that we have placed vertices and bends up to index i. To get a correct polar ordering around $v_{i-1,\star}$ we need to have the next bends above the line through $v_{i-1,\star}$ and $v_{i,\star}$ thus we place the next bend at $z_{i+1,0} = z_i + \frac{1}{2}(z_i - z_{i-1}) + 1$ and the following bends on edges going to $v_{i+1,\star}$ at $z_{i+1,j} = z_i + \frac{1}{2}(z_i - z_{i-1}) + 1 + j$.

The vertex $v_{i+1,\star}$ is placed at the same polar angle about $v_{i,\star}$ as $e_{\star,\star \to i+1,\frac{n}{k}-1}$ which gives $z_{i+1} = z_i + 2(z_{i+1,\frac{n}{k}-1} - z_i) = z_i + 2(\frac{z_i - z_{i-1}}{2} + \frac{n}{k}) = 2z_i - z_{i-1} + 2\frac{n}{k}$; solving this recurrence[1] yields $z_i = i(i+1)\frac{n}{k}$. Then we obtain $z_{i,j} = z_{i-1} + \frac{1}{2}(z_{i-1} - z_{i-2}) + 1 + j = (i-1)(i+1)\frac{n}{k} + 1 + j$. To summarize, the coordinates of the vertices and bends are

$$v_{i,j} = \left(2i, \; i\frac{n}{k} + j, \; i(i+1)\frac{n}{k} \right)$$

$$e_{i_1,j_1 \to i_2,j_2} = \left(2i_2 - 1, \; i_1\frac{n}{k} + j_1, \; (i_2^2 - 1)\frac{n}{k} + 1 + j_2 \right)$$

Proof of Lemma 2. The correct polar ordering of the $v_{i,\star}$ viewed from $v_{i_0,\star}$ is guaranteed since all these points are ordered on a convex curve (i.e. a parabola). Let L_i be the line through $v_{i,\star}$ and $v_{i+1,\star}$. The correct polar ordering of $v_{i,\star}$, the $e_{\star,\star \to i+1,j}$ and $v_{i+1,\star}$, viewed from $v_{i,\star}$, comes directly from the construction; moreover, this ordering is the same for all viewpoints $v_{i_0,\star}$, $i_0 < i$, since these viewpoints lie above L_{i-1} (see Figure 1(b)). ∎

4   Proof of Correctness

We say that two edges cross if their relative interiors intersect. We prove in this section that no two edges of our construction cross. We first show that the edges joining vertices within the same packet induce no crossing. Then, we show that there is no crossing between two outgoing segments, two incoming segments, and finally one outgoing and one incoming segment.

Edges joining vertices within packets. We use the same technique as in the Morin-Wood construction [10] to ensure that no two edges joining vertices

---

[1] An inductive verification is easy since with this formula we have:
$2z_i - z_{i-1} + 2\frac{n}{k} = \frac{n}{k}[2i(i+1) - i(i-1) + 2] = \frac{n}{k}[i^2 + 3i + 2] = (i+1)(i+2)\frac{n}{k} = z_{i+1}$.

within a given packet cross. An edge joining vertices within a packet crosses no other edge joining vertices within another packet since the projection onto the Y-axis of the bounding boxes of the Morin-Wood constructions do not intersect. Finally, the bounding box of the Morin-Wood construction properly intersects no edge joining distinct packets since they do not properly intersect in XZ-projection (see Fig. 1(c)). Hence, edges joining vertices within a packet cross no other edge.

Outgoing-outgoing segments. If two outgoing segments start from different vertices, they lie in two different planes parallel to the XZ-plane. Otherwise, by Lemma 2, the two segments only share their starting point. Hence no two outgoing segments cross.

Incoming-incoming segments. Note that an incoming segment joining $e_{\star,\star\to i,j}$ to $v_{i,j}$ lies in the plane $P_{i,j}$ through the two lines parallel to the Y-axis and containing, respectively, all the $e_{\star,\star\to i,j}$ and all the $v_{i,\star}$. For a pair of incoming segments, we consider three cases according to whether both segments finish at the same vertex, at distinct vertices of the same packet, or at vertices of different packets. In the first case, the segments live in a plane $P_{i,j}$; they start at different bends and end at the same vertex, hence they do not cross. In the second case, the two segments live in two planes $P_{i,j}$ and $P_{i,j'}$ whose intersection is the line $v_{i,j}v_{i,j'}$. The segments end there and thus cannot cross. In the third case, the segments do not overlap in the X-direction, thus they do not cross.

Incoming-outgoing segments. Consider an outgoing segment joining vertex $v_{i_1,\star}$ to bend $e_{i_1,\star\to i_3,\star}$ and an incoming segment joining bend $e_{\star,\star\to i_2,\star}$ to vertex $v_{i_2,\star}$, where $\star$ can be any value (see Figure 2). The ranges over the X-axis of the two segments are $[2i_1, 2i_3-1]$ and $[2i_2-1, 2i_2]$. They overlap only if $i_1 < i_2 < i_3$, and, in such a case, Lemma 2 yields that, viewed from $v_{i_1,\star}$, the points satisfy the polar ordering $e_{\star,\star\to i_2,\star} \preceq_{i_1} v_{i_2,\star} \prec_{i_1} e_{i_1,\star\to i_3,\star}$. This implies that, in projection onto the XZ-plane, points $e_{\star,\star\to i_2,\star}$ and $v_{i_2,\star}$ are below the line segment joining $v_{i_1,\star}$ and $e_{i_1,\star\to i_3,\star}$. Hence the two segments do not cross.

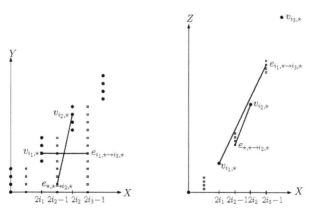

Fig. 2. Incoming and outgoing segments in XY and XZ-projections

5   Volume Analysis

The dimension of the bounding box of our construction for edges between packets has size smaller than $2k \times n \times kn$ since the highest vertex has Z-coordinate $z_{k-1} = k(k-1)\frac{n}{k}$. The complete subgraphs within packets have size $k \times \frac{n}{k} \times \left\lceil \frac{\pi^2}{3}\frac{n^2}{k^3} \right\rceil$ and thus our complete construction fits in a box of size $O(k) \times n \times O(\frac{n^2}{k^3} + kn)$. To balance the increasing and decreasing terms of the Z-dimension we choose k such that $\frac{n^2}{k^3} = kn$ that is $k = n^{\frac{1}{4}}$. Recall that we assumed for simplicity that k and $\frac{n}{k}$ were integers; for any n we can apply our construction with $\lceil n^{\frac{1}{4}} \rceil^4$ vertices and then remove the extra vertices and edges. We thus have the following result.

Theorem 1. *Every complete graph K_n has a one-bend drawing in an axis-parallel box of dimensions $O(n^{\frac{1}{4}}) \times n \times O(n^{\frac{1}{4}})$ and volume $O(n^{2.5})$.*

Remark 1. An alternative for applying Lemma 1 with boxes that match the X-dimension of our construction (choosing $q = k$ with $m = \frac{n}{k}$) is to take boxes whose size matches the Z-dimension of our construction (choosing $q = kn$ with $m = \frac{n}{k}$). Then the dimension of the bounding box of the Morin-Wood construction for interpacket edges is $O(\frac{n^2}{k^2} \cdot \frac{1}{kn}) \times \frac{n}{k} \times kn$ which gives a total size for our construction of $O(k + \frac{n}{k^3}) \times n \times 2kn$. This is still optimal for $k = n^{\frac{1}{4}}$ but it offers a trade-off between volume and aspect ratio of the box for $k \in [1, n^{\frac{1}{4}}]$.

References

1. P. Bose, J. Czyzowicz, P. Morin and D. R. Wood. The maximum number of edges in a three-dimensional grid-drawing, *JGAA*, 8(1):21–26, 2004.
2. T. Calamoneri and A. Sterbini. 3D straight-line grid drawing of 4-colorable graphs, Information Processing Letters 63(2):97–102, 1997.
3. R. F. Cohen, P. Eades, T. Lin, and F. Ruskey. Three-dimensional graph drawing, *Algorithmica*, 17:199–208, 1997.
4. V. Dujmović, P. Morin, and D. R. Wood. Layout of graphs with bounded tree-width, *SIAM J. of Computing*, 34(3)553-579, 2005.
5. V. Dujmović, and D. R. Wood. Three-dimensional grid drawings with sub-quadratic volume, (GD '03), *LNCS* 2912:190–201, Springer-Verlag, 2004.
6. V. Dujmović, and D. R. Wood. Stacks, Queues and Tracks: Layouts of Graph Subdivisions, *Discrete Math and Theoretical Computer Science*, 7:155-202, 2005.
7. B. Dyck, J. Joevenazzo, E. Nickle, J. Wilsdon, and S. Wismath. Drawing K_n in 3D with 2 Bends Per Edge, U. of Lethbridge Tech Rep #CS-01-04: 2–7, Jan 2004.
8. S. Felsner, G. Liotta, S. Wismath. Straight-line drawings on restricted integer grids in two and three dimensions, *JGAA*, 7(4):363–398, 2003.
9. M. Kaufmann and R. Wiese. Embedding vertices at points: few bends suffice for planar graphs, *JGAA*, 6(1):115–129, 2002.
10. P. Morin and D. R. Wood. Three-dimensional 1-bend graph drawings, *JGAA*, 8(3), 2004.
11. J. Pach, T. Thiele, and G. Tóth. Three-dimensional grid drawings of graphs, (GD '97), *LNCS*, 1353:47–51, Springer-Verlag, 1997.

Small Area Drawings of Outerplanar Graphs*
(Extended Abstract)

Giuseppe Di Battista and Fabrizio Frati

Dipartimento di Informatica e Automazione, Università di Roma Tre
gdb@dia.uniroma3.it, fabriziofrati@tiscali.it

Abstract. We show three linear time algorithms for constructing planar straight-line grid drawings of outerplanar graphs. The first and the second algorithm are for balanced outerplanar graphs. Both require linear area. The drawings produced by the first algorithm are not outerplanar while those produced by the second algorithm are. On the other hand, the first algorithm constructs drawings with better angular resolution. The third algorithm constructs outerplanar drawings of general outerplanar graphs with $O(n^{1.48})$ area. Further, we study the interplay between the area requirements of the drawings of an outerplanar graph and the area requirements of a special class of drawings of its dual tree.

1 Introduction

Straight-line drawings of planar graphs have been studied by several authors and constitute one of the main fields of investigation in Graph Drawing. Groundbreaking works of the end of the 20th Century [5, 13, 4] have shown that a planar graph with n vertices has a planar straight-line drawing with integer coordinates ("grid" drawing) with $O(n^2)$ area. Further, it has been shown [12] that there exist graphs that, for such drawings, require quadratic area.

Planar straight-line grid drawings have also been studied for subclasses of planar graphs, looking for subquadratic area bounds. For example a linear area algorithm for drawing binary trees with arbitrary aspect ratio has been shown in [8].

Another subclass of planar graphs that attracted research work in this field is the one of the outerplanar graphs. An outerplanar graph is a planar graph that has a planar drawing such that all its vertices are on the outer face. The dual graph of an outerplanar graph is a tree (but for the outer face). Garg and Rusu [9] proved that an n-vertex outerplanar graph has a planar straight-line grid drawing with $O(d \cdot n^{1.48})$ area, where d is the maximum degree of the vertices of the graph. Biedl [1] conjectured that $O(n \lg n)$ area is sufficient for such graphs.

In [10, 2] are presented algorithms for constructing straight-line drawings with vertices in general position.

Outerplanar graphs have been studied also with respect to other types of drawings. In [1] and in [11] are presented algorithms to construct planar polyline drawings with

---

* Work partially supported by EC - Fet Project DELIS - Contract no 001907, by "Project ALGO-NEXT: Algorithms for the Next Generation Internet and Web: Methodologies, Design, and Experiments", MIUR Programmi di Ricerca Scientifica di Rilevante Interesse Nazionale, and by "MAIS Project", MIUR–FIRB.

P. Healy and N.S. Nikolov (Eds.): GD 2005, LNCS 3843, pp. 89–100, 2005.

$O(n \log n)$ area and $O(d \cdot n)$ area, respectively. An algorithm for constructing in three dimensions straight-line drawings with linear volume is presented in [7].

In this paper we present the following results. They always refer to planar straight-line grid drawings. We show (Section 3) a linear time algorithm for constructing non-outerplanar drawings of balanced outerplanar graphs in linear area and with angular resolution $\geq \frac{c}{\sqrt{n}}$, with c constant. A balanced outerplanar graph is such that its dual tree is balanced. We define a new type of drawings of binary trees, called star-shaped drawings (Section 4). We show that, given a drawing of an outerplanar graph it can be found a star-shaped drawing of its dual tree with the same area bound. Conversely, given a star-shaped drawing of a binary tree it can be found a drawing of its dual outerplanar graph with the same area bound but for the placement of two special vertices. Based on such correspondence, we show a linear time algorithm for drawing a balanced outerplanar graph in linear area (Section 4). The drawings obtained with this algorithm are outerplanar, but the angular resolution is worse with respect to the algorithm of Section 3. Again, based on the above correspondence and exploiting a decomposition technique of binary trees presented in [3], we show a linear time algorithm for constructing outerplanar drawings of general outerplanar graphs with $O(n^{1.48})$ area (Section 5).

2   Preliminaries

We assume familiarity with Graph Drawing (see e.g. [6]).

An *outerplanar graph* is a planar graph that has a planar drawing with all its vertices on the same (say outer) face. Such a drawing is called *outerplanar drawing*. In this paper we deal with outerplanar graphs that are also biconnected. However, this is not a limitation since an outerplanar graph can be always augmented with a linear number of extra edges to a biconnected outerplanar graph. Hence, the algorithms and theorems we present can be applied also to general outerplanar graphs after a simple preprocessing step that does not alter the number of vertices of the graph.

We define the *dual graph* of an outerplanar graph G as follows. The vertices of the dual graph are the faces of G, with the exception of the outer face that is not associated to any vertex of the dual of G. Two vertices f_1 and f_2 of the dual graph sharing an edge of G are connected, in the dual graph, by edge (f_1, f_2). The dual graph of an outerplanar graph is always a tree. Hence, in the following we call it *dual tree*.

A *maximal outerplanar graph* is an outerplanar graph such that all its faces but, eventually, the outer face are composed by three edges. Note that any outerplanar graph can be augmented to a maximal outerplanar by adding extra edges. The vertices of the dual graph of a maximal outerplanar graph have degree at most three. From now on, unless otherwise specified, we assume that outerplanar graphs are maximal.

We can select an edge (u, v) of the outer face of an outerplanar graph G and root the dual tree T of G at the internal face r containing (u, v). Let w be the third vertex of r. We call vertices u and v *poles* and vertex w *central vertex*. We also call u *left vertex* and v *right vertex*. Consider a face f of T and suppose that f is composed in G by edges (v_1, v_2), (v_2, v_3), and (v_3, v_1), in this clockwise order around f. Also, suppose that the parent of f in T and f share edge (v_1, v_2) or that (f is the root) $(v_1, v_2) = (u, v)$. The face sharing with f (if any) edge (v_3, v_1) is the *left child* of f, while the face sharing with f (if any) edge (v_2, v_3) is the *right child* of f. We obtain a binary tree.

A *balanced outerplanar graph* is an outerplanar graph whose dual tree can be rooted to a balanced binary tree. The *height* of an outerplanar graph is the number of nodes on the longest path of its dual tree from the root to a leaf. A *complete outerplanar graph* is an outerplanar graph whose dual tree is a complete binary tree. A *grid drawing* of a graph is such that all its vertices have integer coordinates. A *straight-line drawing* is such that all edges are rectilinear segments. Let Γ be a straight-line grid drawing and consider the smallest rectangle $B(\Gamma)$ with sides parallel to the x- and y-axes that covers Γ completely. We call $B(\Gamma)$ the *bounding box* of Γ. We denote with $b(\Gamma)$, $t(\Gamma)$, $l(\Gamma)$ and $r(\Gamma)$ the bottom, top, left and right side of $B(\Gamma)$, respectively. The *height* (*width*) of Γ is one plus the height (width) of $B(\Gamma)$. The *area* of Γ is the height of Γ multiplied by its width.

3   Non-outerplanar Drawings of Balanced Outerplanar Graphs

We call G_h a complete outerplanar graph with height h, T_h its dual tree, and Γ_h its planar straight-line grid drawing. Let also u_h, v_h and w_h be the left vertex, the right vertex and the central vertex of G_h, respectively.

We show an inductive algorithm to draw complete outerplanar graphs. **Base case:** if $h = 1$, then place u_1 in $(0,0)$, v_1 in $(1,1)$ and w_1 in $(1,0)$. **Inductive case:** if $h > 1$, suppose you have drawn Γ_{h-1}; let r be the line through v_{h-1} and w_{h-1}, let b be the line through u_{h-1} and v_{h-1} and let a be the line parallel to and at horizontal distance one unit from r, in the opposite side of the drawing with respect to r. Shift u_{h-1} and v_{h-1} along b of one horizontal unit, moving away from Γ_{h-1}. Now mirror the modified drawing Γ_{h-1} with respect to a. Insert the edge from u_{h-1} to its symmetric vertex, say z. Let $u_h = u_{h-1}$, $v_h = z$ and $w_h = v_{h-1}$. Examples of the drawings produced by the algorithm are shown in Fig. 1. Showing the planarity of the obtained drawings is trivially done by induction. Now we analyze their area requirement. Let $height_h$ and $width_h$ be the height and the width of Γ_h, respectively. We distinguish two cases. **h is even:** it's easy to see that $height_{h-1} = 2 \cdot height_{h-2} + 1$ and that $height_h = height_{h-1} + 2$. So we have $height_h = 2 \cdot height_{h-2} + 3$. Hence we obtain:

$$height_h = \ldots \underbrace{(((height_2 \cdot 2 + 3) \cdot 2 + 3) \ldots \cdot 2 + 3}_{\frac{h-2}{2} \, times} = height_2 \cdot 2^{\frac{h-2}{2}} + 3 \cdot 2^{\frac{h-4}{2}} +$$

$3 \cdot 2^{\frac{h-6}{2}} + \ldots + 3$. Let $m = \frac{h-2}{2}$; replacing $height_2$ with its value 4 we obtain: $height_h = 4 \cdot 2^m + 3 \cdot 2^{m-1} + 3 \cdot 2^{m-2} + \ldots + 3 = 4 \cdot 2^m + 3 \cdot (2^{m-1} + 2^{m-2} + \ldots + 1) = 4 \cdot 2^m + 3 \cdot (2^m - 1) = 7 \cdot 2^m - 3 = 7 \cdot 2^{\frac{h-2}{2}} - 3 = \frac{7}{2} \cdot 2^{\frac{h}{2}} - 3 = \frac{7}{2} \cdot 2^{(\lg n)\frac{1}{2}} - 3 = \frac{7}{2} \cdot n^{\frac{1}{2}} - 3 = \frac{7}{2} \cdot \sqrt{n} - 3 = O(n^{\frac{1}{2}})$. It's easy to see that: $width_h = 2 \cdot height_h - 1 = 7 \cdot \sqrt{n} - 7 = O(n^{\frac{1}{2}})$. If **h is odd**, using $height_h = 2 \cdot height_{h-1} + 1$ we obtain: $height_h = \left(\frac{7}{2} \cdot 2^{\frac{h-1}{2}} - 3 \right) \cdot 2 + 1 = \frac{7}{\sqrt{2}} \cdot 2^{\frac{h}{2}} - 5 = \frac{7}{\sqrt{2}} \cdot 2^{(\lg n)\frac{1}{2}} - 5 = \frac{7}{\sqrt{2}} \cdot \sqrt{n} - 5 = O(n^{\frac{1}{2}})$. It's easy to see that the width is equal to the height, hence we have: $width_h = \frac{7}{\sqrt{2}} \cdot \sqrt{n} - 5 = O(n^{\frac{1}{2}})$.

About the angular resolution, let u_h be the left vertex of G_h. Recall that $u_{h-1} = u_h$. Passing from G_{h-1} to G_h the number of the neighbours of u_{h-1} increases by one,

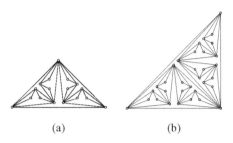

Fig. 1. Applications of the algorithm in *Section 3*. (a) Γ_4. (b) Γ_5.

beacuse of the insertion in Γ_h of the symmetric vertex of u_{h-1}. Let $m\ (t)$ be the largest odd (even) integer $\leq h$. We can prove by induction that the smallest angle in the drawing is ϕ, the angle between half-lines a and b starting at u_h and passing respectively through v_a, the neighbour of u_h inserted in Γ_m, and through v_b, the neighbour of u_h inserted in Γ_{m-2}. Let c be the half-line starting at u_h and passing through v_c, the neighbour of u_h inserted in Γ_t. Let v_d be the intersection point between a and the line through v_b and orthogonal to a. Let v_e be the intersection point between c and the line through v_b and orthogonal to c. Let R_1 be the triangle whose vertices are u_h, v_b and v_e. We denote with a_h, b_h and c_h the lengths of the segments $\overline{v_b v_e}$, $\overline{u_h v_e}$ and $\overline{u_h v_b}$, respectively. Note that, by construction, $b_h = a_h + 1$. Now suppose h is odd; following the construction of the algorithm we obtain $a_h = 2 \cdot a_{h-2} + 4$ and solving the recurrence equation we obtain $a_h = \frac{7}{2} \cdot 2^{\frac{h-1}{2}} - 4$. Hence $b_h = \frac{7}{2} \cdot 2^{\frac{h-1}{2}} - 3$. Applying the Pythagorean theorem to R_1 we obtain $c_h = \sqrt{a_h^2 + b_h^2} = \sqrt{\frac{49}{4} \cdot 2^h - 49 \cdot 2^{\frac{h-1}{2}} + 25}$. Observing that $\overline{v_b v_d} = \frac{\sqrt{2}}{2}$, for every h, we finally obtain: $\phi \approx \sin \phi = \dfrac{\sqrt{2}}{2\sqrt{\frac{49}{4} \cdot 2^h - 49 \cdot 2^{\frac{h-1}{2}} + 25}} > c \cdot 2^{-h/2} = \frac{c}{\sqrt{n}}$, where c is a constant. If h is even, in a similar way we obtain: $\phi \approx \sin \phi = \dfrac{\sqrt{2}}{2\sqrt{\frac{49}{8} \cdot 2^h - \frac{35}{2} \cdot 2^{\frac{h}{2}} + 13}} > c \cdot 2^{-h/2} = \frac{c}{\sqrt{n}}$, where c is a constant. From the above discussion and from the fact that a balanced outerplanar graph can be augmented to complete without altering its height we have:

Theorem 1. *Given an n-vertex balanced outerplanar graph G with height h, there exists an $O(n)$ time algorithm that constructs a planar straight-line grid drawing Γ of G such that: (i) if h is even, then the height of Γ is $\frac{7}{2}\sqrt{n} - 3$ and its width is $7\sqrt{n} - 7$; (ii) if h is odd, then the height of Γ is $\frac{7}{\sqrt{2}}\sqrt{n} - 5$ and its width is $\frac{7}{\sqrt{2}}\sqrt{n} - 5$; (iii) the angular resolution of Γ is greater than $\frac{c}{\sqrt{n}}$, with c constant; (iv) if G is complete, then isomorphic subgraphs of G have congruent drawings in Γ up to a translation and a reflection; and (v) if G is complete, then Γ is axially symmetric.*

4   Outerplanar Drawings and Star-Shaped Drawings

Let T be a binary tree rooted at r. The *leftmost (rightmost) path* of T is the path $v_0, v_1, \ldots, v_m$ such that $v_0 = r$, v_{i+1} is the left (right) child of v_i, $\forall i$ such that $0 \leq$

$i \leq m - 1$, and v_m doesn't have a left (right) child. The *outer-left set* (*outer-right set*) of a planar straight-line drawing Γ of T is the set of points with integer coordinates from which we can draw edges to each one of the nodes of the leftmost (rightmost) path of T without crossing Γ. The *left-right* (*right-left*) *path* of a node $n \in T$ is the path $v_0, v_1, \ldots, v_m$ such that $v_0 = n$, v_1 is the left (right) child of v_0, v_{i+1} is the right (left) child of v_i, $\forall i$ such that $1 \leq i \leq m - 1$, and v_m doesn't have a right (left) child. The *left polygon of the neighbours* (*right polygon of the neighbours*) of a node $n \in T$ is the polygon of the segments representing in Γ the edges of the left-right path (of the right-left path) plus an extra segment connecting v_m and v_0.

A planar straight-line order-preserving drawing Γ of T is *star-shaped* if all the following conditions are satisfied. (1) For each node $n \in T$ its left (right) polygon of neighbours $P_l = (n, v_1, \ldots, v_m)$ ($P_r = (n, v_1, \ldots, v_m)$) is a simple polygon and each segment $(n, v_i), 2 \leq i \leq m - 1$ belongs to the interior of P_l (P_r), but for its endpoints n and v_i. (2) For each pair of nodes $n_1, n_2 \in T$ the left polygon of neighbours or the right polygon of neighbours of n_1 does not intersect with the left polygon of neighbours or with the right polygon of neighbours of n_2, but, possibly, at common endpoints or at common edges. (3) There exist point p_l in the outer-left set of T and point p_r in the outer-right set of T such that segment (p_l, p_r) doesn't intersect any edge of Γ.

Given a drawing Γ of an outerplanar graph we call *internal subdrawing* the drawing obtained by deleting from Γ its poles and their incident edges.

Lemma 1. *Let G be an n-vertex outerplanar graph such that its dual tree T has a star-shaped drawing with $f(n)$ area. We have that G has an outerplanar straight-line drawing such that the area of its internal subdrawing is $f(n)$.*

Lemma 2. *Let G be an n-vertex outerplanar graph that has an outerplanar straight-line drawing with $f(n)$ area. We have that its dual tree T has a planar star-shaped straight-line drawing with an area that is at most $f(n)$.*

To prove the above lemmas we first establish a correspondence γ between the vertices of G and the nodes of T, so that for each node $n \in T$ there is one and only one vertex v of G such that $\gamma(n) = v$ and for each vertex $v \in G$, but for the poles, there is one and only one node $n \in T$ such that $\gamma^{-1}(v) = n$. Consider a subtree of T rooted at n. Suppose that (v_l, v_c) is the edge of G dual to the edge connecting n to its left child (if any). Analogously, suppose that (v_r, v_c) is the edge of G dual to the edge connecting n to its right child (if any). We set $\gamma(n) = v_c$. Now, suppose you have a planar star-shaped straight-line grid drawing Γ of T. Map each vertex v of G, but for its poles, to the point where the node n such that $\gamma^{-1}(v) = n$ is drawn. Map the left vertex u_l of G to a point p_l of the outer-left set and the right vertex v_r of G in a point p_r of the outer-right set so that the edge (p_l, p_r) doesn't intersect any of the edges of T. Draw the edges from u_l to each vertex on the leftmost path of T and the edges from v_r to each vertex on the rightmost path of T. Draw the edge (u_l, v_r). By Condition (3) in the definition of star-shaped drawing and by the definitions of outer-left set and of outer-right set, p_l and p_r exist and their incident edges don't intersect Γ. For each node n (and so for each vertex $v = \gamma(n)$) draw edges to each vertex on its left-right path and to each vertex on its right-left path. Because of Condition 1 and 2 in the definition of star-shaped drawing each of such segments doesn't intersect any other segment of the drawing. The drawing obtained after these insertions is an outerplanar straight-line grid drawing of G as a consequence of the construction and of the correspondence between vertices of G and nodes of T. We have also just seen that each step preserves the initial planarity. The area bound of Lemma 1 is easily obtained by observing that the vertices of G (but the poles) and the nodes of T have exactly the same coordinates.

Now we can start from an outerplanar drawing Φ of G, then we can use again the correspondence between vertices of G and nodes of T to obtain a star-shaped drawing of T. Remove from Φ the poles of G. For each vertex v let n be the node of T such that $\gamma^{-1}(v) = n$ and let n_l and n_r be the left and the right child of n, respectively. Remove all edges incident on v, but those whose second endpoint is a vertex z such that $\gamma^{-1}(z) = n_l$ or $\gamma^{-1}(z) = n_r$. We obtain a star-shaped drawing of T: it's easy to see that the drawing is planar, straight-line, grid and order-preserving and that all the conditions of a star-shaped drawing are verified, since the initial drawing Φ is a planar straight-line grid drawing of G. Again, the area bound of Lemma 2 is easily obtained by observing that the vertices of G (but the poles) and the nodes of T have exactly the same coordinates.

We apply the above lemmas to construct a linear area drawing of a complete outerplanar graph. We denote with T_h a complete binary tree, r_h its root, and Γ_h its drawing. What follows is an inductive algorithm to construct a star-shaped drawing of a complete binary tree. **Base case:** if $h = 1$, then place r_1 in $(0,0)$. **Inductive case:** if $h > 1$, suppose you have drawn Γ_{h-1}. Now we distiguish two subcases. **h is even:** let r be the highest horizontal line such that r intersects Γ_{h-1}. Let a be the line above r parallel to and at vertical distance one unit from r. Let b be the lowest line with slope $\frac{\pi}{4}$ with respect to the x-axis and such that b intersects Γ_{h-1}. Mirror Γ_{h-1} with respect to a. Place r_h at the intersection between a and b. Insert the edges from r_h to its children. If **h is odd** let r be the highest line with slope $\frac{3\pi}{4}$ with respect to the x-axis and such that r intersects Γ_{h-1}. Let a be the line above r parallel to and at vertical distance two units from r. Let b be the lowest line with slope $\frac{\pi}{4}$ with respect to the x-axis and such that b intersects Γ_{h-1}. Mirror Γ_{h-1} respect to a. Translate the new part of the drawing by a vector $(-1,0)$. Place r_h at the intersection between a and b. Insert the edges from r_h to its children. A drawing produced by the algorithm is shown in Fig. 2.a.

It is easy to see, by induction, that the resulting drawing is star-shaped. Now we analyze the area requirements of the above algorithm. Let $height_h$ and $width_h$ be the height and the width of Γ_h, respectively. We distinguish two cases. **h is even:** it's easy to see that $height_{h-1} = height_{h-2} + 2$ and that $height_h = 2 \cdot height_{h-1} + 1$. So we have $height_h = 2 \cdot height_{h-2} + 5$. Hence we obtain:

$$height_h = \ldots(((height_2 \cdot 2 + 5) \cdot 2 + 5) \ldots \cdot 2 + 5 = height_2 \cdot 2^{\frac{h-2}{2}} + 5 \cdot 2^{\frac{h-4}{2}} +$$

$$\underbrace{}_{\frac{h-2}{2} \, times}$$

$5 \cdot 2^{\frac{h-6}{2}} + \ldots + 5$. Let $m = \frac{h-2}{2}$; replacing $height_2$ with its value 3 we obtain: $height_h = 3 \cdot 2^m + 5 \cdot 2^{m-1} + 5 \cdot 2^{m-2} + \ldots + 5 = 3 \cdot 2^m + 5 \cdot (2^{m-1} + 2^{m-2} + \ldots + 1)$ $= 3 \cdot 2^m + 5 \cdot (2^m - 1) = 8 \cdot 2^m - 5 = 8 \cdot 2^{\frac{h-2}{2}} - 5 = 4 \cdot 2^{\frac{h}{2}} - 5 = 4 \cdot 2^{(\lg n)\frac{1}{2}} - 5 = 4 \cdot n^{\frac{1}{2}} - 5 = 4 \cdot \sqrt{n} - 5 = O(n^{\frac{1}{2}})$. It's easy to see that $width_h = \frac{height_h + 1}{2} = 2 \cdot \sqrt{n} - 2 = O(n^{\frac{1}{2}})$. If **h is odd**, using $height_h = height_{h-1} + 2$ we obtain: $height_h$

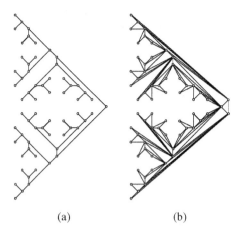

(a)   (b)

Fig. 2. Application of the algorithm in *Section 4*. (a) Γ_6. (b) The drawing of G_6 built over Γ_6.

$$= \left(4 \cdot 2^{\frac{h-1}{2}} - 5\right) + 2 = \frac{4}{\sqrt{2}} \cdot 2^{\frac{h}{2}} - 3 = \frac{4}{\sqrt{2}} \cdot 2^{(\lg n)^{\frac{1}{2}}} - 3 = \frac{4}{\sqrt{2}} \cdot \sqrt{n} - 3 = O(n^{\frac{1}{2}}).$$ It's
easy to see that $width_h = height_h - 1 = \left(\frac{4}{\sqrt{2}} \cdot \sqrt{n} - 3\right) - 1 = \frac{4}{\sqrt{2}} \cdot \sqrt{n} - 4 = O(n^{\frac{1}{2}})$.

We exploit the above algorithm and Lemma 1 to prove the following theorem.

Theorem 2. *Given an n-vertex balanced outerplanar graph G with height h, there exists an $O(n)$ time algorithm that constructs an outerplanar straight-line grid drawing Γ of G such that: (i) if h is even, then the height of Γ is $4\sqrt{n} - 5$ and its width is $2\sqrt{n} - 1$; (ii) if h is odd, then the height of Γ is $\frac{4}{\sqrt{2}}\sqrt{n} - 3$ and its width is $\frac{4}{\sqrt{2}}\sqrt{n} - 3$; (iii) the angular resolution of Γ is less than $\frac{c}{n}$, with c constant; (iv) if G is complete, then isomorphic subgraphs of G have congruent drawings in Γ up to a translation and a reflection; and (v) if G is complete, then Γ is axially symmetric.*

Proof. Γ is constructed as follows. First, we add to G dummy vertices and edges to make it complete without altering h. Second, we draw star-shaped its dual tree T. Third, using the correspondence between the vertices of G and the nodes of T introduced in the proof of Lemmas 1 and 2, we build a drawing Γ' of the internal subgraph of G. Finally, we place the poles of G and their incident edges, obtaining Γ. This is done as follows. We place the left vertex on the same line of $b(\Gamma')$, one unit to the right of $r(\Gamma')$ and we place the right vertex on the same line of $t(\Gamma')$ one unit to the right of $r(\Gamma')$. This placement allows to draw edges from the left vertex to each node of the leftmost path of T and from the right vertex to each node of the rightmost path of T without crossings. Furthermore, this placement increases by one unit the width without altering the height of Γ. Note that similar but different placements of the poles, as the one in Fig. 2.b, are also possible.

The bounds on height and width of Γ descend from the bounds given for star-shaped drawings. Now we analyze the angular resolution. Namely, we show that there is an angle that decreases faster than $\frac{1}{n}$. If h is odd let v_1 be the root of T, else (h even) let v_1 be the left child of the root of T. Let $(v_1, w_0, w_1, \ldots, w_m)$ be the left-right path of

v_1. Let ϕ be the angle between the half-lines a and b starting at v_1 and passing through w_{m-1} and w_m, respectively. From trigonometry we have:

$\sin\phi = \sin(\widehat{w_m v_1 w_0})\cos(\widehat{w_{m-1}v_1 w_0}) - \sin(\widehat{w_{m-1}v_1 w_0})\cos(\widehat{w_m v_1 w_0})$. Observe that $\overline{v_1 w_0} = \sqrt{2}, \forall h$. Let k be the biggest even integer $\leq h-1$. We have $\overline{w_0 w_m} = \sqrt{2}(2\cdot 2^{\frac{k}{2}} - 3)$; moreover $\overline{w_0 w_{m-1}} = \overline{w_0 w_m} - \sqrt{2}$, since $\overline{w_{m-1}w_m} = \sqrt{2}, \forall h$; hence $\overline{w_0 w_{m-1}} = \sqrt{2}(2\cdot 2^{\frac{k}{2}} - 4)$. Using the Pythagorean theorem we obtain:

$$\sin\phi = \frac{\overline{w_0 w_m}}{\sqrt{\overline{w_0 w_m}^2 + \overline{v_1 w_0}^2}} \cdot \frac{\overline{v_1 w_0}}{\sqrt{\overline{w_0 w_{m-1}}^2 + \overline{v_1 w_0}^2}} +$$
$$+ \frac{\overline{v_1 w_0}}{\sqrt{\overline{w_0 w_m}^2 + \overline{v_1 w_0}^2}} \cdot \frac{\overline{w_0 w_{m-1}}}{\sqrt{\overline{w_0 w_{m-1}}^2 + \overline{v_1 w_0}^2}} =$$
$$= \frac{\overline{v_1 w_0}(\overline{w_0 w_m} - \overline{w_0 w_{m-1}})}{\sqrt{\overline{w_0 w_m}^2 + \overline{v_1 w_0}^2}\sqrt{\overline{w_0 w_{m-1}}^2 + \overline{v_1 w_0}^2}} =$$
$$= \frac{\sqrt{2}\sqrt{2}}{\sqrt{2(2\cdot 2^{\frac{k}{2}} - 3)^2}\sqrt{2(2\cdot 2^{\frac{k}{2}} - 4)^2}} =$$
$$= \frac{2}{2\sqrt{4\cdot 2^k - 12\cdot 2^{\frac{k}{2}} + 10}\sqrt{4\cdot 2^k - 16\cdot 2^{\frac{k}{2}} + 17}}.$$

Hence $\phi \approx \sin\phi < c\cdot(2^{-k})$ and since $k = O(h)$, we have $\phi < \frac{c}{n}$, with c constant.

5 Outerplanar Drawings of General Outerplanar Graphs

This section is devoted to the proof of the following theorem. The main ingredients of the proof are: (i) a recursive algorithm for constructing a star-shaped drawing of a binary tree, (ii) Lemma 1, and (iii) Lemma 3 presented by Chan in [3].

Theorem 3. *Given an n-vertex outerplanar graph G, there exists an $O(n)$ time algorithm that constructs an $O(n^{1.48})$ area outerplanar straight-line grid drawing of G.*

Lemma 3. *[3] Let $p = 0.48$. Given any binary tree T of size n, there exists a root-to-leaf path π such that for any left subree α and right subtree β of π, $|\alpha|^p + |\beta|^p \leq (1-\delta)n^p$, for some constant $\delta > 0$.*

First, we show two techniques, called Constructions 1–2, for constructing a star-shaped drawing Γ_i, with $i \in \{1,2\}$, of a general binary tree T with n nodes. Each one is defined in terms of itself and of the other one. In the following we call *spine* a root-to-leaf path $S = (v_0, v_1, \ldots, v_m)$ of T. Let s_i be the non spine child of v_i and let $T(s_i)$ be the subtree of T rooted at s_i. We denote with $W_i(n)$ the width of Γ_i, with $W_{i,l}(n)$ $(W_{i,r}(n))$ the width of the part of Γ_i that is to the left (to the right) of S and with $n(t)$ the number of nodes in the subtree of T rooted at t.

Now we show **Construction** 1. First, we draw each $v_i \in S$ together with $T(s_i)$, obtaining $\Gamma(v_i)$; then we put all the $\Gamma(v_i)$ together to obtain Γ_1. Construction 1 has four subcases, labelled $1xy$, $x \in \{t,b\}$ and $y \in \{l,r\}$. Index x states that S is drawn

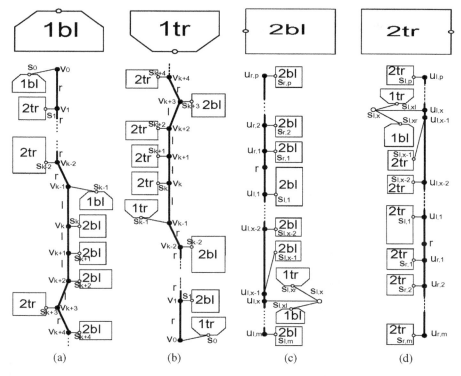

Fig. 3. Constructions (a)$1bl$, (b)$1tr$, (c)$2bl$ and (d)$2tr$. The edges (v_i, v_{i+1}) labelled r (l) are such that v_{i+1} is the right (resp. left) child of v_i. The thick edges show the spine (in Construction 1) and the leftmost and the rightmost paths (in Construction 2) of T.

going towards the top ($x = t$) or towards the bottom ($x = b$) of Γ_1. Index y states that the leftmost path ($y = l$), or the rightmost path ($y = r$), is drawn going towards the left. In the following we show the details of Construction $1bl$, while the others are easily obtained from $1bl$ after a reflection with respect to the x-axis and/or a switch of the left with the right and vice-versa. Constructions $1bl$ and $1tr$ are shown in Fig. 3.

Suppose v_1 is the left (right) child of v_0. Let k be the first index such that v_k is the right (left) child of v_{k-1}. In the following we denote the subtree $T(s_{k-1})$ ($T(s_0)$) also with $T(s_l)$ and we denote the subtree $T(s_0)$ ($T(s_{k-1})$) also with $T(s_r)$. Draw $T(s_0)$ and $T(s_{k-1})$ with Construction $1bl$, obtaining $\Gamma(s_0)$ and $\Gamma(s_{k-1})$, respectively. Draw v_0 one unit above and one unit to the left (right) of $B(\Gamma(s_0))$, obtaining $\Gamma(v_0)$. Draw v_{k-1} one unit above and one unit to the right (left) of $B(\Gamma(s_{k-1}))$, obtaining $\Gamma(v_{k-1})$. Draw any other left (right) subtree with Construction $2tr$ (with Construction $2bl$), obtaining $\Gamma(s_i)$. If s_i is the left (right) child of v_i, draw v_i on the same horizontal channel and one unit to the right (to the left) of s_i, obtaining $\Gamma(v_i)$.

Now we put together all the $\Gamma(v_i)$, $0 \leq i \leq m$ as follows. Place $\Gamma(v_0)$ anywhere in the plane. For $1 \leq i \leq m$, if v_i is the left child (right child) of v_{i-1} and v_{i+1} is the left child (right child) of v_i or v_i is a leaf ($i = m$), then draw $\Gamma(v_i)$ so that v_i is on the same vertical channel of v_{i-1} and so that $b(\Gamma(v_{i-1}))$ is one unit above the $t(\Gamma(v_i))$.

Else (v_i is the left child (right child) of v_{i-1} and v_{i+1} is the right child (left child) of v_i) if v_i is a left child (right child) draw v_i on the vertical channel one unit to the left (to the right) with respect to the vertical channel of v_{i-1} and so that $b(\Gamma(v_{i-1}))$ is one unit above $t(\Gamma(v_i))$.

Property 1. Construction $1bl$ guarantees that all the vertices of the leftmost (rightmost) path of T are visible from any point that is above and to the left (right) of $B(\Gamma_1)$.

Property 2. Suppose that the drawing of Construction $2tr$ is star-shaped and that it places the leftmost and the rightmost paths of the tree on the right side of its bounding box. Suppose also that the drawing of Construction $2bl$ is star-shaped and that it places the leftmost and the rightmost paths of the tree on the left side of its bounding box. We have that the drawing obtained with Construction $1bl$ is star-shaped.

Property 3. $W_{1,l}(n) = \max(W_1(n(s_l)), \max_i(W_2(n(s_i))))$, where i is such that s_i is the left child of v_i. $W_{1,r}(n) = \max(W_1(n(s_r)), \max_i(W_2(n(s_i))))$, where i is such that s_i is the right child of v_i.

Analogous properties hold for Constructions $1br$, $1tl$, and $1tr$.

Construction 2 is as follows. We have four subcases, say $2xy$, where $x \in \{t, b\}$ and $y \in \{l, r\}$. Index x states that the leftmost path is drawn going towards the top ($x = t$) or going towards the bottom ($x = b$) of Γ_2. Index y states that the root is drawn on the right side ($y = r$) or on the left side ($y = l$) of Γ_2. In the following we show Construction $2bl$, while the other cases are easily obtained from $2bl$ after a reflection with respect to the y-axis and/or a switch of the left with the right and vice-versa. Constructions $2bl$ and $2tr$ are shown in Fig. 3. Let r be the root of T, let $C_l = (u_{l,0}, u_{l,1}, \ldots, u_{l,m})$ ($C_r = (u_{r,0}, u_{r,1}, \ldots, u_{r,p})$) be the leftmost (rightmost) path of T, with $u_{l,0} = u_{r,0} = r$. Let $s_{l,i}$ ($s_{r,i}$) be the right (left) child of a node $u_{l,i} \in C_l$ ($u_{r,i} \in C_r$); we call $T(s_{l,i})$ ($T(s_{r,i})$) the subtree of T rooted in $s_{l,i}$ ($s_{r,i}$). First, we draw each $u_{l,i} \in C_l$ together with $T(s_{l,i})$ and each $u_{r,i} \in C_r$ together with $T(s_{r,i})$, obtaining $\Gamma(u_{l,i})$ and $\Gamma(u_{r,i})$ respectively; then we put all the $\Gamma(u_{l,i})$ and the $\Gamma(u_{r,i})$ together to obtain Γ_2.

Let k and j be two indexes such that $k, j \in \{l, r\}$ and let x such that $1 \leq x \leq m$ if $k = l$ and such that $1 \leq x \leq p$ if $k = r$. Find the heaviest subtree $T(s_{k,x})$ among all the subtrees $T(s_{j,i})$. Let $T(s_{k,xl})$ and $T(s_{k,xr})$ be the left and the right subtree of $s_{k,x}$, with root $s_{k,xl}$ and $s_{k,xr}$, respectively. Draw $T(s_{k,xl})$ with Construction $1bl$ and draw $T(s_{k,xr})$ with Construction $1tr$, obtaining $\Gamma(s_{k,xl})$ and $\Gamma(s_{k,xr})$, respectively. Draw any other subtree $T(s_{j,i})$ with Construction $2bl$, obtaining $\Gamma(s_{j,i})$.

Place $\Gamma(s_{k,xl})$ anywhere in the plane. Place $\Gamma(s_{k,xr})$ so that $b(\Gamma(s_{k,xr}))$ is three vertical units above $t(\Gamma(s_{k,xl}))$ and so that $l(\Gamma(s_{k,xr}))$ is on the same vertical channel of $l(\Gamma(s_{k,xl}))$. Place $s_{k,x}$ one unit above $t(\Gamma(s_{k,xl}))$ and one unit to the right of the rightmost boundary between $r(\Gamma(s_{k,xl}))$ and $r(\Gamma(s_{k,xr}))$. Draw $u_{k,x}$ on the same horizontal channel of $s_{k,x}$, one unit to the left of $l(\Gamma(s_{k,xl}))$. If $k = l$ ($k = r$) draw $u_{k,x-1}$ one unit above (one unit below) $u_{k,x}$. Place $\Gamma(s_{k,x-1})$ so that $l(\Gamma(s_{k,x-1}))$ is on the same vertical channel of $l(\Gamma(s_{k,x}))$ and so that (if $k = l$) $b(\Gamma(s_{k,x-1}))$ is one unit above $t(\Gamma(s_{k,x}))$ or (if $k = r$) $t(\Gamma(s_{k,x-1}))$ is one unit below $b(\Gamma(s_{k,x}))$, obtaining $\Gamma(u_{k,x-1})$. For each $\Gamma(s_{j,i})$, but for $\Gamma(s_{k,x-1})$ and $\Gamma(s_{k,x})$, place $u_{j,i}$ one unit to the left of $s_{j,i}$, obtaining $\Gamma(u_{j,i})$. Finally place all the $\Gamma(u_{j,i})$ (and so also $\Gamma(u_{k,x-1})$)

so that all $u_{j,i}$ are on the same vertical channel, so that $b(\Gamma(u_{r,i}))$ is one unit above $t(\Gamma(u_{r,i-1}))$, $2 \le i \le p$, so that $t(\Gamma(u_{l,i}))$ is one unit below $b(\Gamma(u_{l,i-1}))$, $2 \le i \le m$, and so that $t(\Gamma(u_{l,1}))$ is one unit below $b(\Gamma(u_{r,1}))$.

Property 4. Construction $2bl$ guarantees that all the vertices of the leftmost (rightmost) path of T are on the left side of the bounding box of Γ_2.

Property 5. Suppose that the drawing of Constructions $1tr$ and $1bl$ are star-shaped. Suppose that Construction $1tr$ is such that the leftmost (the rightmost) path of $T(s_{k,xr})$ is visible from any point that is below and to the right (to the left) of $B(\Gamma(s_{k,xr}))$. Suppose also that Construction $1bl$ is such that the leftmost (the rightmost) path of $T(s_{k,xl})$ is visible from any point that is above and to the left (to the right) of $B(\Gamma(s_{k,xl}))$. We have that the drawing obtained with Construction $2bl$ is star-shaped.

Property 6. $W_2(n) = \max(2 + W_1(n(s_{k,xl})), 2 + W_1(n(s_{k,xr})), \max(1 + W_2(n(s_{j,i}))))$, where $j \in \{l, r\}$ and i is not equal to x.

Analogous properties hold for Constructions $2br$, $2tl$, and $2tr$.

We can use Constructions 1–2 for constructing a star-shaped drawing Γ of a binary tree T as follows. First, we select any spine. Second, we apply Construction $1bl$. Third, we recursively apply all the constructions in the appropriate cases. From the above properties we have that Γ is star-shaped.

At this point we can draw a general outerplanar graph G with dual tree T as follows. First, we draw T with the above algorithm. Second, we apply Lemma 1 to construct an outerplanar drawing of the internal subgraph of G with the same height and width of T. Third, exploiting Property 1 we place the poles of G obtaining a drawing that has the same height and width plus one unit.

Now we analyze the height and the width of Γ. About the height, it's easy to see that there is at least one vertex for each horizontal line that intersects Γ. So we immediately obtain that the height of Γ is $O(n)$. About the width $W(n)$, let n_1 (n_2) be the number of vertices of the heaviest left (right) subtree of the spine S. We want to show that $W(n) \le W(n_1) + W(n_2) + 6$.

We focus on $W_{1,l}(n)$ to show that $W_{1,l}(n) \le W(n_1) + 2$. For this purpose we start from the expression of $W_{1,l}(n)$ as a function of $W_1(n)$ and of $W_2(n)$, then we substitute $W_2(n)$ with its definition as function of $W_1(n)$ and of $W_2(n)$. We repeat this substitution until we have obtained that $W_{1,l}(n)$ is defined only in terms of $W_1(n)$.

Let $n(s_j^*)$ be the maximum number of nodes of a subtree recursively drawn with Construction 2, after that j substitutions of $W_2(n)$ with its definition (as a function of $W_1(n)$ and of $W_2(n)$) have been made. Let $n(s_{j,l}^*)$ and $n(s_{j,r}^*)$ be the number of nodes of the left and the right subtrees of s_j^*, respectively.

By Property 3 we have $W_{1,l}(n) = \max(W_1(n(s_l)), \max(W_2(n(s_i))))$, with i such that $T(s_i)$ is the left subtree of a spine node v_i. By applying several times Property 6 to the above equation we have: $W_{1,l}(n) = \max(W_1(n(s_l)), 2 + W_1(n(s_{k,xl})), 2 + W_1(n(s_{k,xr})), 1 + W_2(n(s_1^*))) \le \max(W_1(n(s_l)), 2 + W_1(n(s_{k,xl})), 2 + W_1(n(s_{k,xr})), 3 + W_1(n(s_{1,l}^*)), 3 + W_1(n(s_{1,r}^*)), 2 + W_2(n(s_2^*))) \le \max(W_1(n(s_l)), 2 + W_1(n(s_{k,xl})), 2 + W_1(n(s_{k,xr})), 3 + W_1(n(s_{1,l}^*)), 3 + W_1(n(s_{1,r}^*)), 4 + W_1(n(s_{2,l}^*)), 4 + W_1(n(s_{2,r}^*)), 3 + W_2(n(s_3^*))) \le \dots \le \max(W_1(n(s_l)), 2 + W_1(n(s_{k,xl})), 2 + W_1(n(s_{k,xr})), 3 +$

$W_1(n(s_{1,l}^*)), 3 + W_1(n(s_{1,r}^*)), 4 + W_1(n(s_{2,l}^*)), 4 + W_1(n(s_{2,r}^*)), 5 + W_1(n(s_{3,l}^*)), 5 + W_1(n(s_{3,r}^*)), \ldots)$

Observe that $n(s_{j+1}^*) \le \frac{1}{2}n(s_j^*)$, since we draw the heaviest subtree T' of $T(s_j^*)$ with Construction 1 and a subtree T'' with size greater than $\frac{1}{2}n(s_j^*)$ implies $n(T') + n(T'') > n(s_j^*)$, that is impossible by definition. Hence, assuming $W_1(n) > \lg n$, we obtain $W_{1,l}(n) \le \max(W_1(n(s_l)), 2 + W_1(n(s_{k,xl})), 2 + W_1(n(s_{k,xr}))) \le 2 + W_1(n_1)$. With similar arguments we obtain $W_{1,r}(n) \le 2 + W_1(n_2)$. Observing that S is drawn on two adjacent vertical channels we have $W_1(n) = W_{1,l}(n) + W_{1,r}(n) + 2$, hence we obtain $W(n) = W_1(n) \le W_1(n_1) + W_1(n_2) + 6 \le W(n_1) + W(n_2) + 6$. As done in [3], we can choose in linear time a spine of T by maintaining the invariance that $n_1{}^p + n_2{}^p \le (1 - \delta)n^p$. Observe that $W(n) \le \max_{n_l^p + n_r^p \le (1-\delta)n^p}(W(n_l) + W(n_r) + 6)$, for any left (right) subtree of S with n_l (n_r) nodes; by induction this solves to $W(n) = O(n^p)$ and applying Lemma 3, we can complete the analysis of the width of Γ concluding that is possible to get $W(n) = O(n^{0.48})$.

From the results on the height and on the width, we obtain the $O(n^{1.48})$ area bound on Γ. It is easy to see that the algorithm can be implemented to run in linear time.

References

1. T. Biedl. Drawing outer-planar graphs in $O(n \log n)$ area. In M. Goodrich, editor, *Graph Drawing (Proc. GD '02)*, volume 2528 of *Lecture Notes Comput. Sci.*, pages 54–65. Springer-Verlag, 1997.
2. P. Bose. On embedding an outer-planar graph in a point set. In G. Di Battista, editor, *Graph Drawing (Proc. GD '97)*, volume 1353 of *Lecture Notes Comput. Sci.*, pages 25–36. Springer-Verlag, 1997.
3. T.M. Chan. A near-linear area bound for drawing binary trees. *Algorithmica*, 34(1), 2002.
4. M. Chrobak and T. H. Payne. A linear-time algorithm for drawing a planar graph on a grid. *IPL*, 54(4):241–246, 1995.
5. H. de Fraysseix, J. Pach, and R. Pollack. How to draw a planar graph on a grid. *Combinatorica*, 10(1):41–51, 1990.
6. G. Di Battista, P. Eades, R. Tamassia, and I. G. Tollis. *Graph Drawing*. Prentice Hall, Upper Saddle River, NJ, 1999.
7. S. Felsner, G. Liotta, and S. K. Wismath. Straight-line drawings on restricted integer grids in two and three dimensions. *J. Graph Algorithms Appl.*, 7(4):363–398, 2003.
8. A. Garg and A. Rusu. Straight-line drawings of binary trees with linear area and arbitrary aspect ratio. In M. Goodrich, editor, *Graph Drawing (Proc. GD '02)*, volume 2528 of *LNCS*, pages 320–331. Springer-Verlag, 2002.
9. A. Garg and A. Rusu. Area-efficient planar straight-line grid drawings of outerplanar graphs. In G.Liotta, editor, *Graph Drawing (Proc. GD '03)*, volume 2912 of *Lecture Notes Comput. Sci.*, pages 129–134. Springer-Verlag, 2003.
10. P. Gritzmann, B. Mohar, J. Pach, and R. Pollack. Embedding a planar triangulation with vertices at specified points. *Amer. Math. Monthly*, 98(2):165–166, 1991.
11. C. E. Leiserson. Area-efficient graph layouts (for VLSI). In *Proc. 21st Annu. IEEE Sympos. Found. Comput. Sci.*, pages 270–281, 1980.
12. S. Malitz and A. Papakostas. On the angular resolution of planar graphs. *SIAM J. Discrete Math.*, 7:172–183, 1994.
13. W. Schnyder. Embedding planar graphs on the grid. In *Proc. 1st ACM-SIAM Sympos. Discr. Alg.*, pages 138–148, 1990.

Volume Requirements of 3D Upward Drawings[*]

Emilio Di Giacomo[1], Giuseppe Liotta[1],
Henk Meijer[2], and Stephen K. Wismath[3]

[1] Dip. di Ing. Elettronica e dell'Informazione, Università degli Studi di Perugia, Italy
{digiacomo, liotta}@diei.unipg.it
[2] School of Computing, Queen's University, Kingston, Ontario, Canada
henk@cs.queensu.ca
[3] Dept. of Math. and Computer Science, University of Lethbridge, Canada
wismath@cs.uleth.ca

Abstract. This paper studies the problem of drawing directed acyclic graphs in three dimensions in the straight-line grid model, and so that all directed edges are oriented in a common (upward) direction. We show that there exists a family of outerplanar directed acyclic graphs whose volume requirement is super-linear. We also prove that for the special case of rooted trees a linear volume upper bound is achievable.

1 Introduction

The problem of computing 3D grid drawings of graphs so that the vertices are represented at integer grid-points, the edges are crossing-free straight-line segments, and the volume is small, has received a lot of attention in the graph drawing literature (e.g., [4, 5, 7, 8, 9, 12, 13]). While the interested reader is referred to the exhaustive introduction and list of references of [9] for reasons of space, we recall in this extended abstract some of the more recent results on the subject. In what follows, n denotes the number of vertices, and m the number of edges of a graph.

Dujmović and Wood [12] proved that drawings on an integer grid with an $O(n^{1.5})$ volume can be obtained for planar graphs, graphs with bounded degree, graphs with bounded genus, and graphs with no K_h (h constant) as a minor. Bose et al. [3] proved that the maximum number of edges in a grid drawing of dimensions $X \times Y \times Z$ is $(2X-1)(2Y-1)(2Z-1) - XYZ$, which implies a lower bound of $\frac{m+n}{8}$ on the volume of a 3D grid drawing of any graph. Felsner et al. [13] initiated the study of restricted integer grids, where all vertices are drawn on a small set of parallel grid lines, called tracks and proved that outerplanar graphs can be drawn by using three tracks on an integer grid of size $O(1) \times O(1) \times O(n)$. Dujmović, Morin, and Wood [9] showed that a graph G admits a drawing on an integer grid of size $O(1) \times O(1) \times O(n)$ if and only if G admits a drawing

---

[*] Research partially supported by MIUR under Project "ALGO-NEXT(Algorithms for the Next Generation Internet and Web: Methodologies, Design and Experiments)", and the Natural Sciences and Engineering Research Council of Canada.

P. Healy and N.S. Nikolov (Eds.): GD 2005, LNCS 3843, pp. 101–110, 2005.
© Springer-Verlag Berlin Heidelberg 2005

on an integer grid consisting of a constant number of tracks. Dujmović, Morin, and Wood used this result to show in [9] that graphs of bounded tree-width (including, for example, series-parallel graphs and k-outerplanar graphs with constant k) have 3D straight-line grid drawings of $O(n)$ volume. Some of the constant factors in the volume bounds of [9] are improved in [8]. As far as we know, the question of whether all planar graphs admit a 3D straight-line grid drawing of $O(n)$ volume remains a fascinating open problem.

This paper studies the problem of computing 3D straight-line grid drawings of directed acyclic graphs so that all edges are drawn oriented in a common direction; such drawings are called *3D upward drawings* in the remainder of the paper. Recall that 2D straight-line grid drawings of directed acyclic graphs such that all edges are drawn upward are a classical subject of investigation in the graph drawing literature (see, e.g. [1, 2, 14, 18]). Little is known about volume requirements of 3D upward drawings. Poranen [19] presented an algorithm to compute a 3D upward drawing of an arbitrary series-parallel digraph in $O(n^3)$ volume. This bound can be improved to $O(n^2)$ and $O(n)$ if the series-parallel digraph has some additional properties. The major contributions of the present paper can be listed as follows.

- We introduce and study the notion of *upward track layout*, which extends a similar concept studied by Dujmović, Morin, and Wood (see, e.g. [9, 10, 11, 12]). We relate upward track layouts to upward queue layouts and use this relationship to prove some of our volume bounds.
- We show that there exist outerplanar directed acyclic graphs which have a $\Omega(n^{1.5})$ volume lower bound. This result could be regarded as the 3D counterpart of a theorem in [6], which proves that upward grid drawings in 2D can require area exponential in the number of vertices. Note however that the class of graphs that we use for our lower bound has an $O(n^2)$ upward drawing in 2D. Also note that undirected outerplanar graphs admit a 3D grid drawing in optimal $O(n)$ volume [13].
- Motivated by the above super-linear lower bound, we investigate families of outerplanar graphs which admit upward 3D drawings of linear volume. In particular, we show that every tree has an upward 3D drawing on a grid of size $O(1) \times O(1) \times O(n)$.

The remainder of this paper is organized as follows. Preliminaries can be found in Section 2. The definition of upward track layout, and the volume lower bound for 3D upward drawings of outerplanar graphs are in Section 3. How to compute linear-volume 3D upward drawings of trees is the subject of Section 4. Other families of graphs and gaps on the volume are discussed in Section 5. Some proofs are sketched or omitted for reasons of space.

2   Preliminaries

Let G be a directed acyclic graph (DAG). The *underlying undirected graph* $\widehat{G}$ of G is the undirected graph obtained by ignoring the directions of the edges of G.

A *3D straight-line grid drawing* Γ of an *undirected* graph G maps each vertex of G to a distinct point of $\mathbb{Z}^3$ and each edge of G to the straight-line segment between its vertices. we denote the x-, y- and z-coordinates of p by $x(v)$, $y(v)$ and $z(v)$. A *crossing-free straight-line grid drawing* is a straight-line grid drawing such that edges intersect only at shared end-vertices and an edge only intersect a vertex that is an end-vertex of that edge.

A (crossing-free) straight-line grid drawing of a *DAG G* is a (crossing-free) straight-line grid drawing of the underlying undirected graph $\widehat{G}$ of G. A 3D straight-line grid drawing of G is *upward* if for each directed edge $(u, v) \in G$ we have $z(u) < z(v)$.

The *bounding box* of a straight-line grid drawing Γ of a graph G is the minimum axis-aligned box containing Γ. If the sides of the bounding box of a 3D straight-line grid drawing Γ parallel to the x-, y-, and z-axis have lengths $W - 1$, $D - 1$ and $H - 1$, respectively, we say that Γ has *width* W, *depth* D and *height* H. We also say that Γ has *size* $W \times D \times H$ and *volume* $W \cdot D \cdot H$.

3   Volume Requirements of 3D Upward Drawings

In this section we present a super-linear lower bound on the volume of 3D upward drawings of outerplanar DAGs. In order to do this, we start by introducing and studying the concept of an *upward track layout*, which extends the (undirected) notion of an improper track layout as defined by Dujmović et al. [10].

3.1   Upward Track Layouts

Let $G = (V, E)$ be an undirected graph. A *t-track assignment* γ of G consists of a partition of V into t sets $V_0, V_1, \ldots, V_{t-1}$ and a total order $\leq_i$ for each set V_i. We write $u <_i w$ if $u \leq_i w$ and $u \neq w$. There is an *overlap* if there exist three vertices u, v, w such that $u, v, w \in V_i$, $(u, w) \in E$ and $u <_i v <_i w$. There is an *X-crossing* if there exist two edges (u, v) and (w, z) such that $u, w \in V_i$, $v, z \in V_j$, with $i \neq j$, and $u <_i w$ and $z <_j v$. A *t-track layout* of G is a t-track assignment of G without overlaps and X-crossings. The minimum value of t such that G has an t-track layout is called the *track number* of G and is denoted as $\mathsf{tn}(G)$.

Definition 1. *Let $G = (V, E)$ be a DAG. An upward t-track layout of G is a partition of V into t sets $V_0, V_1, \ldots, V_{t-1}$, called* tracks, *a total order $\leq_i$ for each track V_i and a partial order $\preccurlyeq$ on V such that there is no overlap, there is no X-crossing, if $(u, v) \in E$ then $u \preccurlyeq v$ and if $u \leq_i v$ for some i then $u \preccurlyeq v$.*

The minimum value of t such that G has an upward t-track layout is called the *upward track number* of G and is denoted as $\mathsf{utn}(G)$.

We complete this section by studying the relationship between upward track layout and another well-known graph parameter, namely the upward queue-number [15, 16, 17].

Let $G = (V, E)$ be an undirected graph. A *q-queue layout* of G consists of a total ordering $\leq_\sigma$ of V and a partition of E into q sets, called *queues*, such that there are no two edges (u, v) and (w, z) in the same queue such that $u <_\sigma w <_\sigma z <_\sigma v$, where $u <_\sigma w$ means $u \leq_\sigma w$ and $u \neq w$. The minimum value of q such that G has a q-queue layout is called the *queue number* of G and is denoted as $\mathsf{qn}(G)$.

Let $G = (V, E)$ be a DAG. An *upward q-queue layout* of G consists of a total ordering $\leq_\sigma$ of V and a partition of E into q sets, called *queues*, such that it is a q-queue layout for the undirected underlying graph $\hat{G}$ of G and for each edge $(u, v) \in E$, we have $u <_\sigma v$. The minimum value of q such that G has an upward q-queue layout is called the *upward queue number* of G and is denoted as $\mathsf{uqn}(G)$.

Lemma 1. *Let G be a DAG. Then*

$$\mathsf{uqn}(G) \leq \binom{\mathsf{utn}(G)}{2} + \mathsf{utn}(G).$$

Sketch of Proof. The total ordering σ of the queue layout is a total order that respects the partial order $\preceq$ of the track layout. All edges between any pair of tracks can be put in a queue. All edges on a track can be put in a queue.   □

Note that in the undirected case Dujmović et al. [10] proved that $\mathsf{qn}(G) \leq \mathsf{tn}(G)$, for every graph G. As the following lemma shows, the relationship stated by Lemma 1 can be asimptotically tight for DAGs.

Lemma 2. *For all n there exists a DAG G with at least n vertices such that $\mathsf{uqn}(G) \geq (\mathsf{utn}(G) - 2)^2/2$.*

Proof. Let k be the smallest integer such that there is a value t for which $t(t+1) = 2k \geq n$. Consider the graph $G = (V, E)$. The set V is $V_u \cup V_v$ where $V_u = \{u_0, u_1, \ldots, u_{k-1}\}$ and $V_v = \{v_0, v_1, \ldots, v_{k-1}\}$. The set of edges E is $E_u \cup E_v \cup E_{uv}$ where $E_u = \{(u_i, u_{i+1}) \mid 0 \leq i < k-1\}$, $E_v = \{(v_i, v_{i+1}) \mid 0 \leq i < k-1\}$ and $E_{uv} = \{(u_i, v_{k-1-i}) \mid 0 \leq i < k\}$. The graph G contains the Hamiltonian path consisting of $E_u \cup E_v$ plus the edge (u_{k-1}, v_0). The order of the vertices of G in this Hamiltonian path is the unique topological sort of G and therefore it must be the total order for the upward queue layout. Since no two edges from E_{uv} can belong to the same queue, it follows that $\mathsf{uqn}(G) \geq k$. It is not hard to see that in fact $\mathsf{uqn}(G) = k$.

Consider the following upward layout of G on $t + 2$ tracks. Place all vertices of V on the tracks in the order given below. For an illustration see Figure 1, where $k = 10$ and $t = 4$. Place vertices $u_0, u_1, \ldots, u_{t-1}$ on track 0. Then place the next $t - 1$ vertices of V_u on track 1, the next $t - 2$ vertices of V_u on track 2, etc. So track $t - 1$ contains the vertex u_{k-1}. Place v_0 on track $t + 1$. Place v_1 and v_2 on tracks $t + 1$ and t respectively. Then place the next three vertices of V_v on tracks $t + 1$, t and $t - 1$, etc. So the last group of vertices placed is $\{v_{k-t}, v_{k-t-1}, \ldots, v_{k-1}\}$, and they lie on tracks $t + 1, t, \ldots, 2$. It can easily be verified that the edges of E do not form an X-crossing. So $\mathsf{utn}(G) \leq t + 2$. We have $\mathsf{uqn}(G) = k = t(t+1)/2 \geq (\mathsf{utn}(G) - 2)^2/2$, so the lemma holds.   □

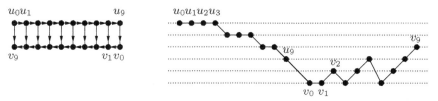

Fig. 1. A graph G and a partial 5-track layout of G

3.2   Volume Requirement

We next show a super-linear volume lower bound by using the results in the previous subsection and the following lemma.

Lemma 3. *Let G be a DAG and let Γ be a 3D straight-line upward grid drawing of G such that the sides of the bounding box of Γ parallel to the x-, y-, and z-axis have length W, D, and H, respectively. Then $\mathsf{utn}(G) \leq W \cdot D$.*

Sketch of Proof. All lines in Γ parallel to the z-axis are the tracks of the track layout. The total ordering $\leq_i$ on each track V_i and the partial order $\prec$ for the track layout can be defined according to the z-coordinates of the vertices in Γ. □

Theorem 1. *There exists an outerplanar DAG G with n vertices such that any crossing-free 3D straight-line upward grid drawing of G requires $\Omega(n^{1.5})$ volume.*

Proof. Consider the DAG $G = (V, E)$ with $m = 3n/2 - 2$ edges as defined in the proof of Lemma 2 and illustrated in Figure 1 with $n = 20$.

As we saw in the proof of Lemma 2, $\mathsf{uqn}(G) = k = n/2$. Assume for contradiction that there exists a 3D straight-line upward grid drawing Γ of G with volume $o(n^{1.5})$. Let W, D, and H be the width, depth, and height of Γ. Since Γ is upward, we have $z(u_0) < z(u_1) < \cdots < z(u_{k-1}) < z(v_0) < z(v_1) < \cdots < z(v_{k-1})$. This implies that $H \geq n$. In order to have a volume of $o(n^{1.5})$ it must be that $W \cdot D = o(n^{\frac{1}{2}})$. By Lemma 3 this would imply $\mathsf{utn}(G) = o(n^{\frac{1}{2}})$. By Lemma 1, we have $\mathsf{uqn}(G) = O(\mathsf{utn}(G)^2)$ and therefore it would be $\mathsf{uqn}(G) = o(n)$, but this is impossible because we proved that $\mathsf{uqn}(G) = \Omega(n)$. □

Note that in contrast to Theorem 1, undirected outerplanar graphs admit a crossing-free 3D straight-line upward grid drawing in optimal $O(n)$ volume [13]. Theorem 1 can be regarded as the three-dimensional counterpart of well-known results which show that in two-dimensions, undirected and directed planar graphs have different area requirements [6].

4   Compact 3D Upward Drawings of Trees

Based on the result of Theorem 1 we next investigate whether there exist meaningful families of outerplanar DAGs with $o(n^{1.5})$ volume upper bounds. In this section we study compact 3D upward drawings of trees and paths. We recall that Heath et al. [17] proved that every tree DAG has an upward 2-queue layout and that every path DAG has an upward 1-queue layout.

Definition 2. *Let $G = (V, E)$ be a DAG. A 3D upward straight-line grid drawing Γ of G on t lines is a drawing of G with vertices placed on t lines parallel to the z-axis such that the drawing induced by the vertices on two of the t lines is crossing-free.*

Recall that in an upward grid drawing, we also have $z(u) < z(v)$ for all edges (u, v).

Lemma 4. *If DAG G has a 3D upward straight-line grid drawing on t-lines of height H, then G has a 3D crossing-free straight-line upward grid drawing of size $t \times p \times p \cdot H$ and volume $O(t^3 \cdot H)$, where p is the smallest prime number such that $p \geq t$.*

The lemma follows directly from a similar result in [9].

Corollary 1. *Let G be a DAG with n vertices. G has a 3D crossing-free straight-line upward grid drawing of size $\mathsf{utn}(G) \times p \times p \cdot n$ and volume $O(\mathsf{utn}(G)^3 \cdot n)$, where p is the smallest prime number such that $p \geq \mathsf{utn}(G)$.*

Lemma 5. *Let T be a directed tree with n vertices. Then T admits an upward straight-line grid drawing on 7 lines.*

Sketch of Proof. Let v be a vertex of T. The set of edges of T is E. We use $T^+(v)$ to denote the subtree of T induced by all vertices w for which there is a directed path of length ≥ 0 from v to w. Similarly, $T^-(v)$ is the subtree of T induced by all vertices w for which there is a directed path of length ≥ 0 from w to v.

Let r be a vertex of T that has no incoming edges, i.e. there are no edges (v, r) in E. Let F_0 be $T^+(r)$. Let $F_1 = \{T^-(w) \mid v \in F_0, w \notin F_0, (w, v) \in E\}$. In other words F_1 is a forest of trees $T^-(w)$ for all nodes w for which there is an edge (w, v) in E with $v \in F_0$ and $w \notin F_0$. Similarly, let $F_2 = \{T^+(w) \mid v \in F_1, w \notin F_0 \cup F_1, (v, w) \in E\}$, $F_3 = \{T^-(w) \mid v \in F_2, w \notin F_1 \cup F_2, (w, v) \in E\}$, etc. Since T is connected, it follows that each vertex v of T belongs to some F_i.

We first draw the single tree of F_0, i.e. the tree $T^+(r)$, on tracks 0, 1 and 2 using the wrap-around algorithm described in [13]. We then place the roots

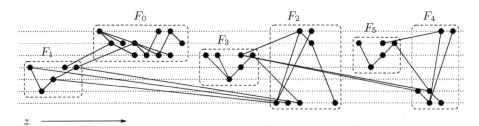

Fig. 2. Drawing of a tree decomposed into 6 forests

of the trees in F_1 on track 3. We then use the algorithm of [13] again to place the remaining vertices of F_1 on tracks 3, 4 and 5, but now wrapping the trees from high z values to smaller z values. Suppose we have placed F_i on tracks $j, j + 1, j + 2$. If i is odd we place the roots of the forest F_{i+1} on track $j + 3$, sufficiently far above F_{i-1} to leave room for F_{i+2}. We then use the wrap-around algorithm to place the remaining vertices of F_{i+1} on tracks $j+3, j+4, j+5$. If i is even we place the roots of the forest F_{i+1} on track $j + 3$, below all vertices of F_i, but above F_{i-2}. We then use the wrap-around algorithm to place the remaining vertices of F_{i+1} on tracks $j + 3, j + 4, j + 5$, so that all vertices of F_{i+1} are above the vertices of F_{i-2}. It can be shown that the resulting drawing has no overlaps and no X-crossings. For an illustration, see Figure 2. $\qquad\square$

Theorem 2. *Every directed tree T with n vertices admits a 3D crossing-free straight-line upward grid drawing of size $7 \times 7 \times 7 \cdot n$ and volume $O(n)$.*

An immediate consequence of Lemma 5 is that for every tree T $\mathsf{utn}(T) \leq 7$. It is possible to prove that there exists a directed tree T such that $\mathsf{utn}(T) \geq 4$. Therefore the following theorem holds.

Theorem 3. *Let T be a directed tree. Then $4 \leq \mathsf{utn}(T) \leq 7$.*

For the special case of a path, the result of Theorem 2 can be further improved as shown in the following.

Theorem 4. *Every directed path P with n vertices admits a 3D crossing-free straight-line upward grid drawing of size $2 \times 2 \times n$ and volume $O(n)$.*

Sketch of Proof. Let P be a directed path with vertices $v_0, \ldots, v_{n-1}$. We assume without loss of generality that the first edge from v_0 to v_1 is directed in the direction from v_0 to v_1. Decompose the path into k chains of consecutive edges that are similarly directed. We refer to the vertices where the path changes direction as $w_0 = v_0, w_1, w_2, ..., w_k = v_{n-1}$, and the directed chains as $W_0 = w_0 \to w_1$, $W_1 = w_1 \leftarrow w_2$, $W_2 = w_2 \to w_3$, etcetera.

These chains alternate in direction and our goal is to draw them on three tracks in the order $0, 1, 2, 0, 1, \ldots$. The algorithm to layout the chains is straight-forward except that some care is required if there is a long down chain that

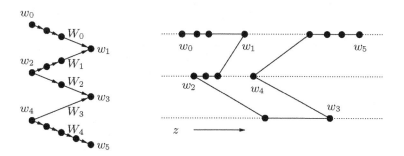

Fig. 3. Paths on 3 tracks

might interfere with previously placed vertices. To avoid this, we maintain an invariant that places vertices of the form w_{2i+1} sufficiently high. See Figure 3 for the general technique. Finally, the tracks can be drawn in 3D (non-coplanarly) in an upward manner and achieving the stated size and volume.   □

5   Extensions to Other Families of DAGs

Let $G = (V, E)$ be a DAG. A *vertex c-colouring* of G is a partition $\{V_i : 1 \le i \le c\}$, such that for every edge $(u, v) \in E$, if $u \in V_i$ and $v \in V_j$, then $i \ne j$. The minimum value of c such that G has a vertex c-colouring is called the *chromatic number* and is denoted by $\chi(G)$. A *strong star colouring* of a graph G is a vertex colouring of G such that each bichromatic subgraph consists of a star and possibly some isolated vertices. The minimum value of c such that G has a strong star colouring with c colours is called the *strong star chromatic number* and is denoted by $\chi_{sst}(G)$. Th definition of strong star chromatic number is due to Dujmović and Wood [12] who observed that track number is at most strong star chromatic number, i.e. $\mathsf{tn}(G) \le \chi_{sst}(G)$. It is easy to prove that also $\mathsf{utn}(G) \le \chi_{sst}(G)$.

In [12] it has been proven that every graph G with m edges and maximum degree $\Delta \ge 1$ has strong star chromatic number $\chi_{sst}(G) < 14\sqrt{\Delta m}$ and $\chi_{sst}(G) < 15m^{2/3}$. Consequences of these results are that every planar graph has upward track number $O(n^{2/3})$ and that this bound reduces to $O(\sqrt{n})$ if the planar graph has bounded degree. This allows us to find upper bounds on the volume of a 3D crossing-free straight-line upward grid drawing of several families of graphs. In particular, outerplanar graphs and Halin Graphs as special cases of planar graphs with unbounded degree, have upward track number $O(n^{2/3})$ and by Corollary 1, volume $O(n^3)$. On the other hand k-planar graphs (i.e. planar graphs with maximum vertex degree at most k) and X-trees as examples of planar graphs with bounded degree have upward track number $O(\sqrt{n})$ and by Corollary 1, volume $O(n^{2.5})$. It is easy to construct an X-tree and a Halin graph that contains the graph of Figure 1, which is outerplanar, planar and k-planar

Table 1. Upper and Lower Bounds on the Volume of a 3D crossing-free straight-line upward grid drawing of different families of graphs

Family of DAGs	Volume Upper Bound	Volume Lower Bound
Trees	$O(n)$ ($7 \times 7 \times 7 \cdot n$)	$\Omega(n)$
Paths	$O(n)$ ($2 \times 2 \times n$)	$\Omega(n)$
X-trees	$O(n^{2.5})$	$\Omega(n^{1.5})$
Halin	$O(n^3)$	$\Omega(n^{1.5})$
Outerplanar	$O(n^3)$	$\Omega(n^{1.5})$
Planar	$O(n^3)$	$\Omega(n^{1.5})$
k-planar	$O(n^{2.5})$	$\Omega(n^{1.5})$
arbitrary	$O(n^4)$	$\Omega(n^{1.5})$

for each $k \geq 3$. It follows that a lower bound of $\Omega(n^{1.5})$ on the volume of a 3D crossing-free straight-line upward grid drawing can be established for all these families of graphs. We conclude by observing that a trivial upper bound on the upward track number of an arbitrary graph G is $O(n)$ and hence by Corollary 1 a trivial upper bound on the volume is $O(n^4)$. Table 1 summarizes these upper and lower bounds on the volume.

References

1. P. Bertolazzi, G. Di Battista, G. Liotta, and C. Mannino. Upward drawings of triconnected digraphs. *Algorithmica*, 6(12):476–497, 1994.
2. P. Bertolazzi, G. Di Battista, C. Mannino, and R. Tamassia. Optimal upward planarity testing of single-source digraphs. *SIAM J. Computing*, 27(1):132–169, 1998.
3. P. Bose, J.Czyzowicz, P. Morin, and D. R. Wood. The maximum number of edges in a three-dimensional grid-drawing. *J. of Graph Algorithms and Applications*, 8(1):21–26, 2004.
4. T. Calamoneri and A. Sterbini. 3D straight-line grid drawing of 4-colorable graphs. *Information Processing Letters*, 63(2):97–102, 1997.
5. R. F. Cohen, P. Eades, T. Lin, and F. Ruskey. Three-dimensional graph drawing. *Algorithmica*, 17(2):199–208, 1997.
6. G. Di Battista, R. Tamassia, and I. G. Tollis. Area requirement and symmetry display of planar upward drawings. *Discrete and Computational Geometry*, 7:381–401, 1992.
7. E. Di Giacomo. Drawing series-parallel graphs on restricted integer 3D grids. In G. Liotta, editor, *Proc. of 11th International Symposium on Graph Drawing, GD 2003*, volume 2912 of *Lecture Notes Computer Science*, pages 238–246. Springer-Verlag, 2004.
8. E. Di Giacomo, G. Liotta, and H. Meijer. Computing straight-line 3D grid drawings of graphs in linear volume. *Computational Geometry*. to appear.
9. V. Dujmović, P. Morin, and D. R. Wood. Layout of graphs with bounded tree-width. *SIAM J. on Computing*, 34(3):553–579, 2005.
10. V. Dujmović, A. Por, and D. R. Wood. Track layouts of graphs. *Discrete Math. Theor. Comput. Sci.*, 6(2):497–522, 2004.
11. V. Dujmović and D. R. Wood. Layouts of graph subdivisions. In J. Pach, editor, *Proc. 12th International Symposium on Graph Drawing, GD 2004*, volume 3383 of *Lecture Notes in Computer Science*, pages 133–143. Springer, 2004.
12. V. Dujmović and D. R. Wood. Three-dimensional grid drawings with sub-quadratic volume. In J. Pach, editor, *Towards a Theory of Geometric Graphs*, volume 342, pages 55–66. 2004.
13. S. Felsner, G.Liotta, and S. K. Wismath. Straight-line drawings on restricted integer grids in two and three dimensions. *J. of Graph Algorithms and Applications*, 7(4):363–398, 2003.
14. A. Garg and R. Tamassia. On the computational complexity of upward and recti-linear planarity testing. *SIAM J. Computing*, 31(2):601–625, 2001.
15. L. S. Heath and S. V. Pemmaraju. Stack and queue layouts of posets. *SIAM J. on Discrete Mathematics*, 10:599–625, 1997.
16. L. S. Heath and S. V. Pemmaraju. Stack and queue layouts of directed acyclic graphs: Part II. *SIAM J. on Computing*, 28:1588–1626, 1999.

17. L. S. Heath, S. V. Pemmaraju, and A. Trenk. Stack and queue layouts of directed acyclic graphs: Part I. *SIAM J. on Computing*, 28:1510–1539, 1999.
18. M. D. Hutton and A. Lubiw. Upward planning of single-source acyclic digraphs. *SIAM J. on Computing*, 25(2):291–311, 1996.
19. T. Poranen. A new algorithm for drawing series-parallel digraphs in 3D. Technical Report A-2000-16, Dept. of Computer and Information Sciences, University of Tampere, Finland, 2000.

How to Embed a Path onto Two Sets of Points

Emilio Di Giacomo, Giuseppe Liotta, and Francesco Trotta

Dipartimento di Ingegneria Elettronica e dell'Informazione,
Università degli Studi di Perugia
{digiacomo, liotta, francesco.trotta}@diei.unipg.it

Abstract. Let R and B be two sets of points such that the points of R are colored red and the points of B are colored blue. Let P be a path such that $|R|$ vertices of P are red and $|B|$ vertices of P are blue. We study the problem of computing a crossing-free drawing of P such that each blue vertex is represented as a point of B and each red vertex of P is represented as a point of R. We show that such a drawing can always be realized by using at most one bend per edge.

1 Introduction

Let G be a planar graph such that each vertex of G is colored with either the red or the blue color. Let R and B be two distinct sets of red and blue points in the plane, respectively, such that $|R|$ equals the number of red vertices of G and $|B|$ equals the number of blue vertices of G. A *bichromatic point-set embedding of G onto $R \cup B$* is a crossing-free drawing such that those vertices that are blue in G are mapped to points of B and those vertices that are red in G are mapped to points of R. The mapping of each blue/red vertex of G to a corresponding blue/red point of $R \cup B$ is not part of the input.

The problem of computing bichromatic point-set embeddings for different subclasses of planar graphs has attracted considerable interest during the last fifteen years. We briefly recall here only some of the most relevant results concerning the case that G is a simple path, since this is the main subject of this short paper. For an exhaustive survey see [5]. In what follows we shall denote with S the set $R \cup B$ and implicitly assume that the red (blue) points of S are always as many as the red (blue) vertices of the bi-colored input path P.

Akiyama and Urrutia [2] exhibit a set S of sixteen points in convex position on which a proper 2-colored path P does not admit a straight-line bichromatic point-set embedding, and present an $O(n^2)$-time algorithm to test whether a proper 2-colored path has a straight-line bichromatic point-set embedding on a given set of points. Abellanas et al. [1] also study straight-line point-set embeddings for a path P with a proper 2-coloring. They show that if either the convex hull of S consists of all red points and no blue points or S is a linearly separable bipartition (i.e. there exists a line that separates all blue points from the red ones), then P has a straight-line point-set embedding onto S. Finally, a recent paper by Kaneko, Kano, and Suzuki [4] provides a complete characterization of those paths with a proper 2-coloring that admit a straight-line bichromatic

P. Healy and N.S. Nikolov (Eds.): GD 2005, LNCS 3843, pp. 111–116, 2005.

point-set embedding onto any set of points S in general position: If P has at most twelve vertices or if it has exactly fourteen vertices, then P always admits a straight-line bichromatic point-set embedding onto S; for all other cases, there exist configurations of S for which P does not admit a straight-line bichromatic point-set embedding.

Motivated by the result of Kaneko, Kano, and Suzuki [4], we study the problem of constructing a bichromatic point-set embedding of a 2-colored path by removing the restriction that no three points of S are collinear and by not assuming that the given 2-coloring is proper. We observe that allowing collinearities naturally leads to bichromatic point set embeddings whose edges can contain bends. The main contribution of this paper is the following theorem.

Theorem 1. *Let P be a simple path such that each vertex of P is colored with either the red or the blue color. Let R and B be two distinct sets of points in the plane such that $|R|$ equals the number of red vertices of P and $|B|$ equals the number of blue vertices of P. Then P admits a bichromatic point-set embedding onto $R \cup B$ with at most one bend per edge.*

The proof of Theorem 1 is based on showing that a 2-colored path admits a bichromatic point-set embedding onto any given set S if and only if it has a suitably defined 2-page bichromatic book embedding (see Section 2).

2   Preliminaries

Let $G = (V, E)$ be a planar graph. A 2-*coloring* of G is a partition of V into 2 disjoint sets V_b and V_r. We call *blue vertices* the vertices of V_b and *red vertices* the vertices of V_r. A 2-coloring is *proper* if for every edge $(u, v) \in E$ we have $u \in V_b$ and $v \in V_r$. Given a vertex v we denote by $c(v)$ the color of v. If a graph G has a 2-coloring we say that it is 2-*colored*, if the 2-coloring is proper we say that G is *properly 2-colored*.

Let G be a planar 2-colored graph and let $S = B \cup R$ be a set of points in the plane, such that $|B| = |V_b|$ and $|R| = |V_r|$. We call *blue points* the points of B and *red points* the points of R. A *point-set embedding* onto S of G is a planar drawing Γ such that the vertices of G are drawn in Γ on the points of S, and each edge of G is drawn as a polyline in Γ (Kaufmann and Wiese [6] show that any planar graph admits a points-set embedding). G has a *bichromatic point-set embedding* onto S if G has a point-set embedding onto S such that every blue vertex is drawn on a blue point, and every red vertex is drawn on a red point. A planar 2-colored graph G is *bichromatic point-set embeddable* if for any set of points, $S = R \cup B$ such that $|B| = |V_b|$ and $|R| = |V_r|$, G has a bichromatic point-set embedding onto S.

Let G be a planar graph. An *h-page book embedding* of G consists of a linear ordering λ of the vertices of G and a partition of the edges of G into h disjoint sets, called *pages*, such that there are no two edges (u, v) and (w, z) in the same page with $u < w < v < z$ in λ. A different but equivalent definition of an *h-page* book embedding is the following. An *h-page book embedding* of G is a drawing of G such that all the vertices of G are drawn as points of a straight line l called *spine*, each edge is drawn on one of h half-planes, called *pages*, having l as a common boundary, and no two edges in the same page cross. According to this second definition, a book embedding is a drawing rather than a combinatorial object. In the following we shall always refer to this "geometric" definition rather than to the "combinatorial" one. In the special case when $h = 2$ we have that a 2-page book embedding of G is a planar drawing such that all the vertices are drawn as points of a straight line l, and each edge is drawn on one of the two half-planes defined by l.

A *red-blue sequence* σ is a sequence of points along a straight line l such that each point $p \in \sigma$ is either red or blue. Given a point p of σ, we denote by $c(p)$ the color of p. Let n_r and n_b be the number of red and blue points in a red-blue sequence σ, respectively, and let G be a planar 2-colored graph such that $|V_b| = n_b$ and $|V_r| = n_r$. An h-page book embedding of G *consistent with* σ is an h-page book embedding of G such that each vertex v of G is represented by a point p of σ and $c(v) = c(p)$. Notice that the exact position of the points of σ on the line l is not relevant for the existence of the book embedding, and only their relative order is important. A planar 2-colored graph G is *h-page bichromatic book embeddable* if, for any red-blue sequence σ with $|V_b| = n_b$ and $|V_r| = n_r$, G has an h-page book embedding consistent with σ. Let γ be an h-page book embedding of G, and let v be a vertex of G. We say that v is *accessible* from a page π if there is no edge (u, w) in π such that $u < v < w$ in the linear ordering of γ. Analogously we say that a point $p \in \sigma$ is *accessible* from a page π if there is no edge (u, w) in π such that $u < p < w$ in the linear ordering of γ. Two vertices/points accessible from a common page can be connected by an edge without creating any crossings.

In [3] it has been proved that there is a strong connection between point-set embeddability and book embeddability. More precisely, the following lemma is an immediate consequence of [3].

Lemma 1. [3] *Let G be a planar graph. G admits a 2-page book embedding if and only if G admits a point-set embedding with at most 1 bend per edge on any set of points.*

The following theorem shows that the result can be extended to the case of bichromatic point-set embedding and bichromatic book embedding. The proof is omitted for reasons of space.

Theorem 2. *Let G be a planar 2-colored graph. Then G is bichromatic point-set embeddable with at most 1 bend per edge if and only if it is 2-page bichromatic book embeddable.*

3   Bichromatic Point-Set Embedding of Paths

In this section we prove Theorem 1 and apply it to the bichromatic point-set embeddability of cycles. Based on Theorem 2, it suffices to prove the following.

Theorem 3. *Let P be a 2-colored path, and let σ be any red-blue sequence. Then P has a 2-page bichromatic book-embedding that is consistent with σ.*

Proof. Let V_b and V_r be the set of blue and red vertices of P respectively, and let σ be any red-blue sequence such that $n_b = |V_b|$ and $n_r = |V_r|$, where n_b and n_r are the number of blue and red points of σ, respectively. Denote as $p_0, p_1, \ldots p_{n-1}$ the points of σ in the order they have in σ. We describe how to construct a 2-page bichromatic book embedding of P consistent with σ. We shall denote with P_k the sub-path of P induced by the first $k + 1$ vertices of P. The $k + 1$ vertices of P_k are denoted as $v_0, v_1, \ldots, v_k$.

The proof is constructive and adds one vertex and one edge per step to the bichromatic book embedding. At step k all vertices of P_{k-1} have already been added to the bichromatic book embedding, and we add vertex v_k and edge (v_{k-1}, v_k). We denote by $\sigma_k \subseteq \sigma$ the red-blue sequence consisting of all points representing the vertices of P_k. We prove by induction that at the end of step k the following invariants hold:

Property 1. Let p_i be the rightmost point of σ_k. Denote as NB_k the set of all points of $\sigma \setminus \sigma_k$ that precede p_i in σ. All points in NB_k have the same color and are all accessible from the same page π. Furthermore, vertex v_k is accessible from π.

Property 2. Let p_j be the point of σ_k representing vertex v_k, and let p_i be the rightmost point of σ_k. Either $i = j$, or if $j \neq i$ then $c(p_{i+1}) \neq c(p_j)$.

At step $k = 0$ we choose the leftmost point p_i of σ such that $c(p_i) = c(v_0)$. Properties 1 and 2 trivially hold in this case. At step $k > 0$ vertex v_k and edge (v_{k-1}, v_k) are added according to the following cases, which depend on the position of the point representing v_{k-1} in σ_{k-1}.

Case 1. v_{k-1} **is represented as the rightmost point** p_i **of** σ_{k-1}. There are three sub-cases (see also Figure 1):

Case 1.a. If $c(p_{i+1}) = c(v_k)$ then map v_k to p_{i+1}, and arbitrarily assign (v_{k-1}, v_k) to one of the two pages. No crossing is created by adding edge (v_{k-1}, v_k) because v_{k-1} and v_k are represented as consecutive points in the sequence. Properties 1 and 2 hold in this case. Namely, $NB_k = NB_{k-1}$ because there is no point between p_i and p_{i+1}. Hence all points in NB_k have the same color and are all accessible from a same page π by induction. Also, v_k is represented as the rightmost point of σ_k, and hence it is accessible from both pages.

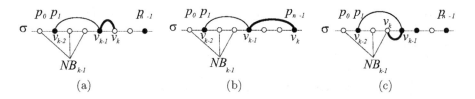

(a)   (b)   (c)

Fig. 1. Illustrations for Theorem 3 (a) Case 1.a (b) Case 1.b (c) Case 1.c

It follows that Property 1 holds. Concerning the statement of Property 2, we observe that in this case the point of σ_k representing vertex v_k is the rightmost point.

Case 1.b. Neither p_{i+1} nor the vertices in NB_{k-1} have the same color as v_k. Map v_k to the first vertex p_j to the right of p_i that has the same color as v_k, i.e. $j = \min\{h \mid h > i \ \wedge \ c(p_h) = c(v_k)\}$. By induction all points in NB_{k-1} are accessible from a same page π. We assign edge (v_{k-1}, v_k) to the other page (the one different from π). The addition of edge (v_{k-1}, v_k) does not introduce any crossings, because there is no other edge with an endvertex mapped on a point between p_i and p_j. We have that $NB_k = NB_{k-1} \cup \{p_{i+1}, p_{i+2}, \ldots, p_{j-1}\}$, and that $c(p_{i+1}) = c(p_{i+2}) = \cdots = c(p_{j-1}) \neq c(v_k)$, because p_j is the first point after p_i such that $c(p_j) = c(p_i)$. It follows that all vertices of NB_k have the same color. Also, they are all accessible from π because we assign edge (v_{k-1}, v_k) to the page different from π. Hence the invariant expressed by Property 1 is maintained. Concerning the statement of Property 2, we observe that also in this case the point of σ_k representing vertex v_k is the rightmost point.

Case 1.c. $c(p_{i+1}) \neq c(v_k)$, $NB_{k-1} \neq \emptyset$, and the vertices of NB_{k-1} have the same color as v_k. We map v_k to the rightmost point p_j of NB_{k-1}, i.e. $j = \max\{h \mid p_h \in NB_{k-1}\}$. By induction all vertices of NB_{k-1} are accessible from a page π, and we assign edge (v_{k-1}, v_k) to π. The addition of edge (v_{k-1}, v_k) does not create a crossing because, by Property 1, v_{k-1} and p_j are accessible from a common page. We have that $NB_k = NB_{k-1} \setminus \{p_j\}$. It follows that the vertices of NB_k all have the same color and are all accessible from a page π by induction. Point p_j is accessible from π by induction, and it remains accessible also after that edge (v_{k-1}, v_k) is drawn on π. Thus Property 1 holds. Property 2 holds since $c(p_{i+1}) \neq c(v_k)$.

Case 2. v_{k-1} **is not represented as the rightmost point** p_i **of** σ_{k-1}. We distinguish three sub-cases (see also Figure 2):

Case 2.a. $c(v_k) = c(v_{k-1})$ and $NB_{k-1} \neq \emptyset$. By induction all points of NB_{k-1} have the same color. Also, note that the points of NB_{k-1} plus the point representing v_{k-1} all belong to NB_{k-2} by induction, and hence they all have the same color as v_k. We map v_k to the rightmost point p_j of NB_{k-1}, i.e. $j = \max\{h \mid p_h \in NB_{k-1}\}$. By induction the vertices of NB_{k-1} are accessible from a page π; we assign edge (v_{k-1}, v_k) to π. The addition of (v_{k-1}, v_k) does not create a crossing

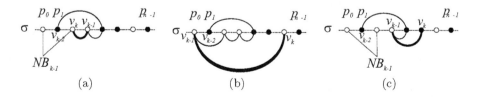

Fig. 2. Illustrations for Theorem 3 (a) Case 2.a (b) Case 2.b (c) Case 2.c

because, by Property 1, v_{k-1} and p_j are accessible from a common page. We have that $NB_k = NB_{k-1} \setminus \{p_j\}$, therefore all points of NB_k have the same color and are all accessible from page π by induction. Point p_j was accessible from π by induction, and it remains accessible also after edge (v_{k-1}, v_k) is drawn on π. Thus, Property 1 holds. Property 2 holds because by induction $c(p_{i+1}) \neq c(v_{k-1})$ and $c(v_k) = c(v_{k-1})$.

Case 2.b. $c(v_k) = c(v_{k-1})$ and $NB_{k-1} = \emptyset$. Choose the first vertex p_j to the right of p_i such that p_j has the same color as v_k, i.e. $j = \min\{h \mid h > i \wedge c(p_h) = c(v_k)\}$. Since the point representing v_{k-1} is an element of NB_{k-2}, this point is accessible from a page π by induction. We assign edge (v_{k-1}, v_k) to π. Since point p_j is to the right of p_i, p_j is accessible from both pages, and therefore the addition of edge (v_{k-1}, v_k) does not create a crossing. We have that $NB_k = p_{i+1}, p_{i+2}, \ldots, p_{j-1}$. Notice that $c(p_{i+1}) = c(p_{i+2}) = \cdots = c(p_{j-1}) \neq c(v_k)$ because p_j is the first point after p_i such that $c(p_j) = c(p_i)$. It follows that all points of NB_k have the same color. Also, they are all accessible from the page different from π. Vertex v_k is accessible from both pages because it is drawn on the rightmost point of σ_k. Therefore the invariants of Property 1 holds. Property 2 trivially holds since v_k is represented as the rightmost point of σ_k.

Case 2.c. $c(v_k) \neq c(v_{k-1})$. By Property 2 we have that $c(p_{i+1}) = c(v_k)$. Map v_k to p_{i+1}. Since the point representing v_{k-1} is an element of NB_{k-2}, it is accessible from a page π. We assign edge (v_{k-1}, v_k) to π. Since point p_{i+1} is to the right of p_i, it is accessible from both pages, and therefore the addition of edge (v_{k-1}, v_k) does not create a crossing. We have that $NB_k = NB_{k-1}$ because there is no point between p_i and p_{i+1}. Hence all points in NB_k have the same color, and are all accessible from a same page by induction. Also v_k is represented as the rightmost point of σ_k and hence it is accessible from both pages. It follows that both the invariants expressed by Properties 1 and 2 are maintained.

This concludes the proof of this theorem and hence of Theorem 1.   □

References

1. M. Abellanas, J. Garcia-Lopez, G. Hernández-Peñver, M. Noy, and P. A. Ramos. Bipartite embeddings of trees in the plane. *Discrete Applied Mathematics*, 93(2-3):141–148, 1999.
2. J. Akiyama and J. Urrutia. Simple alternating path problem. *Discrete Mathematics*, 84:101–103, 1990.
3. E. Di Giacomo, W. Didimo, G. Liotta, and S. K. Wismath. Book-embeddability of series-parallel digraphs. *Algorithmica*. to appear.
4. A. Kaneko, M. Kano, and K. Suzuki. Path coverings of two sets of points in the plane. In J. Pach, editor, *Towards a Theory of Geometric Graph*, volume 342 of Contempory Mathematics. American Mathematical Society, Providence, 2004.
5. A. Kanenko and M. Kano. Discrete geometry on red and blue points in the plane - a survey -. In *Discrete and Computational Geometry*, volume 25 of *Algorithms and Combinatories*. Springer, 2003.
6. M. Kaufmann and R. Wiese. Embedding vertices at points: Few bends suffice for planar graphs. *Journal of Graph Algorithms and Applications*, 6(1):115–129, 2002.

Upward Spirality and Upward Planarity Testing*

Walter Didimo, Francesco Giordano, and Giuseppe Liotta

Università di Perugia, Italy
{didimo, giordano, liotta}@diei.unipg.it

Abstract. The upward planarity testing problem is known to be NP-hard. We describe an $O(n^4)$-time upward planarity testing and embedding algorithm for the class of digraphs that do not contain rigid triconnected components. We also present a new FPT algorithm that solves the upward planarity testing and embedding problem for general digraphs.

1 Introduction

An *upward planar drawing* of a planar digraph G is a crossing-free drawing of G such that the vertices of G are mapped to points of the plane and the edges of G are drawn as simple curves that are monotone in the upward direction. A digraph that admits an upward planar drawing is an *upward planar digraph*. Unfortunately, not all planar digraphs are upward planar. The digraph of Figure 1(a) is not upward planar independent of the choice of its planar embedding. The upward planarity testing problem asks whether a planar digraph G has an upward planar drawing.

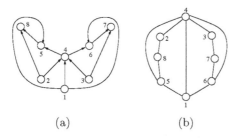

(a)   (b)

Fig. 1. (a) A digraph G that is not upward planar. (b) The underlying undirected graph of G is a series-parallel graph, i.e., it does not have rigid components.

The upward planarity testing problem is a classical subject of investigation in the graph drawing literature, and many papers have been devoted to this subject during the last decade. Bertolazzi et al. [1] present an $O(n^2)$-time algorithm that tests whether a digraph with a given planar embedding is upward planar. Garg and Tamassia [9] show that the problem in the variable embedding setting is NP-complete. Papakostas [14] presents an $O(n^2)$-time algorithm for testing the

* This work is partially supported by the MIUR Project ALGO-NEXT: Algorithms for the Next Generation Internet and Web: Methodologies, Design and Applications.

P. Healy and N.S. Nikolov (Eds.): GD 2005, LNCS 3843, pp. 117–128, 2005.

upward planarity of outerplanar digraphs. Hutton and Lubiw [13] describe an $O(n^2)$-time testing algorithm for digraphs that have a single source. Bertolazzi et al. [3] improve this last result by showing an optimal $O(n)$ testing algorithm for the same class of digraphs studied by Hutton and Lubiw. Bertolazzi et al. [2] describe a branch-and-bound testing algorithm for biconnected planar digraphs. Recently, fixed parameter tractable (FPT) algorithms have also been designed: Chan [4] presents an $O(t! \cdot 8^t \cdot n^3 + (2 \cdot t)^{3 \cdot 2^c} t! \cdot 8^t \cdot n)$-time algorithm where c and t are the number of cut-vertices and the number of triconnected components of G, respectively. Healy and Lynch [12] improve Chan's result by giving an $O(2^t \cdot t! \cdot n^2)$-time algorithm; in the same paper, Healy and Lynch describe a second upward planarity testing algorithm whose time complexity is $O(n^2 + k^4(2k+1)!)$, with $k = |E| - |V|$.

In this paper we describe a polynomial time algorithm and a new FPT algorithm for the upward planarity testing problem in the variable embedding setting. More precisely:

- We introduce and study the concept of *upward spirality* (Section 3), which is a measure of how much a component of a digraph is "rolled-up" in an upward planar drawing. A similar concept was introduced in the literature in the context of orthogonal drawings [6].
- We describe an $O(n^4)$-time upward planarity testing and embedding algorithm for the class of series-parallel digraphs, i.e. biconnected digraphs whose $SPQR$-tree does not have any R-node (Section 4). Our algorithm still runs in polynomial time even if the digraph is not biconnected and any block is a series-parallel digraph.
- Using the above results, we design a new FPT algorithm for upward planarity testing of general digraphs whose time complexity is $O(d^t \cdot n^3 + d \cdot t^2 \cdot n + d^2 \cdot n^2)$, where d is the maximum diameter of any split component of G and t is the number of (non-trivial) triconnected components of G (Section 5).

For reasons of space, all proofs are omitted and some sections are sketched. Details can be found in [8].

2   Preliminaries

We assume familiarity with basic concepts of graph drawing and graph planarity [5]. Let G be a planar digraph with a given planar embedding. A vertex of G is *bimodal* if the circular list of its incident edges can be partitioned into two (possibly empty) lists, one consisting of incoming edges and the other consisting of outgoing edges. If all vertices of G are bimodal then G and its embedding are called *bimodal*. Acyclicity and bimodality are necessary conditions for the upward planar drawability of an embedded planar digraph [1]. However, they are not sufficient conditions.

Let f be a face of an embedded planar bimodal digraph G and suppose that the boundary of f is visited clockwise if f is internal, and counterclockwise if f is external. Let $a = (e_1, v, e_2)$ be a triplet such that v is a vertex of the boundary of f and e_1, e_2 are incident edges of v that are consecutive on the boundary of f.

Triplet a is called an *angle of f*. Also, a is a *switch angle of f* if the direction of e_1 is opposite to the direction of e_2 (note that e_1 and e_2 may coincide if G is not biconnected). If e_1 and e_2 are both incoming in v, then a is a *sink-switch of f*; if they are both outgoing, a is a *source-switch of f*. A source or a sink of G is called a *switch vertex* of G; a vertex that is not a switch vertex is called an *internal vertex* of G.

Let Γ be an upward planar drawing of G and let a be an angle of G. Label a with a letter L (resp. a letter S) if it is a switch angle and has in Γ a value greater (resp. less) than π. Label a with a letter F if it is not a switch angle. The labeled embedded digraph U_G so obtained is called an *upward planar representation* of G, and can be viewed as the equivalence class of all (embedding preserving) upward planar drawings of G that induce the same angle labeling on G. Drawing Γ is also said to be an upward planar drawing that *preserves U_G*.

Now, consider an embedded planar digraph G and a labeling of its angles with labels L, S, and F. If v is a vertex of G, we denote by $L(v)$, $S(v)$, and $F(v)$ the number of angles at v that are labeled L, S, and F, respectively. The *degree* of v is defined as the number of angles at v, and is denoted as $deg(v)$. Also, if f is a face of G, $L(f)$, $S(f)$, and $F(f)$ denote the number of angles of f that are labeled L, S, and F, respectively. The following result is a restatement of the results in [1].

Lemma 1. *Let G be an acyclic planar bimodal embedded digraph with angle labels L, S, F. G and its labeling define an upward planar representation if and only if the following properties hold:* **(UP1)** *If v is a switch vertex of G then: $L(v) = 1$, $S(v) = deg(v) - 1$, $F(v) = 0$;* **(UP2)** *If v is not a switch vertex of G then: $L(v) = 0$, $S(v) = deg(v) - 2$, $F(v) = 2$;* **(UP3)** *If f is a face of G then: $L(f) = S(f) - 2$ if f is internal and $L(f) = S(f) + 2$ if f is external.*

From an upward planar representation U_G it is always possible to construct in linear time an upward planar drawing of G that preserves U_G, where each edge is drawn as a straight-line segment or as a polyline. Figure 2 shows an embedded planar digraph G, an upward planar representation U_G of G, and an upward planar drawing of G within U_G. Given an upward planar representation U_G, the angles labeled L, S, and F are called *large, small,* and *flat* angles, respectively. If

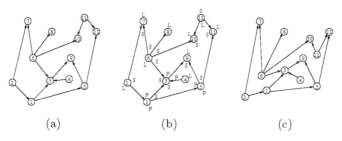

(a)   (b)   (c)

Fig. 2. (a) A planar embedded bimodal digraph G. (b) An upward planar representation U_G of G. (c) An upward planar drawing of G within U_G.

Fig. 3. Transformation of an $SPQR$-tree into its canonical form

G' is a subgraph of G, then G' has an *upward planar representation* $U_{G'}$ induced by U_G, which is defined as follows. Let $a = (e_1, v, e_2)$ be an angle of G', and let A be the counterclockwise sequence of angles of U_G between e_1 and e_2. Angle a in $U_{G'}$ is labeled: L if A either contains one large angle or two flat angles; F if A contains only one flat angle; S otherwise.

Let G be a biconnected graph and let $e = \{s, t\}$ be any edge of G, called *reference edge*. The $SPQR$-*tree* of G with respect to e describes a decomposition of G in terms of its triconnected components, and implicitly represents all planar embeddings of G with e on the external face. We assume familiarity with all formal definitions about $SPQR$-trees [7]. Suppose that G is given with an *st*-numbering of its vertices, such that the source and the sink of this numbering are the end-vertices s, t of the reference edge of G. If T is the $SPQR$-tree of G with respect to e, given any node μ of T, let u and v be the two poles of μ, so that u precedes v in the *st*-numbering. We call u and v the *first pole* and the *second pole* of the pertinent graph G_μ of μ. If G has a fixed planar embedding with reference edge e on the external face, the *right face* of G_μ is the face to the right of G_μ in G, while moving from u to v. The *left face* of G_μ is the face to the left of G_μ in G, while moving from u to v. The path on the right face of G_μ, going from u to v, is called the *right path* of G_μ. The path on the left face of G_μ, going from u to v, is called the *left path* of G_μ.

In the remainder of the paper, we consider $SPQR$-trees of directed graphs (digraphs) G. In this case, the computation of the decomposition tree is done exactly as for undirected graphs, by ignoring the orientation of the edges of G. Notice that, there is no connection between the orientation of the edges of G and the definition of first and second poles of the pertinent digraphs. In order to simplify the description of our upward planarity testing algorithm, we use *canonical SPQR-trees*, i.e., $SPQR$-trees where each S-node has always two children. A canonical $SPQR$-tree T of G can be constructed from an $SPQR$-tree of G by applying on every S-node the transformation illustrated in Figure 3. A canonical $SPQR$-tree of G has a number of nodes that is still linear in the number of vertices of G.

We say that a biconnected digraph G is a *series-parallel digraph* if its $SPQR$-tree only consists of Q-, S-, and P-nodes.

3   Upward Spirality

In the following, we assume that G is a biconnected digraph, T an $SPQR$-tree of G, U_G an upward representation of G, and G_μ the pertinent digraph of a node μ of T, with first pole u and second pole v.

Let $P = < v_1, e_1, v_2, \ldots, v_i, e_i, \ldots, e_{k-1}, v_k >$ be any simple (undirected) path (possibly a simple cycle) in G, and let U_P be the upward planar representation of P induced by U_G. Consider a vertex v_i ($i \in \{2, \ldots, k-1\}$) that is a switch of P, and denote by $a = (e_{i-1}, v_i, e_i)$, $a' = (e_i, v_i, e_{i-1})$ the two angles at v_i in U_P. Walking on P from v_1 to v_k, we say that v_i is a *left turn* (resp. *right turn*) of U_P if a (resp. a') is large. We denote by $n(U_P)$ the number of right turns minus the number of left turns of U_P, and we call $n(U_P)$ the *turn number* of P in U_G, or simpler, the *turn number* of U_P. Similarly, if P is a simple cycle, i.e. $v_1 = v_k$, and we walk clockwise on P, we say that we encounter a left turn (resp. right turn) of U_P on any switches of P that has a large angle (resp. small angle) inside the cycle. Because of Lemma 1, if P is a simple cycle of U_G, then its turn number is $n(U_P) = 2$.

Denoted by $w \in \{u, v\}$ any of the two poles of G_μ, we want to classify w on the basis of the labeling of the angles at w in U_G. The label of the angle at w in the right face (resp. in the left face) of G_μ is called the *right inter-label* (resp. the *left inter-label*) of w. An *intra-label* of w is any label of an angle at w internal at G_μ. We assign to each angle label an integer weight, in such a way that labels S, F, and L have weight 0, 1, and 2, respectively. The *intra-labeling weight* of w is the sum of the weights of all intra-labels of w. From properties UP1 and UP2 of Lemma 1, the intra-labeling weight of w ranges from 0 to 2.

In U_G, we describe the angles labeling of the pole w of G_μ, by using a string $t_w = XY\lambda$, such that X is the left inter-label of w, Y is the right inter-label of w, and λ is the intra-labeling weight of w. We say that t_w is the *pole category* of w. We remark that, since U_G is an upward planar representation, not all categories $XY\lambda$ ($X, Y \in \{S, F, L\}, \lambda \in \{0, 1, 2\}$) are possible for a pole w of a pertinent digraph of G. Indeed, as also observed above, the sum of all angle labels at w must verify UP1 and UP2, and w must be bimodal. Hence, the following lemma immediately follows (see also Figure 4):

Lemma 2. *The possible pole categories of any pole of G_μ in U_G are: $SS0$, $SS1$, $SS2$, $SF0$, $SF1$, $FS0$, $FS1$, $FF0$, $SL0$, $LS0$.*

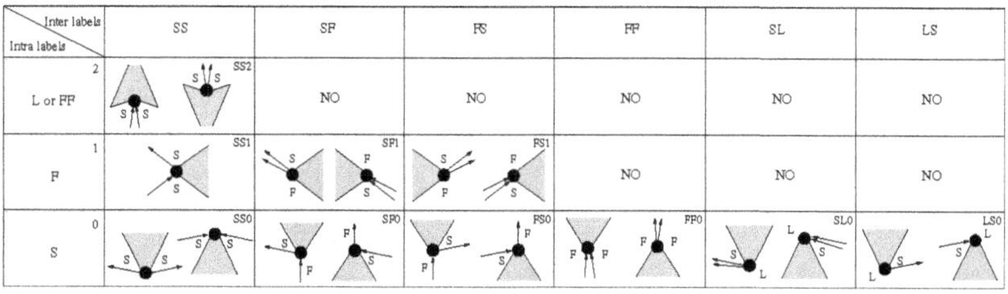

Fig. 4. Illustration of the pole categories for the first pole of a pertinent digraph within an upward planar representation. Grey portions are the internal parts of the pertinent digraph. The two labels around the pole are the inter-labels of the pole. The illustration for the second pole is symmetric.

In order to introduce the notion of upward spirality we need to identify two suitable vertices that we call the *left external vertex of w*, denoted as w_l, and the *right external vertex of w*, denoted as w_r, where w is still any of the two poles of G_μ. The right and the left external vertices of w are defined based on the pole category t_w of w, with respect to G_μ in U_G. More precisely, let e_l be the edge incident on w, that is on the left path of G_μ and that does not belong to G_μ; let e_r be the edge incident on w, that is on the right path of G_μ and that does not belong to G_μ. Also, let x be the end-vertex of e_l other than w and let y be the end-vertex of e_r other than w. The external vertices w_l and w_r of w are defined as follows: (**Case 1**) One of the following three subcases is verified: (i) $t_w \in \{SS0, SF0, FS0, FF0\}$; (ii) $t_w = SL0$ and w is the first pole of G_μ; (iii) $t_w = LS0$ and w is the second pole of G_μ. In this case $w_l = w_r = w$. (**Case 2**) One of the following two subcases is verified: (i) $t_w \in \{FS1, SF1, SS1, SS2\}$; (ii) $t_w = SL0$ and w is the second pole of G_μ; (iii) $t_w = LS0$ and w is the first pole of G_μ. In this case $w_l = x$ and $w_r = y$.

Let u_l, u_r be the left and the right external vertices of the first pole u of G_μ and let v_l, v_r be the left and the right external vertices of the second pole v of G_μ. Let P_{uv} be an (undirected) path from u to v in G_μ. The undirected path $P_l = (u_l, u) \cup P_{uv} \cup (v, v_l)$ is called a *left spine of G_μ*. The path $P_r = (u_r, u) \cup P_{uv} \cup (v, v_r)$ is called a *right spine of G_μ*. For example, the left spine and the right spine of a pertinent digraph are highlighted in Figure 5.

The following lemma shows that the turn number of a spine of a pertinent digraph of an upward representation is an invariant property of the upward representation itself.

Lemma 3. *Let P'_r, P''_r be two distinct right spines of G_μ and let P'_l, P''_l be two distinct left spines of G_μ. Then $n(U_{P'_r}) = n(U_{P''_r})$ and $n(U_{P'_l}) = n(U_{P''_l})$.*

For example, in Figure 5, $G_{\mu'}$ has only two left spines, that also concide with the right spines. The turn number of these spines is -1. Based on Lemma 3, we can denote by $n_l(U_{G_\mu})$ the turn number of any left spine of G_μ in U_G, without

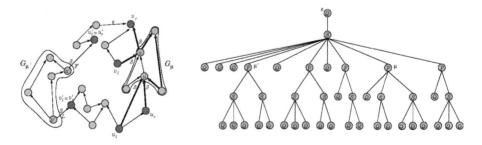

Fig. 5. An upward planar representation of a series-parallel digraph G, and an $SPQR$-tree T of G rooted at edge e. G_μ and $G_{\mu'}$ are the pertinent digraphs of nodes μ and μ' of T, with poles u, v, u', v', respectively. The pole categories of u and v are $SS2$ and $SS1$, respectively. The ones of u' and v' are $FS1$ and $SL0$, respectively. The left and the right spines of G_μ constructed on the right path of G_μ are highlighted.

ambiguity; similarly, $n_r(U_{G_\mu})$ denotes the turn number of any right spine of G_μ. The *upward spirality* of G_μ within U_G (or simpler, the *upward spirality* of U_{G_μ}), is denoted as $\sigma(U_{G_\mu})$ and is defined as follows: $\sigma(U_{G_\mu}) = \frac{n_l(U_{G_\mu})+n_r(U_{G_\mu})}{2}$.

For example, in Figure 5, $\sigma(U_{G_\mu}) = -1/2$, and $\sigma(U_{G_{\mu'}}) = -1$. Suppose now that P_l and P_r are a left spine and a right spine of G_μ, constructed using the same path $P_{uv} = <u, w_1, w_2, \ldots, w_k, v>$ between the poles u, v of G_μ. We can rewrite the turn number of the spines as follows: $n(U_{P_l}) = n(U_{P_{uv}}) + a_{u_l} + a_{v_l}$, $n(U_{P_r}) = n(U_{P_{uv}}) + a_{u_r} + a_{v_r}$, where $a_{u_l} = n(U_{P_{u_l}})$, $a_{u_r} = n(U_{P_{u_r}})$, $a_{v_l} = n(U_{P_{v_l}})$, $a_{v_r} = n(U_{P_{v_r}})$, and $P_{u_l} =<u_l, u, w_1>$, $P_{u_r} =<u_r, u, w_1>$, $P_{v_l} =<w_k, v, v_l>$, $P_{v_r} =<w_k, v, v_r>$. Of course, $a_{u_l}, a_{u_r}, a_{v_l}, a_{v_r} \in \{-1, 0, 1\}$. From the invariant property of Lemma 3, the upward spirality of U_{G_μ}, can be rewritten as follows:

$$\sigma(U_{G_\mu}) = n(U_{P_{uv}}) + \frac{(a_{u_l} + a_{u_r})}{2} + \frac{(a_{v_l} + a_{v_r})}{2} \tag{1}$$

In order to uniquely refer to the values $a_{u_l}, a_{u_r}, a_{v_l}, a_{v_r}$ for the upward spirality of U_{G_μ}, we aim at rewriting $\sigma(U_{G_\mu})$ in a kind of canonical form, choosing always a "special" path P_{uv}. We define the following equivalence relationship between any two paths P'_{uv}, P''_{uv} of G_μ, within a given upward representation U_G of G. We say that P'_{uv}, P''_{uv} are *turn equivalent* if $n(U_{P'_{uv}}) = n(U_{P''_{uv}})$, i.e, if they have the same turn number. Since $\sigma(U_{G_\mu})$ assumes the same value if we use P'_{uv} or P''_{uv} in Formula (1), and since $a_{u_l}, a_{u_r}, a_{v_l}, a_{v_r} \in \{-1, 0, 1\}$, then the turn-equivalence relationship partitions the set of the undirected paths of U_{G_μ}, from the first to the second pole, into a finite set of equivalence classes. The following lemma gives a useful property of the paths of G_μ.

Lemma 4. *Let P^r_{uv} be a path of G_μ that is turn-equivalent to the right path of G_μ, and let P^l_{uv} be a path of G_μ that is turn-equivalent to the left path of G_μ. If P_{uv} is any path of G_μ between u and v, then $n(U_{P^l_{uv}}) \geq n(U_{P_{uv}}) \geq n(U_{P^r_{uv}})$.*

In Formula (1) we now choose as path P_{uv} any path P^r_{uv} that is turn-equivalent to the right path of G_μ, and we consider the corresponding values $(a_{u_l} + a_{u_r})/2$ and $(a_{u_l} + a_{u_r})/2$. Denote $n(U_{P^r_{uv}})$ by $\alpha(U_{G_\mu})$, and denote $(a_{u_l} + a_{u_r})/2$, $(a_{u_l} + a_{u_r})/2$ by $\alpha_u(U_{G_\mu})$ and $\alpha_v(U_{G_\mu})$, respectively.

The upward spirality of U_{G_μ} can be rewritten in the following canonical form: $\sigma(U_{G_\mu}) = \alpha(U_{G_\mu}) + \alpha_u(U_{G_\mu}) + \alpha_v(U_{G_\mu})$. We call $\alpha(U_{G_\mu})$ the *internal spirality* of U_{G_μ}, and $\alpha_u(U_{G_\mu})$, $\alpha_v(U_{G_\mu})$ the *first-pole spirality* and the *second-pole spirality*, respectively. From Lemma 4, each of the terms $a_{u_l}, a_{u_r}, a_{v_l}, a_{v_r}$ in Formula (1) takes the maximum possible value when $P_{uv} = P^r_{uv}$. This also implies that, for any choice of P_{uv}, $(a_{u_l} + a_{u_r})/2 \leq \alpha_u(U_{G_\mu})$ and $(a_{v_l} + a_{v_r})/2 \leq \alpha_v(U_{G_\mu})$. Therefore, for each pole category, it is possible to determine the exact value of the two pole spiralities, since we know that they take the maximum possible value and since we know what are the two external vertices. The next results prove that the upward spirality can only take a linear number of values.

Lemma 5. *Let $\bar{n}$ be the minimum number of switches on any path between the poles u and v of G_μ. Then, $-\bar{n}-2 \leq \sigma(U_{G_\mu}) \leq \bar{n}+2$. Also, $\alpha_u(U_{G_\mu}) + \alpha_v(U_{G_\mu}) \in \{-1, -1/2, 0, 1/2, 1, 3/2, 2\}$.*

Theorem 1. *Let G be a digraph with n vertices, T an $SPQR$-tree of G, and G_μ the pertinent digraph of a node μ of T. There are at most $O(n)$ values for the upward spirality of G_μ within any upward planar representation of G.*

The following lemmas describe the relationships between the upward spiralities of series and parallel compositions, and the ones of their components.

Lemma 6. *Let μ be an S-node of T with children μ_1 and μ_2. Let G_μ be the pertinent digraph of μ, with poles u and v, and let G_{μ_1}, G_{μ_2} be the pertinent digraphs of μ_1, μ_2, with poles $u_1 = u, v_1$, and $u_2 = v_1, v_2 = v$, respectively. The following relationship holds: $\sigma(U_{G_\mu}) = \sigma(U_{G_{\mu_1}}) + \sigma(U_{G_{\mu_2}})$.*

Lemma 7. *Let μ be a P node of T with children $\mu_1, \dots, \mu_k$, ordered from left to right. Let G_μ be the pertinent digraph of μ and let $G_{\mu_1}, \dots, G_{\mu_k}$ be the pertinent digraphs of $\mu_1, \dots, \mu_k$, respectively. For each $i = 1, \dots, k$, the following relationships hold:* **(1)** $\alpha(U_{G_\mu}) = \alpha(U_{G_{\mu_i}}) + \delta^{(i)}(U_{G_\mu}),\ \delta^{(i)}(U_{G_\mu}) \in \{0, 1, 2, 3, 4\};$ **(2)** $\alpha(U_{G_{\mu_1}}) \geq \alpha(U_{G_{\mu_2}}) \geq \cdots \geq \alpha(U_{G_{\mu_k}}) = \alpha(U_{G_\mu})$.

Consider now the subgraph G' of G consisting of G_μ plus the edges incident on u and v that are external to G_μ, and let $U'_{G'}$ be any upward planar representation of G' such that the planar embedding of the external edges of G_μ and the angle labels between these edges in $U'_{G'}$ are the same as in U_G. Notice that, the planar embedding of G_μ in $U'_{G'}$ can be different from the one in U_G. Denote by $t'_u = X'_u Y'_u \lambda_u$ and $t'_v = X'_v Y'_v \lambda_v$ the pole categories of u and v for U'_{G_μ}. The operation of *substitution* of U_{G_μ} with U'_{G_μ} in U_G defines a new planar embedded digraph $S(U'_{G_\mu}, U_G)$ with angle labels S, F, and L such that: (i) The planar embedding and the labels of the angles of subgraph $G - G_\mu$ are the same as in U_G; (ii) The planar embedding and the labels of the angles of subgraph G_μ are the same as in U'_{G_μ}; (iii) The inter-labels of G_μ at u and at v are X'_u, Y'_u, X'_v, Y'_v, respectively. We say that U_{G_μ} is *substitutable* with U'_{G_μ} in U_G if $S(U'_{G_\mu}, U_G)$ is still an upward planar representation of G. The following theorem is the main result of this section.

Theorem 2. *If U'_{G_μ} and U_{G_μ} have the same upward spirality and the same pole categories (i.e. $t'_u = t_u$, $t'_v = t_v$), then U_{G_μ} is substitutable with U'_{G_μ} in U_G.*

4   Upward Planarity Testing of Series-Parallel Digraphs

The outline of our upward planarity testing and embedding algorithm for series-parallel digraphs is as follows. For each possible choice of an edge e of G, the algorithm computes the $SPQR$-tree T of G with reference edge e. Then, the algorithm visits T from bottom to top, in post-order. Each time a node μ of T is visited, μ is equipped with a set of upward planar representations of G_μ (which we call *feasible set of μ*), such that each upward planar representation is constrained to have assigned pole categories and an assigned value of upward spirality. Using the result of Theorem 2, for each possible combination of pole categories and upward spirality value, the algorithm stores only one constrained upward planar representation, if there exists one. The feasible set of each S-node and P-node of T is computed by considering the feasible sets of its children. In this way, the algorithm incrementally tries to construct an upward planar representation of G with edge e on the external face, from the leaves to the root, while exploring a subset of upward planar representations that is "representative" of the whole set of upward planar representations of G. The algorithm ends if the feasible set of a node is empty or if the feasible sets of all nodes have been successfully computed. In the following we formalize the definition of feasible set and then describe how the feasible sets of the different types of nodes can be computed.

A *feasible tuple* of μ is defined as follows: $\tau_\mu =< U_{G_\mu}, \sigma(U_{G_\mu}), t_u, t_v >$, where U_{G_μ} is an upward planar representation of G_μ with pole categories t_u, t_v and upward spirality $\sigma(U_{G_\mu})$. Let $\tau'_\mu =< U'_{G_\mu}, \sigma(U'_{G_\mu}), t'_u, t'_v >$ and $\tau''_\mu =< U''_{G_\mu}, \sigma(U''_{G_\mu}), t''_u, t''_v >$ be two feasible tuples of μ. We say that U'_{G_μ} and U''_{G_μ} are *spirality equivalent* if $\sigma(U'_{G_\mu}) = \sigma(U''_{G_\mu})$, $t'_u = t''_u$, and $t'_v = t''_v$. In this case, we also say that τ'_μ and τ''_μ are *spirality equivalent*. A *feasible set* $\mathcal{F}_\mu$ of μ is a set of feasible tuples of μ such that there is exactly one representative tuple for each class of spirality equivalent feasible tuples of μ. The next lemma guarantees that our algorithm is able to find an upward planar representation of G with e on the external face, if there exists one.

Lemma 8. *Let G be an upward planar digraph with edge e on the external face, and let T be the $SPQR$-tree of G with respect to e. There exists an upward planar representation U_G of G such that: (i) e is on the external face of U_G; (ii) for each node μ of T, there exists a feasible tuple $\tau_\mu =< U_{G_\mu}, \sigma(U_{G_\mu}), t_u, t_v >$ in the feasible set of μ, where U_{G_μ} is the upward representation of G_μ induced by U_G.*

All the Q-nodes have the same feasible set, which can be computed with a pre-processing step in $O(1)$ time. Namely, if μ is a Q-node, both the internal spirality and the internal-labeling weight of any upward planar representation U_{G_μ} of G_μ are equal to 0. We can only have three upward spirality values for U_{G_μ}: 0, 1, and -1. More precisely, if (u, v) is the (undirected) edge represented by μ, the algorithm inserts in $\mathcal{F}_\mu$ a tuple for each of the following combinations of upward spirality and pole categories: (1) $\sigma(U_{G_\mu}) = 0$, $t_u \in \{SS0, SF0, FS0, FF0, SL0\}$, $t_v \in \{SS0, SF0, FS0, FF0, LS0\}$. (2) $\sigma(U_{G_\mu}) = 0$, $t_u = LS0$ and $t_v = SL0$. (3) $\sigma(U_{G_\mu}) = 1$, $t_u = LS0$, $t_v \in \{SS0, SF0, FS0, FF0, LS0\}$. (4) $\sigma(U_{G_\mu}) = -1$, $t_u \in \{SS0, SF0, FS0, FF0, SL0\}$, $t_v = SL0$. In all these tuples, U_{G_μ} is the edge (u, v) oriented upward.

Let μ be an S-node of T, and let u and v be the first pole and the second pole of G_μ, respectively. Let μ_1, μ_2 be the two children of μ; denote by $u_1 = u$, v_1 the first pole and the second pole of G_{μ_1}; also denote by $u_2 = v_1$, $v_2 = v$ the first pole and the second pole of G_{μ_2}. The feasible set of μ is computed using the relationship of Lemma 6. For each pair of tuples $\tau_1 =< U_{G_{\mu_1}}, \sigma(U_{G_{\mu_1}}), t_{u_1}, t_{v_1} >\in \mathcal{F}_{\mu_1}$, $\tau_2 =< U_{G_{\mu_2}}, \sigma(U_{G_{\mu_2}}), t_{u_2}, t_{v_2} >\in \mathcal{F}_{\mu_2}$, the algorithm checks if the inter-labels of t_{v_1} and t_{u_2} are the same, and if the orientations of the edges incident on

$u_2 = v_1$ in $U_{G_{\mu_1}}$ and $U_{G_{\mu_2}}$ are compatible. In the affirmative case, it constructs a new tuple $\tau = < U_{G_\mu}, \sigma(U_{G_\mu}), t_u, t_v >$, which will be inserted in $\mathcal{F}_\mu$, only if $\mathcal{F}_\mu$ does not already contain a spirality equivalent tuple; τ is defined as follows: $\sigma(U_{G_\mu}) = \sigma(U_{G_{\mu_1}}) + \sigma(U_{G_{\mu_2}})$; $t_u = t_{u_1}$, $t_v = t_{v_2}$; U_{G_μ} is the series composition of $U_{G_{\mu_1}}$ and $U_{G_{\mu_2}}$ on the common vertex $u_2 = v_1$. Since each feasible set has $O(n)$ tuples, the feasible set of an S-node can be computed in $O(n^2)$ time.

The computation of the feasible set of a P-node is a more complicated task, since the skeleton of a P-node with k children has $O(k!)$ possible planar embeddings, and we want to keep the computation polynomial in the number of vertices of the graph. Let μ be a P-node of T, with first pole u and second pole v. Let $\mu_1, \ldots, \mu_k$ be the children of μ. We remark that each G_{μ_i} ($i = 1, \ldots, k$) has $u_i = u$ and $v_i = v$ as the first pole and the second pole, respectively. In order to construct the feasible set of μ, we evaluate the possibility of constructing an upward planar representation U_{G_μ} for each possible way of fixing $\sigma(U_{G_\mu}), t_u$, and t_v. Namely, for each choice of $\sigma(U_{G_\mu}), t_u, t_v$, the algorithm must verify if it is possible to select from the feasible sets of $\mu_1, \ldots, \mu_k$, a subset of upward planar representations $U_{G_{\mu_1}}, \ldots, U_{G_{\mu_k}}$ that can assume a "parallel configuration" compatible with $\sigma(U_{G_\mu}), t_u, t_v$. The conditions of Lemma 7 allow us to limit the number of these configurations, so that it is not needed to consider all permutations of the children of μ in the $skel(\mu)$. Actually, it can be proved that the total number of configurations is constant with respect to the number of vertices of G. The set of possible configurations is defined on the basis of t_u and t_v; each configuration consists of a sequence of groups, such that each group can host a certain number of upward planar representations, all having the same pole categories and the same internal spirality (which also implies the same upward spirality). The groups in the sequence are ordered according to their values of internal spirality. In this way, on the basis of $\sigma(U_{G_\mu})$ and for each configuration above defined, the algorithm tries to select a set of upward representations $U_{G_{\mu_1}}, \ldots, U_{G_{\mu_k}}$ from the feasible sets of $\mu_1, \ldots, \mu_k$ and to assign each of them to a group in the configuration. This assignment problem is solved by searching a feasible flow in a suitable network constructed from the configuration. The formal description of the configurations and the construction of the feasible set using a sequence of flow-based algorithms can be found in [8]. The construction of the feasible sets of all P-nodes can be done in $O(n^3)$ time.

Once all feasible sets have been computed for the nodes of T, the algorithm performs a final step to verify if it is possible to construct an upward planar representation from the feasible set of the root of T (which is a Q-node) and the one of its child. Namely, let μ be the root and let ν be its child. The following lemma holds.

Lemma 9. *G has an upward planar representation U_G if and only if there exist two tuples $\tau_\mu = < U_{G_\mu}, \sigma(U_{G_\mu}), t_{u_\mu}, t_{v_\mu} > \in \mathcal{F}_\mu$, $\tau_\nu = < U_{G_\nu}, \sigma(U_{G_\nu}), t_{u_\nu}, t_{v_\nu} > \in \mathcal{F}_\nu$ such that: (1) $\sigma(U_{G_\mu}) - \sigma(U_{G_\nu}) = 2$; (2) $Y_{u_\mu} = X_{u_\nu}$, $Y_{v_\mu} = X_{v_\nu}$, where $t_w = X_w Y_w \lambda_w$ and $w \in \{u_\mu, u_\nu, v_\mu, v_\nu\}$.*

According to Lemma 9, the algorithm looks for two tuples that verify the conditions (1) and (2) in the statement. If these tuples are found, the final upward planar representation is returned, otherwise the upward planarity testing fails. The next theorem summarizes the main result of this section. The final time complexity of the testing algorithm follows from the above discussion, iterating over all $SPQR$-trees of G (one for each choice of the reference edge).

Theorem 3. *Let G be a biconnected series-parallel digraph with n vertices. There exists an $O(n^4)$-time algorithm that tests if G is upward planar and, if so, that constructs an upward planar drawing of G.*

5   An FPT Algorithm for General Digraphs

To extend the upward planarity testing algorithm above described to general biconnected digraphs, we need to describe how to compute the feasible sets of R-nodes. Unfortunately, to compute the feasible set of an R-node μ, we cannot rely on any relationship between the upward spirality of U_{G_μ} and the upward spirality of its children. Therefore, we simply consider all possible combinations of tuples for each virtual edge of $skel(\mu)$ in constructing U_{G_μ}. Namely, let e_i be a virtual edge of $skel(\mu)$ and let μ_i be the child of μ corresponding to e_i. We substitute to e_i the upward planar representation $U_{G_{\mu_i}}$ of a tuple in the feasible set of μ_i. We repeat this process for each virtual edge, until a "partial candidate" upward planar representation U'_{G_μ} of G_μ is constructed. We then apply on this partial representation the flow-based upward planarity testing algorithm proposed by Bertolazzi et al. [1], where the assignment of the switches to the faces is constrained for the part of the representation that is already fixed. In order to construct the feasible set of μ, we need to run the testing algorithm over all possible combinations of upward spirality and pole categories of U_{G_μ}. For each given value of upward spirality σ and for each choice of pole categories t_u, t_v, we enrich the partial upward representation U'_{G_μ} with a suitable external gadget, that forces U_{G_μ} to have upward spirality σ and pole categories t_u, t_v. This gadget will have a fixed upward planar representation, which is still translated into a set of constraints on the flow network. See [8] for a detailed construction of the external gadgets.

The feasible set of an R-node μ, computed with the above procedure, requires to consider all possible combinations of tuples in the feasible set of the children of μ, and, for each of these combinations, we need to consider all possible values of upward spirality and pole categories. The procedure must be also applied to the two possible planar embeddings of $skel(\mu)$. Denote by t the number of non-trivial triconnected components of G and denote by d the maximum diameter of a split component of G. The feasible set of an R-node of μ can be then computed in $O(d^{t_\mu} \cdot n^2)$ time, where n is the number of vertices of G, and $t_\mu \leq t$ is the number of virtual edges (distinct from the reference edge) of μ. Indeed, the minimum number of switches in any path between the poles of G_μ is at most d, and therefore, from Lemma 5, the upward spirality of U_{G_μ} can take $O(d)$ possible values and the feasible set of any node of T has $O(d)$ tuples. Also, $O(n^2)$ is the complexity of the upward planarity testing of Bertolazzi et al. Hence, the feasible set of all R-nodes can be computed in $O(d^t \cdot n^2)$ time.

Our FPT algorithm can be eventually extended to general planar digraphs, using a recent result of Healy and Lynch [10, 11] about the upward planarity testing of simply connected graphs (refer to [8]). The following theorem holds, by observing that the feasible sets of P- and S-nodes of each $SPQR$-tree T can be computed in $O(d \cdot t^2)$-time and $O(d^2 n)$-time, respectively, and by iterating over all decomposition trees of G.

Theorem 4. *Let G be a connected planar digraph with n vertices. Suppose that each block of G has at most t (non-trivial) triconnected components, and that each split component of a block has a diameter at most d. There exists an $O(d^t \cdot n^3 + d \cdot t^2 \cdot n + d^2 \cdot n^2)$-time algorithm that tests if G is upward planar and, if so, that constructs an upward planar drawing of G.*

Theorem 5. *Let G be a connected planar digraph with n vertices and such that each block is a series-parallel digraph. There exists an $O(n^4)$-time algorithm that tests if G is upward planar and, if so, that constructs an upward planar drawing of G.*

References

1. P. Bertolazzi, G. D. Battista, G. Liotta, and C. Mannino. Upward drawings of triconnected digraphs. *Algorithmica*, 6(12):476–497, 1994.
2. P. Bertolazzi, G. Di Battista, and W. Didimo. Quasi-upward planarity. *Algorithmica*, 32(3):474–506, 2002.
3. P. Bertolazzi, G. Di Battista, C. Mannino, and R. Tamassia. Optimal upward planarity testing of single-source digraphs. *SIAM J. Comput.*, 27:132–169, 1998.
4. H. Chan. A parameterized algorithm for upward planarity testing. In *Proc. ESA '04*, volume 3221 of *LNCS*, pages 157–168, 2004.
5. G. Di Battista, P. Eades, R. Tamassia, and I. G. Tollis. *Graph Drawing*. Prentice Hall, Upper Saddle River, NJ, 1999.
6. G. Di Battista, G. Liotta, and F. Vargiu. Spirality and optimal orthogonal drawings. *SIAM J. Comput.*, 27(6):275–298, 1998.
7. G. Di Battista and R. Tamassia. On-line planarity testing. *SIAM J. Comput.*, 25:956–997, 1996.
8. W. Didimo, F. Giordano, and G. Liotta. Upward spirality and upward planarity testing. Technical report, 2005. RT-006-05, DIEI - Università di Perugia, Italy.
9. A. Garg and R. Tamassia. On the computational complexity of upward and rectilinear planarity testing. *SIAM J. Comput.*, 31(2):601–625, 2001.
10. P. Healy and K. Lynch. Building blocks of upward planar digraphs. In *Proc. GD '04*, volume 3383 of *LNCS*, pages 296–306, 2004.
11. P. Healy and K. Lynch. Building blocks of upward planar digraphs. Technical report, 2005. TR UL-CSIS-05-2.
12. P. Healy and K. Lynch. Fixed-parameter tractable algorithms for testing upward planarity. In *Proc. SOFSEM '05*, volume 3381 of *LNCS*, pages 199–208, 2005.
13. M. D. Hutton and A. Lubiw. Upward planarity testing of single-source acyclic digraphs. *SIAM J. Comput.*, 25(2):291–311, 1996.
14. A. Papakostas. Upward planarity testing of outerplanar dags. In *Proc. GD '95*, volume 894 of *LNCS*, pages 7298–306, 1995.

Graph Treewidth and Geometric Thickness Parameters

Vida Dujmović [1,*] and David R. Wood [2,**]

[1] School of Computer Science, Carleton University, Ottawa, Canada
vida@scs.carleton.ca
[2] Departament de Matemàtica Aplicada II, Universitat Politècnica de Catalunya,
Barcelona, Spain
david.wood@upc.edu

Abstract. Consider a drawing of a graph G in the plane such that crossing edges are coloured differently. The minimum number of colours, taken over all drawings of G, is the classical graph parameter *thickness* $\theta(G)$. By restricting the edges to be straight, we obtain the *geometric thickness* $\bar{\theta}(G)$. By further restricting the vertices to be in convex position, we obtain the *book thickness* $\mathrm{bt}(G)$. This paper studies the relationship between these parameters and the treewidth of G. Let $\theta(\mathcal{T}_k)$ / $\bar{\theta}(\mathcal{T}_k)$ / $\mathrm{bt}(\mathcal{T}_k)$ denote the maximum thickness / geometric thickness / book thickness of a graph with treewidth at most k. We prove that:

- $\theta(\mathcal{T}_k) = \bar{\theta}(\mathcal{T}_k) = \lceil k/2 \rceil$, and
- $\mathrm{bt}(\mathcal{T}_k) = k$ for $k \leq 2$, and $\mathrm{bt}(\mathcal{T}_k) = k + 1$ for $k \geq 3$.

The first result says that the lower bound for thickness can be matched by an upper bound, even in the more restrictive geometric setting. The second result disproves the conjecture of Ganley and Heath [*Discrete Appl. Math.* 2001] that $\mathrm{bt}(\mathcal{T}_k) = k$ for all k. Analogous results are proved for outerthickness, arboricity, and star-arboricity.

1 Introduction

Partitions of the edge set of a graph G into a small number of 'nice' subgraphs is in the mainstream of graph theory. For example, in a proper edge colouring, the subgraphs of the partition are matchings. When the subgraphs are required to be planar (respectively, acyclic), then the minimum number of subgraphs in a partition of G is the *thickness* (*arboricity*) of G. Thickness and arboricity are classical graph parameters that have been studied since the early 1960's. The first results in this paper concern the relationship between treewidth and parameters such as thickness and arboricity. Treewidth is a more modern graph parameter which is particularly important in structural and algorithmic graph theory. For each of thickness and arboricity (and other related parameters), we prove tight bounds on the minimum number of subgraphs in a partition of a graph with treewidth k. These introductory results are presented in Section 2.

---

* Supported by NSERC postdoctoral grant.
** Supported by the Government of Spain grant MEC SB-2003-0270, and by projects MCYT-FEDER BFM2003-00368 and Gen. Cat 2001SGR00224.

P. Healy and N.S. Nikolov (Eds.): GD 2005, LNCS 3843, pp. 129–140, 2005.

The main results of the paper concern partitions of graphs with an additional geometric property. Namely, that there is a drawing of the graph, and each subgraph in the partition is drawn without crossings. This type of drawing has applications in graph visualisation (where each plane subgraph is coloured by a distinct colour), and in multilayer VLSI (where each plane subgraph corresponds to a set of wires that can be routed without crossings in a single layer). When there is no restriction on the edges, the minimum number of plane subgraphs, taken over all drawings of G, is again the thickness of G. By restricting the edges to be straight, we obtain the *geometric thickness* of G. By further restricting the vertices to be in convex position, we obtain the *book thickness* of G. Our main results precisely determine the maximum geometric thickness and maximum book thickness of all graphs with treewidth k. We also determine the analogous value for a number of other related parameters.

The paper is organised as follows. Section 3 formally introduces all of the geometric parameters to be studied. Section 4 states our main results. The proofs of our two main theorems are presented in Sections 5 and 6. The remaining proofs are in the full version of the paper [6].

2   Abstract Graph Parameters

We consider graphs G that are simple, finite, and undirected. Let $V(G)$ and $E(G)$ denote the vertex and edge sets of G. For $A, B \subseteq V(G)$, let $G[A; B]$ denote the bipartite subgraph of G with vertex set $A \cup B$ and edge set $\{vw \in E(G) : v \in A, w \in B\}$. A *graph parameter* is a function f such that $f(G) \in \mathbb{N}$ for all graphs G. For a graph class $\mathcal{G}$, let $f(\mathcal{G}) := \max\{f(G) : G \in \mathcal{G}\}$. If $f(\mathcal{G})$ is unbounded, then let $f(\mathcal{G}) := \infty$.

The *thickness* of a graph G, denoted by $\theta(G)$, is the minimum number of planar subgraphs that partition $E(G)$ (see [11]). A graph is *outerplanar* if it has a plane drawing with all the vertices on the boundary of the outerface. The *outerthickness* of a graph G, denoted by $\theta_o(G)$, is the minimum number of outerplanar subgraphs that partition $E(G)$ (see [8]). The *arboricity* of a graph G, denoted by $\mathsf{a}(G)$, is the minimum number of forests that partition $E(G)$. [12] proved that $\mathsf{a}(G) = \max\{\lceil \frac{|E(H)|}{|V(H)|-1} \rceil : H \subseteq G\}$. A *star-forest* is graph in which every component is a star. The *star-arboricity* of a graph G, denoted by $\mathsf{sa}(G)$, is the minimum number of star-forests that partition $E(G)$ (see [1]). Thickness, outerthickness, arboricity and star-arboricity are always within a constant factor of each other (see [6]).

In the remainder of this section we determine the maximum value of each of the above four parameters for graphs of treewidth k. A set of k pairwise adjacent vertices in a graph G is a *k-clique*. For a vertex v of G, let $N_G(v) := \{w \in V(G) : vw \in E(G)\}$ and $N_G[v] := N_G(v) \cup \{v\}$. We say v is *k-simplicial* if $N_G(v)$ is a k-clique. A *k-tree* is a graph G such that either G is (isomorphic to) the complete graph K_k, or G has a k-simplicial vertex v and $G \setminus v$ is a k-tree. The *treewidth* of a graph G is the minimum $k \in \mathbb{N}$ such that G is a spanning subgraph of a k-tree. Let $\mathcal{T}_k$ denote the class of graphs with treewidth at most k. Many families of graphs have bounded treewidth. $\mathcal{T}_1$ is the class of forests. Graphs in $\mathcal{T}_2$ are obviously planar—a 2-simplicial vertex can always be drawn near the edge connecting its two neighbours. Graphs in $\mathcal{T}_2$ are characterised as those with no K_4-minor, and are sometimes called *series-parallel*.

Theorem 1. $\theta(\mathcal{T}_k) = \lceil k/2 \rceil$

Proof. The upper bound immediately follows from a more general result by [4]. Now for the lower bound. The result is trivial if $k \le 2$. Assume $k \ge 3$. Let $\ell := \lceil k/2 \rceil - 1$. Let G be the k-tree obtained by adding $2\ell^k + 1$ k-simplicial vertices adjacent to each vertex of a k-clique. Suppose that $\theta(G) \le \ell$. In the corresponding edge ℓ-colouring of G, consider the vector of colours on the edges incident to each k-simplicial vertex. There are ℓ^k possible colour vectors. Thus there are at least three k-simplicial vertices x, y, z with the same colour vector. At least $\lceil k/\ell \rceil \ge 3$ of the k edges incident to x are monochromatic. Say these edges are xa, xb, xc. Since y and z have the same colour vector as x, the $K_{3,3}$ subgraph induced by $\{xa, xb, xc, ya, yb, yc, za, zb, zc\}$ is monochromatic. Since $K_{3,3}$ is not planar, $\theta(G) \ge \ell + 1 = \lceil k/2 \rceil$. Therefore $\theta(\mathcal{T}_k) \ge \lceil k/2 \rceil$. □

The proofs of the following two results are similar to that of Theorem 1, and can be found in the full version of the paper [6].

Theorem 2. $\theta_o(\mathcal{T}_k) = \mathsf{a}(\mathcal{T}_k) = k$

Theorem 3. $\mathsf{sa}(\mathcal{T}_k) = k + 1$

3   Geometric Parameters

For our purposes, a *drawing* of a graph represents the vertices by a set of points in the plane in general position (no three collinear), and represents each edge by a simple closed curve between its endpoints, such that the only vertices that an edge intersects are its own endpoints. Two edges *cross* if they intersect at some point other than a common endpoint. A graph drawing with no crossings is *plane*. A plane drawing in which all the vertices are on the outerface is *outerplane*.

The *thickness* of a graph drawing is the minimum $k \in \mathbb{N}$ such that the edges of the drawing can be partitioned into k plane subgraphs; that is, each edge is assigned one of k colours such that monochromatic edges do not cross. Any planar graph can be drawn with its vertices at prespecified locations [9, 13]. Thus a graph with thickness k has a drawing with thickness k [9]. However, in such a representation the edges may be highly curved. This motivates the notion of geometric thickness.

A drawing of a graph is *geometric* if every edge is represented by a straight line-segment. The *geometric thickness* of a graph G, denoted by $\bar{\theta}(G)$, is the minimum $k \in \mathbb{N}$ such that there is a geometric drawing of G with thickness k. [10] first defined geometric thickness under the name of *real linear thickness*, and it has also been called *rectilinear thickness*. By the Fáry-Wagner theorem, a graph has geometric thickness one if and only if it is planar.

We generalise the notion of geometric thickness as follows. The *outerthickness* of a graph drawing is the minimum $k \in \mathbb{N}$ such that the edges of the drawing can be partitioned into k outerplane subgraphs. The *arboricity* and *star-arboricity* of a graph drawing are defined similarly, where it is respectively required that each subgraph be a plane forest or a plane star-forest. Again a graph with outerthickness / arboricity / star-arboricity k has a drawing with outerthickness / arboricity / star-arboricity k [9, 13]. The *geometric outerthickness* / *geometric arboricity* / *geometric star-arboricity* of a graph G, denoted by $\overline{\theta}_{o}(G)$ / $\overline{a}(G)$ / $\overline{sa}(G)$, is the minimum $k \in \mathbb{N}$ such that there is a geometric drawing of G with outerthickness / arboricity / star-arboricity k.

A geometric drawing in which the vertices are in convex position is called a *book embedding*. The *book thickness* of a graph G, denoted by $bt(G)$, is the minimum $k \in \mathbb{N}$ such that there is book embedding of G with thickness k. Note that whether two edges cross in a book embedding is simply determined by the relative positions of their endpoints in the cyclic order of the vertices around the convex hull. One can think of the vertices as being ordered on the spine of a book and each plane subgraph being drawn without crossings on a single page. Book embeddings are ubiquitous structures with a variety of applications; see [5] for a survey with over 50 references. A graph has book thickness one if and only if it is outerplanar [2]. A graph has a book thickness at most two if and only if it is a subgraph of a Hamiltonian planar graph [2]. [15] proved that planar graphs have book thickness at most four.

The *book arboricity* / *book star-arboricity* of a graph G, denoted by $ba(G)$ / $bsa(G)$, is the minimum $k \in \mathbb{N}$ such that there is a book embedding of G with arboricity / star-arboricity k. There is no point in defining "book outerthickness" since it would always equal book thickness.

4   Main Results

In this paper we determine the value of all of the geometric graph parameters defined in Section 3 for $\mathcal{T}_k$. The following theorem, which is proved in Section 6, is the most significant result in the paper. It says that the lower bound for the (abstract) thickness of $\mathcal{T}_k$ (Theorem 1) can be matched by an upper bound, even in the more restrictive setting of geometric thickness.

Theorem 4. $\overline{\theta}(\mathcal{T}_k) = \lceil k/2 \rceil$

We have the following theorem for the geometric outerthickness and geometric arboricity of $\mathcal{T}_k$. It says that the lower bounds for the outerthickness and arboricity of $\mathcal{T}_k$ can be matched by an upper bound on the corresponding geometric parameter. By the lower bound in Theorem 2, to prove Theorem 5, it suffices to show that $\overline{a}(\mathcal{T}_k) \leq k$; we do so in [6].

Theorem 5. $\overline{\theta_o}(\mathcal{T}_k) = \overline{\mathsf{a}}(\mathcal{T}_k) = k$

We have the following theorem for the book thickness and book arboricity of $\mathcal{T}_k$.

Theorem 6. $\mathsf{bt}(\mathcal{T}_k) = \mathsf{ba}(\mathcal{T}_k) = \begin{cases} k & \text{for } k \leq 2 \\ k+1 & \text{for } k \geq 3 \end{cases}$

This theorem gives an example of an abstract parameter that is not matched by its geometric counterpart. In particular, $\mathsf{bt}(\mathcal{T}_k) > \theta_o(\mathcal{T}_k) = k$ for $k \geq 3$. Theorem 6 with $k = 1$ was proved by [2]. That $\mathsf{bt}(\mathcal{T}_2) \leq 2$ was independently proved by [14] and [3]. Note that $\mathsf{bt}(\mathcal{T}_2) = 2$ since there are series parallel graphs that are not outerplanar, $K_{2,3}$ being the primary example. We prove the stronger result that $\mathsf{ba}(\mathcal{T}_2) = 2$ in [6]. [7] proved that every k-tree has a book embedding with thickness at most $k+1$. It is easily seen that each plane subgraph is in fact a star-forest. Thus $\mathsf{bt}(\mathcal{T}_k) \leq \mathsf{ba}(\mathcal{T}_k) \leq \mathsf{bsa}(\mathcal{T}_k) \leq k+1$. We give an alternative proof of this result in [6]. [7] proved a lower bound of $\mathsf{bt}(\mathcal{T}_k) \geq k$, and conjectured that $\mathsf{bt}(\mathcal{T}_k) = k$. Thus Theorem 6 refutes this conjecture. The proof is given in Section 5, where we construct a k-tree G with $\mathsf{bt}(G) > k$.

Finally observe that the upper bound of [7] mentioned above and the lower bound in Theorem 3 prove the following result for the star-arboricity of $\mathcal{T}_k$.

Theorem 7. $\mathsf{sa}(\mathcal{T}_k) = \overline{\mathsf{sa}}(\mathcal{T}_k) = \mathsf{bsa}(\mathcal{T}_k) = k+1$

5   Book Thickness: Proof of Theorem 6 ($k \geq 3$)

By the discussion in Section 4, it suffices to show that for all $k \geq 3$, there is a k-tree G with book thickness $\mathsf{bt}(G) > k$. Define G by the following construction:

- Start with a k-clique V_1.
- Add $k(2k+1)$ k-simplicial vertices adjacent to each vertex in V_1; call this set of vertices V_2.
- For each vertex $v \in V_2$, choose three distinct vertices $x_1, x_2, x_3 \in V_1$, and for each $1 \leq i \leq 3$, add four k-simplicial vertices adjacent to each vertex of the clique $(V_1 \cup \{v\}) \setminus \{x_i\}$. Each set of four vertices is called an i-block of v. Let V_3 be the set of vertices added in this step.

Clearly G is a k-tree. Assume for the sake of contradiction that G has a book embedding with thickness k. Let $\{E_1, E_2, \dots, E_k\}$ be the corresponding partition of the edges. For each ordered pair of vertices $v, w \in V(G)$, let the *arc-set* $V_{\overrightarrow{vw}}$ be the list of vertices in clockwise order from v to w (not including v and w). Say $V_1 = (y_1, y_2, \dots, y_k)$ in anticlockwise order. There are $k(2k+1)$ vertices in V_2. Without loss of generality there are at least $2k+1$ vertices in $V_2 \cap V_{\overrightarrow{y_1 y_k}}$. Let $(v_1, v_2, \dots, v_{2k+1})$ be $2k+1$ vertices in $V_2 \cap V_{\overrightarrow{y_1 y_k}}$ in clockwise order.

Observe that the k edges $\{y_i v_{k-i+1} : 1 \leq i \leq k\}$ are pairwise crossing, and thus receive distinct colours, as illustrated in Figure 1(a). Without loss of generality, each $y_i v_{k-i+1} \in E_i$. As illustrated in Figure 1(b), this implies that

$y_1v_{2k+1} \in E_1$, since y_1v_{2k+1} crosses all of $\{y_iv_{k-i+1} : 2 \leq i \leq k\}$ which are coloured $\{2, 3, \ldots, k\}$. As illustrated in Figure 1(c), this in turn implies $y_2v_{2k} \in E_2$, and so on. By an easy induction, we obtain that $y_iv_{2k+2-i} \in E_i$ for all $1 \leq i \leq k$, as illustrated in Figure 1(d). It follows that for all $1 \leq i \leq k$ and $k - i + 1 \leq j \leq 2k + 2 - i$, the edge $y_iv_j \in E_i$, as illustrated in Figure 1(e). Finally, as illustrated in 1(f), we have:

If $qy_i \in E(G)$ and $q \in V_{\overparen{v_{k-1}v_{k+3}}}$, then $qy_i \in E_i$. $(\star)$

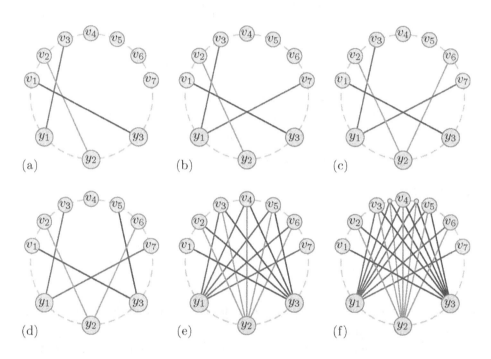

(a)   (b)   (c)

(d)   (e)   (f)

Fig. 1. Example in the proof of Theorem 6 with $k = 3$

Consider any of the twelve vertices $w \in V_3$ that are added onto a clique that contain v_{k+1}. Then w is adjacent to v_{k+1}. Moreover, w is in $V_{\overparen{v_kv_{k+1}}}$ or $V_{\overparen{v_{k+1}v_{k+2}}}$, as otherwise the edge wv_{k+1} crosses k edges of $G[\{v_{k-1}, v_{k+1}\}; V_1]$ that are all coloured differently, which is a contradiction. By the pigeon-hole principle, one of $V_{\overparen{v_kv_{k+1}}}$ and $V_{\overparen{v_{k+1}v_{k+2}}}$ contains at least two vertices from two distinct p-blocks of v_{k+1}. Without loss of generality, $V_{\overparen{v_kv_{k+1}}}$ does. Let these four vertices be (a, b, c, d) in clockwise order.

Each vertex in $\{b, c, d\}$ is adjacent to $k - 1$ vertices of V_1. Not all of b, c, d are adjacent to the same subset of $k - 1$ vertices in V_1, as otherwise all of b, c, d would belong to the same p-block. Hence each vertex in V_1 has a neighbour in $\{b, c, d\}$. By $(\star)$ the edges of $G[\{b, c, d\}, V_1]$ receive all k colours. However, every edge in $G[\{b, c, d\}; V_1]$ crosses the edge av_{k+1}, implying that there is no colour available for av_{k+1}. This contradiction completes the proof.

6   Geometric Thickness: Proof of Theorem 4

The proofs of all of our upper bounds depend upon the following lemma.

Lemma 8. *For every k-tree G, either:*

(1) *there is a (possibly empty) independent set $S \subseteq V(G)$ of k-simplicial vertices in G such that $G \setminus S = K_k$, or*

(2) *there is a nonempty independent set $S \subseteq V(G)$ of k-simplicial vertices in G and a vertex $v \in V(G) \setminus S$, such that:*
   (a) *$G \setminus S$ is a k-tree,*
   (b) *v is k-simplicial in $G \setminus S$,*
   (c) *for every vertex $w \in S$, there is exactly one vertex $u \in N_{G\setminus S}(v)$ such that $N_G(w) = N_{G\setminus S}[v] \setminus \{u\}$,*
   (d) *every k-simplicial vertex of G that is not in S is not adjacent to v.*

Proof. We proceed by induction on $|V(G)|$. If $|V(G)| = k$ then $G = K_k$ and property (1) is satisfied with $S = \emptyset$. If $|V(G)| = k + 1$ then $G = K_{k+1}$ and property (1) is satisfied with $S = \{v\}$ for any vertex v. Now suppose that $|V(G)| \geq k+2$. Let L be the set of k-simplicial vertices of G. Then L is a nonempty independent set, and $G \setminus L$ is a k-tree. Moreover, the neighbourhood of each vertex in L is a k-clique. If $G \setminus L = K_k$, then property (1) is satisfied with $S = L$. Otherwise, $G \setminus L$ has a k-simplicial vertex v. Let S be the set of neighbours of v in L. We claim that property (2) is satisfied. Now $S \neq \emptyset$, as otherwise $v \in L$. Since G is not a clique and each vertex in S is simplicial, $G \setminus S$ is a k-tree. Consider a vertex $w \in S$. Now $N_G(w)$ is a k-clique and $v \in N_G(w)$. Thus $N_G(w) \subseteq N_{G\setminus S}[v]$. Since $|N_G(w)| = k$ and $|N_{G\setminus S}[v]| = k + 1$, there is exactly one vertex $u \in N_{G\setminus S}(v)$ for which $N_G(w) = N_{G\setminus S}[v] \setminus \{u\}$. Part (d) is immediate. □

We now turn to the proof of Theorem 4. The lower bound $\overline{\theta}(\mathcal{T}_k) \geq \lceil k/2 \rceil$ follows from the stronger lower bound $\theta(\mathcal{T}_k) \geq \lceil k/2 \rceil$ in Theorem 1. The theorem is true for all 0-, 1- and 2-trees since they are planar. To prove the upper bound $\overline{\theta}(\mathcal{T}_k) \leq \lceil k/2 \rceil$, it suffices to prove that $\overline{\theta}(2k) \leq k$ for all $k \geq 2$. Let $I := \{i, -i : 1 \leq i \leq k\}$.

Consider a geometric drawing of a 2k-tree G, in which the edges are coloured with k colours. Let v be a 2k-simplicial vertex of G, where $(u_1, u_2, \ldots, u_k, u_{-1}, u_{-2}, \ldots, u_{-k})$ are the neighbours of v in clockwise order around v. Let $F_i(v)$ denote the closed infinite wedge centred at v (but not including v), which is bounded by the ray $\overrightarrow{vu_i}$ and the ray that is opposite to the ray $\overrightarrow{vu_{-i}}$. As illustrated in Figure 2(a), we say that v has the *fan property* if:

– $F_i(v) \cap F_j(v) = \emptyset$ for all distinct $i, j \in I$,
– there are exactly two edges of each colour incident to v, and
– the edges vu_i and vu_{-i} receive the same colour for all $1 \leq i \leq k$.

We proceed by induction on $|V(G)|$ with the hypothesis: "every 2k-tree G has a geometric drawing with thickness k; moreover, if $|V(G)| \geq 2k + 2$, then every 2k-simplicial vertex v of G has the fan property." Let G be a 2k-tree. Apply Lemma 8 to G.

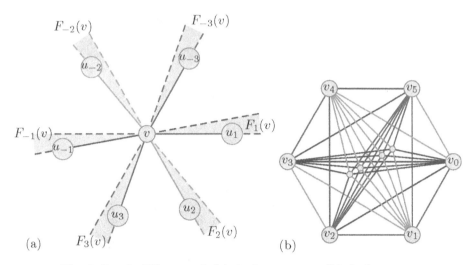

Fig. 2. Proof of Theorem 4: (a) the fan property, (b) the base case

First suppose that Lemma 8 gives a (possibly empty) independent set $S \subseteq V(G)$ of $2k$-simplicial vertices in G such that $G \setminus S = K_{2k}$. Say $V(G \setminus S) = \{v_0, v_1, \ldots, v_{2k-1}\}$. Position $v_0, v_1, \ldots, v_{2k-1}$ evenly spaced on a circle in the plane, and in this order. The edges of $G \setminus S$ can be k-coloured using the standard book embedding of K_{2k} with thickness k, where each edge $v_\alpha v_\beta$ is coloured $\lfloor \frac{1}{2}((\alpha + \beta) \bmod 2k) \rfloor$. Each colour class forms a plane zig-zag pattern. For each vertex $w \in S$ and for all $0 \le i \le k - 1$, colour the edges wv_i and wv_{k+i} by i. As illustrated in Figure 2(b), position the vertices in S in a small enough region near the centre of the circle so that monochromatic edges do not cross, each $w \in S$ has the fan property, and $V(G)$ is in general position. If $|V(G)| \ge 2k+2$, then no vertex in $\{v_0, v_1, \ldots, v_{2k-1}\}$ is $2k$-simplicial in G. Therefore, each $2k$-simplicial vertex of G is in S, and thus has the fan property.

Now suppose that Lemma 8 gives a nonempty independent set $S \subseteq V(G)$ of $2k$-simplicial vertices in G and a vertex $v \in V(G) \setminus S$, such that v is $2k$-simplicial in the k-tree $G \setminus S$. If $|V(G) \setminus S| \ge 2k+2$, then by induction, there is a geometric drawing of $G \setminus S$ with thickness k, in which v has the fan property. Otherwise, $G \setminus S = K_{2k+1}$ and thus the set $S' = \{v\}$ is an independent set of $2k$-simplicial vertices in $G \setminus S$ such that $(G \setminus S) \setminus S' = K_{2k}$. Thus by the construction given above, there is a geometric drawing of $G \setminus S$ with thickness k, in which v has the fan property.

Say $N_{G \setminus S}(v) = (u_1, u_2, \ldots, u_k, u_{-1}, u_{-2}, \ldots, u_{-k})$ in clockwise order about v. Without loss of generality, the edges vu_i and vu_{-i} are coloured i, for all $1 \le i \le k$. Choose a small enough disc D_ϵ centred at v such that:

(a) the only vertices in R_ϵ are $N_{G \setminus S}[v]$,
(b) every edge of $G \setminus S$ that intersects D_ϵ is incident to v (as illustrated in Figure 3), and

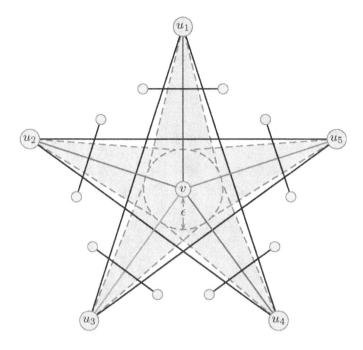

Fig. 3. The 'empty' disc D_ϵ

(c) should a vertex whose neighbourhood is $\{u_1, u_2, \ldots, u_k, u_{-1}, u_{-2}, \ldots, u_{-k}\}$ be placed in D_ϵ, then it would have the fan property.

By Lemma 8, for every vertex $w \in S$, there is exactly one $i \in I$ for which $N_G(w) = N_{G \backslash S}[v] \setminus \{u_i\}$. Let $S_i := \{w \in S : N_G(w) = N_{G \backslash S}[v] \setminus \{u_i\}\}$ for all $i \in I$. Two vertices in S_i have the same neighbourhood in G. For all $i \in I$, choose one vertex $x_i \in S_i$ (if any). We will first draw x_i for all $i \in I$. Once that is completed, we will draw the remaining vertices in S.

As illustrated in 4, for all $i \in I$, colour the edge $x_i v$ by $|i|$, and colour the edge $x_i u_j$ by $|j|$ for all $j \in I \setminus \{i\}$. Now in a drawing of G, for each $i \in I$, $F_i(x_i)$ is the closed infinite wedge bounded by the ray $\overrightarrow{x_i v}$ and the ray that is opposite to $\overrightarrow{x_i u_{-i}}$, and $F_{-i}(x_i)$ is the closed infinite wedge bounded by the ray $\overrightarrow{x_i u_{-i}}$ and the ray that is opposite to $\overrightarrow{x_i v}$. Observe that in a drawing of G, if $x_i \in F_{-i}(v)$ for all $i \in I$, then $v \notin F_\ell(x_i)$ for all $\ell \neq i$. Therefore, for $i \in I$ in some arbitrary order, each vertex x_i can initially be positioned on the line-segment $\overline{vu_{-i}} \cap (D_\epsilon \setminus \{v\})$, so that $x_i \notin \bigcup \{F_\ell(x_j) : \ell \in I \setminus \{j\}\}$ for every $j \in I$. This is possible by the previous observation, since there is always a point close enough to v where x_i can be positioned, so that $x_i \notin \bigcup \{F_\ell(x_j) : \ell \in I \setminus \{j\}\}$ for all the vertices x_j that are drawn before x_i. Observe that each vertex x_i has the fan property in the thus constructed illegal drawing.

Now we move each vertex x_i just off the edge vu_{-i} to obtain a legal drawing. In particular, move each vertex x_i by a small enough distance ϵ' into $F_{-i}(v)$, so that

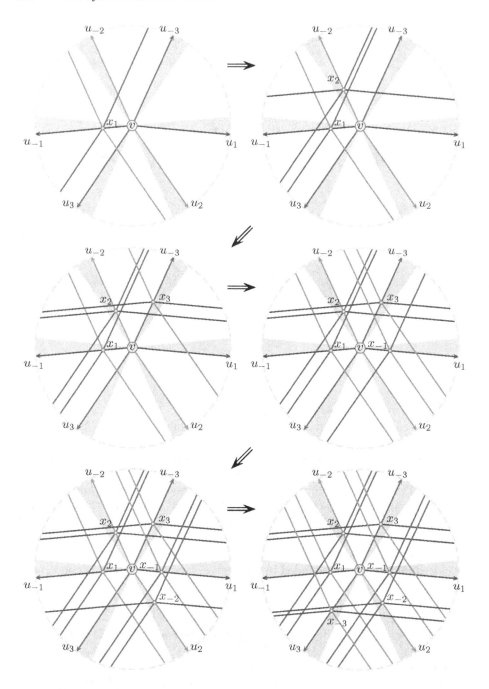

Fig. 4. Placing each x_i on the edge vu_{-i}; the circle D_ϵ is chosen small enough so that the edges incident with u_i are almost parallel

$F_i(x_i)$ does not contain the vertex x_j, for all $j \in I \setminus \{i, -i\}$. This implies that for all distinct $i, j \in I$ with $i \neq -j$, we have that $x_j \notin F_\ell(x_i)$ for all $\ell \in I$.

To prove that monochromatic edges do not cross, we distinguish four types of edges coloured i, where $1 \leq i \leq k$:

1. edges of $G \setminus S$ coloured i,
2. the edges $x_i v$ and $x_{-i} v$,
3. edges $x_j u_i$ for some $j \in I \setminus \{i\}$, and
4. edges $x_\ell u_{-i}$ for some $\ell \in I \setminus \{-i\}$.

First we prove that no type-(1) edge is involved in a monochromatic crossing. No two type-(1) edges cross by induction. Since a type-(2) edge is contained in D_ϵ, by (b) in the choice of ϵ, type-(1) and type-(2) edges do not cross. Suppose that a type-(1) edges e crosses a type-(3) or type-(4) edge. By (a) in the choice of ϵ, e would also cross vu_i. Since vu_i is coloured i, by induction applied to $G \setminus S$, e is not coloured i.

The two type-(2) edges do not cross since they are both incident to v. Type-(3) edges do not cross since they are all incident to u_i. Type-(4) edges do not cross since they are all incident to u_{-i}.

Suppose that a type-(2) edge $x_i v$ crosses a type-(3) edge $x_j u_i$ for some $j \in I \setminus \{i\}$. By construction, $x_i \in F_{-i}(v)$ and $x_j \notin F_{-i}(v)$. Therefore, if $x_j u_i$ crosses $x_i v$, then $x_j u_i$ also crosses the edge vu_{-i}, which is a type-(1) edge of colour $|i|$. Thus this type of crossing was ruled out when type-(1) edges were considered. Now suppose that a type-(2) edge $x_{-i} v$ crosses a type-(3) edge $x_j u_i$ for some $j \in I \setminus \{i\}$. Then $x_j \in F_{-i}(x_{-i})$, which contradicts the placement of x_j. Thus no type-(2) edge crosses a type-(3) edge. By symmetry, no type-(2) edge crosses a type-(4) edge.

If a type-(3) edge $x_{-i} u_i$ crosses a type-(4) edge $x_\ell u_{-i}$ (for some $\ell \in I \setminus \{-i\}$), then $x_\ell u_{-i}$ also crosses the edge vu_i, which is a type-(1) edge coloured $|i|$. Thus this type of crossing was ruled out when type-(1) edges were considered. By symmetry, a type-(4) edge $x_i u_{-i}$ does not cross a type-(3) edge $x_\ell u_i$ (for all $\ell \in I \setminus \{i\}$). Finally, if a type-(3) edge $x_j u_i$ (for some $j \in I \setminus \{i, -i\}$) crosses a type-(4) edge $x_\ell u_{-i}$ (for some $\ell \in I \setminus \{-i, i\}$), then $x_\ell \in F_i(x_j)$ and $x_j \in F_{-i}(x_\ell)$, contradicting our placement of x_ℓ or x_j. Thus type-(3) edges do not cross type-(4) edges.

Each vertex $z \in S_i \setminus \{x_i\}$ can be drawn in a small enough region around x_i, and every edge zu_j coloured with the same colour as $x_i u_j$, so that z has fan property and monochromatic edges do not cross.

It remains to prove that each $2k$-simplicial vertex of G has the fan property whenever $|V(G)| \geq 2k + 2$. By construction that is true for all $2k$-simplicial vertices of G that are in S. The remaining $2k$-simplicial vertices of G are also $2k$-simplicial in the $2k$-tree $G \setminus S$. If $|V(G) \setminus S| \geq 2k + 2$, then by induction, the invariant is also maintained for all $2k$-simplicial vertices of G that are not in S. If $G \setminus S$ is K_{2k+1}, then by Lemma 8(d), there is no $2k$-simplicial vertex of G in $G \setminus S$. Thus the invariant is maintained.

References

[1] YASUKAZU AOKI. The star-arboricity of the complete regular multipartite graphs. *Discrete Math.*, 81(2):115–122, 1990.

[2] FRANK R. BERNHART AND PAUL C. KAINEN. The book thickness of a graph. *J. Combin. Theory Ser. B*, 27(3):320–331, 1979.

[3] EMILIO DI GIACOMO, WALTER DIDIMO, GIUSEPPE LIOTTA, AND STEPHEN K. WISMATH. Book embeddings and point-set embeddings of series-parallel digraphs. In MICHAEL T. GOODRICH AND STEPHEN G. KOBOUROV, eds., *Proc. 10th International Symp. on Graph Drawing* (GD '02), vol. 2528 of *Lecture Notes in Comput. Sci.*, pp. 162–173. Springer, 2002.

[4] GUOLI DING, BOGDAN OPOROWSKI, DANIEL P. SANDERS, AND DIRK VERTIGAN. Partitioning graphs of bounded tree-width. *Combinatorica*, 18(1):1–12, 1998.

[5] VIDA DUJMOVIĆ AND DAVID R. WOOD. On linear layouts of graphs. *Discrete Math. Theor. Comput. Sci.*, 6(2):339–358, 2004.

[6] VIDA DUJMOVIĆ AND DAVID R. WOOD. Graph treewidth and geometric thickness parameters. arXiv.org:math.CO/0503553, 2005.

[7] JOSEPH L. GANLEY AND LENWOOD S. HEATH. The pagenumber of k-trees is $O(k)$. *Discrete Appl. Math.*, 109(3):215–221, 2001.

[8] RICHARD K. GUY. Outerthickness and outercoarseness of graphs. In *Proc. British Combinatorial Conf.*, vol. 13 of *London Math. Soc. Lecture Note Ser.*, pp. 57–60. Cambridge Univ. Press, 1974.

[9] JOHN H. HALTON. On the thickness of graphs of given degree. *Inform. Sci.*, 54(3):219–238, 1991.

[10] PAUL C. KAINEN. Thickness and coarseness of graphs. *Abh. Math. Sem. Univ. Hamburg*, 39:88–95, 1973.

[11] PETRA MUTZEL, THOMAS ODENTHAL, AND MARK SCHARBRODT. The thickness of graphs: a survey. *Graphs Combin.*, 14(1):59–73, 1998.

[12] CRISPIN ST. J. A. NASH-WILLIAMS. Decomposition of finite graphs into forests. *J. London Math. Soc.*, 39:12, 1964.

[13] JÁNOS PACH AND REPHAEL WENGER. Embedding planar graphs at fixed vertex locations. *Graphs Combin.*, 17(4):717–728, 2001.

[14] S. RENGARAJAN AND C. E. VENI MADHAVAN. Stack and queue number of 2-trees. In DING-ZHU DU AND MING LI, eds., *Proc. 1st Annual International Conf. on Computing and Combinatorics* (COCOON '95), vol. 959 of *Lecture Notes in Comput. Sci.*, pp. 203–212. Springer, 1995.

[15] MIHALIS YANNAKAKIS. Embedding planar graphs in four pages. *J. Comput. System Sci.*, 38:36–67, 1986.

Stress Majorization with Orthogonal Ordering Constraints

Tim Dwyer[1], Yehuda Koren[2], and Kim Marriott[1]

[1] School of Comp. Science & Soft. Eng., Monash University, Australia
{tdwyer, marriott}@mail.csse.monash.edu.au
[2] AT&T — Research
yehuda@research.att.com

Abstract. The adoption of the stress-majorization method from multi-dimensional scaling into graph layout has provided an improved mathematical basis and better convergence properties for so-called "force-directed placement" techniques. In this paper we give an algorithm for augmenting such stress-majorization techniques with orthogonal ordering constraints and we demonstrate several graph-drawing applications where this class of constraints can be very useful.

Keywords: graph layout, constrained optimization, separation constraints.

1 Introduction

The family of graph drawing algorithms that attempt to find an embedding of a graph that minimizes some continuous goal function, are variously known as *spring-embedder* or *force-directed placement* algorithms. A popular algorithm in this family has been that of Kamada and Kawai [9] in which squared differences between ideal distances for pairs of nodes and their Euclidean distance in the embedding is minimized. Gansner et al. [6] recently revisited this method and suggested using *functional majorization* — an optimization technique from the field of multidimensional scaling. Functional majorization iteratively improves the drawing by considering a sequence of quadratic forms that bound the stress function from above. They showed that it had distinct advantages over the original algorithm of Kamada and Kawai; particularly, a strictly monotonic decrease in stress and that it could achieve lower values of the cost function in the same running time.

A useful property of the majorization approach is that each iteration involves minimizing a convenient quadratic function. Gansner et al. [6] mentioned that this allows using any available equation solver. In this paper we take advantage of this property, and show how it helps in handling ordering constraints on the nodes. The quadratic nature of the function we minimize in each iteration allows us to efficiently add such linear constraints. In fact, minimizing linearly constrained quadratic functions is known as *quadratic programming*, which is an efficiently solvable problem [13]. However, we have found that general quadratic programming solvers will significantly slow down the stress majorization process. Therefore, we suggest a solver which is crafted especially for our problem, utilizing its unique nature. This solver can deal with ordering constraints without significantly increasing the running time of the layout process. We also demonstrate the utility of imposing this class of constraints — which we call *orthogonal ordering* constraints — to applications such as network layout reflecting the relative positions of an underlying set of coordinates and directed graph drawing.

P. Healy and N.S. Nikolov (Eds.): GD 2005, LNCS 3843, pp. 141–152, 2005.
© Springer-Verlag Berlin Heidelberg 2005

2   Background

We recently introduced the idea of using stress majorization coupled with standard quadratic programming techniques for drawing directed graphs [5]. In the so-called DIG-COLA[1] technique, nodes in the digraph were partitioned into layers based on their hierarchical level and constraints were introduced in the vertical dimension to keep these layers separated. Compared to standard hierarchical graph drawing methods the DIG-COLA algorithm was shown to produce layouts with a much better distribution of edge lengths and for large, dense graphs it was able to find layouts with fewer edge crossings. However, a commercial quadratic programming solver was used to minimize the quadratic forms subject to constraints. This generic approach meant that layout for graphs with hundreds or thousands of nodes could take some minutes to perform.

Another case where orthogonal ordering constraints are useful is when we want to improve the readability of a given layout without significantly changing it. Misue et al. [10] discussed the importance of preserving a user's "mental map" when adjusting graph layouts. One of their models for the mental map focused on preserving *orthogonal ordering* of the nodes in a layout — the relative above/below, left/right positions of the nodes.

The potential for constraint-based, force-directed graph layout was explored by Ryall et al. [11], however their implementation did not use true constraint solving techniques. Rather, they added stiff springs to a standard force-directed model to keep user-selected parts of the diagram roughly spaced as desired. True constraint solving techniques for graph drawing were explored by He and Marriott in [7], where a Kamada-Kawai-based method was extended with an active-set constraint solving technique to provide separation constraints. However, only small examples of fewer than 20 nodes were tested and the scalability of the technique was not tested.

3   Problem Formulation

The general goal function, known as *the stress function*, which we seek to minimize is described by

$$\sum_{i<j} w_{ij}(\|X_i - X_j\| - d_{ij})^2$$

where for each pair of nodes i and j, d_{ij} gives an ideal separation between i and j (usually their graph-theoretical distance), $w_{ij} = d_{ij}^{-2}$ is used as a normalization constant and X is a $n \times d$ matrix of positions for all nodes, where d is the dimensionality of the drawing and n is the number of nodes.

Majorization minimizes this stress function by iteratively minimizing quadratic forms that approximate and bound it from above. Due to its central role in this work, we provide the essential details of the method. Recall that w_{ij} are the normalization constants in the stress function. We use the $n \times n$ matrix A, defined by

$$A_{i,j} = \begin{cases} -w_{ij} & i \neq j \\ \sum_{k \neq i} w_{ik} & i = j \end{cases}. \tag{1}$$

---

[1] Directed Graphs with Constraint-based Layout.

In addition, given an $n \times d$ coordinate matrix Z, we define the $n \times n$ matrix A^Z by

$$A^Z_{i,j} = \begin{cases} -w_{ij} \cdot d_{ij} \cdot \text{inv}(\|Z_i - Z_j\|) & i \neq j \\ -\sum_{k \neq i} A^Z_{i,k} & i = j \end{cases}, \qquad (2)$$

where $\text{inv}(x) = 1/x$ when $x \neq 0$ and 0 otherwise.

It can be shown (see [6]) that the stress function is bounded from above by the quadratic form $F^Z(X)$ defined as

$$F^Z(X) = \sum_{i<j} w_{ij} d^2_{ij} + \sum_{a=1}^{d} \left(\left(X^{(a)}\right)^T A X^{(a)} - 2 \left(X^{(a)}\right)^T A^Z Z^{(a)} \right). \qquad (3)$$

Here, $X^{(a)}$ denotes the a-th column of matrix X. Thus, we have

$$\text{stress}(X) \leqslant F^Z(X) \qquad (4)$$

with equality when $Z = X$.

We differentiate by X and find that the global minima of $F^Z(X)$ are given by solving

$$AX = A^Z Z \qquad (5)$$

This leads to the following iterative optimization process. Given some layout $X(t)$, we compute a layout $X(t+1)$ so that $\text{stress}(X(t+1)) < \text{stress}(X(t))$. We use the function $F^{X(t)}(X)$ which satisfies $F^{X(t)}(X(t)) = \text{stress}(X(t))$. Then, we take $X(t+1)$ as the minimizer of $F^{X(t)}(X)$ by solving (5).

Note that it would be equivalent to consider in each iteration d independent optimization problems, one problem for each axis. Hence the a-th axis of the drawing is determined by minimizing

$$x^T A x - 2x^T A^Z Z^{(a)} \qquad (6)$$

Henceforth, we will work, w.l.o.g., with this 1-D layout formulation as it allows a more convenient notation.

So far we have described the usual, unconstrained stress majorization. In this work we consider a case where we have additional ordering constraints on each axis. Each node i is assigned a level of index $1 \leq lev[i] \leq m$ and variable placement must respect this level. Thus, instead of minimizing (6), we would take the a-th axis of the drawing as the solution of

$$\begin{aligned} \min_x \quad & x^T A x - 2x^T A^Z Z^{(a)} \\ \text{subject to: } & lev[i] < lev[j] \Rightarrow x_i \leq x_j \\ & \text{for all } i, j \in \{1, \dots, n\} \end{aligned} \qquad (7)$$

For brevity henceforth we will replace $2A^Z Z^{(a)}$ with $b \in \mathbb{R}^n$, so the target function is merely $f(x) = x^T A x - x^T b$. We call this the Quadratic Programming with Orthogonal Constraints (QPOC) problem.

It is easy to show that A is positive semi-definite, so the problem has only global minima. Such a quadratic programming problem can be solved in a polynomial time

[13]. However, our experiments show that generic quadratic-programming solvers are much slower than solving an unconstrained problem. To accelerate computation we can utilize two special characteristics of the problem:

1. During the majorization process, we iteratively solve closely related quadratic programs: The constraints and the matrix A are not changed between iterations, while only the vector b is changed. Therefore, the solution of the previous iteration is still a feasible solution for current iteration (satisfying all constraints). Moreover, this previous solution is probably very close to the new optimal solution (e.g., consider that in most iterations the coordinates are only slightly changed). However, such initialization, called "warm-start", is fundamentally not trivial for the barrier (or interior-point) methods used by most commercial solvers.
2. Our constraints are very simple as each of them involve only two variables, being of the form $x_i \leq x_j$. This allows a simple mechanism for guaranteeing the feasibility of the solution.

In the next section we describe an algorithm for solving the QPOC problem.

4   Algorithm

We give an iterative *gradient-projection* algorithm (see Bertsekas [1]) for finding a solution to a QPOC Problem. The algorithm, *solve_QPOC*, is shown in Figure 1. The first step is to decrease $f(x) = x^T Ax + x^T b$, by moving x in the direction of steepest descent, i.e. if the gradient is $g = \nabla f(x) = Ax + b$ this direction is $-g$. While we are guaranteed that — with appropriate selection of step-size s — the energy is decreased by this first step, the new positions may violate the ordering constraints. We correct this by calling the *project* procedure which returns the closest point $\bar{x}$ to x which satisfies the ordering constraints, i.e. it projects x on to the feasible region. Finally, we calculate a vector d from our initial position $\hat{x}$ to $\bar{x}$ and we ensure mono-

---

procedure *solve_QPOC(A, b, lev)*
   $k \leftarrow 0, x \leftarrow initial\_soln()$
  repeat
     $g \leftarrow 2Ax + b$
     $s \leftarrow \frac{g^T g}{g^T Ag}$
     $\hat{x} \leftarrow x$
     $\bar{x} \leftarrow project(\hat{x} - sg, lev)$
     $d \leftarrow \bar{x} - \hat{x}$
     $\alpha \leftarrow \max(\frac{g^T d}{d^T Ad}, 1)$
     $x \leftarrow \hat{x} + \alpha d$
  until $\|\hat{x} - x\|$ sufficiently small
  return x

Fig. 1. Algorithm to find an optimal solution to a QPOC problem with variables $x_1, \ldots, x_n$, symmetric positive-semidefinite matrix A, vector b and $1 \leq lev[i] \leq m + 1$ gives the level for each node i

tonic decrease in stress when moving in this direction by computing a second stepsize $\alpha = \arg\min_{\alpha \in [0,1]} f(x + \alpha d)$ which minimizes stress in this interval.

The procedure *project* is the main technical innovation in this paper. The main difficulty in implementing gradient-projection methods is the need to efficiently project on to the feasible region. Because of the simple nature of the orthogonal ordering constraints we can do this in $O(mn + n \log n)$ time where m is the number of levels and n the number of variables. The *project* procedure (Figure 2) iteratively changes the positions till all constraints are satisifed. In iteration k all constraints involving nodes up to the $(k + 1)$-th level are imposed. More technically, it starts by finding an ordering of the nodes q such that $a = q[i], b = q[i + 1]$ implies either $lev[a] < lev[b]$ or ($lev[a] = lev[b]$ and $x_a \leqslant x_b$). For convenience we also keep an array $1 < p_1, \ldots p_m = n + 1$ of indices for the start of each partition excluding the first (for convenience p_m was set to $n + 1$). When considering partition k, which contains the nodes $above_k = \{u | p_k \leq q[u] < p_{k+1}\}$, we ensure that none of these nodes are assigned positions lower than that of $below_k = \{l | 1 \leq q[l] < p_k\}$. To achieve this we create a minimal set $U_k \subseteq \{j | 1 \leq q[j] < p_{k+1}\}$ that includes nodes violating this condition. To impose the constraints we force all nodes of U_k to lie on a single point $posnU_k$. Since we want to minimize the quadratic function, we take this point as the av-

```
procedure project(x,lev)
    q ← {1 ≤ i ≤ n} sorted by (x_i, lev[i])
    p ← indices to start of each level in q
        s.t. p_1 < ... < p_{m-1} < p_m = n + 1
        and lev[q[p_k]] = lev[q[p_k − 1]] + 1, 1 ≤ k < m
    for 1 ≤ k < m do
        % below_k = {l|1 ≤ q[l] < p_k}, above_k = {u|p_k ≤ q[u] < p_{k+1}}
        % Find U_k = {q[i]|il < i < iu} ⊆ below_k ∪ above_k
        maxiu ← p_{k+1} − 1
        l ← q[p_k − 1], u ← q[p_k]
        sum ← x_l + x_u, w ← 2
        iu ← p_k + 1, il ← p_k − 2
        if x_l > x_u then
            repeat
                finished ← true
                u ← q[iu]
                posnU_k ← sum/w
                if iu ≤ maxiu and x_u < posnU_k then
                    iu ← iu + 1, w ← w + 1
                    sum ← sum + x_u
                    finished ← false
                end if
                l ← q[il]
                if il ≥ 1 and x_l > posnU_k then
                    il ← il − 1, w ← w + 1
                    sum ← sum + x_l
                    finished ← false
                end if
            until finished
            for il < i < iu do
                j ← q[i]
                x_j ← posnU_k
            end for
        end if
    end for
    return x
```

Fig. 2. Algorithm to project variables to the closest position in the feasible region, $1 \leq lev[i] \leq m$ gives the level for each node i

erage of all positions in U_k. The set U_k is minimal in that it does not necessarily include all nodes violating the boundary condition for k, but only the minimal number that need to be moved to $posnU_k$ such that this condition may be satisfied. The following lemma captures this.

Lemma 1. *During execution of* project(x,lev) *after finishing the kth iteration in which U_k and its associated $posnU_k$ are computed*

$$posnU_k = \frac{\sum_{i \in U_k} x_i}{|U_k|} \tag{8}$$

and

$$U_k = \{l \in below_k \mid x_l > posnU_k\} \cup \{u \in above_k \mid x_u < posnU_k\} \tag{9}$$

where the position for x_i is its value before the start of the iteration.

Proof. Equation (8) follows directly from the algorithm and is invariant throughout the loop incrementally building U_k (since whenever U_k is expanded $posnU_k$ is recalculated).

The post-condition (9) implies that U_k includes all nodes that violate the internal constraints among $1, \ldots, p_k - 1$ and $p_k, \ldots, p_{k+1} - 1$. Proof is as follows. The levels are examined in order. When examining level k all nodes in $below_k$ must be sorted by position in q (either by the initial precondition for q or since they have been assigned to a position $posnU_l, l < k$). The precondition for q also ensures that nodes in $above_k$ are sorted by position.

If there is overlap between the tail of $below_k$ and the head of $above_k$ we place these in U_k and set $posnU_k$. We then iteratively examine the successive elements of $below_k$ (from the tail) and $above_k$ (from the head) and add them to U_k until no further overlap is found between these elements and $posnU_k$.

By construction the only elements $l \in below_k$ not placed in U_k are those for which $x_l \leq posnU_k$ (otherwise the loop would not terminate). Dually, for any element $u \in above_k$ not placed in U_k we have that $x_u \geq posnU_k$. Thus

$$U_k \supseteq \{1 \leq q[i] < p_k \mid x_i > posnU_k\} \cup \{p_k \leq i < p_{k+1} \mid x_i < posnU_k\}$$

We now show containment by induction. We prove for $U_k \cap below_k$, while the proof for $U_k \cap above_k$ is analogous. The base case follows from the fact that at the moment we add some $l \in below_k$, it must hold that $x_l > posnU_k$. Now, if later we add $l' \in below_k$, then since $below_k$ is ordered by position, $x_{l'} \leq x_l$. By hypothesis, $x_l > posnU_k$ and since the new $posnU_k$ is the weighted average of x_l' and $posnU_k$, we still have $x_l > posnU_k$. If later we add $u \in above_k$, then since we are adding u we must have $x_u < posnU_k$. Now by hypothesis, $x_l > posnU_k$ and so $x_l > x_u$. Thus as for the previous case $x_l > posnU_k$. □

Corollary 1. *During execution of* project(x,lev) *after finishing the kth iteration in which U_k and its associated $posnU_k$ are computed*

$$posnU_k = \frac{\sum_{i \in U_k} x_i}{|U_k|} \tag{10}$$

where the position of x_i is the input position.

Proof. Notice that unlike Equation (8), the x_i's refer now to the *input* positions, rather than to their values before the current iteration. This makes a difference when we find that $posnU_k < posnU_l, l < k$ and therefore $U_k \supset U_l$ and $posnU_k$ will be calculated from $posnU_l$ for those nodes in U_l rather than their original positions. In this case (10) still holds as

$$
posnU_k = \frac{1}{|U_k|} \left(|U_l| posnU_l + \sum_{i \in U_k \setminus U_l} x_i \right)
$$

$$
= \frac{1}{|U_k|} \left(|U_l| (\frac{1}{|U_l|} \sum_{j \in U_l} x_j) + \sum_{i \in U_k \setminus U_l} x_i \right) = \frac{1}{|U_k|} \sum_{i \in U_k} x_i
$$

$\square$

We now show that this results in a valid gradient-projection method.

Lemma 2. *If the result of the call project(x^0,lev) is x then x is the closest point to x^0 satisfying the ordering constraints defined by lev.*

Proof. (Sketch) We must prove that x minimizes $F(x) = \sum_{i=1}^n (x_i - x_i^0)^2$ subject to satisfying the ordering constraints. It follows from the construction that x satisfies the ordering constraints. Proving optimality is more difficult. Let $u_1, \ldots, u_{m-1}$ be new variables, one for each partition k. We set values to the new variables by setting u_k to be $max\{x_i \mid lev[i] = k\}$.

Recall that if we are minimizing a function F with a set of convex equalities C over variables X, then we can associate a variable λ_c called the Lagrange multiplier with each $c \in C$. Given a solution x we have that this is a minimal solution iff there exist values for the Lagrange multipliers satisfying

$$
\frac{\partial F}{\partial x} = \sum_{c \in C} \lambda_c \frac{\partial c}{\partial x} \tag{11}
$$

for each variable $x \in X$. Furthermore, if we also allow inequalities then the above statement continues to hold as long as $\lambda_c \geq 0$ for all inequalities c of form $c(x) \geq 0$. By definition an inequality c which is not active, i.e., $c(x) > 0$ has $\lambda_c = 0$. These are known as the Karush-Kuhn-Tucker conditions; see [1]. We now prove that x minimizes $F(x)$ subject to, for $k = 1, \ldots, m - 1$:

$$
u_{k-1} \leq u_k \text{ if } k > 1
$$
$$
x_i \leq u_k \text{ for all } i \text{ s.t. } lev[i] = k
$$
$$
x_i \geq u_k \text{ for all } i \text{ s.t. } lev[i] = k + 1
$$

These constraints are equivalent to the ordering constraints.

We show optimality by giving values for all λ_c satisfying Equation (11). An inequality $x_i \leq u_k$ or $x_i \geq u_k$ is active if $i \in U_k \setminus U_{k-1}$. Note that we can have $U_k \subseteq U_{k+1}$, in which case we must be careful to make the right constraint active so as to ensure that each x_i will be involved in no more than one active constraint. For a constraint c of form $x_i \geq u_k$ we set $\lambda_c = \frac{\partial F}{\partial x_i}$ and for c of form $x_i \leq u_k$ we set $\lambda_c = -\frac{\partial F}{\partial x_i}$. The constraint c of form $u_k \leq u_{k+1}$ is active if $U_k \subseteq U_{k+1}$. We set $\lambda_c = -\sum_{i \in U_k} \frac{\partial F}{\partial x_i}$. For all other inequalities c we set $\lambda_c = 0$. We give an extended formal proof of this lemma in [4].

We can now prove the correctness of *solve_QPOC*:

Theorem 1. *solve_QPOC converges to an optimal solution to the input QPOC Problem.*

Proof. Lemma 2 ensures that *solve_QPOC* is a gradient projection method. We now show that a more general proof of convergence for gradient projection methods holds for our specific stepsize calculations. First consider a variant of *solve_QPOC* in which s is always 1 — note that for both constant s and the calculation of s used in Figure 1 the method is equivalent to standard steepest-descent in the case when no active constraints are encountered. With constant $s = 1$ the computation of α implements a Limited Minimization Rule and so from [1–Proposition 2.3.1] every limit point of *solve_QPOC* is a stationary point. Since the original problem is convex any stationary point is an optimal solution. Now consider our computation of s. To ensure convergence we must prove that if $s^k \to 0$ where s^k is the value of s in the kth iteration then the limit point of *solve_QPOC* is a stationary point. But since the computation of s^k is also an example of the Limited Minimization Rule on the unconstrained problem, $s^k \to 0$ only if the limit point of *solve_QPOC* is a stationary point for the unconstrained problem, in which case it must also be a limit point of the constrained problem.

$\square$

4.1 Running Time

The second part of the algorithm, satisfying the constraints, can be performed in $O(mn + n \log n)$ time. However each complete iteration is dominated by computing the desired positions which takes $O(n^2)$ time. This is of course the inherent complexity of the stress function that contains $O(n^2)$ terms. (In fact, this is the same as the complexity of an iteration of the conjugate-gradient method, which is used in the unconstrained majorization algorithm.) In practice only few (5-30) iterations are required to return the optimal solution depending on the threshold on $||x - \hat{x}||$. Running times for graphs with various sizes and with varying numbers of boundaries m are given in Table 1. We compare results for those obtained with the *solve_QPOC* algorithm implemented in C

Table 1. A comparison of results obtained for arranging various graphs with *solve_QPOC* and the *Mosek* interior point method. Times are measured in seconds.

graph	#nodes (n)	#levels (m)	Solve_QPOC		Mosek	
			Time	Stress	Time	Stress
1138bus	1138	231	4.53	74343	209	74374
nos4	100	34	0.14	216.5	2.75	216.8
nos5	468	256	2.172	8517.3	13.0	8614.6
dwa512	512	14	1.23	22464	37.7	22464
dwb512	512	19	1.57	15707	90.8	16418
NSW Rail	312	54/76 (x/y-axis)	4.92	2288	18.6	2274.5
Backbone	2603	2373/1805 (x/y-axis)	55.8	1246960	> 1000	

and the Mosek interior-point quadratic programming solver [14]. Tests were conducted on a 2GHz P4-M notebook PC. As expected, since both solvers return the optimal or near optimal solution, the resulting drawings look identical. However, the dedicated *solve_QPOC* algorithm significantly outperformed the generic solver. The final "stress" value is given as a rough measure of relative quality. Note that this is the final stress value after being monotonically reduced by a number of iterations of the functional-majorization method. Sample graphs were obtained from the Matrix Market [2] (Such as *1138bus* as shown in Figure 4) and some graphs based on geographic coordinates which are shown in Figures 5 and 6.

5   Applications

5.1   Directed Graph Drawing

The method and motivation for drawing directed graphs by constrained majorization is discussed at length in [5]. Generally, a digraph can be said to induce a hierarchical structure on its nodes based on the precedence relationships defined by its directed edges. Consequently, an appropriate depiction of a digraph allocates the y-axis to showing this hierarchy. Thus, if node i precedes node j in the hierarchy, then i will be drawn above j on the y-axis; see, e.g., Sugiyama et al. [12]. This usually leads to the majority of directed edges pointing downwards, thereby showing a clear flow from top to bottom. There are a few possibilities for computing the hierarchical ordering of the nodes. We base our ordering on the "optimal arrangement" suggested by Carmel et al. [3]. Then, we compute the 2-D layout that minimizes the stress, while the y-coordinates of the nodes must obey their hierarchical ordering.

It was shown that this method produces drawings with much more uniform edge lengths making connectivity in large graphs more visible than in drawings produced by standard hierarchical graph drawing techniques.

We reproduce some example graphs drawn in this style and compare performance of our *solve_QPOC* algorithm with that of the solver previously used. Figure 3 illustrates the concept with a small directed graph containing a cycle. Note that since all nodes in

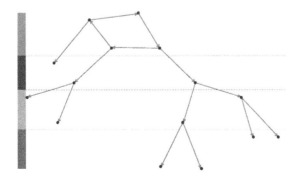

Fig. 3. A directed graph arranged using orthogonal ordering constraints in just the vertical dimension to preserve layering. The color bars on the left side indicate the layer-bands and the faint horizontal lines indicate the boundaries between these layers.

the cycle are in the same hierarchical level they are drawn within the same band. Figure 4 shows a much larger example from the matrix market collection [2].

Fig. 4. The *1138bus* graph (1138 nodes, 1458 edges) from the Matrix market collection[2], displayed as a directed graph

5.2   Layouts Preserving the Orthogonal Ordering

Sometimes a graph has meaningful coordinates. These might be natural physical coordinates associated with the nodes, or just a given layout with which the user is familiar.

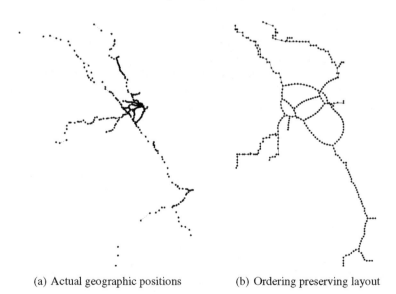

(a) Actual geographic positions   (b) Ordering preserving layout

Fig. 5. The New South Wales rail network (312 nodes, 322 edges) shown with actual geographic positions (left) and then refined using stress minimization with orthogonal ordering constraints (right)

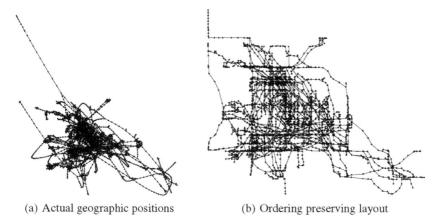

(a) Actual geographic positions   (b) Ordering preserving layout

Fig. 6. A backbone network (2603 nodes, 2931 edges). Left picture is based on the actual geographic coordinates while the right picture is based on ordering-preserving constrained stress minimization.

We want to improve the readability of the given layout while keeping its overall structure, thus preserving the user's mental map and/or natural properties of the layout. A way to achieve these goals is to minimize the stress of the graph, while preserving the original vertical and horizontal ordering of the nodes. These can be achieved by our algorithm. We provide here two examples of refining layouts with meaningful physical coordinates.

The first example involves automatic production of rail network maps. This problem has been tackled as a graph drawing problem by Hong et al. [8]. To produce print quality drawings the authors seek to satisfy quite complex aesthetic requirements such as effective labelling, edges strictly aligned to axes or diagonals and no induced crossings. However, as illustrated in Figure 5, simple orthogonal ordering also goes a long way to improving these diagrams. Note that the underlying geographic relationships are still evident while paths have been straightened and complex sections enlarged.

The second example is an internet backbone network as shown in Figure 6. The layout based on original coordinates contains very dense areas. However, readability is vastly improved by minimizing the stress, while original orthogonal order is preserved.

6   Conclusion and Further Work

We have demonstrated some applications of orthogonal-ordering constraints and that stress majorization can efficiently deal with such constraints. We are currently working on extending the algorithm to work for general separation constraints that may have many more applications, including clustered graph drawing — where we want to separate different clusters — and also cases where we want to restrict portions of the graph to specific rectangular regions. An obvious extension is to allow a wider variety of linear constraints. This would allow restricting portions of the graph to specific convex regions. However solving more general linear constraints requires a more sophisticated algorithm. Active-set techniques [13] may prove promising in this area.

Acknowledgements

Thanks to Damian Merrick for the NSW rail network data and members of the Adaptive Diagrams group at Monash University for their advice and support.

References

1. D. P. Bertsekas, *Nonlinear Programming*, Athena Scientific, 2nd Edition (1999).
2. R. Boisvert, R. Pozo, K. Remington, R. Barrett and J. Dongarra, "The Matrix Market: A web resource for test matrix collections", in *Quality of Numerical Software, Assessment and Enhancement*, Chapman Hall (1997) 125–137.
3. L. Carmel, D. Harel and Y. Koren, "Combining Hierarchy and Energy for Drawing Directed Graphs", *IEEE Trans. Visualization and Computer Graphics* **10** (2004) 46–57.
4. T. Dwyer, Y. Koren and K. Marriott, "Stress Majorization with Orthogonal Ordering Constraints", Technical Report 2005/175, Monash University School of Computer Science and Software Engineering (2005). Available from www.csse.monash.edu.au/~tdwyer
5. T. Dwyer and Y. Koren, "DIG-COLA: Directed Graph Layout through Constrained Energy Minimization", *IEEE Symposium on Information Visualization (Infovis'05)* (To appear 2005).
6. E. Gansner, Y. Koren and S. North, "Graph Drawing by Stress Majorization", *Proc. 12th Int. Symp. Graph Drawing (GD'04)*, LNCS 3383, Springer Verlag (2004) 239–250.
7. W. He and K. Marriott, "Constrained Graph Layout", *Constraints* **3** (1998) 289–314.
8. S. Hong, D. Merrick and H. Nascimento, "The metro map layout problem", *Proc. 12th Int. Symp. Graph Drawing (GD'04)*, LNCS 3383, Springer Verlag (2004) 482–491.
9. T. Kamada and S. Kawai, "An Algorithm for Drawing General Undirected Graphs", *Information Processing Letters* **31** (1989) 7–15.
10. K. Misue, P. Eades, W. Lai, and K. Sugiyama, "Layout Adjustment and the Mental Map", *Journal of Visual Languages and Computing* **6** (1995) 183–210.
11. K. Ryall, J. Marks and S. M. Shieber, "An Interactive Constraint-Based System for Drawing Graphs", *ACM Symposium on User Interface Software and Technology* (1997) 97–104.
12. K. Sugiyama, S. Tagawa and M. Toda, "Methods for Visual Understanding of Hierarchical Systems", *IEEE Trans. Systems, Man, and Cybernetics* **11** (1981) 109–125.
13. J. Nocedal, S. Wright, *Numerical Optimization*, Springer (1999).
14. Mosek Optimization Toolkit V3.2 www.mosek.com.

Fast Node Overlap Removal

Tim Dwyer[1], Kim Marriott[1], and Peter J. Stuckey[2]

[1] School of Comp. Science & Soft. Eng., Monash University, Australia
{tdwyer, marriott}@mail.csse.monash.edu.au
[2] NICTA Victoria Laboratory
Dept. of Comp. Science & Soft. Eng., University of Melbourne, Australia
pjs@cs.mu.oz.au

Abstract. The problem of node overlap removal is to adjust the layout generated by typical graph drawing methods so that nodes of non-zero width and height do not overlap, yet are as close as possible to their original positions. We give an $O(n \log n)$ algorithm for achieving this assuming that the number of nodes overlapping any single node is bounded by some constant. This method has two parts, a constraint generation algorithm which generates a linear number of "separation" constraints and an algorithm for finding a solution to these constraints "close" to the original node placement values. We also extend our constraint solving algorithm to give an active set based algorithm which is guaranteed to find the optimal solution but which has considerably worse theoretical complexity. We compare our method with convex quadratic optimization and force scan approaches and find that it is faster than either, gives results of better quality than force scan methods and similar quality to the quadratic optimisation approach.

Keywords: graph layout, constrained optimization, separation constraints.

1 Introduction

Graph drawing has been extensively studied over the last twenty years [1]. However, most research has dealt with *abstract graph layout* in which nodes are treated as points. Unfortunately, this is inadequate in many applications since nodes frequently have labels or icons and a layout for the abstract graph may lead to overlaps when these are added. While a few attempts have been made at designing layout algorithms that consider node size (e.g. [2, 3, 4]), the approaches are specific to certain layout styles and to the best of the authors' knowledge none are perfect in all situations.

For this reason, a number of papers, e.g. [5, 6, 7, 8, 9, 10], have described algorithms for performing *layout adjustment* in which an initial graph layout is modified so that node overlapping is removed. The underlying assumption is that the initial graph layout is good so that this layout should be preserved when removing the node overlap. Lyons et al.[10] offered a technique based on iteratively moving nodes to the centre of their Voronoi cells until crossings are removed.

P. Healy and N.S. Nikolov (Eds.): GD 2005, LNCS 3843, pp. 153–164, 2005.

Misue et al. [5] propose several models for a user's "mental map" based on *orthogonal ordering, proximity relations* and *topology* and define a simple heuristic *Force Scan* algorithm (FSA) for node-overlap removal that preserves orthogonal ordering. Hayashi et al. [7] propose a variant algorithm (FSA′) that produces more compact drawings while still preserving orthogonal ordering. They also show that this problem is NP-complete. Various other improvements to the FSA method exist and a survey is presented by Li et al. [11]. More recently, Marriott et al. [6] investigated a quadratic programming (QP) approach which minimises displacement of nodes while satisfying non-overlap constraints. Their results demonstrate that the technique offers results that are preferable to FSA in a number of respects, but require significantly more processing time. In this paper we address the last issue.

Our contribution consists of two parts: first, we detail a new algorithm for computing the linear constraints to ensure non-overlap in a single dimension. This has worst case complexity $O(n \log n)$ where n is the number of nodes and generates $O(n)$ non-overlap constraints — assuming that the number of nodes overlapping a single node is bounded by some constant k. Previous approaches have had quadratic or cubic complexity and as far as we are aware it has not been previously realized that only a linear number of non-overlap constraints are required. Each non-overlap constraint has the form $u + a \leq v$ where u and v are variables and $a \geq 0$ is a constant. Such constraints are called separation constraints. Our second contribution is to give a simple algorithm for solving quadratic programming problems of the form: minimize $\sum_{i=1} v_i.weight \times (v_i - v_i.des)^2$ subject to a conjunction of separation constraints over variables $v_1, \ldots, v_n$ where $v_i.des$ is the desired value of variable v_i and $v_i.weight \geq 0$ the relative importance. We show that in practice this algorithm produces optimal solutions to the quadratic program much faster than generic solvers, but also that first part of the algorithm can be run alone to produce near optimal solutions in $O(n \log n)$ time.

2   Background

We assume that we are given a graph G with nodes $V = \{1, \ldots, n\}$, a width, w_v, and height, h_v, for each node $v \in V$,[1] and an initial layout for the graph G, in which each node $v \in V$ is placed at (x_v^0, y_v^0) and $u \neq v \Rightarrow (x_u^0, y_u^0) \neq (x_v^0, y_v^0)$.

We are concerned with layout adjustment: we wish to preserve the initial graph layout as much as possible while removing all node label overlapping. A natural heuristic to use for preserving the initial layout is to require that nodes are moved as little as possible. This corresponds to the Proximity Relations mental map model of Misue et al. [5].

Following [6] we define the *layout adjustment problem* to be the constrained optimization problem: minimize ϕ_{change} subject to C^{no} where the variables of the layout adjustment problem are the x and y coordinates of each node $v \in V$, x_v and y_v respectively, and the objective function minimizes node movement

---

[1] Any extra padding required to ensure a minimal separation between nodes is included in w_v and h_v.

$\phi_{change} = \phi_x + \phi_y = \sum_{v \in V} (x_v - x_v^0)^2 + (y_v - y_v^0)^2$, and the constraints C^{no} ensure that there is no node overlapping. That is, for all $u, v \in V$, $u \neq v$ implies

$$x_v - x_u \geq \tfrac{1}{2}(w_v + w_u) \ (v \text{ right of } u) \vee x_u - x_v \geq \tfrac{1}{2}(w_v + w_u) \ (u \text{ right of } v)$$
$$\vee \ y_v - y_u \geq \tfrac{1}{2}(h_v + h_u) \ (v \text{ above } u) \quad \vee \ y_u - y_v \geq \tfrac{1}{2}(h_v + h_u) \ (u \text{ above } v)$$

A variant of this problem is when we additionally require that the new layout preserves the *orthogonal ordering* of nodes in the original graph, i.e., their relative ordering in the x and y directions. This is a heuristic to preserve more of the original graph's structure. Define $C_x^{oo} = \bigwedge \{x_v \geq x_u \mid x_v^0 \geq x_u^0\}$ and C_y^{oo} equivalently for y. The orthogonal ordering problem adds $C_x^{oo} \wedge C_y^{oo}$ to the constraints to solve.

Our approach to solving the layout adjustment problem is based on [6] where quadratic programming is used to solve a linear approximation of the layout adjustment problem. There are two main ideas behind the quadratic programming approach. The first is to approximate each non-overlap constraint in C^{no} by one of its disjuncts. The second is to separate treatment of the x and y dimensions, by breaking the optimization function and constraint set into two parts. Separating the problem in this way improves efficiency by reducing the number of constraints considered in each problem and if we solve for the x direction first, it allows us to delay the computation of C_y^{no} to take into account the node overlapping which has been removed by the optimization in the x direction.

3   Generating Non-overlap Constraints

We generate the non-overlap constraints in each dimension in $O(|V| \log |V|)$ time using a line-sweep algorithm related to standard rectangle overlap detection methods [12]. First, consider the generation of horizontal constraints. We use a vertical sweep through the nodes, keeping a horizontal "scan line" list of open nodes with each node having references to its closest left and right neighbors (or more exactly the neighbors with which it is currently necessary to generate a non-overlap constraint). When the scan line reaches the top of a new node, this is added to the list and its neighbors computed. When the bottom of a node is reached the the separation constraints for the node are generated and the node is removed from the list.

The detailed algorithm is shown on the left of Figure 1. It uses a vertically sorted list of events to guide the movement of the *scan_line*. An event is a record with three fields, *kind* which is either *open* or *close* respectively indicating whether the top or bottom of the node has been reached, *node* which is the node name, and *posn* which is the vertical position at which this happens.

The *scan_line* stores the currently open nodes. We use a red-black tree to provide $O(\log |V|)$ *insert*, *remove*, *next_left* and *next_right* operations. The functions *new*, *insert* and *remove* create and update the scan line. The functions *next_left(scan_line, v)* and *next_right(scan_line, v)* return the closest neighbors to each side of node v in the scan line.

```
procedure generate_C_x^no(V)
events := { event(open, v, y_v − h_v/2),
                event(close, v, y_v + h_v/2) | v ∈ V}
[e_1, . . . , e_2n] := events sorted by posn
scan_line := new()
for each e_1, . . . , e_2n do
    v := e_i.node
    if e_i.kind = open then
        scan_line := insert(scan_line, v)
        leftv := get_left_nbours(scan_line, v)
        rightv := get_right_nbours(scan_line, v)
        left[v] := leftv
        for each u ∈ leftv do
            right[u] := (right[u] ∪ {v}) \ rightv
        right[v] := rightv
        for each u ∈ rightv do
            left[u] := (left[u] ∪ {v}) \ leftv
    else /* e_i.kind = close */
        for each u ∈ left[v] do
            generate x_u + (w_u + w_v)/2 ≤ x_v
            right[u] := right[u] \ {v}
        for each u ∈ right[v] do
            generate x_v + (w_u + w_v)/2 ≤ x_u
            left[u] := left[u] \ {v}
        scan_line := remove(scan_line, v)
return

function get_left_nbours(scan_line, v)
u := next_left(scan_line, v)
while u ≠ NULL do
    if olap_x(u, v) ≤ 0 then
        leftv := leftv ∪ {u}
        return leftv
    if olap_x(u, v) ≤ olap_y(u, v) then
        leftv := leftv ∪ {u}
    u := next_left(scan_line, u)
return leftv
```

```
procedure satisfy_VPSC(V,C)
[v_1, . . . , v_n] := total_order(V,C)
for i:= 1, . . . , n do
    merge_left(block(v_i))
return [v_1 ← posn(v_1), . . . , v_n ← posn(v_n)]

procedure merge_left(b)
while violation(top(b.in)) > 0 do
    c := top(b.in)
    b.in := remove(c)
    bl := block[left(c)]
    distbltob := offset[left(c)] + gap(c)
                 − offset[right(c)]
    if b.nvars > bl.nvars then
        merge_block(b, c, bl, −distbltob)
    else
        merge_block(bl, c, b, distbltob)
        b := bl
return

procedure merge_block(p, c, b, distptob)
p.wposn := p.wposn + b.wposn −
            distptob × b.weight
p.weight := p.weight + b.weight
p.posn := p.wposn/p.weight
p.active := p.active ∪ b.active ∪ {c}
for v ∈ b.vars do
    block[v] := p
    offset[v] := distptob + offset[v]
p.in := merge(p.in, b.in)
p.vars := p.vars ∪ b.vars
p.nvars := p.nvars + b.nvars
return
```

Fig. 1. Algorithm $generate\_C_x^{no}(V)$ to generate horizontal non-overlap constraints between nodes in V, and algorithm $satisfy\_VPSC(V, C)$ to satisfy the Variable Placement with Separation Constraints (VPSC) problem

The functions $get\_left\_nbours(scan\_line, v)$ and $get\_right\_nbours(scan\_line, v)$ detect the neighbours to each side of node v that require non-overlap constraints. These are heuristics. It seems reasonable to set up a non-overlap constraint with the closest non-overlapping node on each side and a subset of the overlapping nodes. One choice for get_left_nbours is shown in Figure 1. This makes use of the functions $olap_x(u, v) = (w_u + w_v)/2 - |x_u^0 - x_v^0|$ and $olap_y(u, v) = (h_u + h_v)/2 - |y_u^0 - y_v^0|$ which respectively measure the horizontal and vertical overlap between nodes u and v. The main loop iteratively searches left until the first non-overlapping node to the left is found or else there are no more nodes. Each overlapping node u found on the way is collected in $leftv$ if the horizontal overlap between u and v less than the vertical overlap. The arrays $left$ and $right$ detail for each open node v the nodes to each side for which non-overlap constraints should be generated. The only subtlety is that redundant constraints are removed, i.e. if there is currently a non-overlap constraint between any $u \in leftv$ and $u' \in rightv$ then it can be removed since it will be implied by the two new non-overlap constraints between u and v and v and u'.

Theorem 1. *The procedure generate_$C_x^{no}(V)$ has worst-case complexity $O(|V| \cdot k(\log|V| + k)$ where k is the maximum number of nodes overlapping a single node with appropriate choice of heap data structure. Furthermore, it will generate $O(k \cdot |V|)$ constraints.*

Proofs to theorems are provided in the technical report [13]. Assuming k is bounded, the worst case complexity is $O(|V| \log|V|)$.

Theorem 2. *The procedure generate_$C_x^{no}(V)$ generates separation constraints C that ensure that if two nodes do not overlap horizontally in the initial layout then they will not overlap in any solution to C.*

The code for *generate_C_y^{no}*, the procedure to generate vertical non-overlap constraints is essentially dual to that of *generate_C_x^{no}*. The only difference is that any remaining overlap must be removed vertically. This means that we need only find the closest node in the analogue of the functions *get_left_nbours* and *get_right_nbours* since any other nodes in the scan line will be constrained to be above or below these. This means that the number of left and right neighbours is always 1 or less and gives us the following complexity results:

Theorem 3. *The procedure generate_$C_y^{no}(V)$ has worst-case complexity $O(|V| \cdot \log|V|)$. Furthermore, it will generate no more than $2 \cdot |V|$ constraints.*

Theorem 4. *The procedure generate_$C_y^{no}(V)$ generates separation constraints C that ensure that no nodes will overlap in any solution to C.*

4   Solving Separation Constraints

Non-overlap constraints c have the form $u + a \leq v$ where u, v are variables and $a \geq 0$ is the minimum gap between them. We use the notation $left(c)$, $right(c)$ and $gap(c)$ to refer to u, v and a respectively. Such constraints are called *separation constraints*. We must solve the following constrained optimization problem for each dimension:

> *Variable placement with separation constraints (VPSC) problem.* Given n variables $v_1, \ldots, v_n$, a weight $v_i.weight \geq 0$ and a desired value $v_i.des^2$ for each variable and a set of separation constraints C over these variables find an assignment to the variables which minimizes $\sum_{i=1}^{n} v_i.weight \times (v_i - v_i.des)^2$ subject to C.

We can treat a set of separation constraints C over variables V as a weighted directed graph with a node for each $v \in V$ and an edge for each $c \in C$ from $left(c)$ to $right(c)$ with length $gap(c)$. We call this the *constraint graph*. We define $out(v) = \{c \in C \mid left(c) = v\}$ and $in(v) = \{c \in C \mid right(c) = v\}$. Note that edges in this graph are *not* the edges in the original graph.

We restrict attention to VPSC problems in which the constraint graph is acyclic and for which there is at most one edge between any pair of variables.

---

[2] $v_i.des$ is set to x_{vi}^0 or y_{vi}^0 for each dimension, as used in *generate_$C_{\{x|y\}}^{no}$*.

It is possible to transform an arbitrary satisfiable VPSC problem into a problem of this form and our generation algorithm will generate constraints with this property. Since the constraint graph is acyclic it imposes a partial order on the variables: we define $u \preceq_C v$ iff there is a (directed) path from u to v using the edges in separation constraint set C. We will make use of the function $total\_order(V,C)$ which returns a total ordering for the variables in V, i.e. it returns a list $[v_1, \ldots, v_n]$ s.t. for all $j > i$, $v_j \not\preceq_C v_i$.

We first give a fast algorithm for finding a solution to the VPSC algorithm which satisfies the separation constraints and which is "close" to optimal. The algorithm works by merging variables into larger and larger "blocks" of contiguous variables connected by a spanning tree of active constraints, where a constraint $u + a \le v$ is active if at the current position for u and v, $u + a = v$.

The algorithm is shown in Figure 1. It takes as input a set of separation constraints C and a set of variables V. A block b is a record with the following fields: *vars*, the set of variables in the block; *nvars*, the size of *vars*; *active*, the set of constraints between variables in *vars* forming the spanning tree of active constraints; *in*, the set of constraints $\{c \in C \mid right(c) \in b.vars$ and $left(c) \notin b.vars\}$; *out*, out-going constraints defined symmetrically to *in*; *posn*, the position of the block's "reference point"; *wposn*, the sum of the weighted desired locations of variables in the block; and *weight*, the sum of the weights of the variables in the block.

In addition, the algorithm uses two arrays *blocks* and *offset* indexed by variables where $block[v]$ gives the block of variable v and $offset[v]$ gives the distance from v to its block's reference point. Using these we define the function $posn(v) = block(v).posn + offset[v]$ giving the current position of variable v.

The constraints in the field $b.in$ for each block b are stored in a priority queue such that the top constraint in the queue is always the most violated where $violation(c) = left(c) + gap(c) - right(c)$. We use four queue functions: $new()$ which returns a new queue, $add(q, C)$ which inserts the constraints in the set C into the queue q and returns the result, $top(q)$ which returns the constraint in q with maximal violation, $remove(q)$ which deletes the top constraint from q, and $merge(q_1, q_2)$ which returns the queue resulting from merging queues q_1 and q_2. The only slight catch is that some of the constraints in $b.in$ may be *internal* constraints, i.e. constraints which are between variables in the same block. Such internal constraints are removed from the queue when encountered. Another caveat is that when a block is moved *violation* changes value. However, the ordering induced by $violation(c)$ does not change since all variables in the block will be moved by the same amount and so $violation(c)$ will be changed by the same amount for all non-internal constraints. This consistent ordering allows us to implement the priority queues as *pairing heaps* [14] with efficient support for the above operations.

The main procedure, $satisfy\_VPSC$, processes the variables from smallest to greatest based on a total order reflecting the constraint graph. At each stage the invariant is that we have found an assignment to $v_1, .., v_{i-1}$ which satisfies the separation constraints. We process vertex v_i as follows. First, function $block$ is

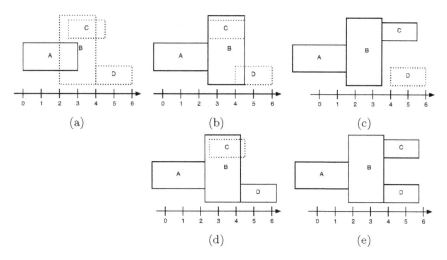

Fig. 2. Example of (non-optimal) algorithm for VPSC problem giving optimal (c) or non-optimal (e) answer

used to create a block b for each v_i setting $b.posn = v_i.des$. Some of the "in" constraints may be violated. If so, we find the most violated constraint c and merge the two blocks connected by c using the function *merge_block*. We repeat this until the block no longer overlaps the preceding block, in which case we have found a solution to $v_1, .., v_i$.

At each step we set $b.posn$ for each block b to the optimum position, i.e. the weighted average of the desired positions: $\frac{\sum_{i=1}^{k} v_i.weight \times (offset[v_i] - v_i.des)}{\sum_{i=1}^{k} v_i.weight}$. By maintaining the fields *wposn* and *weight* we are able to efficiently compute the weighted arithmetic mean when merging two blocks.

Example 1. Consider the example of laying out the boxes A,B,C,D shown in Figure 2(a) each shown at their desired position 1.5, 3, 3.5, and 5 respectively and assuming the weights on the boxes are 1,1,2 and 2 respectively. The constraints generated by *generate_C$_x^{no}$* are $c_1 \equiv v_A + 2.5 \leq v_B$, $c_2 \equiv v_B + 2 \leq v_C$ and $c_3 \equiv v_B + 2 \leq v_D$. Assume the algorithm chooses the total order A,B,C,D. First we add block A, it is placed at its desired position as shown in Figure 2(a). Next we consider block B, $b.in = \{c_1\}$ and the violation of this constraint is 1. We retrieve bl as the block containing A. and calculate *distbltob* as 2.5. We now merge block B into the block containing A. The new block position is 1 as shown in Figure 2(b), and c_1 is added to the active constraints. Next we consider block C, we find it must merge with block AB. The new positions are shown in Figure 2(c). Since there is no violation with the block D, the final position leaves it where it is, i.e. the result is optimal.

Theorem 5. *The assignment to the variables V returned by satisfy_VPSC(V, C) satisfies the separation constraints C.*

```
procedure solve_VPSC(V,C)                          procedure compute_lm()
satisfy_VPSC(V,C)                                  for each c ∈ C do lm[c] := 0 endfor
compute_lm()                                       for each block b do
while exists c ∈ C s.t. lm[c] < 0 do                   choose v ∈ b.vars
    choose c ∈ C s.t. lm[c] < 0                         comp_dfdv(v, b.active, NULL)
    b := block[left(c)]
    lb := restrict_block(b, left(b, c))            function comp_dfdv(v, AC, u)
    rb := restrict_block(b, right(b, c))           dfdv := v.weight × (posn(v) − v.des)
    rb.posn := b.posn                              for each c ∈ AC s.t. v = left(c)
    rb.wposn := rb.posn × rb.weight                         and u ≠ right(c) do
    merge_left(lb)                                     lm[c] := comp_dfdv(right(c), AC, v)
    /* original rb may have been merged */             dfdv := dfdv + lm[c]
    rb := block[right(c)]                          for each c ∈ AC s.t. v = right(c)
    rb.wposn := ∑_{v∈rb} v.weight × (v.des − offset[v])     and u ≠ left(c) do
    rb.posn := rb.wposn/rb.weight                      lm[c] := − comp_dfdv(left(c), AC, v)
    merge_right(rb)                                     dfdv := dfdv − lm[c]
    compute_lm()                                   return dfdv
endwhile
return [v_1 ← posn(v_1), . . . , v_n ← posn(v_n)]
```

Fig. 3. Algorithm to find an optimal solution to a VPSC problem with variables V and separation constraints C

Theorem 6. *The procedure* satisfy_VPSC(V, C) *has worst-case complexity* $O(|V| + |C| \log |C|)$ *with appropriate choice of priority queue data structure.*

Since each block is placed at its optimal position one might hope that the solution returned by *satisfy_VPSC* is also optimal. This was true for the example above. Unfortunately, as the following example shows it is not always true.

Example 2. Consider the same blocks as in Example 1 but with the total order A,B,D,C. The algorithm works identically to the stage shown in Figure 2(b). But now we consider block D, which overlaps with block AB. We merge the blocks to create block ABD which is placed at 0.75, as shown in Figure 2(d). Now block ABD overlaps with block C so we merge the two to the final position 0.166 as shown in Figure 2(e). The result is not optimal.

The solution will be non-optimal if it can be improved by splitting a block. This may happen if a merge becomes "invalidated" by a later merge. It is relatively straight-forward to check if a solution is optimal by computing the Lagrange multiplier λ_c for each constraint c. We must split a block at an active constraint c if λ_c is negative. Because of the simple nature of the separation constraints it is possible to compute λ_c (more exactly $\lambda_c/2$) for the active constraints in each block in linear time. We simply perform a depth-first traversal of the constraints in $b.active$ summing $v.weight \times (posn(v) - v.des)$ for the variables below this variable in the tree. The algorithm is detailed in Figure 3. It assumes the data structures in *satisfy_VPSC* and stores $\lambda_c/2$ in the $lm[c]$ for each $c \in C$. A full justification for this given in [13].

Using this it is relatively simple to extend *satisfy_VPSC* so that it computes an optimal solution. The algorithm is given in Figure 3. This uses *satisfy_VPSC* to find an initial solution to the separation constraints and calls *compute_lm* to compute the Lagrange multipliers. The main while loop checks if the current solution is optimal, i.e. if for all $c \in C$, $\lambda_c \geq 0$, and if so the algorithm terminates. Otherwise one of the constraints $c \in C$ with a negative Lagrange multiplier is chosen (we choose c corresponding to $min\{\lambda_c | \lambda_c < 0, c \in C\}$) and the block b containing c is split into two new blocks, lb and rb populated by $left(b, c)$ and $right(b, c)$ respectively. We define $left(b, c)$ to be the nodes in $b.vars$ connected by a path of constraints from $b.active \setminus \{c\}$ to $left(c)$, i.e. the variables which are in the left sub-block of b if b is split by removing c. We define $right(b, c)$ symmetrically. The split is done by calling the procedure $restrict\_block(b, V)$ which takes a block b and returns a new block restricted to the variables $V \subseteq b.vars$. For space reasons we do not include the (straight-forward) code for this.

Now the new blocks lb and rb are placed in their new positions using the procedures $merge\_left$ and $merge\_right$. First we place lb. Since $lm[c] < 0$, lb wishes to move left and rb wishes to move right. We temporarily place rb at the former position of b and try and place lb at its optimal position. If any of the "in" constraints are violated (since lb wishes to move left the "out" constraints cannot be violated). We remedy this with a call to $merge\_left(lb)$. The placement of rb is totally symmetric, although we must first allow for the possibility that rb has been merged so we update it's reference to the (possibly new) container of $right(c)$ and place it back at its desired position. The code for $merge\_right$ has not been included since it is symmetric to that of $merge\_left$. We have also omitted references to the "out" constraint priority queues used by $merge\_right$. These are managed in an identical fashion to "in" constraints.

Example 3. Consider the case of Example 2. The result of $satisfy\_VPSC$ is shown in Figure 2(d). The Lagrange multipliers calculated for c_1, c_2, c_3 are 1.333, 2.333, and -0.333 respectively. We should split on constraint c_3. We break block ABCD into ABC and D, and placing them at their optimal positions leads to positions shown in Figure 2(c). Since there is no overlap the algorithm terminates.

Theorem 7. *Let θ be the assignment to the variables V returned by solve_VPSC (V, C). Then θ is an optimal solution to the VPSC Problem with variables V and constraints C*

Termination of $solve\_VPSC$ is a little more problematic. $solve\_VPSC$ is an example of an active-set approach to constrained optimization [15]. In practice such methods are fast and lend themselves to incremental re-computation but unfortunately, they may have theoretical exponential worst case behavior and at least in theory may not terminate if the original problem contains constraints that are redundant in the sense that the set of equality constraints corresponding to the separation constraints C, namely $\{u + a = v \mid (u + a \leq v) \in C\}$, contains redundant constraints. Unfortunately, our algorithm for constraint generation may generate equality-redundant constraints. We could remove such redundant separation constraints in a pre-processing step by adding ϵ^i to the gap for the ith separation constraint or else use a variant of lexico-graphic ordering to resolve which constraint to make active in the case of equal violation. We can then show that cycling cannot occur. In practice however we have never found a case of cycling and simply terminate the algorithm after a fixed maximum number of splits.

5   Results

We have compared our method[3] **SAT** = *satisfy_VPSC* and **SOL** = *solve_VPSC* versus **FSA**, the improved Push-Force Scan algorithm [7] and **QP** quadratic programming optimization using the Mosek solver [16]. For SAT, SOL and QP we compare with (**_OO**) and without orthogonal ordering constraints. We did not compare empirically with the Voronoi centering algorithm [10] since it gives very poor results, see Figure 4.

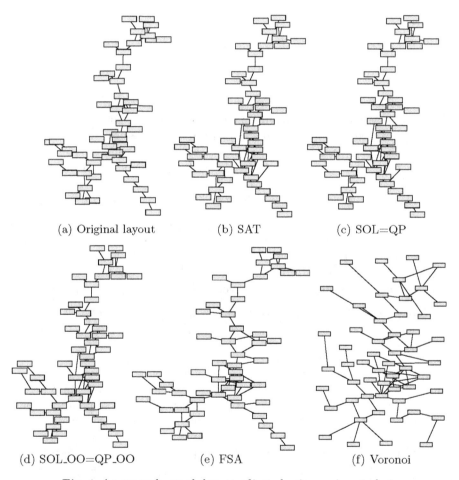

(a) Original layout   (b) SAT   (c) SOL=QP

(d) SOL_OO=QP_OO   (e) FSA   (f) Voronoi

Fig. 4. An example graph layout adjusted using various techniques

Figure 4 shows the initial layout and the results of the various node adjustment algorithms for a realistic example graph. There is little difference between the

---

[3] A C++ implementation of this algorithm is available from http://www.csse. monash.edu.au/~tdwyer.

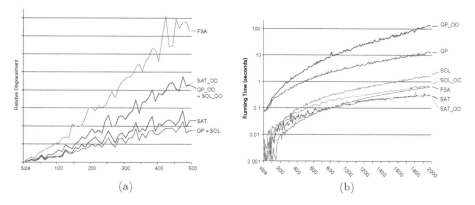

Fig. 5. Comparative (a) total displacement from original positions and (b) times

SAT and SOL results. We include a SOL result with the orthogonal ordering (SOL_OO) constraints which attacks the same problem as FSA. Clearly FSA produces much more spreadout layout. Lastly the Voronoi diagram approach loses most of the structure of the original layout.

Figure 5 gives running times and relative displacement from original position for the different methods on randomly generated sets of overlapping rectangles. We varied the number of rectangles generated but adjusted the size of the rectangles to keep k (the average number of overlaps per rectangle) appoximately constant ($k \approx 10$).

We can see that FSA produces the worst displacements, and that SAT produces very good displacements almost as good as the optimal produced by SOL and QP. We can see that SAT (with or without orthogonal ordering constraints) scales better than FSA. While both SOL and QP are significantly slower, SOL is an order of magnitude faster than QP in the range tested. Adding orthogonal ordering constraints seems to simplify the problem somewhat and SOL_OO requires less splitting than SOL while QP requires more processing time to handle extra constraints. Therefore SOL_OO is significantly faster than QP_OO and SAT_OO returns a solution very near to the optimal while remaining extremely fast. Overall these results show us that SAT is the fastest of all algorithms and gives very close to optimal results.

References

1. Battista, G.D., Eades, P., Tamassia, R., Tollis, I.G.: Graph Drawing: Algorithms for the Visualization of Graphs. Prentice Hall (1999)
2. Harel, D., Koren, Y.: Drawing graphs with non-uniform vertices. In: Proceedings of the Working Conference on Advanced Visual Interfaces (AVI'02), ACM Press (2002) 157–166
3. Friedrich, C., Schreiber, F.: Flexible layering in hierarchical drawings with nodes of arbitrary size. In: Proceedings of the 27th conference on Australasian computer science (ACSC2004). Volume 26., Australian Computer Society (2004) 369–376

4. Marriott, K., Moulder, P., Hope, L., Twardy, C.: Layout of bayesian networks. In: Twenty-Eighth Australasian Computer Science Conference (ACSC2005). Volume 38 of CRPIT., Australian Computer Society (2005) 97–106
5. Misue, K., Eades, P., Lai, W., Sugiyama, K.: Layout adjustment and the mental map. Journal of Visual Languages and Computing **6** (1995) 183–210
6. Marriott, K., Stuckey, P., Tam, V., He, W.: Removing node overlapping in graph layout using constrained optimization. Constraints **8** (2003) 143–171
7. Hayashi, K., Inoue, M., Masuzawa, T., Fujiwara, H.: A layout adjustment problem for disjoint rectangles preserving orthogonal order. In: GD '98: Proceedings of the 6th International Symposium on Graph Drawing, London, UK, Springer-Verlag (1998) 183–197
8. Lai, W., Eades, P.: Removing edge-node intersections in drawings of graphs. Inf. Process. Lett. **81** (2002) 105–110
9. Gansner, E.R., North, S.C.: Improved force-directed layouts. In: GD '98: Proceedings of the 6th International Symposium on Graph Drawing, London, UK, Springer-Verlag (1998) 364–373
10. Lyons, K.A.: Cluster busting in anchored graph drawing. In: CASCON '92: Proceedings of the 1992 conference of the Centre for Advanced Studies on Collaborative research, IBM Press (1992) 327–337
11. Li, W., Eades, P., Nikolov, N.: Using spring algorithms to remove node overlapping. In: Proceedings of the Asia-Pacific Symposium on Information Visualisation (APVIS2005). Volume 45 of CRPIT., Australian Computer Society (2005) 131–140
12. Preparata, F.P., Shamos, M.I. In: Computational Geometry. Springer (1985) 359–365
13. Dwyer, T., Marriott, K., Stuckey, P.J.: Fast node overlap removal. Technical Report 2005/173, Monash University, School of Computer Science and Software Engineering (2005) Available from www.csse.monash.edu.au/~tdwyer.
14. Weiss, M.A.: Data Structures and Algorithm Analysis in Java. Addison Wesley Longman (1999)
15. Fletcher, R.: Practical Methods of Optimization. Chichester: John Wiley & Sons, Inc. (1987)
16. ApS, M.: (Mosek optimisation toolkit v3.2) www.mosek.com.

Delta-Confluent Drawings[*]

David Eppstein, Michael T. Goodrich, and Jeremy Yu Meng

School of Information and Computer Science,
University of California, Irvine,
Irvine, CA 92697, USA
{eppstein, goodrich, ymeng}@ics.uci.edu

Abstract. We generalize the *tree-confluent* graphs to a broader class
of graphs called Δ-*confluent* graphs. This class of graphs and distance-
hereditary graphs, a well-known class of graphs, coincide. Some results
about the visualization of Δ-confluent graphs are also given.

1 Introduction

Confluent Drawing is an approach to visualize non-planar graphs in a planar
way [10]. The idea is simple: we allow groups of edges to be merged together
and drawn as tracks (similar to train tracks). This method allows us to draw, in
a crossing-free manner, graphs that would have many crossings in their normal
drawings. Two examples are shown in Figure. 1. In a confluent drawing, two
nodes are connected if and only if there is a smooth curve path from one to the
other without making sharp turns or double backs, although multiple realizations
of a graph edge in the drawing is allowed.

More formally, a curve is *locally-monotone* if it contains no self intersections
and no sharp turns, that is, it contains no point with left and right tangents that
form an angle less than or equal to 90 degrees. Intuitively, a locally-monotone
curve is like a single train track, which can make no sharp turns. Confluent
drawings are a way to draw graphs in a planar manner by merging edges together
into *tracks*, which are the unions of locally-monotone curves.

An undirected graph G is *confluent* if and only if there exists a drawing A
such that:

- There is a one-to-one mapping between the vertices in G and A, so that, for
  each vertex $v \in V(G)$, there is a corresponding vertex $v' \in A$, which has a
  unique point placement in the plane.
- There is an edge (v_i, v_j) in $E(G)$ if and only if there is a locally-monotone
  curve e' connecting v'_i and v'_j in A.
- A is planar. That is, while locally-monotone curves in A can share overlapping portions, no two can cross.

---

[*] Work by the first author is supported by NSF grant CCR-9912338. Work by the second and the third author is supported by NSF grants CCR-0098068, CCR-0225642,
and DUE-0231467.

P. Healy and N.S. Nikolov (Eds.): GD 2005, LNCS 3843, pp. 165–176, 2005.
© Springer-Verlag Berlin Heidelberg 2005

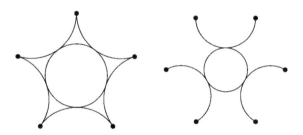

Fig. 1. Confluent drawings of K_5 and $K_{3,3}$

We assume readers have basic knowledge about graph theory and we will use conventional terms and notations of graph theory without defining them. All graphs considered in this paper are simple graphs, i.e., without loop or multi-edge. Confluent graphs are closely related to planar graphs. It is, however, very hard to check whether a given graph can be drawn confluently. The complexity of recognizing confluent graphs is still open and the problem is expected to be hard. Hui, Schaefer and Štefankovič [21] define the notion of *strong confluency* and show that strong confluency can be recognized in **NP**. It is then of interest to study classes of graphs that can or can not be drawn confluently. Several classes of confluent graphs, as well as several classes of non-confluent graphs, have been listed [10].

In this paper we continue in the positive direction of this route. We describe Δ-confluent graphs, a generalization of *tree-confluent* graphs [21]. We discuss problems of embedding trees with internal degree three, including embeddings on the hexagonal grid, which is related to Δ-confluent drawings with large angular resolution, and show that $O(n \log n)$ area is enough for a Δ-confluent drawing of a Δ-confluent graph with n vertices on the hexagonal grid.

Note that although the method of merging groups of edges is also used to reduce crossings in *confluent layered drawings* [14], edge crossings are allowed to exist in a confluent layered drawing.

2 Δ-Confluent Graphs

Hui, Schaefer and Štefankovič [21] introduce the idea of *tree-confluent* graphs. A graph is *tree-confluent* if and only if it is represented by a planar train track system which is topologically a tree. It is also shown in their paper that the class of tree-confluent graphs are equivalent to the class of chordal bipartite graphs.

The class of tree-confluent graphs can be extended into a wider class of graphs if we allow one more powerful type of junctions.

A Δ-*junction* is a structure where three paths are allowed to meet in a three-way complete junction. The connecting point is call a *port* of the junction. A Λ-*junction* is a broken Δ-junction where two of the three ports are disconnected from each other (exactly same as the *track* defined in the tree-confluent draw-ing [21]). The two disconnected paths are called *tails* of the Λ-junction and the remaining one is called *head*.

Fig. 2. Δ-junction and Λ-junction

A Δ-*confluent drawing* is a confluent drawing in which every junction in the drawing is either a Δ-junction, or a Λ-junction, and if we replce every junction in the drawing with a new vertex, we get a tree. A graph G is Δ-confluent if and only if it has a Δ-confluent drawing.

The class of cographs in [10] and the class of tree-confluent graphs in [21] are both included in the class of Δ-confluent graphs. We observe that the class of Δ-confluent graphs are equivalent to the class of distance-hereditary graphs.

2.1   Distance-Hereditary Graphs

A *distance-hereditary* graph is a connected graph in which every induced path is isometric. That is, the distance of any two vertices in an induced path equals their distance in the graph [2]. Other characterizations have been found for distance-hereditary graphs: forbidden subgraphs, properties of cycles, etc. Among them, the following one is most interesting to us:

Theorem 1. [2] *Let G be a finite graph with at least two vertices. Then G is distance-hereditary if and only if G is obtained from K_2 by a sequence of one-vertex extensions: attaching pendant vertices and splitting vertices.*

Here attaching a pendant vertex to x means adding a new vertex x' to G and making it adjacent to x so x' has degree one; and splitting x means adding a new vertex x' to G and making it adjacent to either x and all neighbors of x, or just all neighbors of x. Vertices x and x' forming a split pair are called *true twins* (or *strong siblings*) if they are adjacent, or *false twins* (or *weak siblings*) otherwise.

By reversing the above extension procedure, every finite distance-hereditary graph G can be reduced to K_2 in a sequence of one-vertex operations: either delete a pendant vertex or identify a pair of twins x' and x. Such a sequence is called an *elimination sequence* (or a *pruning sequence*).

In the example distance-hereditary graph G of Figure. 3, the vertices are labelled reversely according to an elimination sequence of G:

17 merged into 16, 16 merged into 15, 15 cut from 3, 14 cut from 2, 13 merged into 5, 12 merged into 6, 10 merged into 8, 11 merged into 7, 9 cut from 8, 8 merged into 7, 7 cut from 6, 6 merged into 0, 5 cut from 0, 4 merged into 1, 3 cut from 1, 2 merged into 1.

The following theorem states that the class of distance hereditary graphs and the class of Δ-confluent graphs are equivalent.

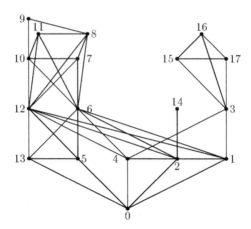

Fig. 3. A distance-hereditary graph G

Theorem 2. *A graph G is distance hereditary if and only if it is Δ-confluent.*

Proof sketch. Assume G is distance hereditary. We can compute the elimination sequence of G, then apply an algorithm, which will be described in Section 2.2, to get a Δ-confluent drawing of G. Thus G is Δ-confluent.

On the other hand, given a Δ-confluent graph G in form of its Δ-confluent drawing A, we can apply the following operations on the drawing A:

1. *contraction.* If two vertices y and y' in A are connected to two ports of a Δ-junction, or y and y' are connected to the two tails of a Λ-junction respectively, then contract y and y' into a new single vertex, and replace the junction with this new vertex.
2. *deletion.* If two vertices y and y' in A are connected by a Λ-junction, y is connected to the head and y' to one tail, remove y' and replace the junction with y.

It is easy to observe that contraction in the drawing A corresponds to identifying a pair of twins in G; and deletion corresponds to removing a pendant vertex in G.

It is always possible to apply an operation on two vertices connected by a junction because the underlying graph is a tree. During each operation one junction is replaced. Since the drawing is finite, the number of junctions is finite. Therefore, we will reach a point at which the last junction is replaced. After that the drawing reduces to a pair of vertices connected by an edge, and the corresponding G reduces to a K_2. Therefore G is a distance-hereditary graph.

This completes the proof of the equivalence between Δ-confluent graphs and distance-hereditary graphs. □

2.2   Elimination Sequence to Δ-Confluent Tree

The recognition problem of distance-hereditary graphs is solvable in linear time (see [2, 20]). The elimination sequence (ordering) can also be computed in linear time. Using the method of, for example, Damiand et al. [9] we can obtain an elimination sequence L for G of Figure. 3:

By using the elimination sequence reversely, we construct a tree structure of the Δ-confluent drawing of G. This tree structure has n leaves and $n-1$ internal nodes. Every internal node has degree of three. The internal nodes represent our Δ- and Λ-junctions. The construction is as follows.

- While L is non-empty do:
  - Get the last *item* from L
  - If *item* is "b merged into a"
    * If edge $(a, b) \in E(G)$, then replace a with a Δ conjunction using any of its three connectors, connect a and b to the other two connectors of the Δ conjunction; otherwise replace a with a Λ conjunction using its head and connect a and b to its two tails.
  - Otherwise *item* is "b cut from a", replace a with a Λ conjunction using one of its tails, connect a to the head and b to the other tail left.

Clearly the structure we obtain is indeed a tree. Once the tree structure is constructed, the Δ-confluent drawing can be computed by visualizing this tree structure with its internal nodes replaced by Δ- and Λ-junctions.

3   Visualizing the Δ-Confluent Graphs

There are many methods to visualize the underlying topological tree of a Δ-confluent drawing. Algorithms for drawing trees have been studied extensively (see [4, 8, 12, 13, 18, 19, 22, 26, 27, 28, 29, 31, 32] for examples). Theoretically all the tree visualization methods can be used to lay out the underlying tree of a Δ-confluent drawing, although free tree drawing techniques might be more suitable. We choose the following two tree drawing approaches that both yield large angular resolution ($\geq \pi/2$), because in drawings with large angular resolution, each junction lies in a center-like position among the nodes connected to it, so junctions are easy to perceive and paths are easy to follow.

3.1   Orthogonal Straight-Line Δ-Confluent Drawings

The first tree drawing method is the orthogonal straight-line tree drawing method. In the drawings by this method, every edge is drawn as a straight-line segment and every node is drawn at a grid position.

Pick an arbitrary leaf node l of the underlying tree as root and make this free tree a rooted tree T (alternatively one can adopt the elimination hierarchy tree of a distance-hereditary graph for use here.) It is easy to see that T is a binary tree because every internal node of the underlying tree has degree three. We can then apply any known orthogonal straight-line drawing algorithm for trees ([e.g.[4, 6, 7, 24, 25, 30]]) on T to obtain a layout. After that, replace drawings of internal nodesz with their corresponding junction drawings.

3.2   Hexagonal Δ-Confluent Drawings

Since all the internal nodes of underlying trees of Δ-confluent graphs have degree three, if uniform-length edges and large angular resolution are desirable, it is then natural to consider the problem of embedding these trees on the hexagonal grid where each grid point has three neighboring grid points and every cell of the grid is a regular hexagon.

Some researchers have studied the problem of hexagonal grid drawing of graphs. Kant [23] presents a linear-time algorithm to draw tri-connected planar graphs of degree three planar on a $n/2 \times n/2$ hexagonal grid. Aziza and Biedl [1] focus on keeping the number of bends small. They give algorithms that achieve $3.5n + 3.5$ bends for all simple graphs, prove optimal lower bounds on number of bends for K_7, and provide asymptotic lower bounds for graph classes of various connectivity. We are not aware of any other result on hexagonal graph drawing, where the grid consists of regular hexagon cells.

In the Δ-confluent drawings on the hexagonal grid, any segment of an edge must lie on one side of a hexagon sub-cell. Thus the slope of any segment is $1/2$, ∞, or $-1/2$. An example drawing for the graph from Figure. 3 is shown in Figure. 4.

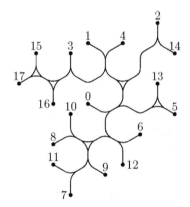

Fig. 4. A hexagonal grid Δ-confluent drawing example

Readers might notice that there are edge bends in the drawing of Figure. 4. Some trees may require a non-constant number of bends per edge to be embedded on a hexagonal grid. Thus it is impossible to embed the tree without edge crossing or edge overlapping, when the bends are limited per edge. However, if unlimited bends are allowed, we show next that Δ-confluent graphs can be embedded in the hexagonal grid of $O(n \log n)$ area in linear time.

The method is to transform an orthogonal straight-line tree embedding into an embedding on the hexagonal grid. We use the results of Chan et al. [6] to obtain an orthogonal straight-line tree drawing. In their paper, a simple "recursive winding" approach is presented for drawing arbitrary binary trees in small area with good aspect ratio. They consider both upward and non-upward cases of orthogonal straight-line drawings. We show that an upward orthogonal straight-line drawing of any binary tree can be easily transformed into a drawing of the same tree on the hexagonal grid.

Figure. 5 (a) exhibits upward orthogonal straight-line drawing for the underlying tree of G in Figure. 3, with node 15 being removed temporarily in order to get a binary tree.

We cover the segments of the hex cell sides with two set of curves: u-curves and v-curves (Figure. 5 (b)). The u-curves (solid) are waving horizontally and the v-curves (dashed) along one of the other two slopes. These two sets of curves are not direct mapping of the lines parallel to x-axis or y-axis in an orthogonal straight-line drawing settings, because the intersection between a u-curve and a v-curve is not a grid point, but a side of the grid cell and it contains two grid points. However this does not matter very much. We choose the lower one of the two grid points in the intersection (overlapping) as our primary point and the

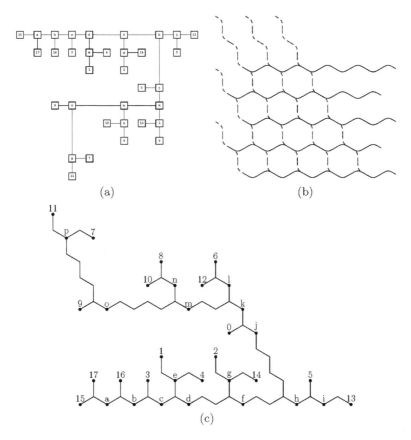

(a)   (b)

(c)

Fig. 5. From upward straight-line orthogonal drawing to hexagonal grid drawing. Internal nodes are labelled with letters and leaves with numbers. (a) orthogonal drawing, generated by Graph Drawing Server (GDS) [5]. (b) u-curves and v-curves. (c) unadjusted result of transformation (mirrored upside-down for a change).

other one as our backup point. So the primary point is at the bottom of a grid cell and its backup is above it to the left. As we can see later, the backup points allow us to do a final adjustment of the node positions.

When doing the transformation from an orthogonal straight-line drawing to a hexagonal grid drawing, we are using only the primary points. So there is a one-to-one mapping between node positions in the orthogonal drawing and the hexagonal grid drawing. However, there are edges overlapping each other in the resultant hexagonal grid drawing of such a direct transformation (e.g. edge (a, b) and edge $(a, 16)$ in Figure. 5 (c)). Now the backup points are used to remove those overlapping portion of edges. Just move a node from a primary point to the point's backup when overlapping happens.

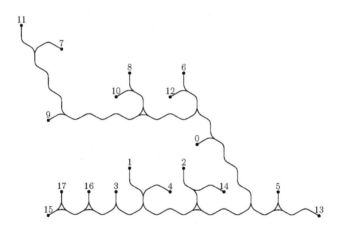

Fig. 6. Final drawing after adjustment

Figure. 6 shows the Δ-confluent drawing of G after overlapping is removed. The drawing does not look compact because the orthogonal drawing from which it is obtained is not tidy in order to have the subtree separation property.

It is not hard to see that backup points are enough for removing all the overlapping portions while the tree structure is still maintained. If wanted, the backup points can be also used to reduce the bends along the edges connecting the tree leaves (e.g. edge connecting node 1). Some bends can be removed as well after junctions are moved (e.g. the subtree of node 8 and 10).

Theorem 3. *Any Δ-confluent graph can be embedded on a grid of size $O(n \log n)$. The representation of its Δ-confluent drawing can be computed in linear time and can be stored using linear space.*

Proof sketch. First the underlying tree of a Δ-confluent graph can be computed in linear time. The transformation runs in linear time as well. It then remains to show that the orthogonal tree drawing can be obtained in linear time. Chan et al.

[6] can realize a upward orthogonal straight-line grid drawing of an arbitrary n-node binary tree T with $O(n \log n)$ area and $O(1)$ aspect ratio. The drawing achieves subtree separation and can be produced in $O(n)$ time.

By using the transformation, we can build a description of the drawing in linear time, which includes the placement of each vertex and representation of each edge. It is straightforward that the drawing has an area of $O(n \log n)$ size. Since the edges are either along u-curves, or along v-curves, we just need to store the two end points for each edge. Note that although some edge might contain $O(\sqrt{n \log n})$ bends (from the "recursive winding" method), constant amount of space is enough to describe each edge. Thus the total space complexity of the representation is $O(n)$. □

In the hexagonal grid drawings for trees, the subtree separation property is retained if the subtree separation in hexagonal grid drawings is defined using u, v area. If different methods of visualizing binary trees on the orthogonal grid are used, various time complexities, area requirements, and other drawing properties for the hexagonal grid Δ-confluent drawing can be derived as well.

4   More About Δ-Confluent Graphs

In this section we discuss a Δ-confluent subgraph problem, and list some topics of possible future work about Δ-confluent graphs.

One way to visualize a non-planar graph is to find a maximum planar subgraph of the original graph, compute a planar drawing of the subgraph, and add the rest of the original graph back on the drawing. An analogous method to visualize a non-Δ-confluent graph would be to find a maximum Δ-confluent subgraph, compute a Δ-confluent drawing, and add the rest back. However, just like the maximum planar subgraph problem, the maximum Δ-confluent subgraph problem is difficult. The problem is defined below, and its complexity is given in Theorem 4.

MAXIMUM Δ-CONFLUENT SUBGRAPH PROBLEM:
INSTANCE: A graph $G = (V, E)$, an integer $K \leq |V|$.
QUESTION: Is there a $V' \subset V$ with $|V'| \geq K$ such that the subgraph of G induced by V' is a Δ-confluent?

Theorem 4. *Maximum Δ-confluent subgraph problem is NP-complete.*

Proof. The proof can be derived easily from Garey and Johnson [17, GT21].

[GT21] INDUCED SUBGRAPH WITH PROPERTY Π:
INSTANCE: A graph $G = (V, E)$, an integer $K \leq |V|$.
QUESTION: Is there a $V' \subset V$ with $|V'| \geq K$ such that the subgraph of G induced by V' has property Π?

It is NP-hard for any property Π that holds for arbitrarily large graphs, does not hold for all graphs, and is hereditary (holds for all induced subgraphs of G whenever it holds for G). If it can be determined in polynomial time whether

Π holds for a graph, then the problem is NP-complete. Examples include "*G* is a clique", "*G* is an independent set", "*G* is planar", "*G* is bipartite", "*G* is chordal."

Δ-confluency is a property that holds for arbitrarily large graphs, does not holds for all graphs, and is hereditary (every induced subgraph of a *Δ*-confluent graph is *Δ*-confluent.) It can be determined in linear time whether a graph is *Δ*-confluent. Thus the maximum *Δ*-confluent subgraph problem is NP-complete.

□

Instead of drawing the maximum subgraph *Δ*-confluently and adding the rest back, We could compute a *Δ*-confluent subgraph cover of the input graph, visualize each subgraph as a *Δ*-confluent drawing, and overlay them together. This leads to the *Δ*-CONFLUENT SUBGRAPH COVERING PROBLEM. Like the maximum *Δ*-confluent subgraph problem, we expect this problem to be hard as well.

This alternative way is related to the concept of *simultaneous embedding* (see [3, 11, 15, 16]). To visualize an overlay of *Δ*-confluent subgraph drawings is to draw trees simultaneously. However *simultaneously embedding* draws only two graphs that share the same vertex set *V*, while a *Δ*-confluent subgraph cover could have a cardinality larger than two. Furthermore, the problem of simultaneously embedding (two) trees hasn't been solved.

Other interesting problems include:

- How to compute the drawing with optimum area (or number of bends, etc.) for a *Δ*-confluent graph?
  Generally hexagonal grid drawings by transforming orthogonal drawings are not area (number of bends, etc.) optimal. If subtree separation is not required, hexagonal grid drawings with more compact area or smaller number of bends can be achieved. Maybe a simple incremental algorithm would work.
- The underlying track system here is topologically a tree. What classes of graphs can we get if other structures are allowed?

References

[1] S. Aziza and T. Biedl. Hexagonal grid drawings: Algorithms and lower bounds. In J. Pach, editor, *Graph Drawing (Proc. GD '04)*, volume 3383 of *Lecture Notes Comput. Sci.*, pages 18–24. Springer-Verlag, 2005.

[2] H. Bandelt and H. M. Mulder. Distance-hereditrary graphs. *J. Combin. Theory Ser. B*, 41:182–208, 1986.

[3] P. Brass, E. Cenek, C. A. Duncan, A. Efrat, C. Erten, D. Ismailescu, S. G. Kobourov, A. Lubiw, and J. S. B. Mitchell. On simultaneous graph embedding. In *8th Workshop on Algorithms and Data Structures*, pages 243–255, 2003.

[4] R. P. Brent and H. T. Kung. On the area of binary tree layouts. *Inform. Process. Lett.*, 11:521–534, 1980.

[5] S. Bridgeman, A. Garg, and R. Tamassia. A graph drawing and translation service on the WWW. In S. C. North, editor, *Graph Drawing (Proc. GD '96)*, volume 1190 of *Lecture Notes Comput. Sci.*, pages 45–52. Springer-Verlag, 1997.

[6] T. M. Chan, M. T. Goodrich, S. R. Kosaraju, and R. Tamassia. Optimizing area and aspect ratio in straight-line orthogonal tree drawings. In S. North, editor, *Graph Drawing (Proc. GD '96)*, volume 1190 of *Lecture Notes Comput. Sci.*, pages 63–75. Springer-Verlag, 1997.

[7] S. Y. Choi. Orthogonal straight line drawing of trees. M.Sc. thesis, Dept. Comput. Sci., Univ. Brown, Providence, RI, 1999.

[8] P. Crescenzi, G. Di Battista, and A. Piperno. A note on optimal area algorithms for upward drawings of binary trees. *Comput. Geom. Theory Appl.*, 2:187–200, 1992.

[9] G. Damiand, M. Habib, and C. Paul. A simple paradigm for graph recognition: Application to cographs and distance hereditary graphs. *Theor. Comput. Sci.*, 263(1–2):99–111, 2001.

[10] M. Dickerson, D. Eppstein, M. T. Goodrich, and J. Y. Meng. Confluent drawing: Visualizing nonplanar diagrams in a planar way. In G. Liotta, editor, *Graph Drawing (Proc. GD '03)*, volume 2912 of *Lecture Notes Comput. Sci.*, pages 1–12. Springer-Verlag, 2004.

[11] C. A. Duncan, D. Eppstein, and S. G. Kobourov. The geometric thickness of low degree graphs. In *20th Annual ACM-SIAM Symposium on Computational Geometry (SCG '04)*, pages 340–346, 2004.

[12] P. Eades, T. Lin, and X. Lin. Two tree drawing conventions. *Internat. J. Comput. Geom. Appl.*, 3:133–153, 1993.

[13] P. D. Eades. Drawing free trees. *Bulletin of the Institute for Combinatorics and its Applications*, 5:10–36, 1992.

[14] D. Eppstein, M. T. Goodrich, and J. Y. Meng. Confluent layered drawings. In J. Pach, editor, *Graph Drawing (Proc. GD '04)*, volume 3383 of *Lecture Notes Comput. Sci.*, pages 184–194. Springer-Verlag, 2005.

[15] C. Erten and S. G. Kobourov. Simultaneous embedding of a planar graph and its dual on the grid. In M. Goodrich and S. Kobourov, editors, *Graph Drawing (Proc. GD '02)*, volume 2518 of *Lecture Notes Comput. Sci.*, pages 575–587. Springer-Verlag, 2003.

[16] C. Erten and S. G. Kobourov. Simultaneous embedding of planar graphs with few bends. In J. Pach, editor, *Graph Drawing (Proc. GD '04)*, volume 3383 of *Lecture Notes Comput. Sci.*, pages 195–205. Springer-Verlag, 2005.

[17] M. R. Garey and D. S. Johnson. *Computers and Intractability: A Guide to the Theory of NP-Completeness.* W. H. Freeman, New York, NY, 1979.

[18] A. Garg, M. T. Goodrich, and R. Tamassia. Area-efficient upward tree drawings. In *Proc. 9th Annu. ACM Sympos. Comput. Geom.*, pages 359–368, 1993.

[19] A. Gregori. Unit length embedding of binary trees on a square grid. *Inform. Process. Lett.*, 31:167–172, 1989.

[20] P. Hammer and F. Maffray. Completely separable graphs. *Discrete Appl. Math.*, 27:85–99, 1990.

[21] P. Hui, M. Schaefer, and D. Štefankovič. Train tracks and confluent drawings. In J. Pach, editor, *Graph Drawing (Proc. GD '04)*, volume 3383 of *Lecture Notes Comput. Sci.*, pages 318–328. Springer-Verlag, 2005.

[22] P. J. Idicula. Drawing trees in grids. Master's thesis, Department of Computer Science, University of Auckland, 1990.

[23] G. Kant. Hexagonal grid drawings. In *Proc. 18th Internat. Workshop Graph-Theoret. Concepts Comput. Sci.*, 1992.

[24] C. E. Leiserson. Area-efficient graph layouts (for VLSI). In *Proc. 21st Annu. IEEE Sympos. Found. Comput. Sci.*, pages 270–281, 1980.

[25] C. E. Leiserson. *Area-efficient graph layouts (for VLSI)*. ACM Doctoral Dissertation Award Series. MIT Press, Cambridge, MA, 1983.

[26] P. T. Metaxas, G. E. Pantziou, and A. Symvonis. Parallel h-v drawings of binary trees. In *Proc. 5th Annu. Internat. Sympos. Algorithms Comput.*, volume 834 of *Lecture Notes Comput. Sci.*, pages 487–495. Springer-Verlag, 1994.

[27] E. Reingold and J. Tilford. Tidier drawing of trees. *IEEE Trans. Softw. Eng.*, SE-7(2):223–228, 1981.

[28] K. J. Supowit and E. M. Reingold. The complexity of drawing trees nicely. *Acta Inform.*, 18:377–392, 1983.

[29] J. S. Tilford. Tree drawing algorithms. Technical Report UIUCDCS-R-81-1055, Department of Computer Science, University of Illinois at Urbana-Champaign, 1981.

[30] L. Valiant. Universality considerations in VLSI circuits. *IEEE Trans. Comput.*, C-30(2):135–140, 1981.

[31] J. Q. Walker II. A node-positioning algorithm for general trees. *Softw. – Pract. Exp.*, 20(7):685–705, 1990.

[32] C. Wetherell and A. Shannon. Tidy drawing of trees. *IEEE Trans. Softw. Eng.*, SE-5(5):514–520, 1979.

Transversal Structures on Triangulations, with Application to Straight-Line Drawing

Éric Fusy

Algorithm Project (INRIA Rocquencourt) and LIX (École Polytechnique)
eric.fusy@inria.fr

Abstract. We define and investigate a structure called transversal edge-partition related to triangulations without non empty triangles, which is equivalent to the regular edge labeling discovered by Kant and He. We study other properties of this structure and show that it gives rise to a new straight-line drawing algorithm for triangulations without non empty triangles, and more generally for 4-connected plane graphs with at least 4 border vertices. Taking uniformly at random such a triangulation with 4 border vertices and n vertices, the size of the grid is almost surely $\frac{11}{27}n \times \frac{11}{27}n$ up to fluctuations of order $\sqrt{n}$, and the half-perimeter is bounded by $n - 1$. The best previously known algorithms for straight-line drawing of such triangulations only guaranteed a grid of size $(\lceil n/2 \rceil - 1) \times \lfloor n/2 \rfloor$. Hence, in comparison, the grid-size of our algorithm is reduced by a factor $\frac{5}{27}$, which can be explained thanks to a new bijection between ternary trees and triangulations of the 4-gon without non empty triangles.

1 Introduction

A plane graph is a connected graph embedded in the plane so that edges do not cross each other. Many algorithms for drawing plane graphs [4, 15, 2, 10] endow the graph with a particular structure, from which it is possible to give coordinates to vertices in a natural way. For example, triangulations, i.e., plane graphs with only faces of degree 3, are characterized by the fact that their inner edges can essentially be partitioned into three spanning trees, called Schnyder Woods, with specific incidence relations [15]. Using these spanning trees it is possible to associate coordinates to each vertex by counting faces on each side of particular paths passing by the vertex. Placing vertices in this way and linking adjacent vertices by segments yields a straight-line drawing algorithm, which can be refined to produce a drawing on a regular grid of size $(n - 2) \times (n - 2)$, see [16].

A plane graph with an outer face of degree k and inner faces of degree 3 is called a triangulation of the k-gon. If the interior of any 3-cycle of edges is a face, the triangulation is *irreducible*. Observe that it implies $k > 3$, unless the graph is reduced to a unique triangle. There exist more compact straight-line drawing algorithms for irreducible triangulations [9, 11], the size of the grid being guaranteed to be $(\lceil n/2 \rceil - 1) \times \lfloor n/2 \rfloor$ in the worst case.

P. Healy and N.S. Nikolov (Eds.): GD 2005, LNCS 3843, pp. 177–188, 2005.

In this extended abstract we concentrate on irreducible triangulations of the 4-gon, which carry a good level of generallity. Indeed many graphs, including 4-connected plane graphs with at least 4 border vertices, can be triangulated (after adding 4 vertices in the outer face) into an irreducible triangulation of the 4-gon, see [1]. By investigating a bijection with ternary trees, we have observed that each irreducible triangulation of the 4-gon can be endowed with a structure, called transversal edge-partition, which can be summarized as follows. Calling S_b, N_r, N_b, S_r (like south-blue, north-red, north-blue, south-red) the 4 border vertices of T in clockwise order, the inner edges of T can be oriented and partitioned into two sets: red edges that "flow" from S_r to N_r, and blue edges that "flow" from S_b to N_b. For those familiar with bipolar orientations [5], i.e. acyclic orientations with two poles, the structure can also be seen as a transversal couple of bipolar orientations, see Section 2.3. As we learned after completing a first draft of this extended abstract, Kant and He used an equivalent structure in [10] and derived nice algorithms of rectangular-dual drawing and of visibility representation. We explore the properties of this structure and show in particular in Theorem 1 that it is of the lattice type.

In Section 3, we derive from the transversal structure a straight-line drawing algorithm of an irreducible triangulation T of the 4-gon. Like drawing algorithms using Schnyder Woods [15, 2], it is based on face counting operations. The first step is to endow T with a particular transversal edge-partition, said *minimal*, which is obtained by application of an iterative algorithm described in Section 2.4. Then the transversal structure is used to associate to each vertex v a path P_r of red edges and a path P_b of blue edges, both passing by v. The abscissa (resp. ordinate) of v is obtained by counting faces on each side of P_r (resp. P_b). Our algorithm outputs a straight line embedding on a regular grid of width W and height H with $W + H \leq n - 1$ if the triangulation has n vertices. This algorithm can be compared to [9] and [11], which produces straight-line drawing on a grid of size $(\lceil n/2 \rceil - 1) \times \lfloor n/2 \rfloor$. However, algorithms of [9] and [11] rely on a particular order of treatment of vertices called canonical ordering, and a step of coordinate-shifting makes them difficult to implement and to carry out by hand. As opposed to that, our algorithm can readily be performed on a piece of paper, because coordinates of vertices can be computed independently with simple face-counting operations. Finally, our algorithm has the nice feature that it respects the structure of transversal edge-partition. Indeed, Theorem 2 ensures that red edges are geometrically oriented from S_r to N_r and blue edges are geometrically oriented from S_b to N_b.

A compact version of the algorithm even ensures that, for a random triangulation with n vertices, the size of the grid is asymptotically almost surely $\frac{11}{27}n \times \frac{11}{27}n$ up to small fluctuations, of order $\sqrt{n}$. Compared to [9] and [11], we do not improve on the size of the grid in the worst case, but improve asymptotically by a reduction-factor 5/27 on the width and height of the grid for a typical (random) object of large size, see Figure 4.2 for an example with $n = 200$. The reduction factor 5/27 can be explained thanks to a new bijection between ternary trees and irreducible triangulations of the 4-gon. This bijection is described in Section 4 and relies on "closure operations", as introduced by G. Schaeffer [13], see also [12] for a bijection with unconstrained triangulations. This bijection has, truth to tell, brought about our discovery of transversal edge-partitions. Indeed, it turns out to "transport" a so-called transversal edge-bicoloration of a ternary tree into the minimal transversal edge-partition of its associated triangulation, in the same way that bijection of [12] transports the structure of Schnyder woods. In addition, the bijection gives a combinatorial way to enumerate rooted 4-connected triangulations, which were already counted by Tutte in [17] using algebraic methods.

2   Definition of Transversal Structures

2.1   Transversal Edge-Partition

Let T be an irreducible triangulation of the 4-gon. Edges and vertices of T are said *inner* or *outer* whether they belong to the outer face or not. A transversal edge-partition of T is a partition of the inner edges of T into two sets, say in blue and in red edge, such that the following conditions are satisfied.

- C1 (Inner vertices): In clockwise order around each inner vertex, its incident edges form: a non empty interval of red edges, a non empty interval of blue edges, a non empty interval of red edges, and a non empty interval of blue edges, see Figure 1a.
- C2 (Border vertices): Writing a_1, a_2, a_3, a_4 for the border vertices of T in clockwise order, all inner edges incident to a_1 and to a_3 are of one color and all inner edges incident to a_2 and to a_4 are of the other color.

Figure 1b gives an example of transversal edge-partition, where we use DARK RED for red edges and LIGHT BLUE for blue edges (the same convention will be used for all figures).

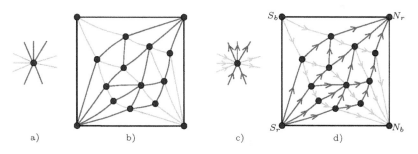

Fig. 1. The structure of transversal edge-partition: local condition (a) and a complete example (b). In parallel, the structure of transversal couple of bipolar orientations: local condition (c) and a complete example (d).

2.2   Lattice Structure

As it is the case with Schnyder Woods and bipolar orientations, the set of transversal edge-partitions of a fixed irreducible triangulation of the 4-gon is a distributive lattice. In addition, the "flip" operation has a nice geometric interpretation. To describe it, we have to introduce some terminology. Given T an irreducible triangulation of the 4-gon endowed with a transversal edge-partition X, we define an *alternating 4-cycle* as a 4-cycle C of inner edges (e_1, e_2, e_3, e_4) of T which are color-alternating (i.e. two adjacent edges of C have different colors). The cycle C is called *essential* if its interior does not properly contain the interior of another alternating 4-cycle. Given a vertex v of C, we call left-edge (resp. right-edge) of v the edge of C starting from v and having the exterior of C on its left (resp. on its right). It can easily be proven that two cases can occur for C: either all edges interior to C and incident to a border vertex v of C have the color of the left-edge of v, then C is called a left alternating 4-cycle; or all edges interior to C and incident to a border vertex v of C have the color of the right-edge of v, then C is called a right alternating 4-cycle.

Theorem 1. *Let T be an irreducible triangulation of the 4-gon. Then the set $\mathcal{E}$ of transversal edge-partitions of T is a non-empty distributive lattice. Given $X \in \mathcal{E}$, the flip operation consists in finding a right alternating 4-cycle C of X and then switching the colors of all edges interior to C, making C a left alternating 4-cycle. The (unique) transversal edge-partition of T without right alternating 4-cycle is said minimal.*

Proof. The non emptiness of $\mathcal{E}$ will be proven constructively in Section 2.4 by providing an algorithm computing the minimal transversal edge-partition of T. The lattice structure follows from the fact that $\mathcal{E}$ is in bijection with the set of orientations of an associated graph (called the angular graph) where each vertex has a fixed outdegree. The set of such orientations with fixed outdegree is well-known to be a distributive lattice, see [6, 7].

2.3   Transversal Couple of Bipolar Orientations

Given a plane graph G and two vertices S (like South) and N (like North) of G incident to the outer face of G, a *bipolar orientation* of G with poles S and N is an acyclic orientation of the edges of G such that, for each vertex v different from S and N, there exists an oriented path from S to N passing by v, see [5] for a detailed decription.

Let T be an irreducible triangulation of the 4-gon. Call N_r, N_b, S_r and S_b the 4 border vertices of T in clockwise order around the outer face of T. A *transversal couple of bipolar orientations* is an orientation and a partition of the inner edges of T into red and blue edges such that the following two conditions are satisfied (see Figure 1d for an example):

- C1' (Inner vertices): In clockwise order around each inner vertex of T, its incident edges form: a non empty interval of outgoing red edges, a non empty

interval of outgoing blue edges, a non empty interval of ingoing red edges, and a non empty interval of ingoing blue edges, see Figure 1c.
- C2' (Border vertices): All inner edges incident to N_b, N_r, S_b and S_r are respectively ingoing blue, ingoing red, outgoing blue, and outgoing red.

This structure is also defined in [10] under the name of regular edge labeling. The following proposition explains the name of transversal couple of bipolar orientations and is also stated in [10]:

Proposition 1. *Let T be an irreducible triangulation of the 4-gon. Given a transversal couple of bipolar orientations of T, the (oriented) red edges induce a bipolar orientation of the plane graph obtained from T by removing S_b, N_b, and all non red edges. Similarly, the blue edges induce a bipolar orientation of T deprived from S_r, N_r and all non blue edges.*

Proposition 2. *To each transversal couple of bipolar orientations of T corresponds a transversal edge-partition of T, obtained by removing the orientation of the edges (Compare Figure 1d and Figure 1b). This correspondence is a bijection.*

Proposition 2 allows us to manipulate equivalently transversal edge-partitions or transversal couples of bipolar orientations. The first point of view is more convenient to describe the lattice structure, the second one will be more convenient to describe the drawing algorithm in Section 3.

2.4   Algorithm Computing the Minimal Transversal Edge-Partition

Let us now describe a simple iterative algorithm to compute transversal edge-partitions. Two different algorithms computing such transversal structures were already presented in [10]. However we need to compute the minimal transversal edge-partition, to be used later in the straight-line drawing algorithm. During the execution, we also orient the edges, so that we compute in fact the underlying transversal couple of bipolar orientations. The algorithm we introduce consists in maintaining and iteratively shrinking a cycle $\mathcal{C}$ of edges of T such that, in particular (we do not detail all invariants here):

- The cycle $\mathcal{C}$ contains the two edges (S_r, S_b) and (S_r, N_b).
- No edge interior to $\mathcal{C}$ connects two vertices of $\mathcal{C}\backslash\{S_r\}$
- All inner edges of T outside of $\mathcal{C}$ are colored and oriented such that Inner-vertex Condition C1' (see Section 2) is satisfied for each inner vertex of T outside of $\mathcal{C}$.

We initialize the cycle $\mathcal{C}$ with vertices S_r, S_b, N_b and all interior neighbours of N_r, color in red all inner edges incident to N_r and orient them toward N_r, see Figure 2b. Observe also that vertices of $\mathcal{C}$ different from S_r can be ordered from left to right with S_b as leftmost and N_b as rightmost vertex. For two vertices v and v' of $\mathcal{C}\backslash\{S_r\}$ with v on the left of v', we write $[v, v']$ for the unique path on $\mathcal{C}$ that goes from v to v' without passing by S_r.

To explain how to update (shrink) $\mathcal{C}$ at each step, we need a few definitions. An *internal path* of $\mathcal{C}$ is a path $\mathcal{P}$ of edges interior to $\mathcal{C}$ and connecting two vertices

v and v' of $\mathcal{C}$. We write $\mathcal{C_P}$ for the cycle constituted by the concatenation of $\mathcal{P}$ and $[v, v']$. The path $\mathcal{P}$ is said *eligible* if the following conditions are satisfied:

- The paths $\mathcal{P}$ and $[v, v']$ have both at least one vertex different from v and v'.
- Each edge interior to $\mathcal{C_P}$ connects a vertex of $\mathcal{P}\backslash\{v, v'\}$ to a vertex of $[v, v']\backslash\{v, v'\}$. In particular, the interior of $\mathcal{C_P}$ contains no vertex.
- The cycle $\mathcal{C}'$ obtained from $\mathcal{C}$ by replacing $[v, v']$ by $\mathcal{P}$ is such that no interior edge of $\mathcal{C}'$ connects two vertices of $\mathcal{C}'\backslash\{S_r\}$.

The update operation is the following: find an eligible internal path $\mathcal{P}$ of $\mathcal{C}$ and write v and v' for its extremities with v on the left of v' (so that v and v' are called respectively left and right extremity of $\mathcal{P}$); then, color each internal edge of $\mathcal{C_P}$ in red and orient it toward $[v, v']\backslash\{v, v'\}$. Color all edges of $[v, v']$ in blue and orient them from v to v'; finally update $\mathcal{C}$ by replacing in $\mathcal{C}$ the path $[v, v']$ by the path $\mathcal{P}$.

It can easily be shown that the absence of non empty triangle on T ensures that the algorithm terminates, i.e. that at each step the cycle $\mathcal{C}$ has an eligible internal path and can be updated (shrinked). After the last update operation, $\mathcal{C}$ is empty. Using all invariants of colors and orientations of edges satisfied by $\mathcal{C}$, it can be shown that the obtained orientation and coloration of inner edges of T is a transversal couple of bipolar orientations. Figure 2 illustrates the complete execution of the algorithm on an example.

This algorithm can easily be adapted to give an algorithm, called COM-PUTEMINIMAL(T), which computes the transversal couple of bipolar orientations associated (by removing orientation of edges) to the minimal transversal edge-partition of T, as defined in Theorem 1. Observe that, at each step of the algorithm, eligible paths of $\mathcal{C}$ can be ordered from left to right, by saying that $\mathcal{P}_1 \geq \mathcal{P}_2$ if the left extremity and the right extremity of $\mathcal{P}_1$ are (weakly) on the

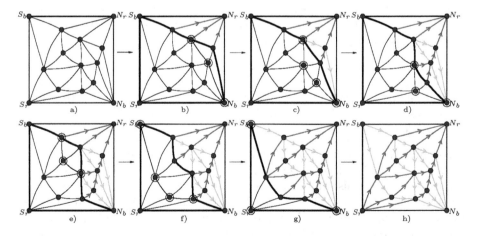

Fig. 2. An example of execution of the algorithm computing the minimal transversal couple of bipolar orientations. Vertices of the rightmost eligible path are surrounded.

left respectively of the left extremity and of the right extremity of $\mathcal{P}_2$. Although this order is only partial, it can easily be shown to admit a unique minimum, called rightmost eligible path of $\mathcal{C}$. Algorithm COMPUTEMINIMAL(T) consists in choosing the rightmost eligible path at each step of the iterative algorithm described above, see also Figure 2, where the execution respects this choice.

Proposition 3. *Given an irreducible triangulation T of the 4-gon, Algorithm* COMPUTEMINIMAL*(T) outputs the transversal couple of bipolar orientations associated to the minimal transversal edge-partition of T (by removing edge orientations). In addition,* COMPUTEMINIMAL*(T) can be implemented to run in linear time.*

3 Application to Straight-Line Drawing

We recall that a straight line drawing of a plane graph G consists in placing all points of G on a regular grid of size $[0, W] \times [0, H]$ and then linking each pair of adjacent vertices of G by a segment, with the condition that two different segments can only meet at their endpoints. The integers W and H are called the width and the height of the grid.

The structure of transversal edge-partition can be used to derive a simple algorithm, called TRANSVERSALDRAW, to perform straight line drawing of an irreducible triangulation T of the 4-gon. First we have to give a few definitions. The plane graph obtained from T by removing all blue (resp. red) edges is called the red-map (resp. blue-map) of T and is denoted by T_r (resp. T_b). We write f_r and f_b for the number of inner faces of T_r and T_b. Given an inner vertex v of T, we define the leftmost outgoing red path of v as the oriented path starting from v and such that each edge of the path is the leftmost outgoing red edge at its origin. As the orientation of red edges is bipolar, this path has no cycle and ends at N_r. We also define the rightmost ingoing red path of v as the path starting from v and such that each edge of the path is the rightmost ingoing red edge at its extremity. This path is also acyclic and ends at S_r. We call separating red path of v the concatenation of these two paths and denote it by $\mathcal{P}_r(v)$. The path $\mathcal{P}_r(v)$ goes from S_r to N_r passing by v, and separates inner faces of T_r into two sets: those on the left of $\mathcal{P}_r(v)$ and those on the right of $\mathcal{P}_r(v)$. Similarly, we define the leftmost outgoing blue path, the rightmost ingoing blue path, and write $\mathcal{P}_b(v)$ for their concatenation, called separating blue path of v.

Algorithm TRANSVERSALDRAW consists of the following steps, see Figure 3 for a complete execution:

- Perform COMPUTEMINIMAL(T) to endow T with its minimal transversal couple of bipolar orientations.
- Take a regular grid of width f_r and height f_b.
- Place the border vertices S_r, S_b, N_b, N_r respectively at coordinates $(0, 0)$, $(0, f_b)$, $(f_r, 0)$ and (f_r, f_b).
- For each inner vertex v of T, place v on the grid in the following way:
  - The abscissa of v is the number of inner faces of T_r on the left of $\mathcal{P}_r(v)$.
  - The ordinate of v is the number of inner faces of T_b on the right of $\mathcal{P}_b(v)$.

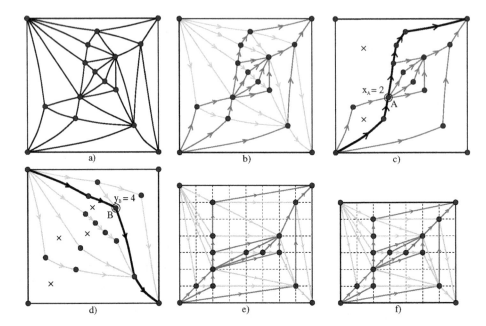

Fig. 3. The execution of Algorithm TRANSVERSALDRAW (a)-(e), and of Algorithm COMPACTTRANSVERSALDRAW (a)-(f) on an example

Algorithm TRANSVERSALDRAW can be enhanced into an algorithm, called COMPACTTRANSVERSALDRAW, giving a more compact drawing. The further step consists in deleting the unused abscissas and ordinates of the drawing computed by TRANSVERSALDRAW. An example is given on Figure 3d, obtained from Figure 3c after having deleted the unused abscissa 3 and the unused ordinate 5.

Theorem 2. *Algorithm* TRANSVERSALDRAW *and Algorithm* COMPACTTRANS VERSALDRAW *can be implemented to run in linear time and compute a straight line drawing of an irreducible triangulation T of the 4-gon such that:*

- *All red edges are oriented from bottom to top and weakly oriented from left to right.*
- *All blue edges are oriented from left to right and weakly oriented from top to bottom.*
- *If T has n vertices, then the width W and height H of the grid of the drawing given by* TRANSVERSALDRAW(T) *verify $W + H = n - 1$.*
- *Let T be taken uniformly at random among irreducible triangulations of the 4-gon with n vertices. The width W_c and the height H_c of the grid of the drawing output by* COMPACTTRANSVERSALDRAW(T) *are asymptotically almost surely equal to $\frac{11}{27}n$, up to fluctuations ϵ_{W_c} and ϵ_{H_c} of order $\sqrt{n}$.*

In fact the transversal structure used to give coordinates to vertices need not to be the minimal one. Using any other transversal couple of bipolar orientations, the three first points of Theorem 2 remain true. However the analysis of the reduction-factor $\frac{5}{27}$ with COMPACTTRANSVERSALDRAW(T) crucially requires that the transversal structure is the minimal one, see Section 4.2.

Corollary 1. *Each 4-connected plane graph G with n vertices and at least 4 vertices on the outer face can be embedded with a straight-line drawing on a regular grid W × H with W + H ≤ n − 1.*

4   Bijection with Ternary Trees and Applications

4.1   Description of the Bijection

A *ternary tree* A is a tree embedded in the plane with nodes of degree 4, called *inner nodes* and nodes of degree 1, called *leaves*. Edges of A connecting two inner nodes are called *inner edges* and edges incident to a leaf are called *stems* (these are "pending" edges). A ternary tree can be rooted by marking one of its leaves, and such rooted ternary trees correspond to the classical definition of ternary trees (i.e. all nodes have either 0 or 3 children).

We describe briefly the bijection (see [8, 12] for detailed descriptions of similar bijections), consisting of three main steps: local closure, partial closure and complete closure. Perform a counterclockwise traversal of A (imagine an ant walking around A with the infinite face on its right). If a stem s and then two inner edges e_1 and e_2 are successively encountered during the traversal, merge the extremity of s with the extremity of e_2, so as to *close* a triangular face. This operation is called *local closure*, see Figure 4b. Now we can restart a counterclockwise traversal around the new Figure F, which is identical to A, except that it contains a triangular face and, more important, the stem s has become an inner edge. Each time we find a succession (stem, edge, edge), we perform a local closure, update the figure, and restart, until no local closure is possible. This greedy execution of local closures is called the *partial closure* of A, see Figure 4c. It can easily be shown that the figure F obtained by partial closure of A does not depend of the order of execution of the local closures. Finally, the last step, called *complete closure* (see Figure 4d), consists in drawing a 4-gon, and then merging the extremity of each unmatched stem with a border vertex, so as to create only

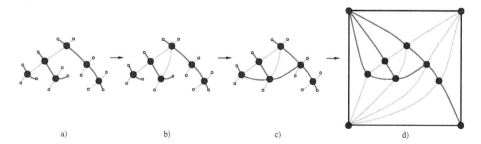

Fig. 4. The execution of the closure on an example

triangular inner faces. It can be shown that the choice of an outer 4-gon is the good one so that this last operation works without conflict.

Observe that the edges of a ternary tree A can be bicolored in blue and red edges so that two successive edges incident to an inner node of A have always different color, see Figure 4a. This bicoloration, unique up to the choice of the colors, is called the transversal edge-bicoloration of A. Observe that Inner Vertex Condition C1 is satisfied on A and remains satisfied throughout the closure.

Theorem 3. *The closure is a bijection between ternary trees with n inner nodes and irreducible triangulations of the 4-gon with n inner vertices.*

The closure transports the transversal edge-bicoloration of a ternary tree into the minimal transversal edge-partition of its image.

Proof. Injectivity can easily be proven by uniqueness of the transversal edge-partition without right alternating 4-cycle. The inverse of the closure consists in computing the minimal transversal edge-partition of T and using the colors to remove some half-edges, so as to leave a ternary tree.

An irreducible triangulation of the 4-gon is *rooted* by choosing one of its 4 border edges and orienting this edge with the infinite face on its right. This well-known operation eliminates symmetries of the triangulation.

Corollary 2. *The closure induces a 4-to-$(2n+2)$ correspondence between the set $\mathcal{A}_n$ of rooted ternary trees with n inner nodes and the set $\mathcal{T}_n$ of rooted irreducible triangulations of the 4-gon with n inner vertices.*

As an enumerative consequence, $|\mathcal{T}_n| = \frac{4}{2n+2}|\mathcal{A}_n| = \frac{4(3n)!}{(2n+2)!n!}$.

Proof. The proof follows easily from the bijection stated in Theorem 3 and from the fact that a ternary tree with n inner nodes has $2n + 2$ leaves and an object of $\mathcal{T}_n$ has 4 edges (the 4 border edges) to carry the root.

4.2   Applications

The closure-bijection has several applications. A first one is a linear-time algorithm to perform uniform random sampling of objects of $\mathcal{T}_n$, using the fact that rooted ternary trees with n inner nodes can readily be uniformly sampled using parenthesis words. A thorough study of such sampling algorithms is given in [14]. In addition, sampled objects of $\mathcal{T}_n$ are naturally endowed, through the closure, with their minimal transversal edge-partition. Hence, we can easily run face-counting algorithms TRANSVERSALDRAW and COMPACTTRANSVERSALDRAW on the sampled objects. Performing simulations on objects of large size ($n \approx 50000$), it was observed that the size of the grid is always approximately $\frac{n}{2} \times \frac{n}{2}$ with TRANSVERSALDRAW and $\frac{n}{2}(1 - \alpha) \times \frac{n}{2}(1 - \alpha)$ with COMPACTTRANSVERSALDRAW, where $\alpha \approx 0.18$. It turns out that the size of the grid can be readily analyzed thanks to our closure-bijection, in the same way that bijection of [12] allowed to analyze parameters of Schnyder woods in [3]. Indeed,

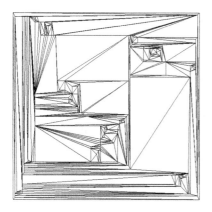 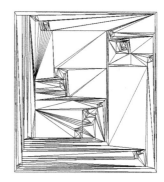

Fig. 5. A random triangulation with 200 vertices embedded with Algorithms TRANSVERSALDRAW and COMPACTTRANSVERSALDRAW

unused abscissas and ordinates of TRANSVERSALDRAW correspond to certain inner edges of the ternary tree, whose number can be proven to be asymptotically almost surely $\frac{5n}{27}$ up to fluctuations of order $\sqrt{n}$.

A second application is counting rooted 4-connected triangulations with n vertices, whose set is denoted by $\mathcal{C}_n$. It is well known that a 4-connected triangulation is a triangulation where each 3-cycle delimits a face. Hence, the operation of removing the root edge of an object of $\mathcal{C}_n$ and carrying the root on the counterclockwise-consecutive edge is an (injective) mapping from $\mathcal{C}_n$ to $\mathcal{T}_{n-4}$. However, given $T \in \mathcal{T}_{n-4}$, the inverse edge-adding-operation can create a separating 3-cycle if there exists an internal path of length 2 connecting the origin of the root of T to the vertex diametrically opposed in the outer face of T. Objects of $\mathcal{T}_{n-4}$ having no such internal path are said *undecomposable* and their set is denoted by $\mathcal{U}_{n-4}$. The above discussion ensures that they are in bijection with $\mathcal{C}_n$. A maximal decomposition of an object T of $\mathcal{T}$ along the above mentioned interior paths of length 2 ensures that T is a sequence of objects of $\mathcal{U}$. After a few simple manipulations and using Corollary 2, we get:

Proposition 4. *The series $C(z)$ counting rooted 4-connected triangulations by their number of inner vertices has the following expression:*

$$C(z) = \frac{z(A(z) - A(z)^2 + 1)}{1 + z(A(z) - A(z)^2 + 1)}$$

where $A(z) = z(1 + A(z))^3$ is the series counting rooted ternary trees by their number of inner nodes.

Acknowledgments. I would like to thank my advisor Gilles Schaeffer. He has greatly helped me to produce this work through numerous discussions, steady encouragment and useful suggestions. I also thank Nicolas Bonichon for fruitful discussions and Thomas Pillot for very efficient implementations of all algorithms presented in this extended abstract.

References

1. Therese C. Biedl, Goos Kant, and Michael Kaufmann. On triangulating planar graphs under the four-connectivity constraint. *Algorithmica*, 19(4):427–446, 1997.
2. N. Bonichon, S. Felsner, and M. Mosbah. Convex drawings of 3-connected plane graphs. In *GD '04: Proceedings of the Symposium on Graph Drawing*, pages 287–299. Springer-Verlag, 2004.
3. N. Bonichon, C. Gavoille, N. Hanusse, D. Poulalhon, and G. Schaeffer. Planar graphs, via well-orderly maps and trees. In 30th *International Workshop, Graph - Theoretic Concepts in Computer Science (WG)*, volume 3353 of *Lecture Notes in Computer Science*, pages 270–284. Springer-Verlag, 2004.
4. H. de Fraysseix, P. Ossona de Mendez, and J. Pach. Representation of planar graphs by segments. *Intuitive Geometry*, 63:109–117, 1991.
5. H. de Fraysseix, P. Ossona de Mendez, and P. Rosenstiehl. Bipolar orientations revisited. *Discrete Appl. Math.*, 56(2-3):157–179, 1995.
6. P. O. de Mendez. *Orientations bipolaires*. PhD thesis, Paris, 1994.
7. Stefan Felsner. Lattice structures from planar graphs. *Electronic Journal of Combinatorics*, (R15):24p., 2004.
8. É. Fusy, D. Poulalhon, and G. Schaeffer. Dissections and trees, with applications to optimal mesh encoding and to random sampling. In *16th Annual ACM-SIAM Symposium on Discrete Algorithms*, January 2005.
9. X. He. Grid embedding of 4-connected plane graphs. In *GD '95: Proceedings of the Symposium on Graph Drawing*, pages 287–299. Springer-Verlag, 1996.
10. G. Kant and Xin He. Regular edge labeling of 4-connected plane graphs and its applications in graph drawing problems. *Theoretical Computer Science*, 172(1-2):175–193, 1997.
11. K. Miura, S. Nakano, and T. Nishizeki. Grid drawings of four-connected plane graphs. *Disc. Comput. Geometry*, 26(2):73–87, 2001.
12. D. Poulalhon and G. Schaeffer. Optimal coding and sampling of triangulations. In *Automata, Languages and Programming. 30th International Colloquium, ICALP 2003, Eindhoven, The Netherlands, June 30 - July 4, 2003. Proceedings*, volume 2719 of *Lecture Notes in Computer Science*, pages 1080–1094. Springer-Verlag, 2003.
13. G. Schaeffer. *Conjugaison d'arbres et cartes combinatoires aléatoires*. PhD thesis, Université Bordeaux I, 1998.
14. G. Schaeffer. Random sampling of large planar maps and convex polyhedra. In *Annual ACM Symposium on Theory of Computing (Atlanta, GA, 1999)*, pages 760–769 (electronic). ACM, New York, 1999.
15. W. Schnyder. Planar graphs and poset dimension. *Order*, 5:323–343, 1989.
16. W. Schnyder. Embedding planar graphs on the grid. In *SODA '90: Proceedings of the first annual ACM-SIAM symposium on Discrete algorithms*, pages 138–148, 1990.
17. W. T. Tutte. A census of planar triangulation. *Canad. J. Math.*, 14:21–38, 1962.

A Hybrid Model for Drawing Dynamic and Evolving Graphs[*]

Marco Gaertler and Dorothea Wagner

Universität Karlsruhe (TH), Faculty of Informatics, 76128 Karlsruhe, Germany
{gaertler, dwagner}@informatik.uni-karlsruhe.de

Abstract. Dynamic processes frequently occur in many applications. Visualizations of dynamically evolving data, for example as part of the data analysis, are typically restricted to a cumulative static view or an animation/sequential view. Both methods have their benefits and are often complementary in their use. In this article, we present a hybrid model that combines the two techniques. This is accomplished by 2.5D drawings which are calculated in an incremental way. The method has been evaluated on collaboration networks.

1 Introduction

Dynamic graphs occur in many applications such as software visualization, animation of graph algorithms or social network analysis. Most of the time a dynamic graph is given by a sequence of graphs that each are snapshots of an ongoing process. While the visualization of individual points in time helps to understand the current situation, a visualization of the whole sequence can reveal information about the evolution in general. So far most visual representations use either a static cumulative view of the sequence or a dynamic animation.

We describe a new hybrid model for dynamic graph drawing that allows a simultaneous representation of both, a cumulative and an animated view. Both views are integrated in such a way that the hybrid layout reveals each of them by changing the perspective or adjusting visual effects, like color or transparency. It is assumed that not only the graph structure but also weights of nodes and edges change over time. A benefit of our approach is the integration of the past evolution of weights by incorporating a cumulative as well as a regressive change, i. e., the weights of nodes and edges reflected in the drawing can also decrease over time. Our approach uses $2.5D$ drawings where time is represented by the third dimension. However, the technique can be generalized to $d.5D$ drawings for arbitrary dimensions d.

Multidimensional visualizations where one or more axes are fixed have been proposed frequently for network data from various applications. Related methods

---

[*] Work partially supported by European Commission - Fet Open project COSIN - COevolution and Self-organisation In dynamical Networks - IST-2001-33555, by European Commission - Fet Open project DELIS - Dynamically Evolving Large Scale Information Systems - Contract no 001907.

P. Healy and N.S. Nikolov (Eds.): GD 2005, LNCS 3843, pp. 189–200, 2005.

use the third dimension to display structural information [3], a hierarchy [4, 8, 1], or an evolution over time [2]. Other visualization techniques for dynamics that are based on conventional 2D or 3D drawings or animations are [9, 11, 5]. See also [6] for an overview and [10] for a more recent work.

The paper is organized as follows. Section 2 introduces our model and the corresponding layout technique. It also includes a short discussion of its benefits and potential drawbacks. The special case of evolving graphs and updating dynamic layouts are topic in Section 3. The results are presented in Section 4. For illustrative purpose, data from the DBLP[1] are used. Finally, Section 5 gives the conclusion.

2   Hybrid Model

In this section, the basic hybrid model is introduced. Section 2.2 and 2.3 provide the description on the model and the algorithmic realization. This is followed by a short discussion of accumulating weights over time while preserving the mental map.

2.1   Notation

A dynamic graph $\mathcal{G}$ is given by a mapping of a time interval $\mathcal{T}$ into the set of weighted graphs. In the following, we assume that there are only finitely many different images of $\mathcal{G}$ and $\mathcal{G}(t) = (V(t), E(t))$ denotes the graph at time $t \in \mathcal{T}$. Without loss of generality we assume that $\mathcal{T}$ can be covered by left-closed and right-open intervals $[t, t'[$ such that $\mathcal{G}$ is fixed on each such interval, and changes on subsequent intervals. For any given point in time $t \in \mathcal{T}$, we denote the earliest time of the left-adjacent corresponding interval with $\mathrm{pred}(t)$ and the earliest time of the right-adjacent corresponding interval with $\mathrm{succ}(t)$. Let $\mathcal{G}(t) = (V(t), E(t))$ be the graph at time $t \in [t_1, t_2[$, then the nodes that have not been in any previous graph are denoted by $V_{\mathrm{new}}(t) := \{v \in V(t) \mid \forall t' < t_1 : v \notin V(t')\}$ and corresponds to 'new' nodes. Similarly $V_{\mathrm{old}}(t) := V(t) \setminus V_{\mathrm{new}}(t)$ denotes the 'old' nodes. Let ω_t denote the weight of a node respectively edge at time t, i.e., $\omega_t \colon V(t) \cup E(t) \to \mathbb{R}_0^+$.

In this way a dynamic graph corresponds to the observations of the (dynamic) process and approximates it with a step function. The changes need not be homogeneously distributed over time and additional observations could be created artifically using interpolation. Furthermore, it reflects the realized changes in the dynamic graph drawing setting.

2.2   Paradigm

The original dynamic graph drawing problem has two realizations: First, the *cumulative view*, which consists of one static layout that emphasizes major trends during the evolution but hides sporadic fluctuations in the graph structure. Secondly, the *animated* or *sequential view*, which requires a static layout for each

---

[1] http://www.informatik.uni-trier.de/~ley/db/

graph that highlights current changes while preserving the general mental map. Both have in common that they are based on the whole sequence. We address a more general problem: Given a dynamic graph $\mathcal{G}$ and its time interval $\mathcal{T} := [0, T[$. For any subinterval $\mathcal{T}' = [t_1, t_2[$ of $\mathcal{T}$ construct suitable layouts that represent the evolution during $\mathcal{T}'$ based on the history of the interval $[0, t_1[$.

Our hybrid model consists of one $2D$ layout for each graph of a sequence embedded layer-wise in $3D$ where the additional dimension represents the time. A sketch of this situation is given in Figure 1. To be more precise, we use one layer for the history $[0, t_1[$ and one layer for each different graph in the interval $[t_1, t_2[$. By tuning the perspective and the individual properties of the layers, the original views, i. e., cumulative and sequential, are obtained: Looking along the time axes (in its negative direction) yields the cumulative view while showing only one layer at a time results in the sequential view. A third

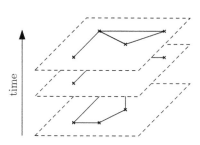

Fig. 1. Hybrid model

kind of view is obtained when identical nodes in different time slots are connected and the perspective is parallel to the layers. It shows the nodes' changes over time. The model is realized by an incremental layout algorithm. First, the history-layer is initialized with a suitable layout obtained by some established algorithm. For every additional layer the nodes are split into two groups, old and new nodes. The old nodes can be easily placed respecting their former positions. The new nodes first need a good initial placement, before the whole layout can be optimized to meet esthetic criteria as well as preserving the mental map.

2.3 Algorithmic Framework

The incremental algorithm, which is associated with the hybrid model, is given in pseudo code in Algorithm 1. It has a large degree of freedom that allows to derive several versions which are optimized for running time, achieved quality, or dependency of temporal knowledge. Especially the last issue also provides means to layout dynamic graphs where only partial information is available during the process. However, a fundamental problem might occur through lack of (future) knowledge, i. e., the position of a connected component which is completely contained in $V_{\mathrm{new}}(t)$ has great influence on the overall quality of a drawing. But the component cannot be properly placed without using information about its future role. Section 3.1 discusses these aspects for dynamic graphs with complete information (evolving graphs).

In the following, some simple methods for Step 1 and 2 are stated. For the initial placement, we suggest a two step approach which combines a barycentric layout with a localized force-directed relaxation. In this way, new nodes are close to their older "anchor" nodes and 'uniformly' spread. This requires that every connected component of $\mathcal{G}(t)$ has at least one node in $V_{\mathrm{old}}(t)$. As for the general optimization step, a modified force-directed approach works well if no further layout properties have to be ensured. The modifications mainly target the incorporation of node and edge weights, i.e., large/heavy nodes should be well distributed, thick/heavy edges should be short.

2.4   Adjustments for Weights and Position

As mentioned in the introduction, the hybrid model is capable of dealing with decreasing weights. In fact, we propose an updating of the weights of nodes and edges to incorporate both their current and their accumulated weight. Basically, there is a tradeoff between old, heavy, and inactive nodes versus young, light, and extremely active nodes. For every interval $[t_2, t_3[$ and its left-adjacent interval $[t_1, t_2[$, we define a semi-cumulative version $\widetilde{\omega}$ of ω_t as shown in Equation (1).

$$\widetilde{\omega_c}(v, t_2) := \frac{\widetilde{\omega}(v, t_1)}{s} + \frac{\omega_{t_2}(v)}{t_2 - t_1} \qquad \text{continuous version}$$

$$\widetilde{\omega_d}(v, t_2) := \frac{\widetilde{\omega}(v, t_1)}{s^{t_2 - t_1}} + \omega_{t_2}(v) \qquad \text{discrete version.}$$

(1)

If there is no left-adjacent interval, then $\widetilde{\omega}(v, t) := \omega_t(v)$. The function $\widetilde{\omega}$ can be extended to a continous function via interpolation. Depending on a scaling parameter s, different behavior is favored, i.e., for $s = 1$ it is the standard cumulative version, for $s > 1$ young and active nodes are preferred over old and inactive nodes while it is vice versa for $0 < s < 1$. The difference between the two versions is the interpretation of time, i.e., the continuous version assumes that the weight $\omega_t(v)$ has been accumulated since the last observation t_1 while the discrete version interprets the weight $\omega_t(v)$ as an instantaneous impulse at time t and that no other impulse has occurred since time t_1. Both models can be justified, the continuous weighting reflects steady growth in contrast to singleton events during an elementary time window that is imitated in the discrete version. In collaboration networks which are restricted to certain publications, like certain conference publications only, one would prefer the discrete version over the continuous one because of the time dependency.

Algorithm 1: Generic hybrid layouter

Input: dynamic graph $\mathcal{G}$ with time interval $[0, T[$ and a subinterval $[t_1, t_2[\subseteq [0, T[$
initialize $\mathcal{G}([0, t_1[)$ with a suitable layout
$t \leftarrow t_1$
while $t \leq t_2$ **do**
    adjust weights on nodes and edges
    for $v \in V_{old}(t)$ **do**
        └ initialize v with its last used position
1   initialize $V_{new}(t)$
2   optimize $\mathcal{G}(t)$ to meet esthetic criteria while preserving the mental map
    └ $t \leftarrow \text{succ}(t)$
project each $G(t)$ to the t.th layer

Similar to other approaches, we introduce additional forces to ensure that the movement of old nodes stays uniform. There are several different ways to anchor a node to its copies in different snapshots. The simplest approach introduces an edge between two identical nodes in consecutive snapshots of ideal length zero. Thus the copies of a node are connected via a path. By introducing additional edges the movement of a node can be further restricted. In the extreme case, all copies of a node are connected to a clique. This type of connection ensures best to preserve the mental map but might slow down the actual layout computation. However, since only one (time) layer is active at a point in time t, there are only T additional positions at which a node might be anchored, where T is the number of previous intervals. Thus even a clique-like connection between identical nodes results in only $\mathcal{O}(T \cdot |V(t)|)$ additional active edges which does not slow down the computation too much. Actually, most of the known techniques to control the movement of nodes can be directly integrated in the hybrid model and its incremental layout method.

3   Extensions of the Hybrid Model

After the basic hybrid model has been introduced in Section 2, two adjustments for specific tasks are presented. First, the modification for evolving graph, i.e., dynamic graph where the whole function is given, and second, a dynamic version of the dynamic graph drawing problem, i.e., given a dynamic graph with a layout, find an extention of this layout if additional time layers are introduced.

3.1   Adjustments for Evolving Graphs

As already mentioned in Section 2.3, incremental layouts cannot find a good position for connected components consisting of only new nodes. Algorithms that are based on the whole sequence avoid this problem through their 'future' dependencies. For example, in [5, 10] identical nodes in consecutive (time-)layers are connected with an edge. During the minimization of the overall forces, a good position of the connected components in early layers is ensured by the position in subsequent layers in which the component has been connected to an already placed part. Thus the relative placement is propagated back in time. A similar scheme can be integrated in the hybrid model: First, the earliest succeeding time is calculated in which the component is connected to some already placed nodes. This layer is then used to estimate the relative position of the component to its anchor nodes and projected back. The optimization step (Algorithm 1, Step 2) treats the components independently and uses the relative placement to ensure that none of them are interfering with each other. If a component has no anchoring nodes, its placement is independent from the remaining graph and can be done arbitrarily. This additional step can be done in linear time plus the time for finding the relative position, which depends on the involved layout technique (Step 1 and 2).

Potential drawbacks are the overhead, if several connected components have anchoring nodes in different time layers, then the relative position for each component involves a 'whole' layout step for each corresponding time-layer. Also, the case where some connected components of a time layer get connected with each other before anchoring nodes appear is a bit problematic. But the above method can be extended to include these cases as well. A different issue is the impact on the overall quality, i. e., because only a relative placement is estimated, certain areas of the layout can become wide-stretched while others are too condensed. This usually happens if the components are rather sparsely connected. By manually adjusting the strength of edges, one can counterweight this effect, however simultaneously diminish the relation between distance in the layout and edge weights.

3.2   Updating Layouts of Dynamic Graphs

In contrast to the connected components, the hybrid models benefits the extension or update of the dynamic graph layout when additional data become available. In other words, given two dynamic graphs $\mathcal{G}$ and $\mathcal{G}'$ such that $\mathcal{G}'$ refines $\mathcal{G}$, i. e., both graphs coincide on some parts of their time interval and differ on the remaining, and a layout $\mathcal{L}$ for $\mathcal{G}$, find a new layout $\mathcal{L}'$ for $\mathcal{G}'$ such that if $\mathcal{G}(t) = \mathcal{G}'(t)$ then also $\mathcal{L}(t) = \mathcal{L}(t')$. This can be interpreted as constraint dynamic graph drawing problem.

Independent of the algorithm, one can always use interpolation of two adjacent fixed time layers for intermediate layers. Using the hybrid model, our approach is to refine the interpolation via bisection, i. e., calculating a rough estimate of the layout for an intermediate layer and using this as an auxiliary layout for the interpolation, more precisely: Let $[t_1, t_{2k}]$ be an interval on which $\mathcal{G}$ is constant and $t_1 < t_2 < \cdots < t_{2k}$ be a subdivision such that $\mathcal{G}'$ is constant on $[t_i, t_{i+1}]$ for $1 \leq i < 2k$ and differs from $\mathcal{G}$. First, a rough placement $L(t_k)$ for time t_k is estimated and afterwards recursively applied to the interval $[t_1, t_k]$. Upon reaching time t_2, an 'exact' placement is calculated instead of the rough placement. Afterwards layouts for $t_3, \ldots, t_k$ are determined in the incremental fashion of the hybrid model. The process is then repeated on the interval $[t_k, t_{2k}]$.

Instead of the bisection approach one could only use the incremental algorithm to interpolate the interval $[t_1, t_{2k}]$, however, if the intervals have many unknown intermediate points or the layout for t_1 and t_{2k} differs a lot, then the overall quality significantly drops.

Consistency. So far, the hybrid model and other dynamic visualization algorithms behaved similar to the update problem. However, there is a difference when comparing the $\mathcal{L}'$ with the layout $\mathcal{L}''$ for $\mathcal{G}'$ ignoring the constraint $\mathcal{L}$. A fully time-dependent algorithm, like the one in [10] can produce very different results for $\mathcal{L}'$ and $\mathcal{L}''$, while general incremental algorithms will give the same partial layouts on the interval $[t_0, t_1]$ where t_0 is the earliest time and t_1 the time of the first deviation of $\mathcal{G}$ and $\mathcal{G}'$.

Also, the hybrid model will produce the same partial layout on $[t_0, t_1]$. Furthermore, if the modifications of the intermediate time slots are small or even consistent with our continual weighting (Section 2.4), then the layouts $\mathcal{L}'$ and $\mathcal{L}''$

of the hybrid model will be very similar. The following observation verifies this claim: If the introduced modifications are small, then both graphs, the original and the modified one, should have similar high-quality layouts. Moreover using a local optimum of force-direction layout of the original graph as initialization for the modified one will quickly convert to a close local optimum.

Thus the similarity of $\mathcal{L}'$ and $\mathcal{L}''$ on a refined interval $[t, t']$ highly depends on the similarity or consistency of the intermediate graphs and the impact on previous modifications but not on succeeding ones. An extreme case would be consistent refinement together with a large modification at the end of the sequence. Traditional algorithms that use the whole available information would produce very different layouts $\mathcal{L}'$ and $\mathcal{L}''$ while incremental and especially the hybrid model would result in very different layouts upto the heavy modification. However, this is paid in terms of achieved overall layout quality.

4 Results

We illustrate some results of our hybrid model for citation networks extracted for the DBLP which is a well-maintained database with approximately 500,000 articles in the area of computer science.

4.1 Data Sets

DBLP maintains information of certain publications. We extracted the overall collaboration graph, i.e., nodes are people and edges connect to nodes if they have common publications. Because the publication activity varies a lot and a single publication can have up to 36 authors, we weighted the edges correspondingly. The weight of a single publication is reciprocal to the number of authors and the weight of all publications in a year is the sum of the individual weights. The weight of a node is the sum of the weights of its incident edges (for a given year).

4.2 Visualizations

In the following, we present several drawings of collaboration networks. In each visualization, there are the following correspondences: node size and cumulative publication weight, node color and time, edge thickness and publication weight, edge color and time. If an edge has a checked pattern, it connects two identical nodes in consecutive snapshots. When speaking of cumulative weight, we always refer to the continuous version shown in Equation (1).

A first example is the authorship of [7], which is one of the first books about graph drawing. Figure 2(a) shows the evolution of their collaboration between 1986 and 2000. It is clearly visible when common publications have occurred, although individual publications are not identifiable. For example the first collaboration between the four authors happened in 1994. Also the node size reflects the continuity of cooperations between authors. However, the visualization

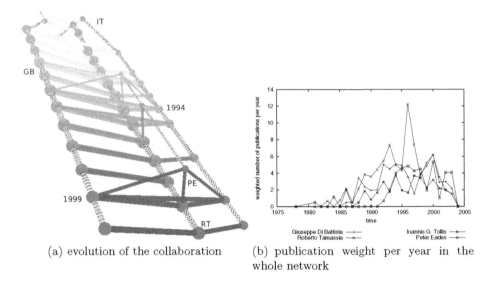

(a) evolution of the collaboration

(b) publication weight per year in the whole network

Fig. 2. Collaboration between Giuseppe Di Battista (GB), Ioannis Tollis (IT), Peter Eades (PE) and Roberto Tamassia (RT) between 1986 and 2000

in Figure 2(a) is limited to the four authors and only their collaboration. Figure 2(b) shows the publication weight of each author within the whole collaboration network. Also, Figure 2(a) illustrates the effect of consistent modifications (Section 3.2) quite well. Between 1991 and 2000, the modifications have been very small, i. e., some reweighting on the nodes and edges, one node appeared while another node disappeared for some time, but the overall layout has been quite stable.

The second example is the collaboration between Ulrik Brandes, Dorothea Wagner and their direct coauthors. Both have published several articles in the graph drawing area and others. Figure 3(a) shows the evolution for 2001 and 2002. The whole network is rather dense, a static cumulative view (without edge weights) is given in Figure 3(c). The static graph has 50 nodes and 161 edges while the time-expanded graph has 206 nodes and 434 edges. Some individual layers are presented in Figure 3(d) and 4 and present the collaboration at specific points in time. In every layer only those nodes that have published something are shown. Again Figure 4(a) and 4(b) clearly indicate that the node size is only relative to the selected network, i. e., nodes like Peter Eades, Joe Marks or Michael Kaufmann who have a large weight in the whole network (see for example Figure 2(b)) have a rather peripheral role in this collaboration network. The balance between old and active nodes and young and active nodes is also visible in Figure 3(a). As shown in Figure 3(b), Ulrik Brandes and Dorothea Wagner have roughly the same amount of weighted publications since 1997. However, Dorothea Wagner has been active since 1989, while Ulrik Brandes started in 1997. However, in the evolutionary view both have a similar size which reflects the similar accumulated publication weight. Using a purely cumulative

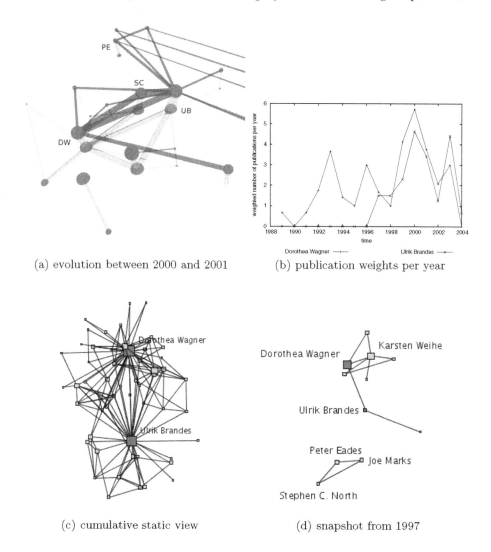

(a) evolution between 2000 and 2001   (b) publication weights per year

(c) cumulative static view   (d) snapshot from 1997

Fig. 3. Collaboration between Ulrik Brandes (UB), Dorothea Wagner (DW) and their direct coauthors between 1989 and 2002. (Other abbreviations are SC for Sabine Cornelsen and PE Peter Eades.)

weight, this would not be the case. Finally, Figure 5 shows the collaboration in a broader sense, i. e., a network with increased number of intermediate coauthors. The visualization shows a sparse connection between the main part that contains both Ulrik Brandes and Dorothea Wagner and a peripheral part.

The final example is a collection of some program committee members of the International Symposium on Graph Drawing. Figure 6(a) shows a 2D project of the evolution between 1986 and 2003 which mask the time axis while a perspective view is given in Figure 6(b). This example reflects both that certain groups

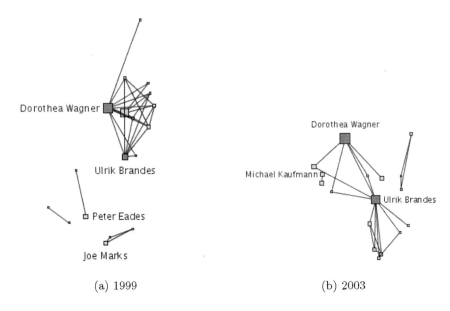

(a) 1999   (b) 2003

Fig. 4. Snapshots of the collaboration between Ulrik Brandes, Dorothea Wagner and their direct coauthors at different points in time

Fig. 5. Partial view of the collaboration between Ulrik Brandes and Dorothea Wagner using more intermediate coauthors

are formed over time that collaborate very closely, but also that occasional collaboration tend to be repeated. Some of the artifacts on the layer 2003 are due to incomplete data.

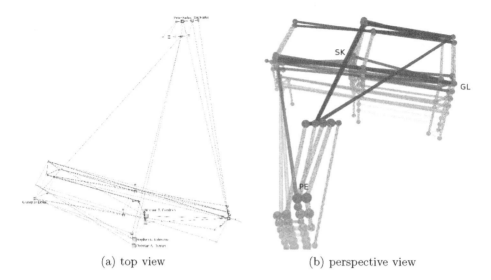

(a) top view   (b) perspective view

Fig. 6. Collaboration between some PC member between 1986 and 2003. (Abbreviations are GL for Guiseppe Liotta, PE Peter Eades and SK for Stephen Kobourov.)

5   Conclusion

We introduced a hybrid model for drawing dynamic and evolving graphs based on 2.5D visualizations. It combines several aspects of static cumulative views and animated/sequential views. The obtained layout permits the general view of the evolution while integrating individual aspects of certain points in time as well as cumulative and regressive changes of weight functions. The method has been evaluated on networks modeling collaboration. In the layouts, structural important nodes were well visible and long-existent nodes did not mask younger nodes. Also sparsely connected components were spatially separated.

References

1. M. Baur, U. Brandes, M. Gaertler, and D. Wagner. Drawing the as graph in 2.5 dimensions. In *Proceedings of the 12th International Symposium on Graph Drawing (GD'04)*, Lecture Notes in Computer Science, pages 43–48, 2005.
2. U. Brandes and S. Corman. Visual unrolling of network evolution and the analysis of dynamic discourse. *Information Visualization*, 2(1):40–50, 2003.
3. U. Brandes and S. Cornelsen. Visual ranking of link structures. *Journal of Graph Algorithms and Applications*, 7(2):181–201, 2003.
4. U. Brandes, T. Dwyer, and F. Schreiber. Visual understanding of metabolic pathways across organisms using layout in two and a half dimensions. *Journal of Integrative Bioinformatics*, 0002, 2004.
5. U. Brandes and D. Wagner. A Bayesian paradigma for dynamic graph layout. In Giuseppe Di Battista, editor, *Proceedings of the 5th Symposium on Graph Drawing (GD'97)*, volume 1353 of *Lecture Notes in Computer Science*, pages 236–247, Rome, Italy, 1997. Springer-Verlag.

6. J. Branke. Dynamic graph drawing. In M. Kaufmann and D. Wagner, editors, *Drawing Graphs*, volume 2025 of *Lecture Notes in Computer Science*, pages 228–246. Springer-Verlag, 2001.
7. Guiseppe di Battista, Peter Eades, Roberto Tamassia, and Ioannis G. Tollis. *Graph Drawing - Algorithms for the Visualization of Graphs*. Prentice Hall, 1999.
8. P. Eades and Q. Feng. Multilevel visualization of clustered graphs. In *Proc. of Graph Drawing*, volume 1190 of *Springer LNCS*, pages 113–128. Springer, 1996.
9. P. Eades, W. Lai, K. Misue, and K. Sugiyama. Preseving the mental map of a diagramm. In *Proceedings of Compugraphics '91*, pages 24–33, 1991.
10. C. Erten, P. Harding, S. Kobourov, K. Wampler, and G. Yee. Graphael: Graph animations with evolving layouts. In *Proceedings of the 11th International Symposium on Graph Drawing (GD'03)*, Lecture Notes in Computer Science, pages 98–110, 2004.
11. S. C. North. Incremental layout with dynadag. In *Proceedings of the 3rd International Symposium on Graph Drawing (GD'95)*, volume 1027 of *Lecture Notes in Computer Science*, pages 409–418. Springer-Verlag, 1996.

Two Trees Which Are Self–intersecting When Drawn Simultaneously

Markus Geyer[1], Michael Kaufmann[1], and Imrich Vrťo[2,⋆]

[1] Universität Tübingen, WSI für Informatik, Sand 13,
72076 Tübingen, Germany
{mk, geyer}@informatik.uni-tuebingen.de
[2] Institute of Mathematics, Slovak Academy of Sciences,
Dúbravská 9, 841 04 Bratislava, Slovakia
vrto@savba.sk

Abstract. An actual topic in the graph drawing is the question how to draw two edge sets on the same vertex set, the so-called simultaneous drawing of graphs. The goal is to simultaneously find a nice drawing for both of the sets. It has been found out that only restricted classes of planar graphs can be drawn simultaneously using straight lines and without crossings within the same edge set. In this paper, we negatively answer one of the most often posted open questions namely whether any two trees with the same vertex set can be drawn simultaneously crossing-free in a straight line way.

1 Introduction

Recently, a new direction in the area of the graph drawing has been opened: Simultaneous planar graph drawing [1, 3, 4, 5, 6]. Consider a set of objects with two different sets of relations. Such structures arise in many applications, e.g. in software engineering, databases, and social networks. The goal is to draw both underlying graphs on the same set of vertices in the plane using straight lines such that each graph alone is displayed as nicely and readable as possible. In case that both graphs are planar, we require that every graph itself is embedded in a plane way. More formally, given two planar graphs $G_1 = (V, E_1)$ and $G_2 = (V, E_2)$, *simultaneous drawing* of G_1 and G_2 is to find their plane straight line drawings D_1 and D_2, such that every vertex is mapped to the same point in both D_1 and D_2. Brass et al. [1] proved that two paths, two cycles and two caterpillars can always be drawn simultaneously. A caterpillar is such a tree that the graph obtained by deleting the leaves is a path. On the other hand, they constructed 2 outerplanar graphs for which the simultaneous drawing is impossible. Erten and Kobourov [5] found an example of a planar graph and a path that do not allow a simultaneous drawing. The most posted open problem in this area is the question whether two trees can always be drawn simultaneously [1, 2, 5]. In this

---

⋆ Supported by a DFG grant "Graphenzeichnen in der Anwendung" No. K812/8-2 and a VEGA grant No. 2/3164/23.

P. Healy and N.S. Nikolov (Eds.): GD 2005, LNCS 3843, pp. 201–210, 2005.

paper we answer this question in negative. Our counterexample consists of two isomorphic trees of depth 2.

2   The Counterexample

The two trees $T_1(n) = (V, E_1)$ and $T_2(n) = (V, E_2)$ are given as follows: $T_1(n)$ and $T_2(n)$ have a common root r with n common children $v_1, \ldots, v_n$. The parameter n will be determined later. The children $v_i, 1 \leq i \leq n$ of r have again children $v_{ij}, 1 \leq i, j \leq n, i \neq j$, s.t. $(v_i, v_{ij}) \in E_1$ and $(v_j, v_{ij}) \in E_2$. We call the edges in $E_1 \cap E_2$ thick black, in $E_1 \setminus E_2$ thin black and those from $E_2 \setminus E_1$ thin gray. We denote the union of the two trees by G_n. A straight line drawing of G_n is called *partially planar* if there is no crossing of 2 edges from E_1 nor crossing of 2 edges from E_2, which is equivalent to the simultaneous drawing of $T_1(n)$ and $T_2(n)$. Fig. 1 shows a partially planar drawing of G_4. Note that this graph class has already been described in [5].

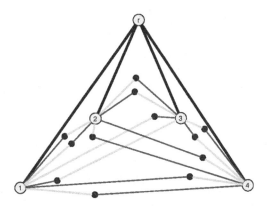

Fig. 1. An example that shows a partially planar drawing of G_4

Theorem 1. *For $n \geq 15$, the simultaneous drawing of G_n is self–intersecting.*

Proof. Since G_{15} is a subgraph of $G_n, n \geq 15$ it is sufficient to prove that any simultaneous drawing for G_{15} is self–intersecting.

Let us assume that there is a partially planar layout L_{15} for G_{15}. We consider such a layout L_{15} and derive a contradiction.

The proof proceeds in three steps:

Lemma 1. *In any partially planar layout L_{15}, there are 8 children of the root such that in the corresponding sublayout L_8 for the subgraph G_8 induced by the root r, the 8 children and the leafs on the connections between them, the root r lies on the outer face of L_8.*

In the following, we only argue on the layout L_8 of the 8 children from the previous lemma and derive a contradiction for L_8. The indexing is done as the children appear in clockwise order seen from the root. We also consider our indexing to be counted $(\bmod\ 8) + 1$ (or $(\bmod\ 5) + 1$ respectively), s.t. after v_8 (or v_5 respectively) we have v_1 again in clockwise order.

Lemma 2. *Let G_5 be any subgraph of the G_8 induced by the root r, by any 5 children $v_1, ..., v_5$ out of the eight children of the root and by the leafs on the connections between these 5 vertices. For all vertices $v_i, 1 \leq i \leq 5$, the two 4-gons defined by the straight-line segments $(v_i, v_{ij}), (v_{ij}, v_j), (v_j, v_{ji})$ and $(v_{ji}, v_i), j = i - 2, i + 2$ do intersect.*

And finally

Lemma 3. *For any layout L_8 of G_8 as defined above there is a vertex v_i, $1 \leq i \leq 8$, such that the two 4-gons defined by the straight-line segments $(v_i, v_{ij}), (v_{ij}, v_j), (v_j, v_{ji})$ and $(v_{ji}, v_i), j = i - 2, i + 2$ do not intersect.*

Lemma 3 is obviously in contradiction to Lemma 2, therefore no such layout could exist. □

In the following section, we provide the proofs of the three lemmata plus all the necessary definitions and useful observations concerning the structures of the layout.

3   The Proofs

3.1   Identifying an Appropriate Subgraph

Lemma 4. *In any partially planar layout L_{15}, there are 8 children of the root such that in the corresponding sublayout L_8 for the subgraph G_8 induced by the root r, the 8 children and the leafs on the connections between them, the root r lies on the outer face of L_8.*

Proof. Let L_{15} be a partially planar layout of graph G_{15}. Let $C = \{v_1, ..., v_{15}\}$ be the children of the root in clockwise order. We identify two children v_i and v_j such that the polygon formed by $(r, v_i), (v_i, v_{ij}), (v_{ij}, v_j)$ and (v_j, r) encloses a maximal number of children of the root. Note that v_i and v_j may not be unique.

Let S be the set of children within the polygon with $|S| = k$. It is easy to see that for the whole subgraph G_k induced by the root r, the children in S and the leafs on the connections between them, the root r lies on the outer face. If $k \geq 8$ we are done and can arbitrarily choose 8 of the children in S to form our subgraph G_8. If $k < 8$, we consider the set $C \setminus S$ of size $l \geq 8$. By the choice of i and j we know that all but one of the connections within $C \setminus S$ lie on the 'same' side of the root such that removing only the two segments (v_i, v_{ij}) and (v_{ij}, v_j) will bring r to the outer face of the layout for the subgraph G_l induced by the root r, the set $C \setminus S$ and the corresponding leafs. For this case, any subset of $C \setminus S$ of size 8 will provide us the desired subgraph G_8. □

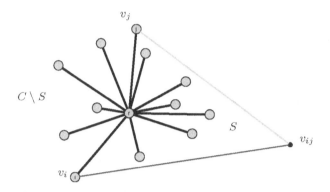

Fig. 2. An example for the choice of v_i and v_j

3.2   Structural Characterizations

After having identified a partially planar layout L for a G_8 such that the root lies at the outer face, we start with some characterizations to prepare the proof of the main lemma.

Take any 5 vertices $v_1, \ldots, v_5$ out of those 8 children in clockwise order. Consider the corresponding graph G_5 induced by the root, by $v_1, \ldots, v_5$ and by $v_{ij}, 1 \leq i, j \leq 5, i \neq j$ and its corresponding layout L_5. Clearly L_5 is partially planar.

- Note that each leaf v_{ij} has actually two adjacent vertices, namely v_i and v_j by a gray edge and by a thin black edge. We can also view each pair of vertices v_i and v_j as being connected by two 2-segment polylines, where one is colored gray/thin black and the other thin black/gray.
- The two connecting 2-segments between each pair v_i and v_j form a 4-gon P_{ij}. It is clear that the participating four segments do not cross.
- None of the 5 vertices lies inside of the P_{ij} and none is enclosed by a sequence of P_{ij}'s. This means that each of them lie on the outer face of the planar subdivision formed by the edges $v_i v_{ij}$ and $v_j v_{ij}$. This is enforced by the black edges from the root to the vertices v_i and by our condition that none of the P_{ij} or a sequence of those enclose the root.
- We say that polygons P_{ij} and P_{lk} are intersecting if a segment of P_{ij} crosses a segment of P_{lk}. Otherwise they are independent.
- Note that a vertex v_i can only have two neighboring vertices v_j, v_k. That are vertices, such that the polygons P_{ij} and P_{ik} are not separating any two vertices. That means the polygons P_{ij} and P_{ik} can be assumed to be independent from each other and from the remaining 4-gons. We also assume from now on, that our numbering reflects this neighbor property and is in clockwise order, e.g. v_i is neighbor to v_{i+1} for $i \in \{1, \ldots, 4\}$ and v_5 is neighbor to v_1.
- The following three configurations for two intersecting polygons P_{ij} and P_{lk} with $i < l < j < k$ are the basics.

Configuration 1: The two leaves incident to P_{ij} lie inside the polygon P_{lk} and the two leaves incident to P_{lk} lie inside the polygon P_{ij}.
Configuration 2: Exactly one leaf of P_{ij} lies inside the polygon P_{lk} and exactly one leaf of P_{lk} lies inside the polygon P_{ij}.
Configuration 3: The two leaves incident to P_{ij} lie outside the polygon P_{lk} and the two leaves incident to P_{lk} lie outside the polygon P_{ij}. See Fig. 3.
- Note that for each polygon the colors can be switched.

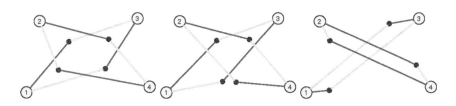

Fig. 3. Configuration 1, 2 and 3

Next we state the main structural lemma, which is identical to our former Lemma 2:

Lemma 5. *In any partially planar drawing of G_5, for each $v_i, 1 \le i \le 5$, there are two 2-segment connections to v_{i+2} and to v_{i-2} that cross.*

Proof. Assume that there is a vertex v_i contradicting the claim. By renumbering, we assume that $v_i = v_1$. This means that the polygons P_{13} and P_{14} are independent.

In what follows we perform a case analysis. On the top level we distinguish two different clockwise orders of the four incident segments of the two polygons attached to v_1:

A) thin black, gray, gray, thin black and
B) thin black, gray, thin black, gray.

Clearly, both polylines of P_{14} separate v_3 from v_5. So P_{14} and P_{35} are intersecting.
Then we discuss for each of A) and B) the Configurations 1,2 and 3, described above, for mutual positions of P_{14} and P_{35}.
And finally, for each of the previous cases we have two subcases:

a) Polygons P_{31} and P_{35} are independent.
b) Polygons P_{31} and P_{35} are intersecting.

Vertex v_2 is on the convex hull between v_1 and v_3. It now has to be connected by two bicolored curves to v_4 as well as to v_5. We describe the possible route of the four paths by the sequence of segments that have to be crossed. Fortunately, this sequence is almost unique.
Before we dive into the case analysis, we formulate some conditions for the solvability:

- incident–segments–condition: Two straight line segments adjacent to the same vertex obviously cannot cross.
- straightness–condition: Two straight line segments cross at most once.
- one–two–condition: Consider a drawing where segments s and s' are adjacent to vertex v, s' forms a double-segment with s'' and in addition s'' crosses s. W.l.o.g. we can assume that there is no such configuration, since any such configuration can easily be redrawn into a configuration where the crossing has been removed.

We assume that the one–two–condition is obeyed in the solution and we always construct a contradiction to the incident–segments condition:

Case A1a: In the Fig. 4(a) consider the two dashed polygonal lines connecting the vertices v_2 to v_4 and v_5 respectively. They indicate the potential route of the corresponding double–segments. Clearly the observation holds that at least one of the two curves have to change its color next to vertex v_2 or within the first polygon P_{13}. If the gray curve changes its color then it follows the thin black curve to v_5 and completely indicates the topological route of the gray-thin black curve from v_2 to v_5. Clearly, this is a contradiction to the incident-segment-condition since the last segment of the gray-thin black curve from v_2 to v_5 intersects the last segment of the thin black-gray double segment from v_3 to v_5. Similarly, if the thin black curve changes its color next to v_2 it follows the gray curve to v_4. As before, we achieve a contradiction to the incident-segment condition since the last segment of the thin black-gray curve from v_2 to v_4 intersects the last segment of the gray-thin black double segment from v_2 to v_4.

Case A1b,A2b: Since the one-two-condition is violated by the thin black-gray double segment between v_3 and v_1 and the first thin black segment between v_3 and v_5, we can safely assume that these cases does not occur.

Case A2a: Analogously as in case A1a, we argue that one of the two curves from Fig. 4(a) has to change colors next to v_2. It therefore indicates one of the routes from v_2 to v_4 or v_5. As in the case A1a, we get a contradiction to the incident-segment condition.

Case A3a: (See Fig. 4(e).) The two curves indicate the similarity to case A1a. One of the curves has to change its color close to v_2 and therefore it produces a violation of the incident–segments–condition.

Case A3b: (See Fig. 4(f).) As before, the curves and the color changing close to v_2 lead to a violation of the incident–segments condition.

Next, we will consider the case B, where we assume that the clockwise order of the edges incident to v_1 is thin black, gray, thin black, gray. The arguments are along the same lines as in case A, but for completeness we consider all the cases:

Case B1a: In Fig. 5(a), we show the two canonical curves one of which has to change its color near v_2 and then follow the other one. Clearly, the same kind of contradiction to the incident–segments condition occurs as in case A1a.

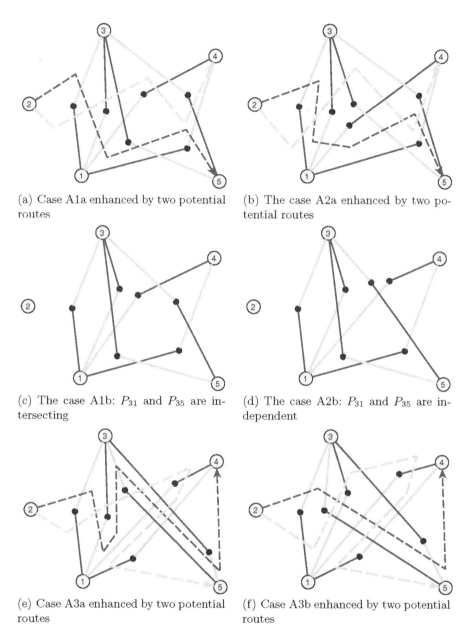

(a) Case A1a enhanced by two potential routes

(b) The case A2a enhanced by two potential routes

(c) The case A1b: P_{31} and P_{35} are intersecting

(d) The case A2b: P_{31} and P_{35} are independent

(e) Case A3a enhanced by two potential routes

(f) Case A3b enhanced by two potential routes

Fig. 4. The different cases for the clockwise ordering thin black, gray, gray, thin black

Case B1b: This case cannot occur at all since there is a crossing of segments of the same color.

Case B2a: (See Fig. 5(c).) As before, the two curves that uniquely indicate the routes induce at least one contradiction to the incident–segments–condition.

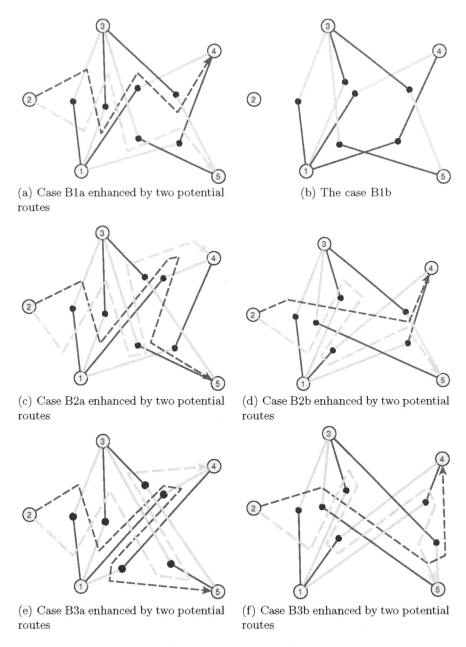

(a) Case B1a enhanced by two potential routes

(b) The case B1b

(c) Case B2a enhanced by two potential routes

(d) Case B2b enhanced by two potential routes

(e) Case B3a enhanced by two potential routes

(f) Case B3b enhanced by two potential routes

Fig. 5. The different cases for the clockwise ordering thin black, gray, thin black, gray

Case B2b: (See Fig. 5(d).) Similar to the case B2a. Although the thin black curve looks promising it violates the incident–segments–condition since it crosses the first gray segment of the double segment from v_4 to v_1.

Case B3a: (See Fig. 5(e).) The snakelike curves immediately lead to a contradiction to the incident–segments–condition.

Case B3b: (See Fig. 5(f).) Analogously to the case B3a. This concludes the proof of the main lemma. □

3.3 The Final Argument

With the next lemma we state a property for any layout of G_8, which is in direct contradiction to a property that has been shown in Lemma 5.

Lemma 6. *For any layout L_8 of G_8 as defined above there is a vertex $v_i, 1 \leq i \leq 8$, such that the two 4-gons defined by the straight-line segments $(v_i, v_{ij}), (v_{ij}, v_j), (v_j, v_{ji})$ and $(v_{ji}, v_i), j = i - 2, i + 2$ (mod 8) do not intersect.*

Proof. Assume the 8 children are numbered in clockwise order, see Fig. 6. By Lemma 5 the polygons P_{13} and P_{17} must intersect. The one–two–condition implies that both polygons lie in the halfplane given by the line $v_3 - v_7$ and the vertex v_1. Symmetrically, the polygons P_{35} and P_{57} lie in the other halfplane. Hence the polygons P_{13} and P_{35} do not intersect. □

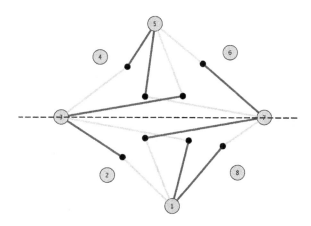

Fig. 6. Polygons P_{31} and P_{35} do not intersect

4 Conclusion

We gave an example of a class of tree pairs that are self-intersecting when drawn simultaneously, but unfortunately the parameter n implies a number of $n^2 + 1$ vertices, our smallest counterexample has size 226. We are optimistic that by more refined arguments this can be improved to $n = 8$ or even $n = 7$.

Another open question is to give a pair of edge-disjoint trees that are self-intersecting when drawn simultaneously. The class G_n can easily be generalized to contain two edge-disjoint trees but our argument for the self-intersection heavily relied on the straight-line edges that are contained in both trees. Finally, it remains an obvious task to generalize these arguments (or find new one) to prove self-intersection for simpler classes of graphs like a tree and a caterpillar.

Acknowledgments

The second author wishes to thank Stephen Kobourov for giving him enough motivation for the final push on the paper during the Dagstuhl-Seminar No. 5191 on Graph Drawing.

References

1. Brass, P., Cenek, E., Duncan, A., Efrat, A., Erten, C., Ismailescu, D., Kobourov, S., Lubiw, A., Mitchell, J.S.B.: On Simultaneous Planar Graph Embeddings. In: Dehne F., Sack, J., Snid, M. (eds.): Workshop on Algorithms and Data Structures, Lecture Notes in Computer Science, Vol. 2748. Springer, Berlin (2002) 243-255
2. Duncan, C.A., Eppstein, D., Kobourov, S.G.: The Geometric Thickness of Low Degree Graphs. In: Boissonant, J.-D., Snoeyink, J. (eds.): 23rd Annual Symp. on Computational geometry. ACM Press, New York (2004) 340-346
3. Erten, C., Kobourov, S.G.: Simultaneous Embedding of a Planar Graph and its Dual on the Grid. In: Bose, P., Morin, P. (eds.): 13th Intl. Symp. on Algorithms & Computation. Lecture Notes in Computer Science, Vol. 2518. Springer, Berlin (2002) 575-587
4. Erten, C., Kobourov, S.G., Le, V., Navabi, A.: Simultaneous Graph Drawing: Layout Algorithms and Visualization Schemes. In: Liotta, G. (ed.): 11th Intl. Symp. on Graph Drawing. Lecture Notes in Computer Science, Vol. 2912. Springer, Berlin (2003) 437-449
5. Erten, C., Kobourov, S.G.: Simultaneous Embedding of Planar Graphs with Few Bends. In: Pach, J. (ed.): 12th Intl. Symp. on Graph Drawing. Lecture Notes in Computer Science, Vol. 3383. Springer, Berlin (2004) 195-205
6. Kobourov, S.G., Pitta, C.: An Interactive Multi-User System for Simultaneous Graph Drawing. In: Pach, J. (ed.): 12th Intl. Symposium on Graph Drawing. Lecture Notes in Computer Science, Vol. 3383. Springer, Berlin (2004) 492-501

C-Planarity of Extrovert Clustered Graphs[*]

Michael T. Goodrich, George S. Lueker, and Jonathan Z. Sun

Department of Computer Science,
Donald Bren School of Information and Computer Sciences,
University of California, Irvine, CA 92697-3435, USA
{goodrich, lueker, zhengsun}(at)ics.uci.edu

Abstract. A clustered graph has its vertices grouped into clusters in a hierarchical way via subset inclusion, thereby imposing a tree structure on the clustering relationship. The c-planarity problem is to determine if such a graph can be drawn in a planar way, with clusters drawn as nested regions and with each edge (drawn as a curve between vertex points) crossing the boundary of each region at most once. Unfortunately, as with the graph isomorphism problem, it is open as to whether the c-planarity problem is NP-complete or in P. In this paper, we show how to solve the c-planarity problem in polynomial time for a new class of clustered graphs, which we call *extrovert* clustered graphs. This class is quite natural (we argue that it captures many clustering relationships that are likely to arise in practice) and includes the clustered graphs tested in previous work by Dahlhaus, as well as Feng, Eades, and Cohen. Interestingly, this class of graphs does not include, nor is it included by, a class studied recently by Gutwenger *et al.*; therefore, this paper offers an alternative advancement in our understanding of the efficient drawability of clustered graphs in a planar way. Our testing algorithm runs in $O(n^3)$ time and implies an embedding algorithm with the same time complexity.

1   Introduction

A *clustered graph* (or *c-graph*) consists of a pair $C = (G, \tau)$, where $G = (V, E)$ is an undirected graph having vertex set V and edge set E, and τ is a rooted tree defining a hierarchy of vertex clusters, which are subsets of V organized hierarchically by subset inclusion. That is, each node of τ represents a cluster that is a subset of V (with the root of τ representing V), and the ancestor-descendant relation of two nodes corresponds to the inclusion relation of two clusters. Any two clusters in this hierarchy are either disjoint or one is completely included in the other. We refer to G and τ as being the *underlying graph* and the *inclusion tree* of C, respectively. Throughout this paper, we reserve the Greek letters ν and μ for clusters and the Roman letters x and y for vertices.

Clustered graphs arise naturally from any context where a hierarchy is imposed on a set of interrelated objects. Naturally, we would like to visualize the hierarchical

---

[*] This is an extended abstract. Work by the first and the third authors is supported by NSF Grants CCR-0225642 and CCR-0312760.

information and relationships that are represented in a clustered graph, and a way of doing so with minimal confusion is to draw the clustered graph in a planar way. Deciding if such a drawing is possible is one of the most interesting problems involving clustered graphs, and was posed by Feng, Eades and Cohen [12]. Formally, this problem, which is called the *c-planarity problem*, asks if C can be drawn (or embedded) in the plane satisfying the following criteria:

1. There is no crossing between any edges of the underlying graph G.
2. Each cluster $\nu \in \tau$ can be enclosed in one simple closed region by a closed curve $b(\nu)$ which is called the *boundary curve* of ν.
3. There is no crossing between boundary curves of any two clusters.
4. There is exactly one crossing between an edge (x, y) and a boundary curve $b(\nu)$ if $x \in \nu$ and $y \notin \nu$. Otherwise there is no crossing between an edge and a boundary curve.

Such a drawing is called a *c-planar drawing*, and a clustered graph is *c-planar* iff it admits such a drawing. Clustered graphs and c-planar drawings bear both significance in theory and interest in practice. For example, if we visualize the communication network of a company such that the vertices, clusters, and edges represent respectively workstations, departments, and communications between two workstations, clearly we want a simple region and single boundary curve for each department and no crossings except those in the above Criterion 4, so that, for any department, we can identify (e.g., for monitoring, blocking, or firewalling) that department's external communications just by looking at the boundary curve of the corresponding cluster. Another example application is in VLSI, where in addition to designing a planar circuit, we might want to piece each functional module together in a hierarchical way.

While the problem of determining if a given graph is planar is well-known to be solvable in linear time (e.g., see [3, 14, 17]), the general c-planarity problem

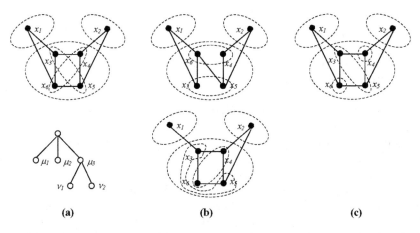

Fig. 1. (a) A c-graph C with 5 clusters $\mu_1 = \{x_1\}$, $\mu_2 = \{x_2\}$, $\mu_3 = \{x_3, x_4, x_5, x_6\}$, $\nu_1 = \{x_3, x_5\}$ and $\nu_2 = \{x_4, x_6\}$, and the inclusion tree τ. C is not c-planar although the underlying graph is planar. (b) Removing any edge of the underlying graph will make C c-planar. (c) Splitting cluster ν_1 or ν_2 will also make C c-planar.

is not known to be solvable in polynomial time. In particular, the existence of boundary curves makes the c-planar testing and embedding significantly harder than simply testing a graph for planarity. In Fig. 1, for example, the c-graph in (a) is not c-planar although the underlying graph is obviously planar, but removing one edge or splitting one cluster in this c-graph will make it c-planar, as illustrated in (b) and (c).

Although a number of papers have addressed the problem of how to draw a c-planar c-graph in the plane [1, 6, 7, 8, 9, 10], very little progress has been made in testing the c-planarity of a given c-graph. Previous work provides effective tests only for a few special classes of c-graphs [4, 5, 12, 13]. In this paper we define and test a new class of c-graphs, which generalizes the result in [5, 12] but is not comparable with [4, 13]. The general problem is still open. So far the testing problem and the embedding problem appear to be equivalent in all solved cases (since each testing algorithm implies an embedding algorithm), so we don't distinguish them unless necessary.

1.1   Previous Results

Let $G(\nu)$ be the subgraph of G induced by cluster ν. (ν is a node in τ and an associated set of vertices in V.) Then ν is a *connected cluster* iff $G(\nu)$ is a connected subgraph. Otherwise $\nu = (\nu^1, \dots, \nu^k)$ is a *disconnected cluster*, where each $G(\nu^i)$ is a connected component of $G(\nu)$.[1] That is, each ν^i is a set of vertices in a connected component of $G(\nu)$. For simplicity, we call ν^i a *chunk* of ν. A connected cluster is considered to have itself as the only chunk, i.e., $\nu = (\nu^1)$ when ν is connected. C is *c-connected* iff all clusters in τ are connected.[2] The c-planarity problem for c-connected c-graphs was solved in $O(n^2)$ time by Feng, Eades and Cohen [12], and then in linear time by Dauhlhaus [5]. For general c-graphs, it is unknown if the problem is NP-hard or not. Gutwenger *et al.* [13] solved in $O(n^2)$ time the case of *almost c-connected* c-graphs, namely, those c-graphs in which either each disconnected cluster $\nu \in \tau$ has its parent and all siblings connected, or all disconnected clusters lie on a path in τ. Cortese *et al.* [4] recently solved in polynomial time another special case, which we call the *cycles of clusters*, where the underlying graph is a cycle and the clusters at each level of the inclusion tree, when contracted into vertices, also form a cycle. To the best of our knowledge, these three classes of c-graphs are the only ones for which c-planarity has been tested in polynomial time.

1.2   Extrovert C-Graphs

We introduce the concept of classifying the disconnected clusters into *extrovert* and *introvert*, and will later solve the c-planarity problem for the case that all disconnected clusters are extrovert.

---

[1] In this paper, we will always use superscripts to denote the partition of an object, and subscripts to distinguish different objects.

[2] The previous papers simply used the term *connected* instead of *c-connected*, but we consider it desirable to introduce the new terminology to distinguish between a connected graph G and a c-connected c-graph C.

An edge $(x, y) \in E(G)$ is called an *extrovert edge* of a cluster ν iff $x \in \nu$ and $y \notin \nu$.[3] We call x an *extrovert vertex* of ν in this case. We denote by E_ν^* and V_ν^* respectively the sets of extrovert edges and extrovert vertices of ν. For a subset $\nu^0 \subset \nu$, we denote by $E_\nu^*(\nu^0)$ and $V_\nu^*(\nu^0)$ respectively the corresponding subsets of E_ν^* and V_ν^*, i.e., $E_\nu^*(\nu^0) = \{e \in E_\nu^* : e \text{ is incident on a vertex of } \nu^0\}$ and $V_\nu^*(\nu^0) = V_\nu^* \cap \nu^0$.

Definition 1. (extrovert chunks, clusters, and c-graphs)

- A *chunk* ν^i of a disconnected cluster $\nu = (\nu^1, \ldots, \nu^k)$ is an *extrovert chunk* iff the parent cluster μ of ν is connected, and $E_\mu^*(\nu^i) \neq \emptyset$.
- A *disconnected cluster* $\nu = (\nu^1, \ldots, \nu^k)$ is an *extrovert cluster* iff each chunk ν^i, $i \in \{1, \ldots, k\}$, is extrovert.
- $C = (G, \tau)$ is an *extrovert c-graph* iff all clusters in τ are either connected or extrovert.

Otherwise the corresponding chunks, clusters and c-graphs are introvert. (See Fig. 1 (a) for example of an extrovert c-graph with extrovert clusters ν_1 and ν_2.)

Like the almost-connected c-graphs of [13], extrovert c-graphs include the class of c-connected c-graphs. Extrovert c-graphs appear to allow a greater degree of disconnectivity than almost-connected c-graphs, since many sibling clusters are allowed to be disconnected. Extrovert c-graphs are also more flexible than the cycles of clusters of [4].

Extrovert c-graphs are a significant generalization of c-connected c-graphs, and we hope they will find use in practice. Intuitively, why might several chunks of a cluster need be drawn together (in the same cluster) when they have no relationships (edges) between them? Perhaps it is because they have similar relationships to entities outside of the cluster. Thus, since our definition requires each chunk of a disconnected cluster ν to have at least one edge going out of the parent cluster of ν, we might expect that this sort of situation arises in practice.

2 Preliminaries

2.1 PQ-Tree and PQ-Reduction

A PQ-tree [3] $T(U)$ is a tree on a set U of n leaves that has two types of internal nodes, P-nodes and Q-nodes, where a P-node can permute its children arbitrarily but a Q-node can only reverse the order of its children. Various combinations of permuting the children of the P-nodes and reversing the order of children of some the Q-nodes result in various permutations of U at the tree leaves. The set of all achievable permutations of the leaves is called the *consistent set* of $T(U)$ and is denoted by CONSISTENT($T(U)$). We say that a subset $S \subseteq U$ of leaves in $T(U)$ is *consecutive* in a permutation π of U if the elements in S appear as a consecutive subsequence in π. PQ-trees support a reduction operation

$$\text{PQ-REDUCE}(T(U), S) \tag{1}$$

---

[3] An *extrovert edge* is called a *virtual edge* in [12].

that returns a new PQ-tree whose consistent set contains exactly those elements of CONSISTENT($T(U)$) in which S is consecutive; if there are no such elements the operation fails.

2.2 Circular Permutations

Suppose we wish to read off the order in which a set of elements appear on the circumference of a circle. Depending on where we start, and whether we read clockwise or counterclockwise, we can obtain various permutations; we will say these permutations are *circularly equivalent*. For example, the permutations $(3, 5, 2, 4, 1, 6)$, $(2, 4, 1, 6, 3, 5)$, and $(5, 3, 6, 1, 4, 2)$ are circularly equivalent. We call an equivalence class of this relation a *circular permutation*. We say any element of the equivalence class is a *representative* of the circular permutation. Informally, a circular permutation represents the order of objects that appear around a circle.

Say a set S is *consecutive* in a circular permutation π if it is consecutive in any representative of π. Informally, this means that the elements of S appear consecutively around the circle.

2.3 PC-Trees and PC-Reduction

PC-trees [15] provide an elegant structure that both simplifies PQ-trees and allows convenient operations on circular permutations. A PC-tree is an unrooted tree with two types of internal nodes, P-nodes and C-nodes, where a P-node can permute its neighbors and a C-node is assigned a cyclic order to its neighbors and can only reverse the order. The *circular consistent set* of a PC-tree $T(U)$ on a set of leaves U, denoted C-CONSISTENT($T(U)$), is the set of all permissible circular permutations of the leaves. Much as with PQ-trees, PC-trees support an operation

$$\text{PC-REDUCE}(T(U), S) \tag{2}$$

that returns a new PC-tree whose circular consistent set contains exactly those circular permutations in C-CONSISTENT($T(U)$) for which the subset S of leaves in $T(U)$ is consecutive; again, if there are no such elements, the operation fails. These trees will be very useful in our algorithm.

It's clear that a PQ-tree is a rooted image of a PC-tree where the Q-nodes correspond to the C-nodes. Therefore the concept of circular consistent set also applies to PQ-trees and the operation PC-REDUCE can take a PQ-tree as input as well. We will not distinguish PQ-trees and PC-trees any more, but use PQ-REDUCE and PC-REDUCE as two operations that can act on the same tree.

2.4 C-Planarity of C-Connected C-Graphs

Let $\mathcal{D}$ be a planar embedding of G. Then for a subgraph H of G we use $\mathcal{D}(H)$ to denote the subembedding of H in $\mathcal{D}$. The boundary of a face in a planar embedding consists of the vertices and edges incident with this face. When ν is a connected cluster, criteria 2–4 in the definition of a c-planar embedding are actually equivalent to requiring that all extrovert vertices of ν are at the boundary of the outer face of $\mathcal{D}(G(\nu))$, and all extrovert edges are in the outer face of $\mathcal{D}(G(\nu))$. The boundary curve $b(\nu)$ can always be obtained by slightly expanding the boundary of the outer face of $\mathcal{D}(G(\nu))$. (See cluster μ_3 in Fig. 1 (a).) In the figures, we use solid lines for the boundary of the outer face of $\mathcal{D}(G(\nu))$ and dashed lines for the boundary curve $b(\nu)$. The definition of c-planarity restricted to c-connected c-graphs then translates into the following property for each cluster.

Property 1 (simple). For a connected cluster ν, a *simple planar embedding of* ν is a planar embedding $\mathcal{D}$ of the graph $(G(\nu) \cup E_\nu^*)$ with the vertices of V_ν^* drawn at the boundary of the outer face of the subembedding $\mathcal{D}(G(\nu))$ and the edges of E_ν^* drawn in the outer face of $\mathcal{D}(G(\nu))$. For a planar embedding $\mathcal{D}$ of G, we say ν is *simple in* $\mathcal{D}$ if the subembedding $\mathcal{D}(G(\nu) \cup E_\nu^*)$ is a simple planar embedding of ν. (In both cases the boundary curve $b(\nu)$ is a slight expansion of the boundary of the outer face of $\mathcal{D}(G(\nu))$.)

The following three lemmas can be deduced from the results of [12]. We summarize and restate them in a particular way to facilitate the presentation of our work. The first lemma is equivalent to Theorem 1 in [12].

Lemma 1. *A c-connected c-graph $C = (G, \tau)$ is c-planar iff G is planar and there exists a planar embedding $\mathcal{D}$ of G in which each $\nu \in \tau$ is simple.*

The next two lemmas are deduced from the testing algorithm of [12]. We provide them without proofs. We also omit the original constructions of [12] that fulfill these procedures.

Lemma 2. *For any connected cluster ν of size m, we can build in $O(m)$ time a PQ-tree on the set of leaves E_ν^*, say $T(E_\nu^*)$, such that the circular consistent set of $(T(E_\nu^*))$ equals the set of circular permutations of the edges of E_ν^* on $b(\nu)$ resulting from all possible simple planar embeddings of ν.*

We write the procedure of building $T(E_\nu^*)$ from ν in [12] as the following operation that converts a subgraph to a PQ-tree.

$$T(E_\nu^*) \leftarrow \text{CONVERT}(\nu). \tag{3}$$

Lemma 3. *For any PQ-tree $T(E_\nu^*)$ resulting from Lemma 2, we can build in $O(m)$ time a representative subgraph R_ν as a replacement of $G(\nu)$ in G with the vertex set r_ν of R_ν being a replacement of the cluster ν, and the extrovert vertex and extrovert edge sets of r_ν remaining V_ν^* and E_ν^*. If we substitute R_ν for $G(\nu)$ in G, then*

- *the circular consistent set of $(T(E_\nu^*))$ equals the set of circular permutations of E_ν^* at $b(r_\nu)$ resulting from all possible simple planar embeddings of r_ν.*
- *G is planar iff G has a planar embedding in which r_ν is simple.*

We write the procedure of building the representative subgraph R_ν from $T(E_\nu^*)$ in [12] as

$$R_\nu \leftarrow \text{REPRESENT}(T(E_\nu^*)), \tag{4}$$

and the procedure of substituting R_ν for $G(\nu)$ in G as

$$G \leftarrow \text{SUBSTITUTE}(G(\nu), R_\nu). \tag{5}$$

The above lemmas characterize the c-planarity of c-connected c-graphs, and provide gadgets to test it. Intuitively, the processes in Lemma 2 and 3 provide that, G has a planar embedding with ν being simple $\Leftrightarrow$ G after the substitution has a planar embedding with r_ν being simple $\Leftrightarrow$ G after the substitution is planar. Then the testing algorithm CPT in [12] traverses τ bottom-up and performs operations (3),(4),(5) for each cluster. After substituting all clusters, G is planar if and only if the original G has a planar embedding that makes all original clusters simple, so that c-planarity testing is converted into planarity testing. The algorithm runs in $O(n^2)$ time.

3   C-Planarity of Extrovert C-Graphs

In this section we characterize the c-planarity of extrovert c-graphs. The following lemma is a straightforward characterization of the c-planarity of c-graphs.

Lemma 4 (Theorem 2 in [12]). *A c-graph $C = (G = (V, E), \tau)$ is c-planar, iff there exists a c-connected c-planar c-graph $C' = (G' = (V, E'), \tau)$ with $E \subset E'$.*

C' is called a *super c-graph* of C. Our idea to characterize the c-planarity of extrovert c-graphs is to treat each chunk of a cluster as a small connected cluster and use the following two properties together with Property 1.

Property 2 (connectable). Let $\nu = (\nu^1, \ldots, \nu^k)$ be a disconnected cluster with its parent cluster μ being connected, and $\mathcal{D}$ be a simple planar embedding of μ in which each chunk ν^i of ν is also simple. We say that ν is connectable in $\mathcal{D}$ iff there is a way to draw $k - 1$ extra edges inside $b(\mu)$ that connect the k chunks of ν into one connected component, without introducing any edge crossings. We call the extra edges *bridges* of ν.

Property 3 (conflict). Let $\nu_l = (\nu_l^1, \ldots, \nu_l^{k_l})$, $l = 1, \ldots,$ be sibling disconnected clusters with their parent cluster μ being connected, and $\mathcal{D}$ be a simple planar embedding of μ in which each chunk ν_l^i of each ν_l is also simple. We say that the ν_l's conflict in $\mathcal{D}$ iff each ν_l is connectable, but there is no way to connect all of the ν_l's inside $b(\mu)$ simultaneously without introducing edge crossings.

Theorem 1. *An extrovert c-graph $C = (G, \tau)$ is c-planar iff G is planar and there exists a planar embedding $\mathcal{D}$ of G such that, each chunk of each cluster is simple in $\mathcal{D}$; each extrovert cluster is connectable in the subembedding of its parent cluster; and no sibling extrovert clusters conflict.*

Proof. [sketch]*Sufficiency.* Assume there is such an embedding $\mathcal{D}$ of G. Since each chunk of cluster is simple, we can add a boundary curve for each chunk in $\mathcal{D}$, which is slightly outside the boundary of the outer face of this chunk. The only thing disqualifying this drawing to be a c-planar drawing is that an extrovert cluster is enclosed in not a single but many regions. Connect each extrovert cluster ν with $k > 1$ chunks by adding $k - 1$ bridges. As required by Property 2 and 3, the bridge between two chunks ν^i and ν^j will cross only with the boundary curves of ν^i and ν^j. So we can merge the k simple closed regions for chunks into one region by "digging tunnels" along the bridges as shown in Fig 2. Since each chunk region is simple and there are only $k-1$ bridges spanning k chunks, the resulting region for the whole cluster is still simple. Doing this for all extrovert clusters gives a c-planar embedding of C.

Fig. 2. Merge two chunk regions and boundary curves by digging a tunnel along a bridge

Necessity. Assume C is c-planar. By Lemma 4, there exists a c-connected super c-graph $C' = (G', \tau)$ of C and it has a c-planar embedding $\mathcal{D}$. We only need to show that in $\mathcal{D}$ of G', each chunk of a cluster is simple; each extrovert cluster is connectable in the subembedding of its parent; and no sibling extrovert clusters conflict. Consider an extrovert cluster $\nu = \{\nu^1, \nu^2, \ldots\}$ with parent cluster μ. Since each chunk ν^i is extrovert, there is an extrovert edge in $E^*_\mu(\nu^i)$ crossing $b(\mu)$. Therefore any chunk ν^i cannot be enclosed in an inner face of another chunk ν^j, so each chunk must be simple. The properties of being connectable and not conflicting are obvious, noting that the extra edges in C' include all the bridges.   □

4   Testing Algorithm

We first convert the inclusion tree τ into τ' by splitting each disconnected cluster into its chunks. Each node $\nu \in \tau$ is a cluster and each node $\nu^i \in \tau'$ is a chunk. (See Fig. 3.) We always use μ for a parent and ν a for child. The frame of our testing algorithm EXTROVERT-CPT is shown in Fig. 4. It inherits the algorithm CPT in [12], except that we process the chunks in τ' instead of the clusters in τ, and insert the following subroutine to filter the permissible circular permutations of extrovert edges at $b(\mu^i)$. (See Fig. 5.)

$$T'(E^*_{\mu^i}) \leftarrow \text{FILTER}(T(E^*_{\mu^i})). \tag{6}$$

Fig. 3. τ and τ' for the c-graph in Fig. 1 (a)

Algorithm EXTROVERT-CPT

1: **for** each $\mu^i \in \tau'$ in postorder **do**
2:    test planarity of $G(\mu^i)$
3:    **if** μ^i is not the root of τ' **then**
4:       $T(E^*_{\mu^i}) \leftarrow$ CONVERT$(G(\mu^i))$
5:       $T'(E^*_{\mu^i}) \leftarrow$ FILTER$(T(E^*_{\mu^i}))$
6:       $R_{\mu^i} \leftarrow$ REPRESENT$(T'(E^*_{\mu^i}))$
7:       $G \leftarrow$ SUBSTITUTE$(G(\mu^i), R_{\mu^i})$

Fig. 4. The frame algorithm for testing c-planarity of an extrovert c-graph. If any subroutine at any moment fails, (either a subgraph is not planar or a reduction is not doable,) then the algorithm stops and returns "not c-planar". Otherwise it returns "c-planar" after passing the planarity test of $G(root(\tau'))$ in Step 2.

Algorithm FILTER$(T(E^*_{\mu^i}))$

for each τ-child ν of μ^i **do**
   contract $S(\nu)$ into a vertex in S
for each ν that is an extrovert τ-child cluster of μ^i **do**
   for each set of vertices $G(\mu^i \backslash \nu)^j$ in $G(\mu^i \backslash \nu)$ that contracts into a connected component $C^j_{S(\mu^i \backslash \nu)}$ in $S(\mu^i \backslash \nu)$ **do**
      $T(E^*_{\mu^i}) \leftarrow$ PC-REDUCE$(T(E^*_{\mu^i}), E^*_{\mu^i}(G(\mu^i \backslash \nu)^j))$
return $T(E^*_{\mu^i})$

Fig. 5. The filter algorithm

Now we describe the filter algorithm FILTER$(T(E^*_{\mu^i}))$ shown in Fig. 5. We call a node $\nu \in \tau$ a τ-child of a node $\mu^i \in \tau'$, and μ^i the τ'-parent of ν, if every chunk ν^i of ν is a child of μ^i in τ'. Since the parent cluster of an extrovert cluster is connected, each $\nu \in \tau$ has exactly one τ'-parent. In addition to G, we maintain a *skeleton* S of G which is initially equal to G but, at the time any μ^i is processed in FILTER$(T(E^*_{\mu^i}))$, contracts every τ-child ν of μ^i into a vertex. We denote by C^j_F the j-th connected component of a disconnected graph F. Let $S(\mu^i \backslash \nu)$, the subgraph resulting from removing ν from $S(\mu^i)$, have connected components $C^1_{S(\mu^i \backslash \nu)}, C^2_{S(\mu^i \backslash \nu)}, \ldots$. Let $G(\mu^i \backslash \nu)^j$ be the part of $G(\mu^i \backslash \nu)$ that contracts into $C^j_{S(\mu^i \backslash \nu)}$ in S (as every τ-child of μ^i is contracted into a vertex). Then we require that in the output of FILTER$(T(E^*_{\mu^i}))$ the extrovert edges of μ^i in each $G(\mu^i \backslash \nu)^j$ are always consecutive among all extrovert edges of μ^i on $b(\mu^i)$. This will be achieved by doing a PC-reduction for each $G(\mu^i \backslash \nu)^j$ as Fig. 5 shows. (Note that $G(\mu^i \backslash \nu)^j$ may not be a connected component of $G(\mu^i \backslash \nu)$, but consist of multiple connected components.)

Recall that by Theorem 1, to qualify a planar embedding of G to be a c-planar embedding of C, we only need to maintain the property of simple for each chunk of each cluster, the property of connectable for each extrovert cluster, and the property of no conflict among all extrovert child clusters in each connected parent cluster. By inheriting the algorithm CPT in [12] but using τ' instead of τ, EXTROVERT-CPT maintains all the diversity of embedding each chunk of cluster to be simple. In addition, when μ^i is the only chunk of a connected cluster μ, FILTER($T(E^*_{\mu^i})$) will further filter the simple planar embeddings of μ, by doing a sequence of PC-reductions for each extrovert child cluster of μ, so that only those in which all extrovert child clusters of μ are connectable and don't conflict are left in the circular consistent set of $T'(E^*_{\mu^i})$. (If μ^i is a chunk of an extrovert cluster, then by the definition of extrovert c-graph all of its τ-children are connected clusters and FILTER($T(E^*_{\mu^i})$) does nothing.) We'll prove in the next section that FILTER($T(E^*_{\mu^i})$) fulfills this purpose by performing a PC-reduction for the extrovert edges of μ^i incident with each $G(\mu^i\backslash\nu)^j$.

5   Proof of Correctness

In this section we show why making some certain sets of extrovert edges consecutive among all extrovert edges at $b(\mu)$ can provide Property 2 and 3 in Sec. 3 to the planar embedding inside $b(\mu)$. In order to prove the main Theorem 2, we first prove the following lemma.

Lemma 5. *Let $\nu = (\nu^1, \ldots, \nu^k)$ be an extrovert cluster with parent cluster μ, and $G(\mu\backslash\nu)$ have connected components $C^1_{G(\mu\backslash\nu)}, C^2_{G(\mu\backslash\nu)}, \ldots$. Let $\mathcal{D}$ be a simple planar embedding of μ in which each ν^i is also simple, and $\pi(E^*_\mu)$ be the circular permutation of E^*_μ at $b(\mu)$. Then ν is connectable in $\mathcal{D}$, iff for each $C^j_{G(\mu\backslash\nu)}$, $j = 1, 2, \ldots$, $E^*_\mu(C^j_{G(\mu\backslash\nu)})$ is consecutive in $\pi(E^*_\mu)$.*

Proof. [sketch]*Necessity.* If there is a $C^j_{G(\mu\backslash\nu)}$ such that $E^*_\mu(C^j_{G(\mu\backslash\nu)})$ is not consecutive in $\pi(E^*_\mu)$, then there are $e_i \in E^*_\mu$, $i \in \{1,2,3,4\}$, such that $e_1, e_2 \in E^*_\mu(C^j_{G(\mu\backslash\nu)})$ are separated by $e_3, e_4 \notin E^*_\mu(C^j_{G(\mu\backslash\nu)})$ at $b(\mu)$. Then there is a path $p \in C^j_{G(\mu\backslash\nu)}$ from e_1 to e_2 cutting $b(\mu)$ into two halves with e_3 and e_4 being on different sides. (See Fig. 6.) We show that each side contains some chunk ν^i of ν, so that ν cannot be connected inside $b(\mu)$ without crossing p. See the side of e_3. If e_3 is incident with some ν^i, then ν^i is on this side. Otherwise e_3 is incident with some $C^{j'}_{G(\mu\backslash\nu)}$ with $j' \neq j$, in which case there must also be some ν^i on this side because $C^j_{G(\mu\backslash\nu)}$ and $C^{j'}_{G(\mu\backslash\nu)}$ were in a connected graph $G(\mu)$ but become disconnected in $G(\mu\backslash\nu)$. Similarly the side of e_4 contains another chunk $\nu^{i'}$ of ν.

Sufficiency. Suppose ν is not connectable inside $b(\mu)$. We greedily connect the chunks of ν until getting a maximal set of connected $\cup\nu^i$ which doesn't contain some $\nu^{i'}$. Then there must be some paths in $G(\mu)$ with two ends $e_1, e_2 \in E^*_\mu$ cutting $b(\mu)$ into two halves and $\cup\nu^i$ and $\nu^{i'}$ on different sides. We can show that among all such paths there is a path p with no vertex of p belonging to ν, which means that p is contained in some connected component $C^j_{G(\mu\backslash\nu)}$ and $E^*_\mu(C^j_{G(\mu\backslash\nu)})$ is not consecutive in $\pi(E^*_\mu)$ since e_1 and e_2 are separated at $b(\mu)$ by the extrovert edges coming from $\cup\nu^i$ and those from $\nu^{i'}$. Details are omitted. □

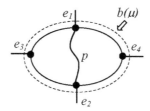

Fig. 6. e_1, e_2 and p are in $C^j_{G(\mu \backslash \nu)}$, and e_3, e_4 are not in $C^j_{G(\mu \backslash \nu)}$. There must be some ν^i on the left of p and another $\nu^{i'}$ on the right.

We conclude with the following two theorems with Theorem 3 showing the correctness and running time of the testing algorithm. Proofs of these theorems are omitted in this extended abstract. An embedding algorithm is implied by the testing algorithm. Details are also omitted.

Theorem 2. *Let* $\nu_l = (\nu_l^1, \ldots, \nu_l^{k_l})$, $l = 1, 2, \ldots$, *be sibling extrovert clusters with connected parent cluster* μ, $\mathcal{D}$ *be a simple planar embedding of* μ *in which each* ν_l^i *is also simple, and* $\pi(E_\mu^*)$ *be the circular permutation of* E_μ^* *at* $b(\mu)$. *Let* S *be the skeleton of* G *in which each child cluster of* μ *is contracted into a vertex,* $S(\mu \backslash \nu_l)$ *have connected components* $C^1_{S(\mu \backslash \nu_l)}, C^2_{S(\mu \backslash \nu_l)}, \ldots$, *and each* $C^j_{S(\mu \backslash \nu_l)}$, $j = 1, 2, \ldots$, *be contracted from a subgraph* $G(\mu \backslash \nu_l)^j$ *of* $G(\mu \backslash \nu_l)$. *Then each* ν_l *is connectable and all of the* ν_l'*s don't conflict in* $\mathcal{D}$, *iff for each* $G(\mu \backslash \nu_l)^j$, $l = 1, 2, \ldots$ *and* $j = 1, 2, \ldots$, $E_\mu^*(G(\mu \backslash \nu_l)^j)$ *is consecutive in* $\pi(E_\mu^*)$.

Theorem 3. *The algorithm EXTROVERT-CPT correctly tests the c-planarity of an extrovert c-graph in* $O(n^3)$ *time.*

References

1. G. D. Battista, W. Didimo, and A. Marcandalli. Planarization of clustered graphs. In *Graph Drawing (GD'01), LNCS 2265*, pages 60–74, 2001.
2. G. D. Battista and R. Tamassia. On-line planarity testing. *SIAM J. Comput.*, 25(5):956–997, 1996.
3. K. Booth and G. Lueker. Testing for the consecutive ones property, interval graphs, and graph planarity using PQ-tree algorithms. *J. Comput. Systems Sci.*, 13(3):335–379, 1976.
4. P. G. Cortese, G. D. Battista, M. Patrignani, and M. Pizzonia. Clustering cycles into cycles of clusters. In *Graph Drawing (GD'04)*, 2004.
5. E. Dahlhaus. A linear time algorithm to recognize clustered planar graphs and its parallelization. In *LATIN'98, LNCS 1380*, pages 239–248, 1998.
6. C. A. Duncan, M. T. Goodrich, and S. G. Kobourov. Planarity-preserving clustering and embedding for large planar graphs. In *Graph Drawing (GD'99), LNCS 1731*, pages 186–196, 1999.
7. P. Eades, Q. Feng, and H. Nagamochi. Drawing clustered graphs on an orthogonal grid. *Journal of Graph Algorithms and Applications*, 3(4):3–29, 1999.

8. P. Eades and Q.-W. Feng. Multilevel visualization of clustered graphs. In *Graph Drawing, GD'96, LNCS 1190*, pages 101–112, 1996.
9. P. Eades, Q.-W. Feng, and X. Lin. Straight-line drawing algorithms for hierarchical graphs and clustered graphs. In *Graph Drawing, GD'96, LNCS 1190*, pages 113–128, 1996.
10. P. Eades and M. L. Huang. Navigating clustered graphs using force-directed methods. *J. Graph Algorithms and Applications: Special Issue on Selected Papers from 1998 Symp. Graph Drawing*, 4(3):157–181, 2000.
11. S. Even and R. E. Tarjan. Computing an st-numbering. *Theoretical Computer Science*, 2(3):339–344, 1976.
12. Q.-W. Feng, P. Eades, and R. F. Cohen. Clustered graphs and C-planarity. In *3rd Annual European Symposium on Algorithms (ESA'95), LNCS 979*, pages 213–226, 1995.
13. C. Gutwenger, M. Jnger, S. Leipert, P. Mutzel, M. Percan, and R. Weiskircher. Advances in C-planarity testing of clustered graphs. In *Graph Drawing (GD'02), LNCS 2528*, pages 220–235, 2002.
14. J. Hopcroft and R. E. Tarjan. Efficient planarity testing. *J. ACM*, 21(4):549–568, 1974.
15. W. Hsu and R.M.McConnell. PC trees and circular-ones arrangements. *Theoretical Computer Science*, 296(1):59–74, 2003.
16. A. Lempel, S. Even, and I. Cederbaum. An algorithm for planarity testing of graphs. In *Theory of graphs: International symposium*, pages 215–232, 1966.
17. T. Nishizeki and N. Chiba. *Planar Graphs: Theory and Algorithms*, volume 32 of *Ann. Discrete Math.* North-Holland, Amsterdam, The Netherlands, 1988.

Non-planar Core Reduction of Graphs

Carsten Gutwenger and Markus Chimani

University of Dortmund, Dortmund, Germany
{carsten.gutwenger, markus.chimani}@cs.uni-dortmund.de

Abstract. We present a reduction method that reduces a graph to a smaller core graph which behaves invariant with respect to planarity measures like crossing number, skewness, and thickness. The core reduction is based on the decomposition of a graph into its triconnected components and can be computed in linear time. It has applications in heuristic and exact optimization algorithms for the planarity measures mentioned above. Experimental results show that this strategy yields a reduction to 2/3 in average for a widely used benchmark set of graphs.

1 Introduction

Graph drawing is concerned with the problem of rendering a given graph on the two-dimensional plane so that the resulting drawing is as readable as possible. Objective criteria for the readability of a drawing depend mostly on the application domain, but achieving a drawing without edge crossings is in general a primary objective. Such a drawing is called a *planar* drawing. However, it is well known that not every graph can be drawn without edge crossings. The famous theorem by Kuratowski [10] shows that a graph is planar if and only if it does not contain a subdivision of $K_{3,3}$ or K_5.

If a graph G is not planar, a question arises naturally: How far away is the graph from planarity? For that reason, various measures for non-planarity have been proposed. The most prominent one is the *crossing number* of a graph which asks for the minimum number of crossings in any drawing of G. Further measures are the *skewness* which is the minimum number of edges we have to remove from G in order to obtain a planar graph, and the *thickness* which is the minimum number of planar subgraphs of G whose union is G. However, finding an optimal drawing with respect to any of these non-planarity measures yields an NP-hard optimization problem [5, 12, 13].

Various heuristic and exact methods for solving these optimization problems have been proposed; please refer to [11, 14, 7] for an overview. It is well known that it is sufficient to consider each biconnected component of the graph separately. We present a new approach based on the triconnectivity structure of the graph which reduces a 2-connected graph to a core that behaves invariant to the above non-planarity measures. We call this core graph the non-planar core $\mathcal{C}$ of G and show that it can be constructed in linear computation time. In order to compute the crossing number, skewness, or thickness of G, any standard algorithm can be applied to $\mathcal{C}$. This approach targets in particular exact

P. Healy and N.S. Nikolov (Eds.): GD 2005, LNCS 3843, pp. 223–234, 2005.

algorithms, since their running times heavily depend on the instance size. It is also constructive in the sense that we can reconstruct a solution for G (e.g., a crossing minimal drawing) from the solution for the core graph $\mathcal{C}$.

This paper is organized as follows. After introducing some basic terminology, the non-planar core is defined in Sect. 3. The next three sections 4–6 apply the new reduction technique to crossing number, skewness and thickness. Section 7 shows that a straight-forward idea to further reduce the size of the core is not possible. We conclude the paper with experimental results.

2   Preliminaries

Let $G = (V, E)$ be a graph. If $(u, v) \in V \times V$, we use $G \cup (u, v)$ as a shorthand for the graph $(V, E \cup (u, v))$. For a subset of the vertices $V' \subseteq V$, we denote with $G[V']$ the vertex induced subgraph (V', E_V), where $E_V \subseteq E$ is the set of edges with both end vertices in V'. If $E' \subseteq E$ is a subset of the edges of G, we denote with $G[E']$ the subgraph induced by the edges in E', that is $G[E'] = (V_E, E')$ with $V_E = \{v \in V \mid v \text{ is incident with an edge in } E'\}$. Suppose that G is planar and let Γ be an embedding of G with face set F. The *dual graph* $\Gamma^* = (F, E^*)$ of Γ contains an edge $e^* = (f, f')$ for every edge $e \in E$ such that e is on the boundary of both f and f'; edge e is also called the *primal edge* of e^*.

2.1   Crossing Number, Skewness, and Thickness

The *crossing number* $\nu(G)$ of a graph $G = (V, E)$ is the minimum number of crossings in any drawing of G. The *skewness* $\mu(G)$ of G is the size of a minimum cardinality edge set F such that $G[E \backslash F]$ is planar, and we call $G[E \backslash F]$ a maximum planar subgraph of G. The *thickness* $\theta(G)$ of G is the minimum number k of planar graphs $G_1, \ldots, G_k$ such that $G_1 \cup \ldots \cup G_k = G$.

We extend the notion of crossing number and skewness to graphs with a given weight function $w : E \to \mathbb{N}$. We call the sum

$$\sum_{\substack{e, f \in E \\ e \text{ crosses } f}} w(e) \cdot w(f)$$

the *crossing weight* of a drawing, and we denote with $\nu(G, w)$ the *weighted crossing number* of G which is the minimum crossing weight of any drawing of G. If $E' \subseteq E$, we define $w(E') := \sum_{e \in E'} w(e)$ to be the weight of E', and we denote with $\mu(G, w)$ the *weighted skewness* of G which is the weight $w(F)$ of a minimum weight edge set F such that $G[E \setminus F]$ is planar.

In the remainder of this paper, we will restrict our attention to 2-connected graphs. However, the results on crossing number, skewness, and thickness can easily be generalized using the following relationships. Let G be a graph and $B_1, \ldots, B_k$ its biconnected components. Then,

$$\nu(G) = \sum_{i=1,\ldots,k} \nu(B_i), \quad \mu(G) = \sum_{i=1,\ldots,k} \mu(B_i), \quad \theta(G) = \max_{i=1,\ldots,k} \theta(B_i).$$

2.2   Minimum Cuts and Traversing Costs

A *cut* in G is a partition $(S, \bar{S})$ of the vertices of G. The *capacity* $c(S, \bar{S})$ of the cut is the cardinality of the set $E(S, \bar{S})$ of all the edges connecting vertices in S with vertices in $\bar{S}$. For two vertices $s, t \in V$, we call $(S, \bar{S})$ an *st-cut* if s and t are in different sets of the cut. A *minimum st-cut* is an *st*-cut of minimum capacity. We denote the capacity of a minimum *st*-cut in G with $\mathrm{mincut}_{s,t}(G)$.

Let $s, t \in V$ and $G \cup (s, t)$ be 2-connected and planar. For an embedding Γ of $G \cup (s, t)$, we define the *traversing costs* of Γ with respect to (s, t) to be the shortest path in the dual graph of Γ that connects the two faces adjacent to (s, t) without using the dual edge of (s, t). We also call the corresponding list of primal edges a *traversing path* for s and t. Gutwenger, Mutzel, and Weiskircher [8] showed that the traversing costs are independent of the choice of the embedding Γ of G. Hence, we define the *traversing costs* of G with respect to (s, t) to be the traversing costs of an arbitrary embedding Γ with respect to (s, t). It is easy to see that a traversing path defines an *st*-cut. The following theorem shows that this *st*-cut is even a minimum *st*-cut.

Theorem 1. *Let* $G = (V, E)$ *be a graph with* $s, t \in V$ *and* $G \cup (s, t)$ *is 2-connected and planar. Then, the traversing costs of* G *with respect to* (s, t) *are equal to* $\mathrm{mincut}_{s,t}(G)$.

We are interested in special subgraphs of a 2-connected, not necessarily planar graph $G = (V, E)$ which we call planar *st*-components. Let $s, t \in V$ be two distinct vertices. We call an edge induced subgraph $C = G[E_C]$ a *planar st-component* of G if $G \cup (s, t)$ is 2-connected and planar, and if $V(C) \cap V' \subseteq \{s, t\}$, where $V' := V(G[E \setminus E_C])$ denotes the vertex set of the graph induced by the edges not contained in C. Obviously, since G is 2-connected, $V(C) \cap V'$ is either empty or contains both s and t.

2.3   SPQR-Trees

SPQR-trees basically represent the decomposition of a biconnected graph into its triconnected components. For a formal definition we refer the reader to [4, 3]. Informally speaking, the nodes of an SPQR-tree $\mathcal{T}$ of a graph G stand for serial (S-nodes), parallel (P-nodes), and triconnected (R-nodes) structures, as well as edges of G (Q-nodes). The respective structure is given by skeleton graphs associated with each node of $\mathcal{T}$, which are either cycles, bundles of parallel edges, or triconnected simple graphs. We denote with *skeleton*(η) the skeleton graph associated with node η. Each edge $e \in skeleton(\eta)$ corresponds to a tree edge $e_{\mathcal{T}} = (\eta, \xi)$ incident with η. We call ξ the *pertinent node* of e. The edge e stands for a subgraph called the *expansion graph* of e that is only attached to the rest of the graph at the two end vertices of e. The expansion graph of e is obtained as follows. Deleting edge $e_{\mathcal{T}}$ splits $\mathcal{T}$ into two connected components. Let $\mathcal{T}_\xi$ be the connected component containing ξ. The *expansion graph* of e (denoted with expansion(e)) is the graph induced by the edges that are represented by the Q-nodes in $\mathcal{T}_\xi$. We further introduce the notation expansion$^+(e)$ for the graph expansion$(e) \cup e$.

For our convenience, we omit Q-nodes and distinguish in skeleton graphs between *real edges* that are skeleton edges whose pertinent node would be a Q-node, and *virtual edges*.

3   The Non-planar Core

Let G be a 2-connected graph and let $\mathcal{T}$ be its SPQR-tree. For a subtree $\mathcal{S}$ of $\mathcal{T}$, we define the *induced graph* $G[\mathcal{S}]$ of $\mathcal{S}$ to be the edge induced subgraph $G[E']$, where E' is the union of all edges in skeletons of nodes of $\mathcal{S}$ that have no corresponding tree edge in $\mathcal{S}$:

$$E' := \bigcup_{\eta \in \mathcal{S}} \{e \in skeleton(\eta) \mid e \text{ has no corresponding tree edge in } \mathcal{S}\}$$

Hence, the induced graph consists of virtual edges representing planar *st*-components and real edges representing edges of G. Analogously to SPQR-trees, we define the expansion graph of a virtual edge in $G[\mathcal{S}]$ and use the notations expansion(e) and expansion$^+(e)$ for a virtual edge e. We can reconstruct G from $G[\mathcal{S}]$ by replacing every virtual edge with its expansion graph. We have in particular $G[\mathcal{T}] = G$.

We define the *non-planar core* of G to be the empty graph if G is planar, and the induced graph of the smallest non-empty subtree $\mathcal{S}$ of $\mathcal{T}$ such that the *expansion*$^+(e)$ is planar for every virtual edge e in $G[\mathcal{S}]$. It is easy to derive the following properties of the non-planar core of G.

Lemma 1. *Let $\mathcal{C} = G[\mathcal{S}]$ be the non-planar core of G.*

(a) $\mathcal{C} = \emptyset \iff G$ is planar
(b) $\mathcal{C} \neq \emptyset \implies$ Every leaf of $\mathcal{S}$ is an R-node with non-planar skeleton.

Proof. The first part follows directly from the definition.

Let $\mathcal{C} \neq \emptyset$ and thus G be non-planar. Then, $\mathcal{S}$ must contain a node with non-planar skeleton. Suppose $\xi \in \mathcal{S}$ is a leaf whose skeleton is planar. Since $\mathcal{S}$ contains at least one further node, ξ has exactly one adjacent node η in $\mathcal{S}$. But then the expansion graph of the virtual edge of ξ in *skeleton*(η) is planar, and hence $\mathcal{S}' := \mathcal{S} - \xi$ is also a subtree of $\mathcal{T}$ with the property that expansion$^+(e)$ is planar for every virtual edge e in $G[\mathcal{S}']$. This is a contradiction to the minimality of $\mathcal{S}$. It follows that every leaf of $\mathcal{S}$ is a node with non-planar skeleton. This must be an R-node, since only R-node skeletons can be non-planar.   □

We extend the non-planar core $\mathcal{C}$ of G by an additional weight function $w :$ $E(\mathcal{C}) \rightarrow \mathbb{N}$. If e is a real edge, then $w(e)$ is 1. Otherwise, let $e = (s, t)$ and we define $w(e) := \text{mincut}_{s,t}(\text{expansion}(e))$. We denote the non-planar core with given edge weights by a pair $(\mathcal{C}, w)$.

Theorem 2. *Let $G = (V, E)$ be a 2-connected graph. Then, the non-planar core of G and the corresponding edge weights can be computed in $\mathcal{O}(|V| + |E|)$ time.*

Algorithm 1. Computation of the non-planar core.

Require: 2-connected graph $G = (V, E)$
Ensure: non-planar core $(\mathcal{C}, w)$ of G

Let $\mathcal{T}$ be the (undirected) SPQR-tree of G

Let *candidates* be an empty stack of nodes
for all $\xi \in \mathcal{T}$ **do**
   $d[\xi] := deg(\xi)$
   if $d[\xi] = 1$ **then**
      candidates.push(ξ)
   end if
end for

$P := \emptyset$
while *candidates* $\neq \emptyset$ **do**
   $\xi := candidates$.pop()
   if $skeleton(\xi)$ is planar **then**
      $P := P \cup \{\xi\}$
      for all $\eta \in Adj(\xi)$ **do**
         $d[\eta] := d[\eta] - 1$
         if $d[\eta] = 1$ **then**
            candidates.push(η)
         end if
      end for
   end if
end while

Let $\mathcal{S}$ be the graph induced by the vertices in $V(\mathcal{T}) \setminus P$
$\mathcal{C} := G[\mathcal{S}]$

for all edges $e \in \mathcal{C}$ **do**
   if e is a virtual edge **then**
      $w(e) :=$ traversing costs of expansion(e) with respect to e
   else
      $w(e) := 1$
   end if
end for

Proof. Algorithm 1 shows a procedure for computing the non-planar core. We achieve linear running time, since constructing an SPQR-tree, testing planarity, and computing traversing costs takes only linear time; see [6, 9, 8].

4   Crossing Number

In this section, we apply the non-planar core reduction to the crossing number problem. The following theorem shows that it is sufficient to compute the crossing number of the non-planar core.

Theorem 3. *Let G be a 2-connected graph, and let (C, w) be its non-planar core. Then,*

$$\nu(G) = \nu(C, w).$$

The proof of Theorem 3 is based on the following lemma which allows us to restrict the crossings in which the edges of a planar st-component may be involved so that we can still obtain a crossing minimal drawing of G. A similar result has been reported by Širáň in [15]. However, as pointed out in [1], the proof given by Širáň is not correct.

Lemma 2. *Let $C = (V_C, E_C)$ be a planar st-component of $G = (V, E)$. Then, there exists a crossing minimal drawing $\mathcal{D}^*$ of G such that the induced drawing $\mathcal{D}_C^*$ of C has the following properties:*

(a) $\mathcal{D}_C^*$ contains no crossings;
(b) s and t lie in a common face f_{st} of $\mathcal{D}_C^*$;
(c) all vertices in $V \setminus V_C$ are drawn in the region of $\mathcal{D}^*$ defined by f_{st};
(d) there is a set $E_s \subseteq E_C$ with $|E_s| = \text{mincut}_{s,t}(C)$ such that any edge $e \in E \setminus E_C$ may only cross through all edges of E_s, or through none of E_C.

Proof. Let $G' = G[E \setminus E_C]$ be the graph that results from cutting C out of G. Let $\mathcal{D}$ be an arbitrary, crossing minimal drawing of G, and let $\mathcal{D}_C$ (resp. $\mathcal{D}'$) be the induced drawing of C (resp. G'). We denote by P the planarized representation of G' induced by $\mathcal{D}'$, i.e. the planar graph obtained from $\mathcal{D}'$ by replacing edge crossings with dummy vertices. Let Γ_P be the corresponding embedding of P and Γ_P^* the dual graph of Γ_P.

Let $p = f_1, \ldots, f_{k+1}$ be a shortest path in Γ_P^* that connects an adjacent face of s with an adjacent face of t. There are $\lambda := \text{mincut}_{s,t}(C)$ edge disjoint paths from s to t in C. Each of these λ paths crosses at least k edges of G' in the drawing $\mathcal{D}$. Hence, there are at least $\lambda \cdot k$ crossings between edges in C and edges in G'. We denote with E_p the set of primal edges of the edges on the path p. Let $\mathcal{D}_C^*$ be a planar drawing of C in which s and t lie in the same face f_{st}, and let E_s be the edges in a traversing path in $\mathcal{D}_C^*$ with respect to s and t. By Theorem 1, there is a minimum st-cut $(S, \bar{S})$ with $E(S, \bar{S}) = E_s$, and thus $|E_s| = \lambda$. We can combine $\mathcal{D}'$ and $\mathcal{D}_C^*$ by placing the drawing of $C[S]$ in face f_1 and the drawing of $C[\bar{S}]$ in f_{k+1}, such that all the edges in E_p cross all the edges in E_s; see Fig. 1. It is easy to verify that the conditions (a)–(d) hold for the resulting drawing $\mathcal{D}^*$. □

We conclude this section with the proof of Theorem 3, i.e. we show that $\nu(G) = \nu(C, w)$.

Proof (of Theorem 3).

"$\leq$" Let $\mathcal{D}_C$ be a drawing of C with minimum crossing weight. For each virtual edge $e = (s, t) \in C$, we replace e by a planar drawing $\mathcal{D}_e$ of the corresponding planar st-component so that all edges that cross e in $\mathcal{D}_C$ cross the edges in a traversing path in $\mathcal{D}_e$ with respect to (s, t). Since w_e is equal to the

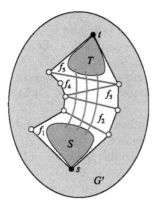

Fig. 1. Final drawing $\mathcal{D}^*$ of G; here, $p = f_1, f_2, f_3, f_4, f_5$ is the shortest path in Γ_P^*

traversing costs of $\mathcal{D}_e$ with respect to (s, t) by definition, replacing all virtual edges in this way leads to a drawing of G with $\nu(\mathcal{C}, w)$ crossings, and hence $\nu(G) \leq \nu(\mathcal{C}, w)$.

"$\geq$" On the other hand, let $\mathcal{D}$ be a crossing minimal drawing of G. For each virtual edge $e = (s, t) \in \mathcal{C}$, we modify $\mathcal{D}$ in the following way. Let C be the planar st-component corresponding to e, and let G' be the rest of the graph. By Lemma 2, we obtain another crossing minimal drawing of G if we replace the drawing of C with a planar drawing $\mathcal{D}_C$ of C such that all edges of G' that cross edges in C will cross the edges in $E(S, \bar{S})$, where $(S, \bar{S})$ is a minimum st-cut in C. If we replace $\mathcal{D}_C$ with an edge $e = (s, t)$ with weight $w(e) := |E(S, \bar{S})| = \mathrm{mincut}_{s,t}(C)$, we obtain a drawing with the same crossing weight.

By replacing all virtual edges in that way, we obtain a drawing of $\mathcal{C}$ whose crossing weight is the crossing number of G. It follows that $\nu(G) \geq \nu(\mathcal{C}, w)$, and hence the theorem holds. □

5   Skewness

We can apply the non-planar core reduction to the skewness of a graph in a rather analogue way. The following lemma establishes our main argument.

Lemma 3. *Let* $G = (V, E)$ *be a 2-connected graph,* $C = (V_C, E_C)$ *a planar st-component of* G, *and* $P = (V, E_P)$ *a maximum planar subgraph of* G. *Then, either* $C \subseteq P$, *or* $|E_C| - |E_P \cap E_C| = \mathrm{mincut}_{s,t}(C)$.

Proof. We distinguish two cases.

Case 1. *There is a path from* s *to* t *in* P *which consists only of edges of* C. Consider an embedding Γ of P. If we cut out C from Γ, then s and t must lie in a common face of the resulting embedding Γ'. On the other other hand, we can construct an embedding Γ_C of C in which s and t lie on the external face. Inserting Γ_C into Γ' yields an embedding of $P \cup C$. Since P is a maximum planar subgraph of G and $C \subseteq G$, it follows that $C \subseteq P$.

Case 2. *There is no such path from s to t.* Let $E' = E_P \cap E_C$ be the edges of C contained in P. It follows that $C' = (V_C, E')$ has at least two connected components, one containing s, and the other containing t. Hence, the number of edges in $E_C \setminus E'$ is at least $\mathrm{mincut}_{s,t}(C)$, which implies $|E_C| - |E_P \cap E_C| \geq \mathrm{mincut}_{s,t}(C)$.

On the other hand, we can construct an embedding of C with s and t on the external face, and remove the $\mathrm{mincut}_{s,t}(C)$ edges in a traversing path of C with respect to (s, t). This yields an embedding Γ with two connected components C_s and C_t with $s \in C_s$ and $t \in C_t$. Let $G' = G[E \setminus E_C]$ be the rest of the graph. Since C_s has only s in common with G' and C_t has only t in common with G', we can insert Γ into any embedding of $G' \cap P$ preserving planarity. This implies that $|E_C| - |E_P \cap E_C| \leq \mathrm{mincut}_{s,t}(C)$ and the lemma holds. □

Using this lemma, we can show that the non-planar core is invariant with respect to skewness.

Theorem 4. *Let G be a 2-connected graph, and let (C, w) be its non-planar core. Then,*

$$\mu(G) = \mu(C, w)$$

Proof. Let $G = (V, E)$ and $C = (V_C, E_C)$.

"$\geq$" Let $P = (V, E_P)$ be a maximum planar subgraph of G. We have $\mu(G) = |E| - |E_P|$. We show that we can construct a planar subgraph $P_C = (V_C, E')$ of C with $w(E_C) - w(E') = \mu(G)$.

Consider a planar st-component C of G. By Lemma 3, we know that either C is completely contained in P, or exactly $\mathrm{mincut}_{s,t}(C)$ many edges of C are not in E_P. In the first case, we know that an st-path is in P, and hence replacing C by the corresponding edge (s, t) preserves planarity. In the second case, the corresponding virtual edge $e = (s, t)$ with weight $w(e) = \mathrm{mincut}_{s,t}(C)$ will not be in P_C.

Constructing P_C in this way obviously yields a planar subgraph (V_C, E') of C with $w(E_C \setminus E') = \mu(G)$.

"$\leq$" Let $P_C = (V_C, E_P)$ be a maximum weight planar subgraph of C, and let D be a drawing of P_C. We have $\mu(C, w) = w(E_C) - w(E_P)$. We show that we can construct a planar subgraph $P' = (V, E')$ of G with $|E| - |E'| = \mu(C, w)$. We again consider a planar st-component C of G. Let $e = (s, t)$ be the corresponding virtual edge, and let D_C be a planar drawing of C in which both s and t lie in the external face. If e is in P_C, we can replace e with the drawing D_C and the resulting drawing remains planar. If e is not in P_C, we remove the edges of a traversing path of C with respect to (s, t) from D_C. This yields a drawing D'_C with two connected components, one containing s, and the other containing t. Obviously, we can add the drawing D'_C to D preserving planarity, and we removed exactly $w(e) = \mathrm{mincut}_{s,t}(C)$ edges from G.

We finally end up with a drawing of a planar subgraph $P = (V, E')$ of G with $|E| - |E'| = \mu(\mathcal{C}, w)$.   □

6   Thickness

For computing the thickness of G, we do not need to consider the weight of edges in the non-planar core $\mathcal{C}$ of G. Instead, we slightly modify $\mathcal{C}$ by splitting every virtual edge (s, t) whose expansion graph does not contain an edge (s, t). We denote the resulting graph with core$^+(G)$.

Theorem 5. *Let G be a 2-connected graph, and let $C' = \mathrm{core}^+(G)$. Then,*

$$\theta(G) = \theta(\mathrm{core}^+(G))$$

Proof. "≥" Let $\theta(G) = k$, and let $G_1, \ldots, G_k$ be k planar graphs with $G_1 \cup \ldots \cup G_k = G$. We consider a planar st-component C. We distinguish two cases:

(i) If there is a graph G_i such that $G_i \cap C$ contains a path from s to t, then we remove all edges and vertices $\neq s, t$ of C from all graphs $G_1, \ldots, G_k$, and we add the edge $e = (s, t)$ to G_i. If C does not contain an edge (s, t), then we also split e.

(ii) Otherwise, we know that $k \geq 2$, and therefore there are two graphs G_i and G_j with $i \neq j$. We add the edges $e_s = (s, d)$ to G_i and $e_t = (d, t)$ to G_j, where d is a new dummy vertex. If any of the end vertices of e_1 (resp. e_2) is not yet contained in G_i (resp. G_j), we also add this vertex.

It follows that we can construct k planar graphs whose union is core$^+(G)$, and thus $\theta(G) \geq \theta(\mathrm{core}^+(G))$.

"≤" Let $\theta(\mathrm{core}^+(G)) = k$, and let $G_1, \ldots, G_k$ be k pairwise edge disjoint planar graphs with $G_1 \cup \ldots \cup G_k = \mathrm{core}^+(G)$. We consider a virtual edge $e = (s, t)$ of the non-planar core of G. Let $C = (V_C, E_C)$ be the expansion graph of e. If C contains an edge (s, t), then e is contained in core$^+(G)$, and thus there is a subgraph, say G_i, containing e. We replace e in G_i by C.

Otherwise, C contains an edge $e = (s, t)$ and e was split into two edges, say $e_1 = (s, d)$ and $e_2 = (d, t)$, in core$^+(G)$. We split C into two edge disjoint graphs C_1 and C_2 in the following way: Let E' be the set of edges incident with s. Then, C_1 is the graph induced by E', and C_2 is the graph induced by $E_C \setminus E'$. Let G_i be the graph containing e_1, and let G_j be the graph containing e_2. If $i = j$, then we replace e_1 and e_2 by C in G_i. Otherwise, we replace e_1 by C_1 in G_i, and e_2 by C_2 in G_j.

It follows that we can construct k planar subgraphs of G whose union is G, and thus $\theta(G) \leq k$.   □

7   Further Reductions

It is a straight-forward idea to try to reduce the computation of crossing number or skewness to the non-planar skeletons of R-nodes. To do this, it would be necessary to be able to merge two components with the following properties:

(a) Both components have exactly two nodes, say s and t, in common.
(b) Each component is – if augmented with a virtual edge (s, t) – non-planar and at least 2-connected.
(c) The crossing number (skewness) of the merged component is the sum of the crossing numbers (skewnesses) of the components.

In the following we will give counterexamples to show that this approach fails.

Crossing Number. Figure 2(a) shows two components and their crossing minimal embedding, with regards to the minimum st-cut of their counterpart, which defines the weight of the virtual edges. The two components have unique minimum st-cuts, denoted by dashed lines. The minimum st-cut of the left component is 7, whereby the minimum st-cut of the right one is 5. The minimum crossing numbers of the left and right components are 10 and 4, respectively; but the minimum crossing number of the merged result is only $2 \cdot 4 + 5 = 13$ (Fig. 2(b)), which is less than the sum $10 + 4 = 14$. The reason is that we have edges that partially cross through the counterpart component.

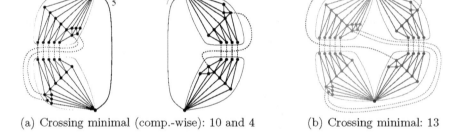

(a) Crossing minimal (comp.-wise): 10 and 4   (b) Crossing minimal: 13

Fig. 2. Calculating only the crossing numbers of the non-planar R-nodes is not correct

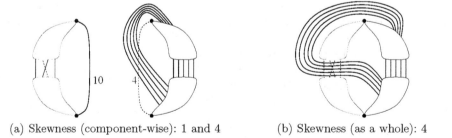

(a) Skewness (component-wise): 1 and 4   (b) Skewness (as a whole): 4

Fig. 3. Calculating only the skewnesses of the non-planar R-nodes is not correct

Skewness. Figure 3(a) shows two components including the virtual edges with the weights of their counterpart's minimum st-cut. The jelly bag cap shaped regions denote dense, crossing-free, 3-connected subgraphs, similar to the ones in Fig. 2. The edges which have to be removed to get a planar subgraph are the

dashed lines. The skewness of the left component is 1 — note that the choice
between the two possibilities is arbitrary. The skewness of the right component
corresponds to removing its virtual edge, and therefore has the value of 4. We can
see that we have one edge that has to be removed for both components, and is
therefore counted twice: the merged drawing has a skewness of only 4, although
the sum of the separate skewnesses would have suggested $1 + 4 = 5$. Note that
we can not even find any set of edges which does not include the virtual edge,
has the size 5, and can be removed in order to get a planar subgraph.

8   Experimental Results and Discussion

We tested the effect of our reduction strategy on a widely used benchmark set
commonly known as the *Rome library* [2]. This library contains over 11.000
graphs ranging from 10 to 100 vertices, which have been generated from a core
set of 112 graphs used in real-life software engineering and database applications.

We found that all non-planar graphs in the library have a single non-planar
biconnected component whose non-planar core is the skeleton of just one R-node.
Fig. 4 shows the average relative size of the non-planar core $\mathcal{C}$ compared to the
non-planar biconnected component (block) and the total graph. Here, the size of
a graph is simply the number of its edges. It turns out that, on average, the size
of the non-planar core is only 2/3 of the size of the non-planar block. Compared
to the whole graph, the size of the non-planar core reduces to about 55% on
average. This shows that the new approach provides a significant improvement
for reducing the size of the graph.

It will be interesting to see the effect the reduction strategy has on the practical performance of heuristics and exact algorithms for computing crossing number, skewness, and thickness. It remains an open problem if we can further reduce

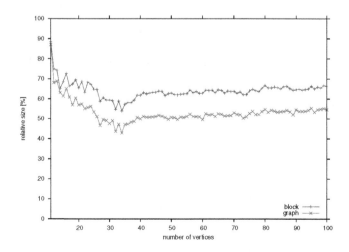

Fig. 4. Relative size of the non-planar core for the Rome graphs

a graph based on its connectivity structure. In particular, there might be the possibility for improvements by considering cut sets with three or more vertices.

References

[1] M. Chimani and C. Gutwenger. On the minimum cut of planarizations. Technical report, University of Dortmund, Germany, 2005.

[2] G. Di Battista, A. Garg, G. Liotta, R. Tamassia, E. Tassinari, and F. Vargiu. An experimental comparison of four graph drawing algorithms. *Comput. Geom. Theory Appl.*, 7:303–325, 1997.

[3] G. Di Battista and R. Tamassia. On-line maintanance of triconnected components with SPQR-trees. *Algorithmica*, 15:302–318, 1996.

[4] G. Di Battista and R. Tamassia. On-line planarity testing. *SIAM J. Comput.*, 25(5):956–997, 1996.

[5] M. R. Garey and D. S. Johnson. Crossing number is NP-complete. *SIAM J. Algebraic Discrete Methods*, 4(3):312–316, 1983.

[6] C. Gutwenger and P. Mutzel. A linear time implementation of SPQR trees. In J. Marks, editor, *Proc. GD 2000*, volume 1984 of *LNCS*, pages 77–90. Springer-Verlag, 2001.

[7] C. Gutwenger and P. Mutzel. An experimental study of crossing minimization heuristics. In G. Liotta, editor, *Proc. GD 2003*, volume 2912 of *LNCS*, pages 13–24. Springer, 2004.

[8] C. Gutwenger, P. Mutzel, and R. Weiskircher. Inserting an edge into a planar graph. *Algorithmica*, 41(4):289–308, 2005.

[9] J. Hopcroft and R. E. Tarjan. Efficient planarity testing. *J. ACM*, 21(4):549–568, 1974.

[10] C. Kuratowski. Sur le problème des courbes gauches en topologie. *Fundamenta Mathematicae*, 15:271–283, 1930.

[11] A. Liebers. Planarizing graphs — A survey and annotated bibliography. *J. Graph Algorithms and Applications*, 5(1):1–74, 2001.

[12] P. C. Liu and R. C. Geldmacher. On the deletion of nonplanar edges of a graph. *Congr. Numerantium*, 24:727–738, 1979.

[13] A. Mansfield. Determining the thickness of graphs is NP-hard. *Math. Proc. Camb. Philos. Soc.*, 93:9–23, 1983.

[14] Petra Mutzel, Thomas Odenthal, and Mark Scharbrodt. The thickness of graphs: A survey. *Graphs and Combinatorics*, 14(1):59–73, 1998.

[15] J. Širáň. Additivity of the crossing number of graphs with connectivity 2. *Periodica Mathematica Hungarica*, 15(4):301–305, 1984.

An Experimental Comparison of Fast Algorithms for Drawing General Large Graphs

Stefan Hachul and Michael Jünger

Universität zu Köln, Institut für Informatik,
Pohligstraße 1, 50969 Köln, Germany
{hachul, mjuenger}@informatik.uni-koeln.de

Abstract. In the last decade several algorithms that generate straight-line drawings of general large graphs have been invented. In this paper we investigate some of these methods that are based on force-directed or algebraic approaches in terms of running time and drawing quality on a big variety of artificial and real-world graphs. Our experiments indicate that there exist significant differences in drawing qualities and running times depending on the classes of tested graphs and algorithms.

1 Introduction

Force-directed graph drawing methods generate drawings of a given general graph $G = (V, E)$ in the plane in which each edge is represented by a straight line connecting its two adjacent nodes. The computation of the drawings is based on associating G with a physical model. Then, an iterative algorithm tries to find a placement of the nodes so that the total energy of the physical system is minimal. Important esthetic criteria are uniformity of edge length, few edge crossings, non-overlapping nodes, and the display of symmetries if some exist.

Classical force-directed algorithms like [5, 15, 7, 4, 6] are used successfully in practice (see e.g. [2]) for drawing general graphs containing few hundreds of vertices. However, in order to generate drawings of graphs that contain thousands or hundreds of thousands of vertices more efficient force-directed techniques have been developed [19, 18, 9, 8, 12, 21, 11, 10]. Besides fast force-directed algorithms other very fast methods for drawing large graphs (see e.g. [13, 16]) have been invented. These methods are based on techniques of linear algebra instead of physical analogies. But they strive for the same esthetic drawing criteria.

Previous experimental tests of these methods are mainly restricted to regular graphs with *grid-like* structures (see e.g. [13, 16, 9, 21, 12]). Since general graphs share these properties quite seldom, and since the test environments of these experiments are different, a standardized comparison of the methods on a wider range of graphs is needed.

In this study we experimentally compare some of the fastest state-of-the-art algorithms for straight-line drawing of general graphs on a big variety of graph classes. In particular, we investigate the force-directed algorithm GRIP of Gajer and Kobourov [9] and Gajer et al. [8], the *Fast Multi-scale Method* (FMS) of

P. Healy and N.S. Nikolov (Eds.): GD 2005, LNCS 3843, pp. 235–250, 2005.

Harel and Koren [12], and the *Fast Multipole Multilevel Method* (FM^3) of Hachul and Jünger [11, 10]. The examined algebraic methods are the algebraic multigrid method ACE of Koren et al. [16] and the *high-dimensional embedding* approach (HDE) by Harel and Koren [13]. Additionally, one of the faster classical force-directed algorithms, namely the *grid-variant algorithm* (GVA) of Fruchterman and Reingold [7], is tested as a benchmark.

After a short description of the tested algorithms in Section 2 and of the experimental framework in Section 3, our results are presented in Section 4.

2   The Algorithms

2.1   The Grid-Variant Algorithm (GVA)

The grid-variant algorithm of Fruchterman and Reingold [7] is based on a model of pairwise repelling charged particles (the nodes) and attracting springs (the edges), similar to the model of the Spring Embedder of Eades [5]. Since a naive exact calculation of the repulsive forces acting between all pairs of charges needs $\Theta(|V^2|)$ time per iteration, GVA does only calculate the repulsive forces acting between nodes that are placed relatively near to each other. Therefore, the rectangular drawing area is subdivided into a regular square grid. The repulsive forces that act on a node v that is contained in a grid box B are approximated by summing up only the repulsive forces that are induced by the nodes contained in B and the nodes in the grid boxes that are neighbors of B. If the number of iterations is assumed to be constant, the best-case running time of the GVA is $\Theta(|V| + |E|)$. The worst-case running time, however, remains $\Theta(|V|^2 + |E|)$.

2.2   The Method GRIP

Gajer et al. [8] and Gajer and Kobourov [9] developed the force-directed multi-level algorithm GRIP. In general, multilevel algorithms are based on two phases. A *coarsening phase*, in which a sequence of coarse graphs with decreasing sizes is computed and a *refinement phase* in which successively drawings of finer graphs are computed, using the drawings of the next coarser graphs and a variant of a suitable force-directed single-level algorithm.

The coarsening phase of GRIP is based on the construction of a *maximum independent set filtration* or *MIS filtration* of the node set V. A MIS filtration is a family of sets $\{V =: V_0, V_1, \ldots, V_k\}$ with $\emptyset \subset V_k \subset V_{k-1} \ldots \subset V_0$ so that each V_i with $i \in \{1, \ldots, k\}$ is a maximal subset of V_{i-1} for which the graph-theoretic distance between any pair of its elements is at least $2^{i-1} + 1$. Gajer and Kobourov [9] use a Spring Embedder-like method as single-level algorithm at each level. The used force vector is similar to that used in the Kamada-Kawai method [15], but is restricted to a suitable chosen subset of V_i.

Other notable specifics of GRIP are that it computes the MIS filtration only and no edge sets of the coarse graphs $G_0, \ldots, G_k$ that are induced by the filtrations. Furthermore, it is designed to place the nodes in an n-dimensional space

$(n \geq 2)$, to draw the graph in this space, and to project it into two or three dimensions.

The asymptotic running time of the algorithm, excluding the time that is needed to construct the MIS filtration, is $\Theta(|V|(\log \mathit{diam}(G)^2))$ for graphs with bounded maximum node degree, where $\mathit{diam}(G)$ denotes the diameter of G.

2.3   The Fast Multi-scale Method (FMS)

In order to create the sequence of coarse graphs in the force-directed multilevel method FMS, Harel and Koren [12] use an $O(k|V|)$ algorithm that finds a 2-approximative solution of the $\mathcal{NP}$-hard *k-center problem*. The node set V_i of a graph G_i in the sequence $G_0, \ldots, G_k$ is determined by the approximative solution of the k_i-center problem on G with $k_i > k_{i+1}$ for all $i \in \{0, \ldots, k-1\}$.

The authors use a variation of the algorithm of Kamada and Kawai [15] as force-directed single-level algorithm. In order to speed up the computation of this method, they modify the energy function of Kamada and Kawai [15] that is associated with a graph G_i with $i \in \{0, \ldots, k-1\}$. The difference to the original energy of Kamada and Kawai [15] is that only some of the $|V(G_i)| - 1$ springs that are connected with a node $v \in V(G_i)$ are considered.

The asymptotic running time of the FMS is $\Theta(|V||E|)$. Additionally, $\Theta(|V|^2)$ memory is needed to store the distances between all pairs of nodes.

2.4   The Fast Multipole Multilevel Method (FM³)

The force-directed multilevel algorithm FM³ has been introduced by Hachul and Jünger [11, 10]. It is based on a combination of an efficient multilevel technique with an $O(|V|\log|V|)$ approximation algorithm to obtain the repulsive forces between all pairs of nodes.

In the coarsening step subgraphs with a small diameter (called *solar systems*) are collapsed to obtain a multilevel representation of the graph. In the used single-level algorithm, the bottleneck of calculating the repulsive forces acting between all pairs of charged particles in the `Spring Embedder`-like force model is overcome by rapidly evaluating potential fields using a novel multipole-based tree-code. The worst-case running time of FM³ is $O(|V|\log|V| + |E|)$ with linear memory requirements.

2.5   The Algebraic Multigrid Method ACE

In the description of their method ACE, Koren et al. [16] define the quadratic optimization problem

$$(P) \qquad \min \ x^T L x \quad \text{so that} \quad x^T x = 1 \quad \text{in the subspace} \quad x^T 1_n = 0 \ .$$

Here $n = |V|$ and L is the *Laplacian* matrix of G.

The minimum of (P) is obtained by the eigenvector that corresponds to the smallest positive eigenvalue of L. The problem of drawing the graph G in two dimensions is reduced to the problem of finding the two eigenvectors of L that are associated with the two smallest eigenvalues.

Instead of calculating the eigenvectors directly, an *algebraic multigrid algorithm* is used. Similar to the force-directed multilevel ideas, the idea is to express the originally high-dimensional problem in lower and lower dimensions, solving the problem at the lowest dimension, and progressively solving a high-dimensional problem by using the solutions of the low-dimensional problems.

The authors do not give an upper bound on the asymptotic running time of ACE in the number of nodes and edges.

2.6   High-Dimensional Embedding (HDE)

The method HDE of Harel and Koren [13] is based on a two phase approach that first generates an embedding of the graph in a very high-dimensional vector space and then projects this drawing into the plane.

The high-dimensional embedding of the graph is generated by first using a linear time algorithm for approximatively solving the k-center problem. A fixed value of $k = 50$ is chosen, and k is also the dimension of the high-dimensional vector space. Then, breadth-first search starting from each of the k center nodes is performed resulting in k $|V|$-dimensional vectors that store the graph-theoretic distances of each $v \in V$ to each of the k centers. These vectors are interpreted as a k-dimensional embedding of the graph.

In order to project the high-dimensional embedding of the graph into the plane, the k vectors are used to define a *covariance* matrix S. The x- and y-coordinates of the two-dimensional drawing are obtained by calculating the two eigenvectors of S that are associated with its two largest eigenvalues. HDE runs in $O(|V| + |E|)$ time.

3   The Experiments

3.1   Test-Environment, Implementations, and Parameter Settings

All experiments were performed on a 2.8 GHz Intel Pentium 4 PC with one gigabyte of memory.

We tested a version of GVA that has been implemented in the framework of AGD [14] by S. Näher and D. Alberts, an implementation of GRIP by R. Yusufov that is available from [22], and implementations of FMS, ACE, HDE by Y. Koren that are available from [17]. Finally, we tested our own implementation of FM³.

In order to obtain a fair comparison, we ran each algorithm with the same set of standard-parameter settings (given by the authors) on each tested graph. However, we are aware that in some cases it might be possible to obtain better results by spending a considerable amount of time with trial-and-error searching for an optimal set of parameters for each algorithm and graph.

3.2   The Set of Test Graphs

Since only few implementations can handle disconnected and weighted graphs, we restrict to connected unweighted graphs, here.

We generated several classes of artificial graphs to examine the scaling of the algorithms on graphs with predefined structures but different sizes.

These are *random grid* graphs that were obtained by first creating regular square grid graphs and then randomly deleting 3% of the nodes. The *sierpinski* graphs were created by associating the *Sierpinski Triangles* with graphs. Furthermore, we generated complete 6-*nary trees*.

The next two classes of artificial graphs were designed to test how well the algorithms can handle highly non-uniform distributions of the nodes and high node degrees. Therefore, we created these graphs in a way so that one can expect that an energy-minimal configuration of the nodes in a drawing that relies on a **Spring Embedder**-like force model induces a tiny subregion of the drawing area which contains $\Theta(|V|)$ nodes. In particular, we constructed trees that contain a root node r with $|V|/4$ neighbors. The other nodes were subdivided into six subtrees of equal size rooted at r. We called these graphs *snowflake* graphs.

Additionally we created *spider* graphs by constructing a circle C containing 25% of the nodes. Each node of C is also adjacent to 12 other nodes of the circle. The remaining nodes were distributed on 8 paths of equal length that were rooted at one node of C. In contrast to the snowflake graphs is that the spider graphs have bounded maximum degree.

The last kind of artificial graphs are graphs with a relatively high edge density $|E|/|V| \geq 14$. We called them *flower* graphs. They are constructed by joining 6 circles of equal length at a single node before replacing each of the nodes by a complete subgraph with 30 nodes (K_{30}).

The rest of the test graphs are taken from real-world applications. In particular, we selected graphs from the *AT&T graph library* [1], from C. Walshaw's graph collection [20], and a graph that describes a social network of 2113 people that we obtained from C. Lipp.

We partitioned the artificial and real-world graphs into two sets. The first set are graphs that consist of few biconnected components, have a constant maximum node degree, and have a low edge density. Furthermore, one can expect that an energy-minimal configuration of the nodes in a **Spring Embedder** drawing of such a graph does not contain $\Theta(|V|)$ nodes in an extremely tiny subregion of the drawing area. Since one can anticipate from previous experiments [13, 16, 9, 12] that the graphs contained in this set do not cause problems for many of the tested algorithms, we call the set of these graphs *kind*. The second set is the complement of the first one, and we call the set of these graphs *challenging*.

3.3   The Criteria of Evaluation

The natural criteria to evaluate a graph-drawing algorithm in practice are the needed running times and the quality of the drawings.

Unlike evaluating the first criterion, evaluating the quality of a drawing is a difficult task. Possible ways are the calculation of the total energy in the underlying force models or the measurement of relevant esthetic criteria (e.g. crossing number, uniformity of the edge lengths). However, one of the most important goals is that an individual user is satisfied with a drawing. Hence, we decided to print the drawings and to comment how well they display the structure of each graph by keeping the modeled esthetic criteria in mind.

4   The Results

4.1   Comparison of the Running Times

Table 1 presents the running times of the methods GVA, FM³, GRIP, FMS, ACE, and HDE for the tested graphs.

Table 1. The test graphs and the running times that are needed by the tested algorithms to draw them. Explanations: (E) No drawing was generated due to an error in the executable. (M) No drawing was generated because the memory is restricted to graphs with $\leq 10,000$ nodes. (T) No drawing was generated within 10 hours of CPU time. B denotes the set of biconnected components of the graphs.

Graph Information							Algorithm Information					
							CPU Time in Seconds					
Type	Name	$|V|$	$|E|$	$|B|$	$\frac{|E|}{|V|}$	max. degree	GVA	FM³	GRIP	FMS	ACE	HDE
Kind Arti- ficial	rnd_grid_032	985	1834	2	1.8	4	12.5	1.9	0.3	1.0	< 0.1	< 0.1
	rnd_grid_100	9497	17849	6	1.8	4	203.4	19.1	4.4	32.0	0.5	0.1
	rnd_grid_320	97359	184532	2	1.9	4	6316.1	215.4	(E)	(M)	4.1	1.3
	sierpinski_06	1095	2187	1	2.0	4	13.1	1.8	0.3	1.0	< 0.1	< 0.1
	sierpinski_08	9843	19683	1	2.0	4	171.7	16.8	4.8	33.0	1.0	0.1
	sierpinski_10	88575	177147	1	2.0	4	3606.4	162.0	(E)	(M)	23.4	1.0
Kind Real World	crack	10240	30380	1	2.9	9	317.5	23.0	6.8	(M)	0.4	0.2
	fe_pwt	36463	144794	55	3.9	15	1869.1	69.0	(E)	(M)	(T)	0.5
	finan_512	74752	261120	1	3.4	54	6319.8	158.2	(E)	(M)	7.5	1.0
	fe_ocean	143437	409593	39	2.8	6	19247.0	355.9	(E)	(M)	4.0	3.4
Chal- lenging Arti- ficial	tree_d_4	1555	1554	1554	1.0	7	14.3	2.6	0.3	2.0	< 0.1	< 0.1
	tree_d_5	9331	9330	9330	1.0	7	130.3	17.7	2.4	43.0	0.5	< 0.1
	tree_d_6	55987	55986	55986	1.0	7	1769.2	121.3	(E)	(M)	4.5	0.5
	snowflake_A	971	970	970	1.0	256	8.0	1.6	0.4	73.0	0.4	< 0.1
	snowflake_B	9701	9700	9700	1.0	2506	143.2	17.4	6.1	3320.0	(T)	< 0.1
	snowflake_C	97001	97000	97000	1.0	25006	14685.7	166.5	(E)	(M)	(T)	0.8
	spider_A	1000	2200	801	2.2	18	17.6	1.9	0.4	1.0	1.1	< 0.1
	spider_B	10000	22000	8001	2.2	18	189.0	17.7	7.2	47.0	8.9	0.1
	spider_C	100000	220000	80001	2.2	18	4568.3	177.2	(E)	(M)	280.7	1.3
	flower_A	930	13521	1	14.5	30	61.7	1.2	0.7	1.0	< 0.1	< 0.1
	flower_B	9030	131241	1	14.5	30	595.1	11.9	19.3	46.0	1.4	0.2
	flower_C	90030	1308441	1	14.5	30	11841.5	121.4	(E)	(M)	(T)	1.4
Chal- lenging Real World	ug_380	1104	3231	27	2.9	856	23.1	2.1	0.4	1.0	< 0.1	< 0.1
	esslingen	2075	5530	867	2.6	97	43.8	4.0	0.5	404.0	1.0	< 0.1
	add_32	4960	9462	951	1.9	31	80.6	12.1	1.6	17.0	0.5	< 0.1
	dg_1087	7602	7601	7601	1.0	6566	624.8	18.1	3.6	5402.0	108.4	< 0.1
	bcsstk_33	8738	291583	1	33.3	140	1494.6	23.8	29.1	6636.0	0.4	0.3
	bcsstk_31	35586	572913	48	16.1	188	4338.4	83.6	(E)	(M)	1.9	0.7
	bcsstk_32	44609	985046	3	22.0	215	6387.1	110.9	(E)	(M)	3.6	0.9

As expected, in most cases **GVA** is the slowest method among the force-directed algorithms. The largest graph fe_ocean is drawn by **GVA** in 5 hours and 20 minutes.

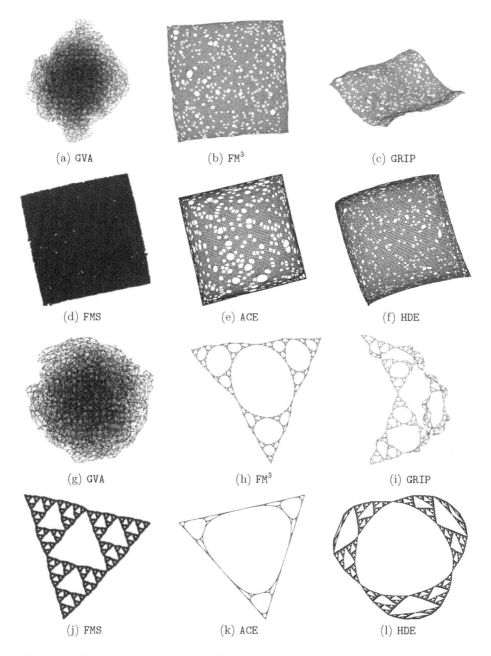

(a) GVA

(b) FM^3

(c) GRIP

(d) FMS

(e) ACE

(f) HDE

(g) GVA

(h) FM^3

(i) GRIP

(j) FMS

(k) ACE

(l) HDE

Fig. 1. (a)-(f) Drawings of rnd_grid_100 and (g)-(l) sierpinski_08 generated by different algorithms

The method FM³ is significantly faster than GVA for all tested graphs. The running times range from less than 2 seconds for the smallest graphs to less than 6 minutes for the largest graph fe_ocean. The subquadratic scaling of FM³ can be experimentally confirmed for all classes of tested graphs.

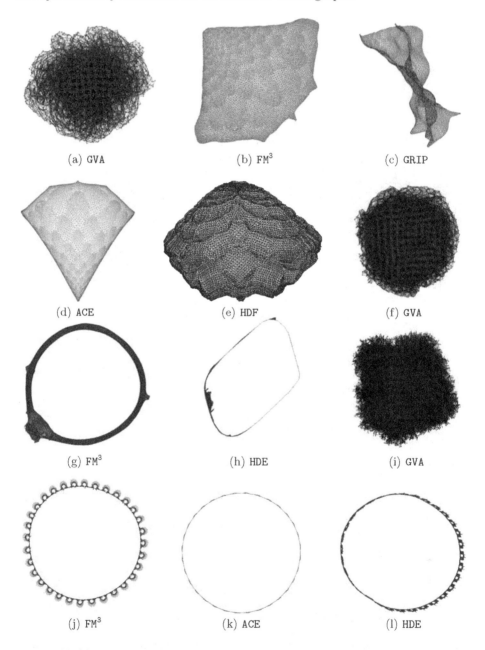

(a) GVA (b) FM³ (c) GRIP

(d) ACE (e) HDF (f) GVA

(g) FM³ (h) HDE (i) GVA

(j) FM³ (k) ACE (l) HDE

Fig. 2. (a)-(e) Drawings of crack, (f)-(h) fe_pwt, and (i)-(l) finan_512 generated by different algorithms

Except for the dense graphs flower_B and bcsstk_33 GRIP is faster than FM^3 (up to a factor 9). Unfortunately, we could not examine the scaling of GRIP for the largest graphs due to an error in the executable.

Since the memory requirement of FMS is quadratic in the size of the graph, the implementation of FMS is restricted to graphs that contain at most 10,000 nodes. The running time of FMS is comparable with that of FM^3 for the smallest and the medium sized kind graphs. In contrast to this, the CPU time of FMS increases

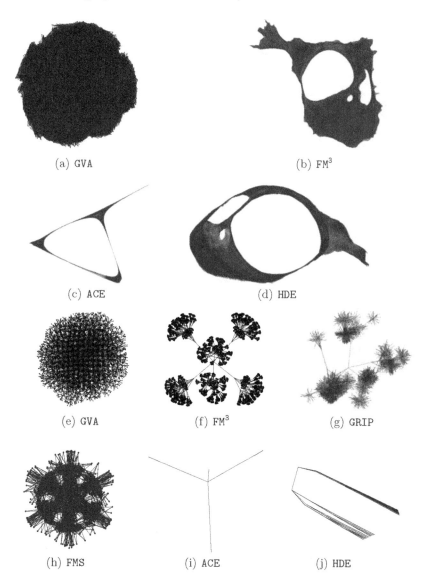

(a) GVA   (b) FM^3

(c) ACE   (d) HDE

(e) GVA   (f) FM^3 (g) GRIP

(h) FMS   (i) ACE   (j) HDE

Fig. 3. (a)-(d) Drawings of fe_ocean and (e)-(j) tree_06_05 generated by different algorithms

drastically for several challenging graphs, in particular for graphs that either contain nodes with a very high degree or have a high edge density.

The algorithm ACE is much faster than the force-directed algorithms for nearly all kind graphs. However, like for FMS, the running times grow extremely when ACE is used to draw several of the challenging graphs.

The linear time method HDE is by far the fastest algorithm. It needs less than 3.4 seconds for drawing even the largest tested graph.

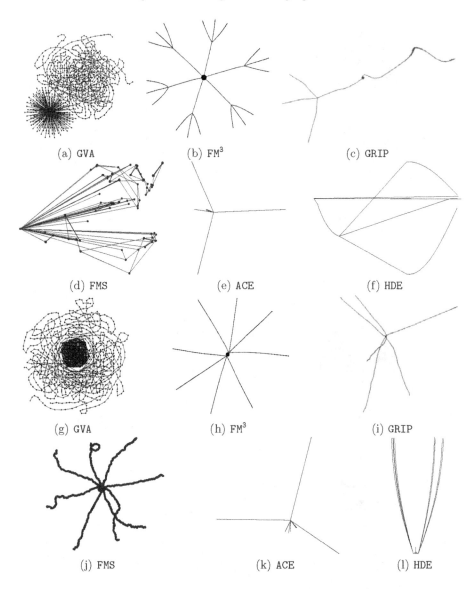

(a) GVA   (b) FM³   (c) GRIP

(d) FMS   (e) ACE   (f) HDE

(g) GVA   (h) FM³   (i) GRIP

(j) FMS   (k) ACE   (l) HDE

Fig. 4. (a)-(f) Drawings of snowflake_A and (g)-(l) spider_A generated by different algorithms

4.2   Comparison of the Drawings

For all kind graphs the classical method GVA does not untangle the drawings
that were induced by the random initial placements.

In contrast to this nearly all algorithms generated comparable pleasing drawings of the kind graphs (see Figure 1, Figure 2, and Figure 3(a)-(d)).

None of the drawings of the complete 6-nary trees (see Figure 3(e)-(j)) is really convincing, since the force-directed algorithms produce many unnecessary

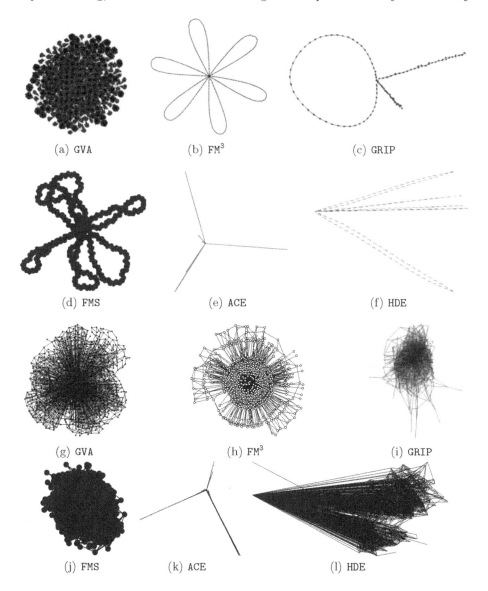

(a) GVA   (b) FM^3 (c) GRIP

(d) FMS   (e) ACE   (f) HDE

(g) GVA   (h) FM^3 (i) GRIP

(j) FMS   (k) ACE   (l) HDE

Fig. 5. (a)-(f) Drawings of flower_B and (g)-(l) ug_380 generated by different algorithms

edge crossings. However, the drawings generated by FM³ and FMS display parts of the regularity of these graphs. The algebraic methods ACE and HDE place many nodes at the same coordinates. In general, this behavior of the algebraic methods can be observed for graphs that consist of many biconnected components. Explanations of the theoretical reasons can be found in [3, 16] and [10].

Except FM³ none of the tested algorithms displays the global structure of the snowflake graphs. Even the drawings of the smallest snowflake graph (see

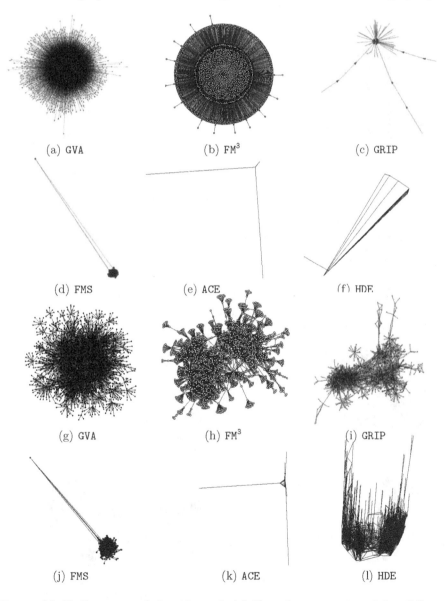

(a) GVA   (b) FM³   (c) GRIP

(d) FMS   (e) ACE   (f) HDE

(g) GVA   (h) FM³   (i) GRIP

(j) FMS   (k) ACE   (l) HDE

Fig. 6. (a)-(f) Drawings of dg_1087 and (g)-(l) esslingen generated by different algorithms

Figure 4(a)-(f)) leave room for improvement. However, **GVA** and **GRIP** visualize parts of its structure in an appropriate way.

The drawings of the spider_A graph (see Figure 4(g)-(l)) that are generated by **GRIP**, **FMS**, and **HDE** are not as symmetric as that generated by FM³. But they display the global structure of the graph. The drawing generated by **GVA** shows the dense subregion, but **GVA** does not untangle the 8 paths. The paths in the

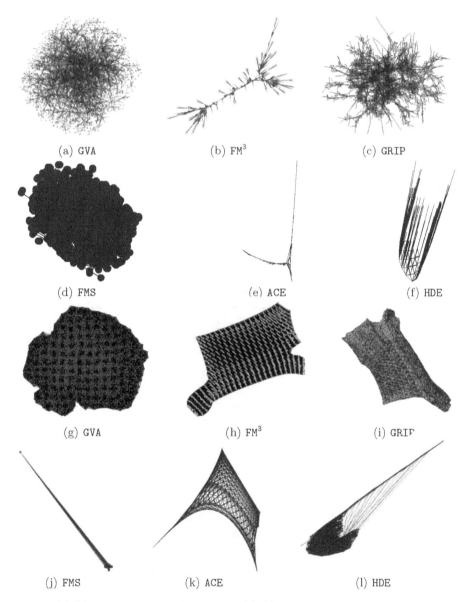

(a) GVA   (b) FM³   (c) GRIP

(d) FMS   (e) ACE   (f) HDE

(g) GVA   (h) FM³   (i) GRIP

(j) FMS   (k) ACE   (l) HDE

Fig. 7. (a)-(f) Drawings of add_32 and (g)-(l) bcsstk_33 generated by different algorithms

drawing of ACE are not displayed in the same length. The drawings of the larger spider graphs are of comparable quality.

The drawings of the flower_B graph (see Figure 5(a)-(f)) that are generated by FMS and HDE display the global structure of the graph but the symmetries are not as clear as in the drawing generated by FM³. The drawings of the other flower graphs are of comparable quality.

We concentrate on the challenging real-world graphs now. The graphs ug_380 and dg_1087 both contain one node with a very high degree. Furthermore, dg_1087 has many biconnected components, since it is a tree. Only the drawings that are generated by GVA, FM³, and GRIP (see Figure 5(g)-(l) and Figure 6(a)-(f)) clearly display the central regions of these graphs. It can be observed that

(a) GVA   (b) FM³   (c) ACE

(d) HDE   (e) GVA   (f) FM³

(g) ACE   (h) HDE

Fig. 8. (a)-(d) Drawings of bcsstk_31_con and (e)-(h) bcsstk_32 generated by different algorithms

the edge lengths of the drawing of dg_1087 that is generated by FM³ are more uniform than in the drawings of dg_1087 that are generated by GVA and GRIP.

The social network esslingen (see Figure 6(g)-(l)) consists of two big well-connected subgraphs. This can be visualized by FM³, GRIP, and HDE. But all drawings contain many edge crossings.

Since add_32 that describes a 32 bit adder contains many biconnected components, we expect that the drawings have a tree-like shape. This structure is visualized by GVA, FM³, GRIP, and ACE (see Figure 7(a)-(f)). The drawings of GVA and GRIP contain many edge crossings, while the drawing of ACE displays the global structure, but hides local details.

Finally, we discuss the drawings of the graphs bcsstk_31_con, bcsstk_32, and bcsstk_33 that have a very high edge density. The drawings of bcsstk_33 (see Figure 7(g)-(l)) that are generated by FM³, GRIP, and ACE are comparable and visualize the regular structure of the graph. The car body that is modeled by the graph bcsstk_31_con (see Figure 8(a)-(d)) is visualized by FM³ and ACE only. All drawings of bcsstk_32 are completely different (see Figure 8(e)-(h)) and an evaluation of the drawings is left to the reader.

5   Conclusion

We can summarize that only GVA, FM³, and HDE generate drawings of all tested graphs. The force-directed multilevel methods and the algebraic methods are — except the methods FMS and ACE for some graphs — much faster than the comparatively slow classical algorithm GVA. HDE, FM³ and GRIP scale well on all tested graphs. FM³ needs few minutes to draw the largest graphs. GRIP is up to factor 9 faster than FM³ but it could not be tested on the largest graphs. All tested methods are much slower than HDE that needs only few seconds to draw even the largest graphs.

As expected, all algorithms, except GVA, generate pleasing drawings of the kind graphs. In contrast to this, the quality of the generated drawings varies a lot depending on the structures of the tested challenging graphs. Only FM³ generates pleasing drawings for the majority of the challenging graphs. But there still remain classes of tested graphs (the complete trees and the social network graph esslingen) for which the drawing quality of all tested algorithms leaves room for improvement.

Acknowledgments. We would like to thank David Alberts, Steven Kobourov, Yehuda Koren, Stefan Näher, and Roman Yusufov for making the implementations of their algorithms available to us. We thank Ulrik Brandes, Carola Lipp and Chris Walshaw for the access to the real-world test graphs.

References

1. The AT&T graph collection: www.graphdrawing.org.
2. F. J. Brandenburg, M. Himsolt, and C. Rohrer. An Experimental Comparison of Force-Directed and Randomized Graph Drawing Methods. In *Graph Drawing 1995*, volume 1027 of *LNCS*, pages 76–87. Springer-Verlag, 1996.

3. U. Brandes and D. Wagner. In *Graph Drawing Software*, volume XII of *Mathematics and Visualization*, chapter visone - Analysis and Visualization of Social Networks, pages 321–340. Springer-Verlag, 2004.
4. R. Davidson and D. Harel. Drawing Graphs Nicely Using Simulated Annealing. *ACM Transactions on Graphics*, 15(4):301–331, 1996.
5. P. Eades. A heuristic for graph drawing. *Congressus Numerantium*, 42:149–160, 1984.
6. A. Frick, A. Ludwig, and H. Mehldau. A Fast Adaptive Layout Algorithm for Undirected Graphs. In *Graph Drawing 1994*, volume 894 of *LNCS*, pages 388–403. Springer-Verlag, 1995.
7. T. M. J. Fruchterman and E. M. Reingold. Graph Drawing by Force-directed Placement. *Software–Practice and Experience*, 21(11):1129–1164, 1991.
8. P. Gajer, M. T. Goodrich, and S. G. Kobourov. A Multi-dimensional Approach to Force-Directed Layouts of Large Graphs. In *Graph Drawing 2000*, volume 1984 of *LNCS*, pages 211–221. Springer-Verlag, 2001.
9. P. Gajer and S. G. Kobourov. GRIP: Graph Drawing with Intelligent Placement. In *Graph Drawing 2000*, volume 1984 of *LNCS*, pages 222–228. Springer-Verlag, 2001.
10. S. Hachul. *A Potential-Field-Based Multilevel Algorithm for Drawing Large Graphs*. PhD thesis, Institut für Informatik, Universität zu Köln, Germany, 2005.
11. S. Hachul and M. Jünger. Drawing Large Graphs with a Potential-Field-Based Multilevel Algorithm (Extended Abstract). In *Graph Drawing 2004*, volume 3383 of *Lecture Notes in Computer Science*, pages 285–295. Springer-Verlag, 2005.
12. D. Harel and Y. Koren. A Fast Multi-scale Method for Drawing Large Graphs. In *Graph Drawing 2000*, volume 1984 of *LNCS*, pages 183–196. Springer-Verlag, 2001.
13. D. Harel and Y. Koren. Graph Drawing by High-Dimensional Embedding. In *Graph Drawing 2002*, volume 2528 of *LNCS*, pages 207–219. Springer-Verlag, 2002.
14. M. Jünger, G. W. Klau, P. Mutzel, and R. Weiskircher. In *Graph Drawing Software*, volume XII of *Mathematics and Visualization*, chapter AGD - A Library of Algorithms for Graph Drawing, pages 149–172. Springer-Verlag, 2004.
15. T. Kamada and S. Kawai. An Algorithm for Drawing General Undirected Graphs. *Information Processing Letters*, 31:7–15, 1989.
16. Y. Koren, L. Carmel, and D. Harel. Drawing Huge Graphs by Algebraic Multigrid Optimization. *Multiscale Modeling and Simulation*, 1(4):645–673, 2003.
17. Y. Koren's algorithms: research.att.com/~yehuda/index_programs.html.
18. A. Quigley and P. Eades. FADE: Graph Drawing, Clustering, and Visual Abstraction. In *Graph Drawing 2000*, volume 1984 of *LNCS*, pages 197–210. Springer-Verlag, 2001.
19. D. Tunkelang. JIGGLE: Java Interactive Graph Layout Environment. In *Graph Drawing 1998*, volume 1547 of *LNCS*, pages 413–422. Springer-Verlag, 1998.
20. C. Walshaw's graph collection: staffweb.cms.gre.ac.uk/~c.walshaw/partition.
21. C. Walshaw. A Multilevel Algorithm for Force-Directed Graph Drawing. In *Graph Drawing 2000*, volume 1984 of *LNCS*, pages 171–182. Springer-Verlag, 2001.
22. R. Yusufov's implementation of GRIP: www.cs.arizona.edu/~kobourov/GRIP.

Hierarchical Layouts of Directed Graphs in Three Dimensions

Seok-Hee Hong[1,2] and Nikola S. Nikolov[1,3]

[1] IMAGEN Program, National ICT Australia Ltd.
[2] School of IT, University of Sydney, NSW, Australia
[3] Department of CSIS, University of Limerick, Limerick, Republic of Ireland
{seokhee.hong, nikola.nikolov}@nicta.com.au

Abstract. We introduce a new graph drawing convention for 3D hierarchical drawings of directed graphs. The vertex set is partitioned into layers of vertices drawn in parallel planes. The vertex set is further partitioned into $k \geq 2$ subsets, called walls. The layout consists of a set of parallel walls which are perpendicular to the set of parallel planes of the layers. We also outline a method for computing such layouts and introduce four alternative algorithms for partitioning the vertex set into walls which address different aesthetic requirements.[1]

1 Introduction

The visual representation of hierarchically organised data has application in areas such as Social Network Analysis, Bioinformatics, Software Engineering, etc. Hierarchies are commonly modeled by directed graphs (digraphs) and thus visualised by algorithms for drawing digraphs. Most of the research effort in this area has been related to improvements of various aspects of the Sugiyama method, the most popular method for creating 2D layered drawings of digraphs [4, 9].

The increasing availability of powerful graphic displays opens new opportunities for developing new methods for 3D graph drawing. There is evidence that 3D graph layouts combined with novel interaction and navigation methods make graphs easier to comprehend by humans and increase the efficiency of task performance on digraphs [10].

However, there has been relatively little research on drawing digraphs in 3D. One of the known approaches is the method of Ostry which consists of computing a layered drawing in 2D and then wrapping it around a cone or a cylinder [8]. Another approach is the method used in the graph drawing system GIOTTO3D [5] which is conceptually different from the Sugiyama method. GIOTTO3D employs a simple 3-phase algorithm for producing 3D layered drawings of digraphs. In the first phase a planarisation method is used to draw the graph in 2D; in the second phase vertices and edges are assigned z-coordinates so that all edges point

---

[1] An online gallery with examples of our 3D hierarchical layouts is available at http://www.cs.usyd.edu.au/~visual/valacon/gallery/3DHL.

P. Healy and N.S. Nikolov (Eds.): GD 2005, LNCS 3843, pp. 251–261, 2005.

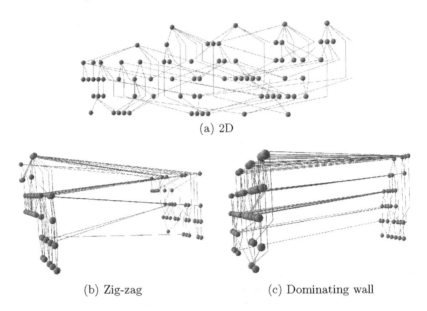

(a) 2D

(b) Zig-zag   (c) Dominating wall

Fig. 1. A 2D layered drawing (a) compared to 3D layered layouts of the same graph with two parallel walls (b) and (c)

into the same direction and the total edge span is minimised; and at the third phase the shape of the vertices and the edges is determined.

This paper presents a method for layered drawing of digraphs in 3D which extends the Sugiyama method. We generalise and extend the work presented in our previous paper [6]. In summary, we propose an extra step after the layer assignment of the 2D Sugiyama method. It consists of partitioning the vertex set into subsets, called *walls*. Any subset of the vertex set can be a wall. We propose that layers occupy parallel planes with all edges pointing in the same direction. Walls also occupy parallel planes which are perpendicular to the layer planes. Each pair of a wall and a layer intersect into a set of vertices placed along the line which is the intersection of the corresponding wall and layer planes. As a result each wall contains a 2D layered drawing. Examples of such a layout can be seen in Figures 1 and 2.

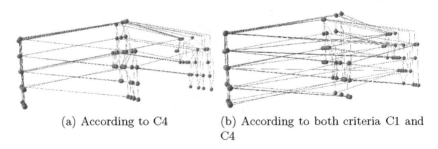

(a) According to C4   (b) According to both criteria C1 and
                                 C4

Fig. 2. Examples of 3D hierarchical layouts with $k > 2$ parallel walls

The motivation behind the proposed drawing convention consists of the following points:

- A 3D layered drawing of a digraph allows the employment of specific 3D navigation and interaction techniques and decreases visual complexity. For example, each wall can be viewed separately, the camera may move along edges between the walls, etc.
- Partitioning the hierarchy into a set of walls each containing a smaller 2D hierarchy allows us to draw the smaller 2D hierarchies efficiently with fast heuristics or even exact algorithms which generally would perform worse if employed for drawing the whole graph as a 2D hierarchy.
- Drawing the 3D hierarchy as a set of 2D hierarchies utilises the extensively developed techniques for drawing hierarchies in 2D.

Our method can be applied to any digraph, such as a class hierarchy that originates from a Software Engineering application, or a hierarchical relationship in a social network, etc. In particular, we report experiments with some of the graphs in the Rome data set [3].

The paper is organised as follows: in the next section we introduce some definitions and an outline of the Sugiyama method for drawing digraphs in 2D. Then in Section 3 we describe our 3D extension to the Sugiyama method. In Section 4 we show some drawings of digraphs computed with our method and we compare them to the corresponding 2D drawings. Finally, in Section 5 we draw some conclusions from this work.

2   Terminology

The Sugiyama method for layered digraph drawing consists of four steps. The first step is to remove all directed cycles from the graph by inverting the direction of some edges. In the second step the vertices of the digraph are partitioned into layers. Let $G = (V, E)$ be a digraph without directed cycles. We denote the set of all immediate predecessors of vertex v by $N^-(v) = \{u : (u, v) \in E\}$, and the set of all its immediate successors by $N^+(v) = \{u : (v, u) \in E\}$. A *layering* of G is defined as an ordered partition $L = \{L_1, L_2, ..., L_h\}$ of its vertex set into h subsets, called *layers*, such that $(u, v) \in E$ with $u \in L_i$ and $v \in L_j$ implies $j < i$. A digraph with a layering is a *layered digraph*. A layering is *proper* if all edges are between vertices in adjacent layers. If this is not the case then after the second step of the Sugiyama method dummy vertices which subdivide long edges, i.e. edges which connect vertices in non-adjacent layers, are introduced. Formally, for each edge $e = (u, v)$ with $u \in L_i$, $v \in L_j$, and $j < i - 1$, we introduce $i - j - 1$ dummy vertices $d^e_{j+1}, d^e_{j+2}, \ldots, d^e_{i-1}$ into layers $L_{j+1}, L_{j+2}, \ldots, L_{i-1}$, respectively. We also replace edge e by edges $(u, d^e_{i-1}), (d^e_{i-1}, d^e_{i-2}), \ldots, (d^e_{j+2}, d^e_{j+1}), (d^e_{j+1}, v)$.

In the third step a linear order is established for the vertices in each layer. And in the last fourth step x- and y-coordinates of all vertices are decided as well as the shape of the edges. Various algorithms, which emphasise different properties of the drawing, have been suggested for each step of the Sugiyama method [2].

3   3D Layered Drawing of Directed Graphs

In this section we propose a 3D extension to the Sugiyama method. In summary, we introduce a new step called *wall assignment* which further partitions the layer into a set of $k > 1$ subsets, called *walls*, after the layering step. Our method is outlined in Algorithm 1.

Algorithm 1. 3D Layered Digraph Drawing

Step 1 (Cycle Removal): Remove all directed cycles by inverting the direction of some edges.

Step 2 (Layer Assignment): Partition the vertex set into h layers, $L_1, L_2, \ldots, L_h$ with $h \geq 2$.

Step 3 (Wall Assignment): Partition the vertices in each layer L_i into k subsets, $L_i^1, L_i^2, \ldots, L_i^k$ with $k \geq 2$.

Step 4 (Vertex Ordering): Set a linear order of the vertices which belong to the same layer and wall.

Step 5 (Coordinate Assignment): Assign x-, y-, and z-coordinates to each vertex. Determine the shape of each edge.

Layers occupy parallel planes and each layer L_i is partitioned into k subsets, $L_i^1, L_i^2, \ldots, L_i^k$. The vertices placed in the jth group of each layer form a *wall*. That is, the set $W^j = \{L_1^j, L_2^j, \ldots, L_h^j\}$ is the jth *wall*. There are k walls in total, each wall occupies a plane, and the walls occupy k parallel planes which are perpendicular to the h planes of the layers. In addition, we require all dummy vertices along the same long edge to be in the same wall in order to avoid more than one bend outside the walls which contain the endpoints of the edge.

Since we perform the wall-assignment step after the introduction of dummy vertices we assume that $G = (V, E)$ is a proper layered digraph with a layering $L = \{L_1, L_2, \ldots, L_h\}$, i.e. each edge connects vertices in adjacent layers. By partitioning the vertices into $k \geq 2$ walls we partition the edge set of a digraph into two subsets: *intra wall edges* and *inter wall edges*. Intra wall edges are edges with both endpoints in the same wall, and inter wall edges are edges with endpoints in different walls. The *span* of an inter wall edge is the absolute value of the difference between the numbers of the two walls which contain the endpoints of that edge. Note that each inter wall edge has at least one endpoint which is not a dummy vertex because we require all dummy vertices along the same long edge to be in the same wall.

The partition of the original vertex set into k walls may originate from the digraph's application domain. They might be the clusters of a given clustered graph. If no such partition is given then the vertex set can be partitioned into k walls according to the following optimisation criteria:

- **C1.** Even distribution of vertices among walls, i.e. balanced partition of the vertex set into walls.
- **C2.** As few as possible inter wall edges for avoiding occlusion in the 3D space.
- **C3.** As few as possible crossings between inter wall edges in the projection of the drawing into a plane which is orthogonal to both the layer planes and the wall planes.
- **C4.** The sum of the spans of inter wall edges should be minimised.
- **C5.** As few as possible crossings between intra wall edges.

These criteria are designed to express the properties of layouts with low visual complexity. They give rise to some hard optimisation problems which require the development of efficient algorithms. In our previous work we have proposed an algorithm for partitioning the vertex set into two walls according to C1 and C2. In the remainder of this section we propose two new methods of partitioning the vertex set into two walls according to C3, and two versions of a method for partitioning the vertex set into k-walls according to C4 and C1.

3.1 Two-Wall Partitions

The following two algorithms for partitioning the vertex set into two walls are designed to have all inter wall edges arranged in a particular pattern such that C3 is satisfied. We call them zig-zag wall partition and dominating wall partition respectively. Both algorithms scan all layers one by one from bottom to top and partition each of them into two subsets. We start with a random balanced partition of the first layer that contains more than one vertex. Each next layer L_i is partitioned into L_i^1 and L_i^2 such that $L_i^1 \cup L_i^2 = L_i$ and $L_i^1 \cap L_i^2 = \phi$ based on the partition of layer L_{i-1}.

The zig-zag wall partition and the dominating wall partition differ in the way layer L_i is divided into two subsets based on the already given partition of layer L_{i-1} for $i \geq 2$. The zig-zag wall partition scans all the vertices in layer L_i and partitions it into L_i^1 and L_i^2 by applying different strategies for even and odd layers. If i is an even number then $u \in L_i$ will be always assigned to L_i^2 except in the case when u has no neighbours in L_{i-1}^2. In that case u will be assigned to L_i^1. That is, a vertex is assigned to L_i^1 if and only if it has all its immediate successors in L_{i-1}^1. Note that this is done after the insertion of dummy vertices and all the immediate successors of u are in layer L_{i-1}. If i is an odd number then $u \in L_i$ is assigned to L_i^2 if and only if all its immediate successors are assigned to L_{i-1}^2. As a result the inter wall edges form a zig-zag pattern between the two walls. The zig-zag wall partition is presented as Algorithm 2. An example layout is shown in Figure 1(b).

The idea of the dominating wall partition is all the inter wall edges have their origin in the same wall, which we call the *dominating wall*. We assume that wall W^1 is the dominating wall, i.e. the wall that consists of $L_1^1, L_2^1, \ldots, L_h^1$. Vertex $u \in L_i$ is assigned to L_i^2 if and only if it has no neighbours in L_{i-1}^1. The dominating wall partition of layer L_i is presented as Algorithm 3. An example layout is shown in Figure 1(c).

Algorithm 2. Zig-zag wall partition of layer L_i

if $i \bmod 2 = 0$ then
    $L_i^1 \leftarrow \{v \in L_i : N^+(v) \cap L_{i-1}^2 = \phi\}$
    $L_i^2 \leftarrow L_i \setminus L_i^1$
else
    $L_i^2 \leftarrow \{v \in L_i : N^+(v) \cap L_{i-1}^1 = \phi\}$
    $L_i^1 \leftarrow L_i \setminus L_i^2$
end if

Algorithm 3. Dominating wall partition of layer L_i

$L_i^2 \leftarrow \{v \in L_i : N^+(v) \cap L_{i-1}^1 = \phi\}$
$L_i^1 \leftarrow L_i \setminus L_i^2$

It is easy to see that both the zig-zag wall partition and the dominating wall partition place all dummy vertices along edge e into the same wall. Both algorithms take $O(|V| + |E|)$ time because each vertex is examined and assigned into a wall and this is done by examining the neighbours of the vertices, i.e. by examining all the edges of the digraph.

Lemma 1. *Both the zig-zag wall partition and the dominating wall partition algorithms assign all dummy vertices along an edge to the same wall and partition the vertex set of the graph into two subsets in linear time.*

3.2   k-Wall Partitions

We have designed a third algorithm for partitioning the vertex set into $k \geq 2$ walls according to C4. That is, the sum of spans of inter wall edges is kept small.

Similar to the two algorithms described above all the layers are scanned one by one from bottom to top. The first layer which contains more than one vertex is partitioned randomly and each following layer L_i is partitioned on the basis of the partition of layer L_{i-1}.

For partitioning layer L_i into k subsets we apply Algorithm 4. In summary, for each vertex $u \in L_i$ all its immediate successors are considered, and u is placed in the wall whose number is the closest integer to the average of the wall numbers of the immediate successors of u. In other words, the wall u is placed in the barycenter of the walls its immediate successors are placed in. An example layout is shown in Figure 2(a).

In order to achieve a more even distribution of vertices between the walls, i.e. to satisfy C1, we can keep track on the number of vertices currently assigned to a wall and when we compute b take this into account giving preference to the walls with fewer number of vertices. The implementation of such a procedure is presented in Algorithm 5 which is a generalised version of Algorithm 4. Now the wall for vertex u is computed as

$$b = \left\lfloor \frac{\sum_{j=1}^{k} j * max\{0, neighbours[j] - |L_i^j|\}}{\sum_{j=1}^{k} max\{0, neighbours[j] - |L_i^j|\}} + 0.5 \right\rfloor . \tag{1}$$

Algorithm 4. k-wall partition of layer L_i

---

for all $j = 1..k$ do
    $L_i^j \leftarrow \phi$
end for
for all $u \in L_i$ do
    if $|N^+(u)| = 0$ then
        Let b be a number such that $|L_i^b| = min\{|L_i^1|, |L_i^2|, \ldots, |L_i^k|\}$
    else
        for all $j = 1..k$ do
            $neighbours[j] \leftarrow |N^+(u) \cap L_{i-1}^j|$
        end for
        $b \leftarrow \left\lfloor \frac{\sum_{j=1}^k j * neighbours[j]}{\sum_{j=1}^k neighbours[j]} + 0.5 \right\rfloor$
    end if
    $L_i^b \leftarrow L_i^b \cup \{u\}$
end for

Algorithm 5. k-wall balanced partition of layer L_i

---

for all $j = 1..k$ do
    $L_i^j \leftarrow \phi$
end for
for all $u \in L_i$ do
    if u is a dummy vertex then
        Let b be the number such that the only immediate successor of u is assigned to L_{i-1}^b.
    else
        for all $j = 1..k$ do
            $neighbours[j] \leftarrow |N^+(v) \cap L_{i-1}^j|$
        end for
        if $\sum_{j=1}^k max\{0, neighbours[j] - |L_i^j|\} > 0$ then
            $b \leftarrow \left\lfloor \frac{\sum_{j=1}^k j * max\{0, neighbours[j] - |L_i^j|\}}{\sum_{j=1}^k max\{0, neighbours[j] - |L_i^j|\}} + 0.5 \right\rfloor$
        else
            Let b be a number such that $|L_i^b| = min\{|L_i^1|, |L_i^2|, \ldots, |L_i^k|\}$
        end if
    end if
    $L_i^b \leftarrow L_i^b \cup \{u\}$
end for

---

in the case $\sum_{j=1}^k max\{0, neighbours[j] - |L_i^j|\} > 0$. Otherwise u is placed in the wall with currently the fewest number of vertices in the current layer.

It is easy to see that Algorithm 4 guarantees that all dummy vertices along an edge belong to the same wall. However, this is no longer guaranteed with the proposed balancing technique. Thus, when applying the balancing technique we first need to check whether u is dummy and if it is then to place it into the wall of its immediate successor.

The time complexity of the proposed k-wall partitioning algorithm is also $O(|V|+|E|)$ because similar to the two-wall partitioning algorithms each vertex and each edge are scanned once.

Lemma 2. *Both versions of the k-wall partition algorithm assign all dummy vertices along an edge to the same wall and partition the vertex set of the graph into $k \geq 2$ subsets in linear time.*

4   Computational Results

We applied the presented layout technique for computing 3D hierarchical layouts of graphs taken from the Rome dataset [3]. We randomly chose ten graphs with 75 vertices. The number of edges of each of them is close to twice the number of vertices. In the remainder of this section we compare the different wall assignment techniques:

- (ZZ) Zig-zag wall partition.
- (DW) Dominating wall partition.
- (MC) Balanced min-cut wall partition introduced in [6].
- (KW) k-wall partition.
- (BW) Balanced k-wall partition.
- (SW) Single wall, i.e. a 2D hierarchical layout.

MC is a version of the balanced min-cut wall partition introduced in [6]. Note that we have slightly modified the original algorithm to make sure all dummy vertices along an edge go to the same wall which we did not consider in our previous work.

We assumed that the input graphs are undirected and we assigned direction to each edge from the vertex with the higher degree to the vertex with the lower degree. The reason for doing this is because we had to assign the direction of the edges by using the same method in all experimental graphs. It also allows us to test how our 3D hierarchical drawings can be used for emphasising centrality in large and complex networks, in particular the degree centrality, by having a loose connection between the layers and the centrality values. If both endpoints of an edge have the same degree then the direction is assigned randomly. We applied Algorithm 1 six times, once for each wall assignment technique. Each time we used the same algorithms for the other four steps: cycle removal, layer assignment, vertex ordering, and coordinate assignment.

We remove directed cycles by reversing the direction of the back edges in a DFS tree of the digraph. For layer assignment we used the longest-path algorithm followed by an improvement heuristic [7]. The vertex ordering step is performed with a layer-by-layer sweep and the barycenter heuristic taking into account the wall partition. Details about the vertex ordering with 2 walls can be found in [6]. We used a trivial extension of the same method for $k > 2$ walls. For the coordinate assignment step we applied the Brandes-Köpf algorithm for each wall independently [1]. The z-coordinates are given by the wall numbers.

Table 1. Vertex and edge distribution for two-wall partitions

graph	ZZ			DW			MC		
	vertices	edges	inter wall edges	vertices	edges	inter wall edges	vertices	edges	inter wall edges
1	77 30	75 21	24	42 65	44 48	28	55 52	51 48	21
2	55 73	47 75	32	44 84	43 70	41	66 62	63 64	27
3	63 46	57 36	46	41 68	41 47	51	54 55	53 53	33
4	65 56	61 51	33	41 80	40 67	38	62 59	62 57	26
5	79 58	73 54	31	49 88	51 77	30	70 67	73 61	24
6	80 43	83 37	34	43 80	47 65	42	65 58	65 56	33
7	107 53	119 46	30	67 93	77 80	38	82 78	85 79	31
8	73 51	67 48	30	60 64	63 48	34	62 62	59 66	20
9	97 43	104 38	26	65 75	69 55	44	72 68	70 69	29
10	71 77	65 78	43	51 97	57 81	48	75 73	79 77	30

Table 2. Vertex and edge distribution for k-wall partitions

graph	number of walls	KW		BW	
		vertices	edges	vertices	edges
1	3	28 39 40	22 35 30	42 34 31	26 22 18
2	3	46 42 40	42 35 39	46 46 36	31 35 25
3	2	48 61	35 67	58 51	52 46
4	3	38 46 37	29 39 27	43 45 33	30 33 20
5	4	34 39 34 30	26 33 28 22	41 29 33 34	28 18 27 23
6	4	33 51 20 19	26 56 13 12	42 28 31 22	30 16 24 10
7	4	34 53 43 30	22 52 36 24	46 42 45 27	30 30 32 18
8	3	40 44 40	27 38 31	39 40 45	23 26 36
9	3	36 58 46	26 59 35	52 42 44	40 33 29
10	3	46 62 40	37 67 35	54 53 41	37 44 30

First we compare the two-wall partitioning algorithms. Table 1 shows the vertex and edge distribution between the walls as well as the number of inter wall edges. It can be observed that MC does really find wall partitions which have more even distribution of vertices and edges between the two walls. The MC partitions typically have the minimum number of inter wall edges, which is what we expected. The advantage of ZZ and DW is the special arrangement of inter wall edges they guarantee. We can also see that the distribution of vertices and edges between the two walls in ZZ and DW is not necessarily unbalanced.

Table 2 compares the vertex and edge distribution for KW and BW. The number of walls k is a half of the number of layers h, i.e. $k = \lfloor \frac{h}{2} \rfloor$. Thus, different graphs have a different number of walls. We can observe that BW does find more balanced distribution of vertices between the walls only if the KW distribution is not balanced (e.g., graph 9). When the KW distribution is relatively well balanced then BW may perform worse than KW (e.g., graph 8).

Table 3. Total span of inter wall edges in k-wall partitions

graph	number of walls	KW		BW	
		number of inter wall edges	span of inter wall edges	number of inter wall edges	span of inter wall edges
1	3	33	33	54	68
2	3	47	47	63	78
3	2	37	37	41	41
4	3	50	50	62	77
5	4	49	57	62	94
6	4	47	57	74	105
7	4	61	69	85	118
8	3	49	49	60	73
9	3	48	48	66	80
10	3	47	47	75	93

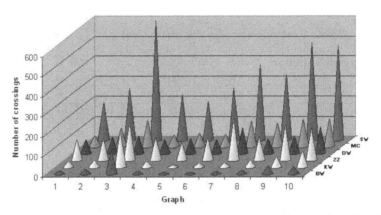

Fig. 3. Total number of edge crossings between intra wall edges (in all walls)

Table 3 compares KW and BW in terms of sum of the span of inter wall edges. KW performs better than BW. We observed that the KW wall partitions have very few inter wall edges between non-adjacent walls.

We have also computed the total number of crossings between intra wall edges. The results are presented in Figure 4. As expected, the bigger the number of walls the fewer crossings between intra wall edges. The best of the two-wall partitions is DW which typically results in a fewer number of intra wall edge crossings than ZZ and MC. The fewest number of intra wall edge crossings is reached in BW which is a result we did not expect. However, the BW partitions have a larger number of inter wall edges and a larger sum of spans of inter wall edges than KW (see Table 3).

5   Conclusions and Current Work

We introduced a 3D extension to the Sugiyama method. We propose to partition the vertex set into $k \geq 2$ walls; each wall contains a 2D drawing of a layered digraph. This is done by introducing an additional wall assignment step into the Sugiyama method after the layer assignment step. We propose three different wall assignment algorithms which meet different optimisation criteria.

The computational results suggest that the proposed 3D graph drawing convention results in reduced visual complexity. The choice of a particular wall assignment algorithm may highly depend on the interaction and navigation techniques. It is also possible to develop new vertex ordering heuristics specific for the 3D layered graph drawings with k walls. We are also looking at the possibility of defining new optimisation problems arising from 3D drawing aesthetic criteria.

Acknowledgement. The authors would like to thank Michael Forster for the valuable discussion and suggestions as well as for implementing the Brandes-Köpf algorithm for coordinate assignment which we used in the drawings included in the paper.

References

1. U. Brandes and B. Köpf. Fast and simple horizontal coordinate assignment. In P. Mutzel, M. Jünger, and S. Leipert, editors, *Graph Drawing: Proceedings of 9th International Symposium, GD 2001*, volume 2265 of *Lecture Notes in Computer Science*, pages 31–44. Springer-Verlag, 2002.
2. G. Di Battista, P. Eades, R. Tamassia, and I. G. Tollis. *Graph Drawing*. Prentice Hall, 1999.
3. G. Di Battista, A. Garg, G. Liotta, R. Tamassia, E. Tassinari, and F. Vargiu. An experimental comparison of four graph drawing algorithms. *Computational Geometry: Theory and Applications*, 7:303–316, 1997.
4. P. Eades and K. Sugiyama. How to draw a directed graph. *Journal of Information Processing*, 13(4):424–437, 1990.
5. A. Garg and R. Tamassia. GIOTTO: A system for visualizing hierarchical structures in 3D. In S. North, editor, *Graph Drawing: Symposium on Graph Drawing, GD '96*, volume 1190 of *Lecture Notes in Computer Science*, pages 193–200. Springer-Verlag, 1997.
6. S.-H. Hong and N. S. Nikolov. Layered drawings of directed graphs in three dimensions. In S.-H. Hong, editor, *Information Visualisation 2005: Asia-Pacific Symposium on Information Visualisation (APVIS2005)*, volume 45, pages 69–74. CRPIT, 2005.
7. N. S. Nikolov and A. Tarassov. Graph layering by promotion of nodes. *Special issue of Discrete Applied Mathematics associated with the IV ALIO/EURO Workshop on Applied Combinatorial Optimization*, to appear.
8. D. Ostry. Some three-dimensional graph drawing algorithms. Master's thesis, University of Newcastle, 1996.
9. K. Sugiyama, S. Tagawa, and M. Toda. Methods for visual understanding of hierarchical system structures. *IEEE Transaction on Systems, Man, and Cybernetics*, 11(2):109–125, February 1981.
10. C. Ware and G. Franck. Viewing a graph in a virtual reality display is three times as good as a 2D diagram. In *IEEE Conference on Visual Languages*, pages 182–183, 1994.

Layout Effects on Sociogram Perception

Weidong Huang[1,2], Seok-Hee Hong[1,2], and Peter Eades[1,2]

[1] IMAGEN Program, National ICT Australia Ltd.
[2] School of Information Technologies, University of Sydney, Australia
{weidong.huang, seokhee.hong, peter.eades}@nicta.com.au

Abstract. This paper describes a within-subjects experiment in which we compare the relative effectiveness of five sociogram drawing conventions in communicating underlying network substance, based on user task performance and usability preference, in order to examine effects of different spatial layout formats on human sociogram perception. We also explore the impact of edge crossings, a widely accepted readability aesthetic. Subjective data were gathered based on the methodology of Purchase et al. [14]. Objective data were collected through an online system.

We found that both edge crossings and conventions pose significant affects on user preference and task performance of finding groups, but either has little impact on the perception of actor status. On the other hand, the node positioning and angular resolution might be more important in perceiving actor status. In visualizing social networks, it is important to note that the techniques that are highly preferred by users do not necessarily lead to best task performance.

1 Introduction

Social networks can be modeled as graphs, and visualized as node-edge diagrams where nodes represent *actors*, and edges represent *relationships* between them. With advances in display media, the use of node-edge diagrams or *sociograms* (see Figure 1 for an example) has been increasingly important and popular in social network analysis. Sociograms serve as simple visual illustrations in helping people to explore and understand network structure, and to communicate specific information about network characteristics to others.

One of the major concerns in network visualization is *effectiveness*. There are two issues involved here: one is readability. Readability can be affected by not only intrinsic network characteristics [6], but also layout. In particular, *edge crossings* has long been widely accepted a major aesthetic [13,4]. Purchase [16], in her pioneering work of a user study which compared the relative effects of five aesthetic criteria (bends, crosses, angels, orthogonality and symmetry) on abstract graphs, also concluded that edge crossings has greatest impact on human graph understanding. Subsequently the aesthetic of edge crossings was validated on UML diagrams [14]. The remark of Purchase et al. [14] that there is no guarantee that results of domain-independent experiments could automatically apply to domain-specific diagrams motivates this paper. We are not aware of any previous user studies in examining edge crossings impact on the perception of social networks when domain-specific tasks are performed.

P. Healy and N.S. Nikolov (Eds.): GD 2005, LNCS 3843, pp. 262–273, 2005.

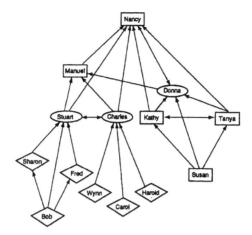

Fig. 1. Advice network formed by an auditing team. Courtesy of Krackhardt [9]. Ellipses represent managers; diamonds represent staff auditors and boxes represent secretaries. A line from Donna to Nancy indicates that Donna seeks advice from Nancy.

The other is communication. Good readability does not necessarily lead to effective underlying semantics communication. When a social network is mapped to a sociogram, what matters is relationship patterns, not the physical positioning of nodes [17]. However, previous studies [10, 11] revealed that the spatial layout of nodes in a sociogram does affect viewers in perceiving social network characteristics. Therefore there is practical need for investigating the actual communication effectiveness of a particular visualization method. Surprisingly, although considerable amount of fancy techniques have been proposed in literature, very little empirical evidence is available to support their effectiveness in communicating network structure to humans [12].

To address the above questions, we conducted a user study. In this study, we compared the communication effectiveness of five sociogram drawing conventions, and investigated the impact of edge crossings under each convention, based on user preference and task performance. Subjective data were gathered based on the methodology of Purchase et al. [14]. Objective data were collected through an online system.

1.1 Sociogram Constructing Conventions

Many visualization techniques are aimed to highlight one or two aspects of the network structure, and confirm to some aesthetics to improve the readability. Of our particular interest are the following five conventions.

1. Circular layout: all nodes are placed on a circle [17].
2. Hierarchical layout: nodes are arranged by mapping actors' status scores to the nodes' vertical coordinates [2].
3. Radial Layout: all nodes are laid on circumference of circles in a way that their distances from the center exactly reflect their centrality levels [3].

4. Group layout: nodes are separated into different groups with nodes in the same group close to one another. See [5] for a review.
5. Free layout: nodes are arranged without any particular purpose.

For more details about drawing conventions and background, see [7].

2   Experiment

2.1   Subjects

Twenty-three subjects were recruited from a student population in computer science on a completely voluntary basis. All the subjects were postgraduates and computer literate. All had node-edge diagram experience such as UML or ER, associated with their study units; six of them were graph drawing research students. All had neither academic nor working experience related to social networks. They were reimbursed $20 each for their time upon the completion of their tasks.

2.2   Design

Networks. For this experiment, two networks are used. One is Krackhardt's *advice network* [9]. The network is modeled as a directed graph with 14 nodes and 23 edges as shown in Figure 1. The network has three groups in a sense that we discuss later in this section.

The Katz status scores [8] of the actors are shown in Table 1. In this study, Katz status score was used as the index of importance. We expected subjects to perceive the importance in accordance with, and we measured their perceptions against the actors' Katz status scores.

The second one is a fictionalized network which was produced from the first one by eliminating all the directions. This gives an undirected graph with 14 nodes and 23 edges, which we call a *collaboration network*. A line between A and B means that A and B collaborate with each other.

Sociograms. We used a total of 12 drawings: 1) for each of the five conventions, two drawings of the advice network - one with minimum crossings and one with many crossings (see Figure 2); 2) for the collaboration network, a free convention drawing with minimum crossings and a free convention drawing with many crossings.

Table 1. Actors' Katz status scores

Nancy	1.00	Fred	0.02
Donna	0.66	Sharon	0.02
Manuel	0.57	Harold	0.00
Stuart	0.19	Wynn	0.00
Charles	0.17	Susan	0.00
Kathy	0.08	Bob	0.00
Tanya	0.08	Carol	0.00

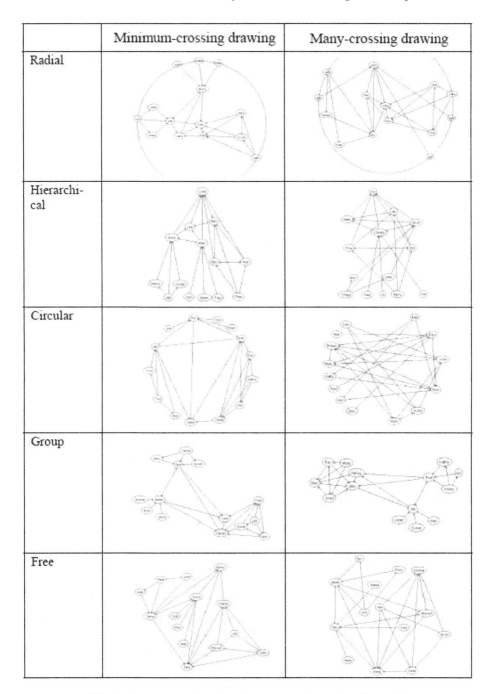

	Minimum-crossing drawing	Many-crossing drawing
Radial		
Hierarchi-cal		
Circular		
Group		
Free		

Fig. 2. Sociograms for the advice network used in the study

All nodes were labeled with different names; in every drawing, each node was mapped to a new name. By providing a context and background for each network, and names for actors, subjects were expected to perform tasks from the real world social network perspective [10]. However, subjects were not made aware that the drawings had the same graph structure.

Tasks. For this experiment, we considered two common social network measures which are frequently highlighted in sociograms: one is *importance* or status of actors; the other is the *presence of social groups*, in which the connections among actors are relatively dense. The whole session included 3 main tasks:

1. Online tasks
   (a) find 3 most important actors and rate them according to their importance levels; and
   (b) determine how many groups are in the network, and separate the 4 highlighted actors according to their group membership, given the condition that one actor should not belong to more than one group, and one group should not include only one actor. In formal tests, the same 4 nodes (actors) across all drawings were highlighted in red rectangles.
2. Subjective rating tasks
   (a) Usability acceptance rating: with one page showing all 6 many-crossing drawings, and the other page showing all 6 minimum-crossing drawings, subjects were required to rate their usability based on a scale from -3 (completely unacceptable) to +3 (completely acceptable) for importance tasks and group tasks, respectively.
   (b) Crossing preference rating: each many-crossing (A) and minimum-crossing (B) pair was shown one by one, 6 pairs in total. Subjects needed to indicate their preferences for importance tasks and group tasks, respectively, based on a scale from -2 (strongly A) to +2 (strongly B), where, for example, "Strongly A"means A is strongly preferred over B.
   (c) Overall usability ranking: with all 10 advice network drawings being shown in one page, subjects needed to choose 3 drawings that they least preferred and 3 drawings that they most preferred for their overall usability, using a scale from -3 to -1 and from 1 to 3, respectively.
3. Questionnaires: there were 2 questionnaires with each having a different focus, and to be presented to subjects before and after they were debriefed about edge crossings and drawing conventions, respectively. The first questionnaire asked subjects information about their study background, experience with node-edge diagrams and social networks, how they interpret sociograms, and any network structure and sociogram features that they think may influence their graph perceptions. The second questionnaire asked about their thoughts about conventions and edge crossings.

For the above rating tasks, subjects were also required to write down a short explanation for each answer.

Online System. Sociograms were displayed by a custom-built online experimental system. The system was designed so that:

- A question is shown first, a button on the screen is pressed, then the corresponding drawing is shown; after writing down the answer on the answer sheet provided, the button is pressed and the next question is shown, and so on.
- Each subject's response time for each drawing is logged. This starts once a drawing is completely displayed and ends once the button is pressed.

The study employed a within-subjects design. For online tasks, 10 of the 12 drawings were randomly chosen and shown to comply with the time schedule and to reduce fatigue. The order of group and importance questions for each drawing was also random. Subjects were told they could have breaks during the question viewing periods if they wished. There was no time limit on task completion, although they were recommended to answer each question in one minute. During the preparation time, subjects were instructed to answer each question in the context of the underlying network and as quickly as possible without compromising accuracy.

2.3 Procedure

A pilot study had been conducted with another four subjects who did not have any social network background to check our methodology. They showed that they quickly understood the questions and felt comfortable with the experiment. They related the visual network representations with their daily social experiences when performing tasks.

The formal tests took place in a computer laboratory, in which all PCs had the same specifications. Before starting the experiment, subjects were asked to read the information sheet, sign the consent form, read through and understand the tutorial material, ask questions and practice with the online system. The drawings used for practice were quite different from the ones used in formal tests, since the practice was only for familiarization with the procedure and system, not for them to get experienced with sociogram reading.

Once ready to start, subjects indicated to the experimenter, and started running the online system performing tasks formally. After the online reading tasks followed by a short break which was to refresh subjects' memory, they proceeded with the rating tasks, and the first questionnaire. Then, after being given a debriefing document explaining the nature of the study, edge crossings and drawing conventions, subjects were asked to do the rating tasks (a) and (b) again, and finally finished with the second questionnaire. Subjects were also encouraged to verbalize any thoughts and feelings about the experiment. The whole session took about 60 minutes.

3   Results

The data of three subjects were discarded due to the failure of following instructions. Since the collaboration network sociograms did not create distinct difference with their counterparts, these data have been omitted in our analysis.

For simplicity, we use C, R, G, H, and F to represent circular, radial, group, hierarchical and free drawing conventions, respectively, and use P for minimum-crossing drawing, and C for many-crossing drawing. Therefore CP denotes circular minimum-crossing drawing; CC denotes circular many-crossing drawing, and so on.

3.1   User Preference Data

Usability Acceptance. Subjects' usability rating scale data are illustrated in Figures 3-4 and analyzed using ANOVA (with Fisher's PLSD, for pairwise comparisons).

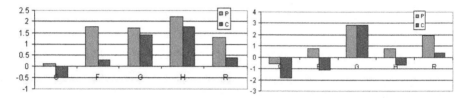

Fig. 3. Mean importance usability scores   **Fig. 4.** Mean group usability scores

Table 2. Pre- and post-debriefing scores (only data with significant changes are shown)

Importance	HP	HC	RP	RC	Group	RP	CC	FC	HC
Pre	2.21	1.75	1.29	0.42	Pre	1.96	-1.83	-1.13	-0.67
Post	2.75	2.29	1.87	1.63	Post	1.50	-1.43	-0.54	-0.08
P Value	0.04	0.05	0.00	0.00	P Value	0.02	0.03	0.02	0.02

For importance tasks, the minimum-crossing drawing was generally rated higher than the many-crossing one for each convention. There was a significant crossings effect between all minimum-crossing and all many-crossing drawings (p=0.00). Also there was a significant convention effect among all drawings (p=0.00). The pairwise comparisons revealed that the user acceptance for CP was significantly different for GP, HP and FP, respectively; the user acceptance for RP was significantly different for CP and HP, respectively. Furthermore, *paired t tests* showed that after debriefing, the drawings of both hierarchical and radial conventions were rated significantly higher (see Table 2). In particular, the mean scores of the hierarchical convention pair were higher than all others, showing that the positioning of nodes was perceived more important than edge crossings for importance tasks.

For group tasks, GC and GP were rated much higher than others; in fact the others were perceived as having little usefulness. Analysis found a significant crossings affect for each pair of drawings for all conventions (p=0.00) except group convention. Also, conventions produced a significant difference among all drawings (p=0.00). Pairwise comparisons showed that all pairs were significantly different except between HP and FP. In subjects' post-debriefing ratings, there was a significant change for RP, CC, FC and HC (see Table 2).

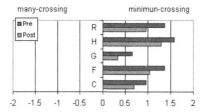
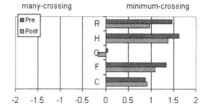

Fig. 5. Mean importance preference scores   **Fig. 6.** Mean group preference scores

Crossing Preference. As can be seen from Figures 5-6, generally, subjects preferred the minimum-crossing drawing more for each convention. The *1-sample t
tests* against the hypothesized mean (=0) revealed that for all conventions except
group convention, subjects' preference for the minimum-crossing drawings over
the corresponding many-crossing drawings was statistically significant (p<0.01).

The *paired t tests* revealed that there were no significant changes between the
pre- and post-debriefing ratings, although the post-debriefing preferences were
generally weaker than pre-debriefing preferences for both importance and group
tasks.

Overall Usability Ranking. Subjects' ranking values for each drawing were
summed as a weighted value, and weighted values for all drawings are shown
in Figure 7. It can be seen that generally, the many-crossing drawings were
less preferred except GC, which was ranked the highest for its overall usability,
followed by GP, then HP. Both CC and FC had the lowest weighted value,
indicating that they were considered having little overall practical utility.

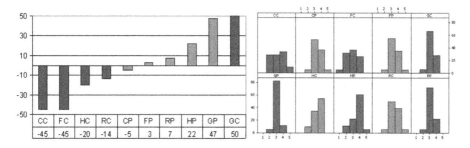

Fig. 7. Overall usability values

Fig. 8. Distributions of reported group
number

3.2 User Performance Data

Response Time. Subjects' response time data are illustrated in Figures 9-10
and analyzed using the non-parametric method of Kruskal-Wallis.

For importance tasks, subjects spent shorter time with the minimum-crossing
drawing than the many-crossing drawing for each convention in general. Among
all minimum-crossing drawings, the shortest time was spent with GP, followed

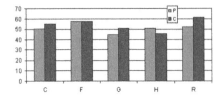

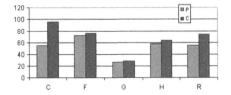

Fig. 9. Median time (sec.) for importance   **Fig. 10.** Median time (sec.) for group

by CP, HP, RP, and finally FP. However, statistical tests revealed that these differences of response time were not statistically significant in terms of the effect of either edge crossings or drawing conventions.

For group tasks, following a similar pattern, again shorter time was spent with the minimum-crossing drawing than the many-crossing one for each convention. Among all minimum-crossing drawings, the shortest time was spent with GP, followed by RP, CP, HP, and finally FP. Analysis showed that there was a significant crossings effect between the minimum-crossing and many-crossing pair of circular convention (p=0.012), and between all minimum-crossing and all many-crossing drawings (p=0.021), and a significant convention difference among all drawings (p=0.000). Pairwise comparisons showed that subjects spent significantly shorter time with GP than all others at the level of 0.01.

Reported Group Number and Member Group Assignment. Figure 8 illustrates the distribution of the reported group number for each drawing. As can be seen, GP had largest proportion of subjects (82.4%) responded "correctly" (3 as expected). An analysis of variance of the reported group number for all drawings showed there was a significant difference at the level of 0.066.

Also, at dyad level, the member group assignment task was to investigate edge crossings and convention impact on the perception of actors' co-memberships. As can be seen from Figure 11, a relatively larger proportion of subjects performed this task correctly on the minimum-crossing drawing than on the many-crossing drawing for each convention except free convention. Among all minimum-crossing drawings, GP yielded the highest correctness rate (76.5%).

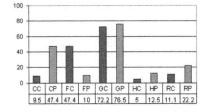

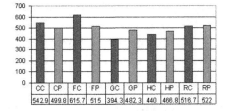

Fig. 11. Group assignment correctness(%)   **Fig. 12.** Weighted values for identifying important actors

Identifying Most Important Actors. Figure 12 shows the weighted values for all drawings. The weighted value is to measure a drawing's overall effectiveness of conveying information about importance, and calculated in the following way: First we gave an index of 5 to the most important actor, 2 to the second and 1 to the third; then the productions of indices and corresponding correctness percentages were summed as a weighted value for each drawing. It can be clearly seen that FC had the highest weighted value, followed by CC, RP, and then RC; GC had the lowest.

4   Qualitative Data and Discussions

The analysis of subjects' responses to questionnaires and interviews revealed that subjects had a strong preference of placing nodes on the top or in the center to highlight importance, and clustering nodes in the same group and separating groups to highlight groups. They had tendency to believe that nodes in the center or on the top are more important, and nodes in close proximity belong to the same group.

There was strong evidence that edge crossings contribute to the significant difference in user preference, usability acceptance, and group task performance. Edge crossings not only affect the ease of reading, but also affect the understanding of network structures.

With respect to drawing convention, for importance tasks, hierarchical convention was strongly preferred, while for group tasks, group convention was strongly preferred. Users achieved the highest response accuracy with group convention for group tasks. However, the highest response accuracy did not come with hierarchical convention for importance tasks. For overall usability, group convention was the one for which the usability was rated high and user performance was well as well.

Quite surprisingly, subjects were overwhelmingly in favor of hierarchical convention for importance tasks; they spent relatively short time with HP, but obtained the lowest correctness rate among all minimum-crossing drawings. On the other hand, FC obtained the highest correctness rate, but relatively long time was spent with it. We realized that some subjects had complained in questionnaires and verbally that in some drawings, edges were incident to nodes too closely to clearly identify arrow directions. Visual inspection revealed that indeed, free convention drawings had very good angular resolution, while hierarchical convention made angular resolution relatively low, where edges had to be crowded in one side of nodes. In addition, subjects spent longer time with FC, which might actually allow them to have better chance to understand network structure better.

In summary, no obvious evidence was found that either edge crossings or conventions pose significant impact on user importance task performance. Users generally performed better when they took longer time. We conjecture that only those tasks which are closely related to edges and involve edge tracing can be significantly affected, such as finding groupings. On the other hand, for communicating information about actor status, the angular resolution and node positioning in a sociogram might be more important, compared to drawing conventions and reducing the number of edge crossings.

For detailed discussion of user responses, the recommendations for sociogram design, and some further hypotheses derived about human graph perception, see [7].

5   Conclusions

This study, together with previous research [10, 11], has demonstrated that how sensitive the human sociogram perception is to spatial layout, and how important it is to have visualization techniques evaluated for their actual effectiveness in communication from human understanding point of view. It should be noted that visualization techniques, which are highly preferred by users, do not necessarily always produce best task performance, as demonstrated in this study.

The findings from this study should be interpreted within the limitations of the given experimental settings. In this study we had only investigated the relative effectiveness of five "explanatory visualization" [2] conventions and edge crossings impact under each convention, in communicating actor status and subgroup information to novice audience. Their usability in assisting professionals to explore and understand social networks remains untouched and is beyond the scope of this study. For a comprehensive overview in this field, see [1].

Additional studies are needed to empirically identify and investigate the impact of other possible variables of human sociogram perception in a more controllable manner.

Acknowledgements. The authors are grateful to Mr Kelvin Cheng, and Mr Le Song for their helpful comments on questionnaire design, and to students who willingly took part in the experiment. Ethical clearance for this study was granted by the University of Sydney, December 2004.

References

1. Card, S., Mackinlay, J., and Shneiderman, B. (1999): Readings in information visualization: using vision to think. Morgan Kaufmann Publishers, San Francisco, CA.
2. Brandes, U., Raab, J. and Wagner. D. (2001): Exploratory Network Visualization: Simultaneous Display of Actor Status and Connections. Journal of Social Structure 2(4).
3. Brandes, U., Kenis, P. and Wagner, D. (2003): Communicating Centrality in Policy Network Drawings. IEEE Trans. on Visualization and Computer Graphics 9(2): 241-253.
4. Di Battista, G., Eades, P., Tamassia, R. and Tollis, I. (1998): Graph drawing: algorithms for the visualisation of graphs, Prentice Hall.
5. Freeman, L. (1999): Visualizing Social Groups. American Statistical Association, Proceedings of the Section on Statistical Graphics, 47-54.

6. Ghoniem, M., Fekete, J. and Castagliola, P. (2004): A Comparison of the Readability of Graphs Using Node-Link and Matrix-Based Representations. Proceedings of the 10th IEEE InfoVis'04, Austin, TX. IEEE Press. pp. 17-24.

7. Huang, W., Hong, S., and Eades, P. (2005): Layout Effects: Comparison of Sociogram Drawing Conventions. http://www.it.usyd.edu.au/~whua5569/ex/.

8. Katz, L. (1953): A new status index derived from sociometric analysis. Psychometrika, 18: 39-43.

9. Krackhardt, D. (1996): Social Networks and Liability of Newness for Managers. In C. L. Cooper and D. M. Rousseau (eds.), John Wiley Sons, Ltd. New York, NY. Trends in Organizational Behavior, Volume 3, pp. 159-173.

10. McGrath, C., Blythe, J. and Krackhardt, D. (1997): The effect of spatial arrangement on judgments and errors in interpreting graphs. Social Networks, 19(3):223-242.

11. McGrath, C., Blythe, J. and Krackhardt, D. (1996): Seeing Groups in Graph Layout. Connections, 19(2): 22-29.

12. McGrath, C., Krackhardt, D. and Blythe, J. (2003): Visualizing Complexity in Networks: Seeing Both the Forest and the Trees. Connections, 25(1): 37-47.

13. Moreno, J. L. (1953): Who shall survive: Foundations of Sociometry, Group Psychother-apy, and Sociodrama. Beacon House Inc.

14. Purchase, H., Allder, J. Carrington, D. (2002): Graph Layout Aesthetics in UML Diagrams: User Preferences. J. Graph Algorithms Appl. 6(3): 255-279.

15. Purchase, H., Cohen R., James M. (1995): Validating graph drawing aesthetics. Proceedings of the Graph Drawing Symposium, Passau, Germany (GD '95), 435-446, Springer-Verlag.

16. Purchase, H. (1997): Which aesthetic has the greatest effect on human understanding? Proceedings of the 5th International Symposium on Graph Drawing (GD '97), 248-261. Springer-Verlag.

17. Scott, J. (2000): Social Network Analysis: A Handbook. Sage Publications, 2nd edition.

On Edges Crossing Few Other Edges in Simple Topological Complete Graphs

Jan Kynčl and Pavel Valtr

Department of Applied Mathematics and Institute for Theoretical Computer Science (ITI), Charles University, Malostranské nám. 25, 118 00 Praha 1, Czech Republic

Abstract. We study the existence of edges having few crossings with the other edges in drawings of the complete graph (more precisely, in simple topological complete graphs). A *topological graph* $T = (V, E)$ is a graph drawn in the plane with vertices represented by distinct points and edges represented by Jordan curves connecting the corresponding pairs of points (vertices), passing through no other vertices, and having the property that any intersection point of two edges is either a common end-point or a point where the two edges properly cross. A topological graph is *simple*, if any two edges meet in at most one common point.

Let $h = h(n)$ be the smallest integer such that every simple topological complete graph on n vertices contains an edge crossing at most h other edges. We show that $\Omega(n^{3/2}) \leq h(n) \leq O(n^2/\log^{1/4} n)$. We also show that the analogous function on other surfaces (torus, Klein bottle) grows as cn^2.

1 Introduction

A *topological graph* $T = (V, E)$ is a graph drawn in the plane with vertices represented by distinct points and edges represented by Jordan curves connecting the corresponding pairs of points (vertices), passing through no other vertices, and having the property that any intersection point of two edges is either a common end-point or a point where the two edges properly cross. A topological graph is *simple*, if any two edges meet in at most one common point.

One of the traditional themes in the area of graph drawings is to realize a given abstract graph as a topological graph so that the number of edge crossings is minimized. Here we consider a variant of a "dual" problem. We study realizations of the complete graph where each edge crosses "many" other edges.

Consider a network model drawn as a topological graph where the edge crossings are used for the exchange of some commodities (or information) between the two crossing edges. In any such model, edges with few crossings can exchange only small amounts of the commodities with the other edges within a time unit. This leads to the question about the existence of drawings in which each edge crosses "many" other edges.

If we can choose the underlying abstract graph on n vertices, then we can realize it with each edge crossing $\Omega(n^2)$ other edges. E.g., take the vertices of a regular n-gon and connect each vertex by straight-line segments with the $\approx n/3$

P. Healy and N.S. Nikolov (Eds.): GD 2005, LNCS 3843, pp. 274–284, 2005.

opposite vertices. Each edge in the obtained topological graph crosses at least $\approx n^2/9$ other edges. Moreover, each edge is realized by a straight-line segment, thus it is a so-called *rectilinear drawing* (sometimes also called a *geometric graph*). If the underlying graph is fixed then the situation is much more complicated. In this paper we restrict our attention to topological complete graphs, i.e., to realizations of (abstract) complete graphs. We are not aware of any result for other classes of graphs.

If any two edges are allowed to cross each other at most twice, then there are various realizations of the complete graph with each edge crossing $\Omega(n^2)$ other edges. E.g., take n points (vertices) on a short horizontal segment s and for any two vertices a, b, connect a and b by an arc constructed as follows. Let $U(a, b)$ be the unit circle going through a and b and having the center above s. Then the edge ab is drawn as the arc obtained from $U(a, b)$ by removing the part below the segment ab. Then any two edges with no common vertex cross once or twice. A different example of such a drawing is described in [14]. In this paper we show that the situation is different for simple topological (complete) graphs.

According to the so-called crossing lemma [1,9], if T is a topological graph with n vertices and $e \geq (3 + \varepsilon)n$ edges then its crossing number is at least $\Omega(e^3/n^2)$, (i.e., it contains at least $\Omega(e^3/n^2)$ crossing pairs of edges). It follows that if T is a topological complete graph then its crossing number is $\Omega(n^4)$ (this has also a quite easy direct proof). If T is simple then there are at most $\binom{n-2}{2} = O(n^2)$ crossings on each edge. It follows that a simple topological complete graph on n vertices contains $\Omega(n^2)$ edges each of which crosses $\Omega(n^2)$ other edges.

We study the existence of edges with (much) fewer than cn^2 crossings. Let us remark that in any rectilinear drawing of K_n the edges on the boundary of the convex hull do not cross any other edge. On the other hand, Harborth and Thürmann [8] found a simple topological complete graph in which each edge crosses some other edges.

Let $h = h(n)$ be the smallest integer such that every simple topological complete graph on n vertices contains an edge crossing at most h other edges. Harborth and Thürmann [8] proved $h(n) > (\frac{3}{4}+o(1))n$. Other related questions were studied e.g. in [5,6,7,16,17]. It has been asked in the preliminary version of the book [2] whether $h(n) = O(n)$, and the final version of [2] contains a conjecture that $h(n) = o(n^2)$. In this paper we show that $h(n)$ grows much faster and we also give the first subquadratic upper bound on $h(n)$:

Theorem 1.
$$\Omega(n^{3/2}) \leq h(n) \leq O(n^2/\log^{1/4} n).$$

We describe two essentially different constructions giving the lower bound. We present both of them, since they may help in closing the gap between the bounds given in Theorem 1. We conjecture that the lower bound is closer to the asymptotic behavior of $h(n)$ than the upper bound, and maybe even $h(n) = \Theta(n^{3/2})$. We remark that our proof gives a reasonable constant involved in the Ω−notation in the lower bound in Theorem 1. For simplicity of presentation, we do not compute the constants.

It is interesting that for other surfaces (torus, Klein bottle, real projective plane) it is possible to find simple topological complete graphs with each edge crossing $\Omega(n^2)$ other edges. This is discussed in the last section of the paper.

Brass, Moser, and Pach [2] describe a connection between the function $h(n)$ and the maximum number of disjoint edges in a topological graph. They have suggested the following greedy procedure: Select an edge intersecting the smallest number of other edges, delete these edges, and repeat the procedure. The lower bound in Theorem 1 indicates limits of this procedure in some cases. We remark that finding many disjoint edges and various similar questions on topological and geometric graphs have recently received a lot of attention, e.g. see [3, 4, 10, 11, 12, 13, 14, 15].

2   The Lower Bound

2.1   First Construction

Let $\mathcal{S}$ be the unit sphere in $\mathbf{R}^3$. Our topological complete graph giving the lower bound in Theorem 1 will be drawn on $\mathcal{S}$ by choosing an appropriate set P_n of n points on $\mathcal{S}$ and then connecting each pair of points of P_n by the shortest arc contained in $\mathcal{S}$. The points of P_n will be "well distributed" on $\mathcal{S}$ and *in general position*, meaning that no two points of P_n are antipodal and no three points of P_n lie on a common great circle of $\mathcal{S}$.

The crucial requirement on P_n is the following condition:

(C) If $d = d(P_n)$ denotes the minimum (Euclidean) distance of a pair of points of P_n then for any point $q \in \mathcal{S}$, the $1.1d$-neighborhood of q contains a point of P_n.

The set P_n is constructed as follows. First, we inductively construct n auxiliary points $a_1, \ldots, a_n$. Choose a point $a_1 \in \mathcal{S}$ arbitrarily. Now, let $i \in \{1, \ldots, n-1\}$ and suppose that $a_1, \ldots, a_i$ have already been selected. Then we choose a_{i+1} as a point on $\mathcal{S}$ maximizing the quantity $\min\{\|a_1 - a_{i+1}\|, \|a_2 - a_{i+1}\|, \ldots, \|a_i - a_{i+1}\|\}$. Clearly, we can slightly perturb the constructed set $\{a_1, \ldots, a_n\}$ so that the perturbed set, P_n, is in general position and satisfies condition (C).

Observe that $(d =) d(P_n) = \Theta(1/\sqrt{n})$ follows from the following three facts by a simple counting argument: (i) the area of $\mathcal{S}$ is $\Theta(1)$, (ii) the $1.1d$-neighborhoods of the points of P_n cover $\mathcal{S}$, and (iii) the $0.49d$-neighborhoods of the points of P_n are pairwise disjoint.

Let $T = T_n$ be the simple topological complete graph on $\mathcal{S}$ such that $V(T) = P_n$ and that $E(T)$ consists of the shortest curves on $\mathcal{S}$ connecting the pairs of vertices. We have to show that every edge in T crosses $\Omega(n^{3/2})$ other edges.

We use the notions *equator, northern/southern hemisphere* of $\mathcal{S}$ in the obvious way. Clearly, for any two vertices a, b, the edge ab is a portion of the great circle containing a, b. Thus, it suffices to show that if a portion I of a great circle of $\mathcal{S}$ has length $|I| = d$ then it is intersected by at least $\Omega(n^{3/2})$ edges of T. We may suppose that I is a portion of the equator. We denote the end-points of I by s

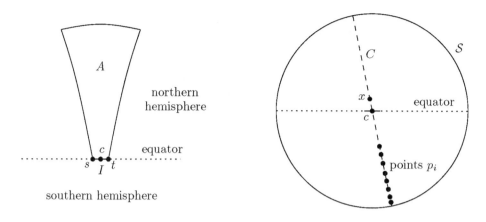

Fig. 1. The region A (left) and the points p_i (right)

and t. For a point $x \in S$ not lying on the equator, *the spherical triangle stx* is the region on S bounded by I and by the two shortest arcs contained in S and joining x with the points s and t, respectively.

Let c be the mid-point of the arc I. We consider the region A on S of the points x on the northern hemisphere such that $||x - c|| < \frac{1}{100}$ and that the spherical triangle stx has the inner angles at s and t each at most 0.6π (see Figure 1). The region A is bounded by I and by three arcs of length $\Theta(1)$. Clearly, its area is $\Theta(1)$ and it contains $\Theta(n)$ points of P_n. It suffices to show that any point of $A \cap P_n$ is an end-point of $\Omega(\sqrt{n})$ edges intersecting I.

Let $x \in A \cap P_n$. Consider the great circle C going through the points x and c (see Figure 1). Since $d = \Theta(1/\sqrt{n})$, it is possible to select $\Theta(\sqrt{n})$ points $p_1, p_2, \ldots, p_t$ in the intersection of C with the southern hemisphere such that $\frac{1}{10} < ||c - p_i|| < \sqrt{2}$ (for each i) and $||p_i - p_j|| > 2.2d$ (for any $i \neq j$). In general, the points p_i do not lie in P_n. However, the $1.1d$-neighborhood of each p_i contains a point $p_i' \in P_n$. By the choice of the points p_i, the points p_i' are pairwise distinct. It is not difficult to verify that each of the $\Theta(\sqrt{n})$ edges xp_i' intersects the arc I. This completes the proof that any edge in $T = T_n$ crosses $\Omega(n^{3/2})$ other edges.

2.2   Second Construction

Our second construction giving the lower bound in Theorem 1 is only briefly outlined in this extended abstract. We start with any fixed simple topological complete graph T in which each edge has at least one crossing, e.g with the drawing on Fig. 2. Let $V(T) = \{v_1, v_2, \ldots, v_t\}$. Let $n \geq t$ and suppose for simplicity that $\sqrt{n/t}$ is an integer. We replace each vertex v_i by a set V_i of n/t vertices placed in a square lattice $\sqrt{n/t} \times \sqrt{n/t}$ of a very small diameter. Any two vertices in distinct sets $V_i, V_j, i \neq j$, will be connected by an edge contained in a small neighborhood of the edge $v_i v_j$ of T. Let $i \in \{1, \ldots, t\}$ and suppose that the edges in T incident to v_i leave the vertex v_i in a counterclockwise order

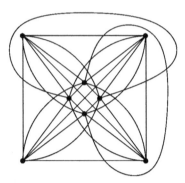

Fig. 2. A simple topological complete graph on 8 vertices in which each edge crosses another edge

$v_i v_{j_1}, v_i v_{j_2}, \ldots, v_i v_{j_{t-1}}$. In a small neighborhood of the convex hull of V_i we draw the edges leaving from the vertices of V_i so that any edge connecting a vertex of V_i with a vertex of V_{j_1} leaves the vertex of V_i along a vector parallel to some vector $(1, \varepsilon)$, where $\varepsilon > 0$ is very small (and different for different edges), and similarly any edge connecting any vertex of V_i with a vertex of V_{j_2}, V_{j_3}, or $V_{j_4} \cup \ldots \cup V_{j_{t-1}}$ (respectively) leaves the vertex of V_i along a vector parallel to some vector $(\varepsilon, 1)$, $(-1, \varepsilon)$, or $(\varepsilon, -1)$ (respectively). This ensures that after a very tiny perturbation of V_i and after connecting any two vertices of V_i by a straight-line segment, each such segment (edge) will be intersected by at least $\left(\sqrt{n/t} - 1 \right) n/t = \Theta(n^{3/2})$ edges connecting vertices of V_i with the vertices of $V_{j_1} \cup V_{j_2} \cup V_{j_3} \cup V_{j_4}$. It is not too difficult to check that the whole construction can be done so that the resulting drawing is a simple topological (complete) graph. Moreover, any edge connecting vertices from distinct sets V_i, V_j has $(n/t)^2 = \Theta(n^2)$ crossings in a small neighborhood of the point where the edge $v_i v_j$ crosses another edge of the graph T. Thus the obtained topological graph gives the lower bound in Theorem 1.

3   The Upper Bound

Topological graphs G, H are said to be *weakly isomorphic*, if there exists an incidence preserving one-to-one correspondence between $(V(G), E(G))$ and $(V(H), E(H))$ such that two edges of G intersect if and only if the corresponding two edges of H do. Let C_m denote a complete convex geometric graph with m vertices (note that all such graphs are weakly isomorphic to each other). A simple topological complete graph with m vertices is called *twisted* and denoted by T_m, if there exists a *canonical* ordering of its vertices $v_1, v_2, \ldots, v_m$ such that for every $i < j$ and $k < l$ two edges $v_i v_j, v_k v_l$ cross if and only if $i < k < l < j$ or $k < i < j < l$ (see Figure 3). Figure 4 shows an equivalent drawing of T_m on the cylindric surface. If G, H are topological graphs, we say that G *contains* H, if G has a topological subgraph weakly isomorphic to H.

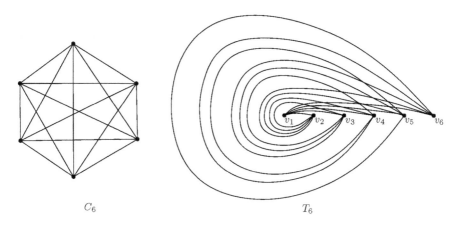

Fig. 3. The convex geometric graph C_6 and the twisted graph T_6

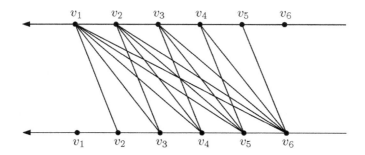

Fig. 4. Drawing of the twisted graph T_6 on the cylindric surface

In the proof of the upper bound, we will use the following asymmetric form of the result of Pach, Solymosi and Tóth [13]:

Theorem 2. [13] *There exists a $c > 0$ such that for all positive integers n, m_1, m_2 satisfying $m_1 m_2 \leq c \log^{1/4} n$ every simple topological complete graph with n vertices contains C_{m_1} or T_{m_2}.*

We will use this theorem for $m_1 = c' \log^{1/4} n$ and m_2 constant.

Now we prove two lemmas, the first one related to the complete geometric graph C_m, the second one related to the twisted drawing T_5.

Lemma 1. *Let G be a simple topological complete graph with n vertices. If G contains C_m, then there exists an edge in G which crosses at most $2n^2/m$ other edges.*

Proof. Let H be a topological complete subgraph of G with m vertices weakly isomorphic to C_m. H has a face F that is bounded by a non-crossing Hamiltonian cycle C consisting of m edges. Without loss of generality, suppose that F is the outer face of H. Then all edges of H lie inside the region bounded by the cycle C. We denote this region by R.

Claim. Let c be a simple continuous curve which starts and ends inside F, does not go through any vertex of H and crosses each edge of H at most once. Then c crosses at most two edges of the cycle C.

Proof. For contradiction, suppose that c crosses more than two edges of C. Then the intersection of c with R consists of $k \geq 2$ disjoint arcs $c_1, c_2, \ldots, c_k$ (see Figure 5). In the region R, the arcs c_1, c_2 separate two portions of C, denoted by α, β, from each other (see Figure 5). Since $c \supseteq c_1 \cup c_2$ intersects each edge of C at most once, each of the arcs α, β contains a vertex of G. However, any edge e connecting a vertex on α with a vertex on β intersects both c_1 and c_2. Thus, it intersects c more than once — a contradiction.

Let c be an arbitrary edge of G and let k be the number of edges of C that are crossed by c. First, we delete from c a small neighborhood of its end-points, receiving a curve c' that is disjoint with all vertices of H and crosses the same edges as c does. If some of the end-points of c' lies inside the region R, we delete from c' the initial part between the end-point and the first point a, at which c' crosses C, including a small neighborhood of a. We receive a curve c'' that has both its end-points inside F and crosses at least $k - 2$ edges from C. By the previous claim, c'' crosses at most 2 edges from C, thus $k \leq 4$.

G has less than $\frac{n^2}{2}$ edges, thus there are at most $2n^2$ crossings between the edges of G and the edges of C. By the pigeon-hole principle, among the m edges of C there is an edge, which crosses at most $2n^2/m$ edges of G.

Consider a simple topological complete graph H weakly isomorphic to the twisted graph T_m with the canonical ordering $v_1, v_2, \ldots, v_m$ of its vertices. The face incident with the vertices v_{m-1} and v_m only is called an *outer* face of H (it coincides with the outer face of the drawing of T_m at Figure 3), similarly the face incident with the vertices v_1 and v_2 only is called an *inner* face of H.

Lemma 2. *Let H be a simple topological complete graph weakly isomorphic to T_5. There does not exist a simple continuous curve c, which crosses each edge of H at most once, does not go through any vertex of H, begins and ends inside the outer face of H and intersects the inner face of H.*

Proof. Let $v_1, v_2, \ldots, v_5$ be the canonical ordering of the vertices of H. Consider a Hamiltonian cycle H_5, which is a subgraph of H with the edge set $E(H_5) =$

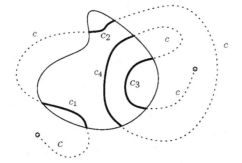
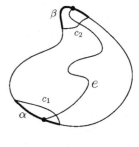

Fig. 5. The arcs c_i (left) and the arcs α, β (right)

$\{v_1v_2, v_2v_3, v_3v_4, v_4v_5, v_5v_1\}$. Let F_1, F_2, F_3, F_4 be the four faces of H_5 such that F_1 is incident with the vertices v_1 and v_2 only and F_i borders with F_{i+1}, $i = 1, 2, 3$ (see Figure 6). Note that F_1 (F_4) is the inner (outer) face of H and that F_i does not border with F_j if $|i - j| \geq 2$.

For contradiction, suppose that there exists a simple continuous curve c starting and ending inside F_4 and passing through F_1, avoiding all vertices of H and crossing each edge of H at most once. Choose a point $p \in c \cap F_1$. By the previous observation, between the starting point and p, c has to pass through the faces F_2 and F_3, so it must cross at least three edges of H_5. Similarly, c crosses at least three edges of H_5 between the point p and its end-point. But H_5 has only five edges, thus at least one of them is crossed by c more than once, a contradiction.

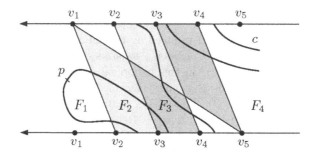

Fig. 6. The graph H_5 and a curve c with six crossings

The following theorem gives the upper bound in Theorem 1.

Theorem 3. *There exists a $c > 0$ such that in every simple topological complete graph with n vertices there exists an edge that crosses at most $cn^2/\log^{1/4} n$ other edges.*

Proof. Let G be a simple topological complete graph with n vertices. By Theorem 2, every induced subgraph of G with at least $n^{1/8}$ vertices contains T_{20} or $C_{\frac{c'}{2}\log^{1/4} n}$. If G contains $C_{\frac{c'}{2}\log^{1/4} n}$ then, by Lemma 1, G has an edge which crosses at most $\frac{4}{c'}n^2/\log^{1/4} n$ other edges. For the rest of the proof, suppose that G does not contain $C_{\frac{c'}{2}\log^{1/4} n}$, thus every induced subgraph of G with at least $n^{1/8}$ vertices contains T_{20}.

Let T_{20}^1 be a complete subgraph of G with 20 vertices weakly isomorphic to T_{20} and let $v_1^1, v_2^1, \ldots, v_{20}^1$ be a canonical ordering of its vertices. Consider a graph H^1 with the vertex set $V(H^1) = V(T_{20}^1)$ and the edge set $E(H^1) = \{v_1^1v_2^1, v_2^1v_3^1, \ldots, v_{19}^1v_{20}^1, v_1^1v_5^1, v_6^1v_{10}^1\}$ (see Figure 7). Denote the faces of H^1 as $F_1^1, F_2^1, \ldots, F_7^1$ such that F_1^1 is the inner face of T_{20}^1, F_7^1 contains the outer face of T_{20}^1 and F_i^1 borders with F_{i+1}^1, $i = 1, 2, \ldots, 6$ (as on the Figure 7).

Applying Lemma 2 on the twisted induced subgraph of T_{20}^1 with the vertices $v_1^1, v_2^1, \ldots, v_5^1$ we get that every edge of G, which crosses $v_1^1v_2^1$, has at least one

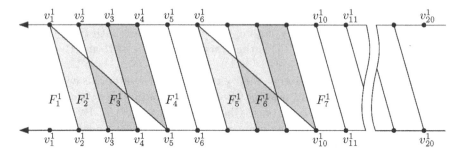

Fig. 7. The graph H^1 and its seven faces

end-point in the set $A^1 = F_1^1 \cup F_2^1 \cup F_3^1 \cup \{v_3^1, v_4^1, v_5^1\}$. Denote $a^1 = |A^1 \cap V(G)|$. If $a^1 < n^{1/8}$, then there are at most $n \cdot n^{1/8} = n^{9/8}$ edges with one end-point in A^1, thus at most $n^{9/8}$ edges cross the edge $v_1^1 v_2^1$. In the other case, the complete subgraph of G induced by the set $A^1 \cap V(G)$ has a subgraph T_{20}^2 weakly isomorphic to T_{20}. Consider a twisted subgraph H of T_{20}^1 induced by the vertices $v_{10}^1, v_9^1, v_8^1, v_7^1, v_6^1$ (in this canonical ordering). Every edge of T_{20}^2 has both its end-points inside the outer face of H, so it cannot intersect the inner face of H (by Lemma 2). This yields that all edges of T_{20}^2 lie in the set $B^1 = F_1^1 \cup F_2^1 \cup \ldots \cup F_6^1 \cup \{v_3^1, v_4^1, \ldots, v_{10}^1\}$. Denote $b_1 = |B^1 \cap V(G)|$. Note that $A^1 \subseteq B^1$, thus $a_1 \leq b_1$. It follows that at most one face of the graph T_{20}^2 does not lie in B^1. So we can choose a canonical ordering $v_1^2, v_2^2, \ldots, v_{20}^2$ of the vertices of T_{20}^2 such that the faces $F_1^2, F_2^2, \ldots, F_6^2$ of the graph H^2 (defined analogically as H^1 and its faces F_i^1) lie in B^1. We define sets A^2, B^2 and numbers a^2, b^2 analogically as A^1, B^1, a^1, b^1. B^2 is a proper subset of B^1, since all vertices of T_{20}^2 and faces $F_1^2, F_2^2, \ldots, F_6^2$ are contained in B^1, but, for example, vertex v_{11}^2 does not lie in B^2. It yields that $b_2 < b_1$. If $a_2 \geq n^{1/8}$, then there exists a twisted complete subgraph T_{20}^3 of G induced by some 20 vertices of the set $A^2 \cap V(G)$. Further we proceed by induction, similarly as above. In the i-th step, assuming that $a_{i-1} \geq n^{1/8}$, we find a twisted complete subgraph T_{20}^i of G with 20 vertices and define two integers a_i, b_i satisfying $0 \leq a_i \leq b_i < b_{i-i}$. After finitely many steps, we get a number a_i, which is less than $n^{1/8}$. It means that the edge $v_1^i v_2^i$ in the graph T_{20}^i is crossed by less than $n^{9/8} < n^2/\log^{1/4} n$ other edges of G.

4   Other Surfaces

Here we show that an analogue of the function $h(n)$ is quadratic for the torus and for the Klein bottle[1]:

---

[1] The same result for the projective plane has been recently found by Attila Pór (personal communication). We describe Pór's construction at the end of this section. Since any drawing of a finite graph on the projective plane can be easily transformed to a drawing on the Klein bottle, Pór's construction can be used to obtain an alternative proof of Proposition 1 for the Klein bottle.

Proposition 1. *On the torus and on the Klein bottle, there exists a simple topological complete graph with each edge having at least cn^2 crossings.*

Proof. Consider a rectangle from which, after gluing its opposite sides, we get a torus. Place the vertices $v_1, v_2, \ldots v_n$ along its upper and lower side in this order. We draw the edges the following way: if $j - i$ mod $n \leq \lfloor \frac{n-1}{2} \rfloor$, or if $j - i$ mod $n = \frac{n}{2}$ and $i \leq \frac{n}{2}$, we represent the edge $v_i v_j$ as a segment starting at the upper vertex v_i, directing down and to the right, possibly leaving the rectangle on the right-hand side and entering on the left-hand side and ending at the lower vertex v_j. At Figure 8, you can see the representation of the edges incident to one vertex. It is clear that in this drawing each two edges intersect at most once and that every edge crosses at least cn^2 other edges.

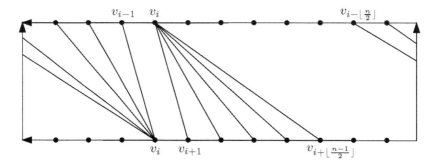

Fig. 8. Edges incident to the vertex v_i in the drawing of K_n on the torus

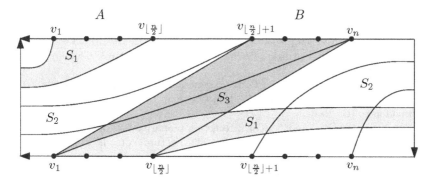

Fig. 9. Drawing of K_n on the Klein bottle

For the drawing on the Klein bottle, divide the vertices into two sets $A = \{v_1, v_2, \ldots, v_{\lfloor \frac{n}{2} \rfloor}\}$ and $B = \{v_{\lfloor \frac{n}{2} \rfloor + 1}, \ldots, v_n\}$ and place all the edges into three strips S_1, S_2, S_3. S_1 contains all edges among the vertices of A, S_2 contains all edges among the vertices of B, and S_3 all edges between A and B (see Figure 9). Clearly, we can draw the edges such that no two of them intersect more than once. It is not difficult to verify that each edge crosses at least cn^2 other edges.

We now describe Attila Pór's construction of a simple topological complete graph on n vertices with each edge intersecting at least $\Omega(n^2)$ other edges. The projective plane can be obtained by adding a line at infinity to the real plane. We place the vertices of the constructed topological graph in the vertices of a regular n-gon P. Any two vertices are connected by the portion of the line through the two vertices outside of the polygon P. It is easy to see that any edge is intersected by $\Omega(n^2)$ other edges.

References

1. M. Ajtai, V. Chvátal, M. Newborn, and E. Szemerédi: Crossing-free subgraphs. Annals of Discrete Mathematics **12** (1982), 9–12
2. P. Brass, W. Moser, and J. Pach: Research problems in discrete geometry. Springer (2005)
3. G. Cairns and Y. Nikolayevsky: Bounds for generalized thrackles. Discrete Comput. Geom. **23** (2000), 191–206
4. J. Černý: Geometric graphs with no three disjoint edges. Discrete Comput. Geom. (to appear)
5. H. Harborth: Crossings on edges in drawings of complete multipartite graphs. Colloquia Math. Soc. János Bolyai **18** (1978), 539–551
6. H. Harborth and M. Mengersen: Edges without crossings in drawings of complete graphs. J. Comb. Theory, Ser. B **17** (1974) 299–311
7. H. Harborth and M. Mengersen: Drawings of the complete graph with maximum number of crossings. Congr. Numerantium **88** (1992), 225–228
8. H. Harborth and C. Thürmann: Minimum number of edges with at most s crossings in drawings of the complete graph. Congr. Numerantium **102** (1994), 83–90
9. F.T. Leighton: New lower bound techniques for VLSI. Math. Systems Theory **17** (1984), 47–70
10. J. Pach, R. Pinchasi, G. Tardos, and G. Tóth: Geometric graphs with no self-intersecting path of length three. Graph drawing 2002, 295–311, Lecture Notes in Comput. Sci., 2528, Springer, Berlin, 2002; also European J. Combin. **25** (2004), no. 6, 793–811
11. J. Pach, R Radoičić, and G. Tóth: A generalization of quasi-planarity. Towards a theory of geometric graphs, 177–183, Contemp. Math., 342, Amer. Math. Soc., Providence, RI, 2004
12. J. Pach, R Radoičić, and G. Tóth: Relaxing planarity for topological graphs. Discrete and computational geometry, 221–232, Lecture Notes in Comput. Sci. **2866**, Springer, Berlin, 2003
13. J. Pach, J. Solymosi and G. Tóth: Unavoidable configurations in topological complete graphs. Discrete Comput. Geom. **30** (2003), 311 – 320
14. J. Pach and G. Tóth: Disjoint edges in topological graphs. To appear
15. R. Pinchasi and R Radoičić: Topological graphs with no self-intersecting cycle of length 4. Towards a theory of geometric graphs, 233–243, Contemp. Math., 342, Amer. Math. Soc., Providence, RI, 2004
16. G. Ringel: Extremal problems in the theory of graphs. Theory Graphs Appl., Proc. Symp. Smolenice 1963, 85–90 (1964)
17. R.D. Ringeisen, S.K. Stueckle, and B.L. Piazza: Subgraphs and bounds on maximum crossings. Bull. Inst. Comb. Appl. **2** (1991), 33–46

On Balloon Drawings of Rooted Trees

Chun-Cheng Lin and Hsu-Chun Yen*

Dept. of Electrical Engineering, National Taiwan University,
Taipei, Taiwan 106, ROC
sanlin@cobra.ee.ntu.edu.tw, yen@cc.ee.ntu.edu.tw

Abstract. Among various styles of tree drawing, *balloon drawing*, where each subtree is enclosed in a circle, enjoys a desirable feature of displaying tree structures in a rather balanced fashion. We first design an efficient algorithm to optimize angular resolution and aspect ratio for the balloon drawing of rooted unordered trees. For the case of ordered trees for which the center of the enclosing circle of a subtree need not coincide with the root of the subtree, flipping the drawing of a subtree (along the axis from the parent to the root of the subtree) might change both the aspect ratio and the angular resolution of the drawing. We show that optimizing the angular resolution as well as the aspect ratio with respect to this type of rooted ordered trees is reducible to the perfect matching problem for bipartite graphs, which is solvable in polynomial time. Aside from studying balloon drawing from an algorithmic viewpoint, we also propose a local magnetic spring model for producing dynamic balloon drawings with applications to the drawings of galaxy systems, H-trees, and sparse graphs, which are of practical interest.

1 Introduction

Since the majority of algorithms for drawing rooted trees take linear time, rooted tree structures are suited to be used in an environment in which real-time interactions with users are frequent. Among existing algorithms in the literature for drawing rooted trees, triangular tree drawing [8], radial or hyperbolic drawing [5], and balloon drawing [1, 3, 6] with respect to cone trees [9] are popular for visualizing hierarchical graphs. Our concern in this paper is a *balloon drawing* of a rooted tree which is a drawing having the following properties: (1) all the children under the same parent are placed on the circumference of the circle centered at their parent, (2) there exist no edge crossings in the drawing, and (3) with respect to the root, the deeper an edge is, the shorter its drawing length becomes.

Each subtree in the balloon drawing of a tree is enclosed entirely in a circle, which resides in a *wedge* whose end-point is the parent node of the subtree. The radius of each circle is proportional to the number of descendants associated with the root node of the subtree. The ray from the parent to the root of the subtree divides the wedge into two sub-wedges. Depending on whether the two sub-wedge angles are required to be identical or not, a balloon drawing can

---

* Corresponding author. Supported in part by NSC Grant 94-2213-E-002-086, Taiwan.

P. Healy and N.S. Nikolov (Eds.): GD 2005, LNCS 3843, pp. 285–296, 2005.

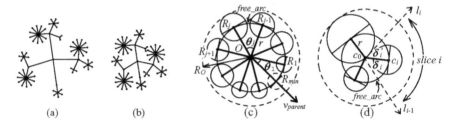

Fig. 1. (a) and (b) illustrate balloon drawings. (c) and (d) illustrate the SNS model. Note that in (c), node O is not the root, and the edge between O and its parent goes through a circle with radius R_{min}; (d) is a star graph centered at c_0.

further be divided into two types: drawings with *even angles* (see Figure 1(a)) and drawing with *uneven angles* (see Figure 1(b)).

The main aesthetic criteria on balloon drawing are *angular resolution* and *aspect ratio*. *Angular resolution* refers to the smallest angle between two adjacent edges incident to the common node in straight-line drawing, whereas *aspect ratio* is defined as the ratio of the largest angle to the smallest angle formed by two adjacent edges incident to the common node in straight-line drawing. A tree layout with a small aspect ratio often enjoys a very balanced view of a tree.

It is not hard to observe that with respect to a rooted *unordered* tree, changing the order in which the children of a node are listed affects the angular resolution as well as the aspect ratio of the drawing. Hence an interesting question arises: *How to find an embedding of a rooted unordered tree such that the balloon drawing of the tree (of the even angle type) has the maximum angular resolution and the minimum aspect ratio?* In the first part of this paper, we demonstrate an efficient algorithm which is guaranteed to yield an optimal balloon drawing in terms of (maximum) angular resolution and (minimum) aspect ratio.

Now consider the case of the uneven angle type. Allowing uneven angles introduces another dimension of flexibility as far as optimizing angular resolution and aspect ratio is concerned. Even if the embedding (ordering) of a tree is given, flipping the drawing of a subtree along the axis going through the parent and the root of the subtree might change the angular resolution and the aspect ratio of the drawing. Notice in the uneven angle case, the angles on the two sides of the axis might not be equal. A related question is: *How to flip uneven angles in the balloon drawing of a rooted ordered tree to achieve optimality in angular resolution and aspect ratio?* Notice in the above, the embedding of the underlying tree is fixed. As it turns out, we are able to reduce the above problem to that of *perfect matching* of bipartite graphs, which admits a polynomial time solution.

Aside from the above two algorithmic issues related to balloon drawing, the second part of this paper deals with the design and implementation of a *force-directed method* (see, e.g., [2, 11]) to provide *dynamic* balloon drawings. Scalability, interactivability, and predictability make dynamic drawing interesting and important in the issues arising from many applications of information visualization. More details about our approach will be given in Section 5.

2   Preliminaries

A tree is a connected acyclic graph. A rooted tree where the order of the subtrees is significant and fixed is called an *ordered tree*; otherwise, it's called *unordered*. A *star graph* with $n+1$ nodes is a rooted tree in which the root node is of degree n and the others are of degree one. Given a drawing of graph G, the *angular resolution* at node v refers to the smallest angle formed by two adjacent edges incident to the common node v in the drawing of G. The *angular resolution* of a drawing of G is defined as the minimum angular resolution among all nodes in G. The *aspect ratio* of a drawing of G is the ratio of the largest angular resolution to the smallest angular resolution in the drawing. The angular resolution (resp., aspect ratio) of a graph is in the range of $(0°, 360°)$ (resp., $[1, \infty)$).

There exist two models in the literature for generating *balloon drawings* of trees. Given a node v, let $r(v)$ be the radius of the drawing circle centered at v. If we require that $r(v) = r(w)$ for arbitrary two nodes v and w that are of the same depth from the root of the tree, then such a drawing is called a balloon drawing under the *fractal model* [4]. Under this model, if r_m and r_{m-1} are the lengths of edges at depths m and $m - 1$, respectively, then

$$r_m = \gamma \times r_{m-1} \qquad (1)$$

where γ is the predefined ratio $(0 < \gamma < 1)$ associated with the fractal drawing.

Unlike the fractal model, the *subtree with nonuniform sizes* (abbreviated as *SNS*) model [1, 3] allows subtrees associated with the same parent to reside in circles of different sizes, and hence, the drawing based on this model often results in a clearer display on large subtrees than that under the fractal model.

Theorem 1 (see [1, 3]). *Given a rooted ordered tree T with n nodes, a balloon drawing under the SNS model can be obtained in $O(n)$ time in a bottom-up fashion with the edge length and the angle between two adjacent edges according to equations (2) and (3) respectively:*

$$r = C/(2\pi) \cong (2\sum\nolimits_{j} R_j)/(2\pi) \qquad (2)$$

$$\theta_j \cong (R_{j-1} + free\_arc + R_j)/r \qquad (3)$$

(see Figure 1(c)(d)) where r is the radius of the inner circle centered at node O; C is the circumference of the inner circle; R_j is the radius of the outer circle enclosing all subtrees of the j-th child of O, and R_O is the radius of the outer circle enclosing all subtrees of O; since there exist the gap between C and the sum of all diameters in Equation (2), we can distribute to every θ_j the gap between them evenly, denoted by $free\_arc$.

Note that the trees considered in [1, 3] are ordered. Since all the angles incident to a common node are the same in the fractal model, changing the ordering of the subtrees of a node at any level does not affect the angular resolution (nor the aspect ratio). Under the SNS model, however, the ordering of subtrees is critical as far as angular resolution and aspect ratio are concerned. Our goal is to devise an algorithm for optimizing the angular resolution and aspect ratio of the balloon drawing of a rooted unordered tree (under the SNS model).

3   Balloon Drawings with Even Angles

First, consider the following way of drawing a star graph with circles of nonuniform size attached to the children of the root (see Figure 1(d) for an example).

Definition 1. *The balloon drawing of a star graph with children of nonuniform size is a drawing in which*

1. *circles associated with different children of the root do not overlap, and*
2. *all the children of the root are placed on the circumference of a circle centered at the root.*

Let S be a star graph with $n + 1$ nodes $\{c_0, c_1, ..., c_n\}$, where c_0 is the root. It can easily be seen from Figure 1(d) that, in a balloon drawing of S, the circle centered at the root is divided into n *wedges* (or *slices*) each of which accommodates a circle associated with a child of c_0. Let δ_i (resp., δ_i') be the angle between rays $\overrightarrow{l_{i-1}}$ and $\overrightarrow{c_0c_i}$ (resp., $\overrightarrow{l_i}$ and $\overrightarrow{c_0c_i}$). The balloon drawing is said to be of *even angle* if $\delta_i = \delta_i'$, for all $1 \leq i \leq n$. That is, $\overrightarrow{c_0c_i}$ divides the respective wedge into two equal sub-wedges; otherwise the drawing is said to be of *uneven angle*. In this section, we only consider balloon drawings of even angle. More will be said about the uneven angle case in Section 4.

Let $\theta_i, 1 \leq i \leq n$, be the degree of the wedge angle enclosing the circle centered at node c_i. (In Figure 1(d) $\theta_i = \delta_i + \delta_i' = 2\delta_i$, assuming the even angle case.) An ordering of the children of c_0 is simply a *permutation* σ of $\{1, ..., n\}$, which specifies the placements of nodes $c_1, ..., c_n$ (and their associated circles) along the circumference of the circle centered at c_0 in the balloon drawing. More precisely, the children are drawn in the order of $c_{\sigma_1}, c_{\sigma_2}, ..., c_{\sigma_n}$, in which c_{σ_i} and $c_{\sigma_{i\oplus 1}}$, $1 \leq i \leq n$, are neighboring nodes. [1] With respect to σ, the degree of the angle between $\overrightarrow{c_0c_i}$ and $\overrightarrow{c_0c_{i\oplus 1}}$ is $(\theta_{\sigma_i} + \theta_{\sigma_{i\oplus 1}})/2$. Hence, the angular resolution (denoted by $AngResl_\sigma$) and the aspect ratio (denoted by $AspRatio_\sigma$) are

$$AngResl_\sigma = \min_{1 \leq i \leq n}\left\{\frac{\theta_{\sigma_i} + \theta_{\sigma_{i\oplus 1}}}{2}\right\}, AspRatio_\sigma = \left\{\frac{\max_{1 \leq i \leq n}\left\{\frac{\theta_{\sigma_i} + \theta_{\sigma_{i\oplus 1}}}{2}\right\}}{\min_{1 \leq i \leq n}\left\{\frac{\theta_{\sigma_i} + \theta_{\sigma_{i\oplus 1}}}{2}\right\}}\right\}. \quad (4)$$

Let Σ be the set of all permutations of $\{1, ..., n\}$. In what follows, we shall design an efficient algorithm to find a permutation that returns

$$optAngResl = \max_{\sigma \in \Sigma}\{AngResl_\sigma\} \textbf{ and } optAspRatio = \min_{\sigma \in \Sigma}\{AspRatio_\sigma\}.$$

The $optAngResl$ is said to *involve* degrees of angles θ_{σ_i} and $\theta_{\sigma_{i\oplus 1}}$ if i is the value minimizing $AngResl_\sigma$ of Equation (4) w. r. t. the optimal permutation σ.

For notational convenience, we order the set of wedge angles $\theta_1, ..., \theta_n$ in ascending order as either

$$m_1, m_2, ..., m_{k-1}, m_k, M_k, M_{k-1}, ..., M_2, M_1 \textbf{ if } n \textbf{ is even,} \quad (5)$$

$$\textbf{or } m_1, m_2, ..., m_{k-1}, m_k, mid, M_k, M_{k-1}, ..., M_2, M_1 \textbf{ if } n \textbf{ is odd,} \quad (6)$$

for some k where m_i (resp. M_i) is the i-th minimum (resp. maximum) among all, and mid is the median if n is odd. We define $\alpha_{ij} = (M_i + m_j)/2, 1 \leq i, j \leq k$.

---

[1] $i \oplus 1$ denotes $(i \bmod n) + 1$.

Procedure 1. OPTBALLOONDRAWING

Input: a star graph S with n child nodes of nonuniform size.

Output: a balloon drawing of S optimizing angular resolution and aspect ratio.

– Sort the set of degrees of the wedge angles (accommodating the n nonuniform circles) into ascending order as mentioned in Equations (5) and (6).
– Output a drawing witnessed by the following circular permutation:
$(M_1, m_2, M_3, m_4, ..., \mu, (, mid), \nu, ..., M_4, m_3, M_2, m_1)$
where $\{\mu, \nu\} = \{M_k, m_k\}$ whose values depends on whether $n = 2k$ or $2k + 1$ and whether k is odd or even. Note that M_1 and m_1 are adjacent.

Recall from Figure 1(c) that, the drawing of the subtree rooted at node O is enclosed in a circle centered at O. By abstracting out the details of each of the subtrees associated with the children of O, the balloon drawing of the subtree at O can always be viewed as a balloon drawing of a star graph with children of nonuniform size rooted at O, regardless of the depth at which O resides. In addition, even if we alter the ordering of the children of O, the size of the outer circle bounding all the children of O remains the same; hence, the optimization of each of the subtrees at depth k does not affect the optimization of their parent at depth $k - 1$. In view of the above, optimizing the angular resolution and the aspect ratio of a balloon drawing of a rooted unordered tree can be carried out in a bottom-up fashion. So, it suffices to investigate how to optimize the angular resolution and the aspect ratio of balloon drawing with respect to star graphs.

Theorem 2. *Procedure 1 achieves optimality in angular resolution as well as in aspect ratio for star graphs.*

Proof. (Sketch) In what follows, we only consider the case

$$\sigma = (M_1, m_2, M_3, m_4, ..., M_{k-1}, m_k, mid, M_k, m_{k-1}, ..., M_4, m_3, M_2, m_1)$$

i.e., $n = 2k + 1$ and k is odd, and assuming that degrees are all distinct; the remaining cases are similar (in fact, simpler).

Recall that $\alpha_{ij} = \frac{M_i + m_j}{2}$. One can easily see the following properties of σ:

Property (1). For each $i \in \{2, ..., k\}$, the angles of degree $(mid + m_k)/2$,
$\alpha_{(i-1)i} = (M_{i-1} + m_i)/2$, $(M_k + mid)/2$, and $\alpha_{i(i-1)} = (M_i + m_{i-1})/2$
are included in σ;

Property (2). The minimum degree of σ must be $(mid + m_k)/2$ or $\alpha_{j(j-1)}$,
while the maximum degree of σ must be $(M_k + mid)/2$ or $\alpha_{(l-1)l}$ for some
$j, l \in \{2, ..., k\}$.

The reason behind Property (2) is that all the angles consecutively appearing in σ have the following ordering relationship:

$$\alpha_{12} > \alpha_{32} < \alpha_{34} > ... > \alpha_{j(j-1)} < ... < \alpha_{(l-1)l} > ... > \alpha_{k(k-1)} < (M_k + mid)/2$$
$$> (mid + m_k)/2 < \alpha_{(k-1)k} > ... > \alpha_{43} < \alpha_{23} > \alpha_{21} < (M_1 + m_1)/2 < \alpha_{12}.$$

Suppose δ is the permutation that witnesses $optAngResl$. ¿From Property (2), the minimum angular resolution of σ must be either $\alpha_{i,i-1}$, for some $2 \leq i \leq k$, or $(mid + m_k)/2$. In what follows, we only need to consider the case when the minimum angular resolution of σ is $\alpha_{i,i-1}$; the other case can be proved similarly.

Now if M_i is a neighbor of m_{i-1} in δ (the optimal permutation), then δ and σ have the same angular resolution and $optAngResl = \alpha_{i,i-1}$ because, otherwise,

- if $optAngResl < \alpha_{i,i-1}$ (which is the angular resolution of σ), then this contradicts that δ is optimal.
- if $optAngResl > \alpha_{i,i-1}$, then this's impossible because δ has an angle of degree $\alpha_{i,i-1} = (M_i + m_{i-1})/2$.

Hence, σ is optimal as well.

On the other hand, suppose x and y $(x < y)$ are the two neighbors of m_{i-1} in δ and neither one is M_i, then both x and y must be greater than M_i; otherwise, the angular resolution of δ is smaller than $(M_i + m_{i-1})/2$ – contradicting δ being optimal. Also note that $optAngResl \geq (m_{i-1} + x)/2$. Now if we look at a partition of the set of wedge angles of S as follows:

$$\underbrace{m_1 < ... < m_{i-1}}_{R_A} < \underbrace{...}_{R_B} < M_i < \underbrace{... < x < ... < y < ... < M_1}_{R_C} . \tag{7}$$

R_A contains $i - 2$ elements, which must be connected to at least $i - 1$ elements of R_C; otherwise, the angular resolution in δ becomes less than $(M_i + m_{i-1})/2$ – a contradiction. R_C originally contains $i - 1$ elements. However, x and y are the two neighbors of m_{i-1} – meaning that together with m_{i-1} they are tied together and cannot be separated. So effectively only '$i-2$' elements of R_C can fill the $i-1$ neighbors of R_A – which is not possible. We have a contradiction. What the above shows is that in the optimal permutation δ, a neighbor of m_{i-1} is M_i. Hence, the angular resolution of δ is $\leq (M_i + m_{i-1})/2$. Since δ witnesses $optAngResl$, $optAngResl = (M_i + m_{i-1})/2$, meaning that σ also produces $optAngResl$.

The above implies that $optAngResl$ must be either $(mid + m_k)/2$ or $\alpha_{i(i-1)}$ for some $i \in \{2, ..., k\}$, which is always included in the circular permutation σ produced by Procedure 1. Similarly, we can prove that the minimum degree of the largest angle of any drawing must be $(M_k + mid)/2$ or $\alpha_{(j-1)j}$ for some $j \in \{2, ..., k\}$, which is also always in σ. Since σ simultaneously possesses both the maximum degree of the smallest angle and the minimum degree of the largest angle of any drawing, σ also witnesses the optimum aspect ratio.   □

Using Procedure 1, the drawing of a rooted unordered tree which achieves optimality in angular resolution and aspect ratio can be constructed efficiently in a bottom-up fashion.

Finally, in order to give a sense perception on the advantages of our algorithm, we implement the algorithm and give a simple experimental result shown in Figure 2. The drawing in (a) based on the fractal model displays that the degrees of angles spanned by adjacent edges are identical, and hence has the best angular resolution and aspect ratio. The major drawback of fractal drawing, as seen

tree information			Angular resolution			Aspect ratio		
node num	max degree	depth	fractal	SNS	our method	fractal	SNS	our method
1000	20	4	18°	3.50°	6.33°	1	13.4961	4.6718

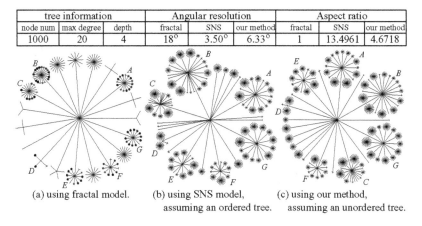

(a) using fractal model.   (b) using SNS model,   (c) using our method,
assuming an ordered tree.   assuming an unordered tree.

Fig. 2. An experimental result. (The above is its statistics.)

in (a), is that visibility deteriorates considerably as we move towards deeper subtrees of complicated structures. For observing more complicated subtrees clearer and easier, one may apply the SNS model (with respect to an ordered tree) to yield (b). (Note that (b) has the same ordering as that in(a).) To a certain extent, the SNS model sacrifices the angular resolution and the aspect ratio in order to gain better visibility for displaying complicated subtree structures. If the ordering of subtrees is allowed to be altered, our optimization algorithm has the ability to optimize the angular resolution and the aspect ratio under the SNS model on the balloon drawing of the rooted unordered tree, as shown in (c).

4   Balloon Drawings with Uneven Angles

The area of a balloon drawing can be measured by the size of the circle enclosing the drawing. Minimizing the area of a drawing is an important issue because any drawing needs to be rendered on a limited region. A careful examination of the approach investigated in Section 3 suggests that the area of balloon drawing generated by the SNS model may not be minimal. Part of the reason is the involvement of the so-called $free\_arc$ described in Theorem 1 and Figure 1(d), serving for the purpose of separating the enclosing circles of two neighboring subtrees. A more subtle point regarding the 'waste' of drawing space is illustrated in Figure 3, in which (a) shows the drawing of a tree under the SNS model. Let T_v be the subtree rooted at v. Based on the approach discussed in Section 3, T_v resides in a circle centered at v and the circle included in a wedge in which the ray from O to v cuts the wedge into two sub-wedges of identical size (i.e., $\theta_1 = \theta_2$). By limiting the drawing to the area formed by two lines (see t_1 and t_2 in (a)) tangent to the outer circle of children of v, the drawing area is reduced, i.e., the new wedge (in which the drawing of T_v resides) is now spanned by lines t_1 and t_2 with the degree of the wedge angle equals $\theta_3 + \theta_4$. Furthermore, the ray $\overrightarrow{Ov}$ cuts this new wedge into two possibly uneven parts (i.e., θ_3 need not be equal

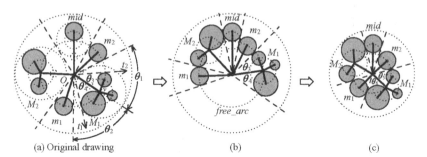

(a) Original drawing   (b)   (c)

Fig. 3. Immediate step of minimizing area where each shaded circle is a star graph

to θ_4). Allowing uneven angles tends to release extra space between the drawings of neighboring subtrees, in comparison with the case discussed in Section 3 when only even angles are permitted. The presence of such extra space allows us to move the position of each subtree inwards (i.e., towards the root node O) which, in turn, reduces the drawing area as (b) shows. The drawing area can further be reduced by shrinking the *free_arc* on the bottom of (b). The final drawing is shown in (c) which obviously has a smaller drawing area (compared with (a)). However, angular resolution and aspect ratio might deteriorate as (c) indicates.

It's interesting to observe that in Figure 3(c) the angular resolution and the aspect ratio might change if we flip the drawing of subtree T_v along the axis $\overrightarrow{Ov}$ (i.e., swapping θ_5 and θ_6). Hence, a natural question is to determine how the two possibly uneven angles associated with a subtree are arranged in order to achieve optimal aspect ratio and angular resolution, assuming the ordering of subtrees in the drawing is fixed. Such a question can be formulated as follows.

Suppose O is the root of a star graph with n subtrees rooted at $A_1, ..., A_n$ which are listed in a counterclockwise fashion. Suppose the degrees of the two angles associated with subtree A_i ($1 \leq i \leq n$) are a_i^0 and a_i^1. Then the sequence of degrees encountered along the circle centered at O can be expressed as:

$$\{\underbrace{a_1^{t_1}, a_1^{t_1'}}_{A_1}, \underbrace{a_2^{t_2}, a_2^{t_2'}}_{A_2}, ..., \underbrace{a_i^{t_i}, a_i^{t_i'}}_{A_i}, ..., \underbrace{a_n^{t_n}, a_n^{t_n'}}_{A_n,}\} \tag{8}$$

where $t_i, t_i' \in \{0,1\}$ and $t_i + t_i' = 1$. With respect to the above, the angular resolution and the aspect ratio can be calculated as follows respectively:

$$AngResl = \min_{1 \leq i \leq n}\{a_i^{t_i'} + a_{i\oplus 1}^{t_{i\oplus 1}}\}, AspRatio = \frac{\max_{1 \leq i \leq n}\{a_i^{t_i'} + a_{i\oplus 1}^{t_{i\oplus 1}}\}}{\min_{1 \leq i \leq n}\{a_i^{t_i'} + a_{i\oplus 1}^{t_{i\oplus 1}}\}}. \tag{9}$$

The problem then boils down to assigning 0 and 1 to t_i and t_i' ($1 \leq i \leq n$) in order to optimize *AngResl* and *AspRatio*. Note that the two values (either ((0, 1) or (1, 0)) of (t_i, t_i') correspond to the two configurations of the drawing of the subtree associated with A_i and one is obtained from the other by flipping along the axis $\overrightarrow{OA_i}$. Consider the following problems:

THE ASPECT RATIO (RESP., ANGULAR RESOLUTION) PROBLEM : Given the
initial drawing of a star graph (with uneven angles) specified by Equation
(8) and a real number r, determining the assignments (0 or 1) for t_i and t'_i
($1 \leq i \leq n$) so that $AspRatio \leq r$ (resp., $AngResl \geq r$); return *false* if no
such assignments exist.

In what follows, we show how the above two problems can be reduced to
perfect matching for bipartite graphs. A *matching* M on a graph G is a set of
edges of G such that any two edges in M shares no common node. A *maximum
matching* of G is a matching of the maximum cardinality. The largest possible
matching on a graph with n nodes consists of $n/2$ edges, and such a matching is
called a *perfect matching*. It is known that the maximum matching problem for
bipartite graphs with n nodes and m edges can be found in $O(\sqrt{m}n)$ time [7].

Theorem 3. *Both the* ASPECT RATIO PROBLEM *and the* ANGULAR RESOLUT-
ION PROBLEM *can be solved in* $O(n^{2.5})$ *time.*

Proof. (Sketch) We consider the *Aspect Ratio* Problem first. Let r be the bound
of the desired aspect ratio. Suppose the set of wedge angles of the star graph is
specified as

$$\{\underbrace{b_1, b'_1}_{A_1}, \underbrace{b_2, b'_2}_{A_2}, ..., \underbrace{b_i, b'_i}_{A_i}, ..., \underbrace{b_n, b'_n}_{A_n}\} \tag{10}$$

Notice that b_i and b'_i are the degrees of the two angles associated with the wedge
in which the drawing of the subtree rooted at A_i resides. b_i (b'_i) can be the
neighbor of one of b_{i-1}, b'_{i-1}, b_{i+1}, and b'_{i+1}, depending on where A_{i-1}, A_i and
A_{i+1} are positioned in the drawing. (For instance, if b_i is paired with b'_{i+1}, the
angle between $\overrightarrow{OA_i}$ and $\overrightarrow{OA_{i+1}}$ becomes $b_i + b'_{i+1}$.) As a result, to determine
whether it is feasible to realize a drawing for which the aspect ratio is less than
or equal to r, our algorithm iteratively selects a pair (x, y) where $x \in \{b_i, b'_i\}$ and
$y \in \{b_{i\oplus1}, b'_{i\oplus1}\}$ so that $x + y$ is assumed to be the '*smallest*' angle in a drawing
respecting the aspect ratio r, if such a drawing exists. Then a bipartite graph
$G_{(x,y)}$ is constructed in such a way that a drawing respecting the aspect ratio r
exists iff $G_{(x,y)}$ has a perfect matching.

To better understand the algorithm, consider the case when $(x, y) = (b_1, b'_n)$.
Let $\phi = b_1 + b'_n$. $G_{(b_1, b'_n)}$ $(=((U, W), E)$, where $U \cup W$ is the set of nodes and
$U \cap W = \emptyset)$ is constructed as follows (assuming $n = 2k$):

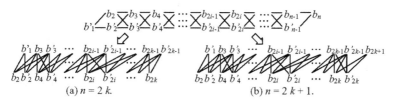

(a) $n = 2k$.   (b) $n = 2k + 1$.

Fig. 4. Illustration of modelling. The nodes with odd (resp. even) index are placed on
the upper (resp. lower) level.

(1) $U = \{b_{2i-1}, b'_{2i-1}; \forall i \in \{1, ..., k\}\} - \{b_1\}; W = \{b_{2i}, b'_{2i}; \forall i \in \{1, ..., k\}\} - \{b'_{2k}\}$

(2) Note that (b_1, b'_{2k}) is the only edge involving b_1 and b'_{2k}, so needs not be considered in computation.

　　(i) For each $i \in \{2, 3, ..., k-1\}$, $(s, t) \in E$, where $s \in \{b_{2i-1}, b'_{2i-1}\}$ and $t \in \{b_{2i-2}, b'_{2i-2}, b_{2i}, b'_{2i}\}$, if $\phi \leq (s+t) \leq r \cdot \phi$, meaning that placing s next to t (inducing an angle of degree $s+t$) respects the aspect ratio, as well as ϕ being the smallest among all the angles in the drawing.

　　(ii) $(b'_1, t) \in E$, where $t \in \{b_2, b'_2\}$, if $\phi \leq (b'_1 + t) \leq r \cdot \phi$.

　　(iii) $(s, t) \in E$, where $s \in \{b_{2k-1}, b'_{2k-1}\}$ and $t \in \{b_{2k-2}, b'_{2k-2}, b_{2k}\}$, if $\phi \leq (s+t) \leq r \cdot \phi$.

See Figure 4(a) for the structure of bipartite graph $G_{(b_1, b'_n)}$ for the case $n = 2k$. The case $n = 2k+1$ is similar (see Figure 4(b)). It is reasonably easy to see that $G_{(b_1, b'_n)}$ has a perfect matching iff there exists a drawing for which $AngResl = b_1 + b'_n$, and $AspRatio \leq r$. By repeatedly selecting a pair (x, y), $x \in \{b_i, b'_i\}$ and $y \in \{b_{i \oplus 1}, b'_{i \oplus 1}\}$, as the one that contributes to $AngResl$ (i.e., the smallest angle), whether a drawing with an aspect ratio $\leq r$ exists or not can be determined.

　　As for the executing time, since every node in the bipartite graph is adjacent to at most four edges, the number of $G_{(x,y)}$ needed to be considered is $O(n)$. For a given m-edge n-node bipartite graph, the perfect matching problem can be solved in $O(\sqrt{m}n)$. Hence, the *Aspect Ratio Problem* can be solved in $O(n \times \sqrt{n}n) = O(n^{2.5})$ time. The solution for the *Angular Resolution Problem* can be performed along a similar line of the proof for the *Aspect Ratio Problem*.　　□

5　Local Magnetic Spring Model

Like [11], our *local magnetic spring model* replaces all edges by *local* magnetized springs and assumes that each node is placed at the center of a *local* polar magnetic field, which can be viewed as a set of vectors radical from the node. Each angle formed by two adjacent radial vectors of the same node is set evenly (resp. according to Equation (3)) if the fractal (resp. SNS) model is applied. So each edge (magnet) is affected by a magnetic torque if it does not align properly in its corresponding magnetic field. From [2, 11], the spring forces acted at each node v and the magnetic torque of v taking v's parent, say $p(v)$, as the reference point of the torque can be calculated respectively according to $F_s = c_s \log(d/r)$ and $\tau = c_\tau \theta^\alpha$, where c_s, c_τ, and α are constants, d is the current length of the spring, r is the natural length of the spring calculated according to Equation (2) (resp. Equation (1)) if the SNS (resp. fractal) model is applied, and θ is the angle formed between the edge $vp(v)$ and its corresponding magnetic field. After setting each spring natural length and the orientation of each magnetic field, our algorithm can output the balloon drawing of a tree automatically (see also an example in Figure 5).

　　Based on the above theory, we develop a prototype system for dynamic balloon drawing of trees, running on a Pentium IV 3.2GHz PC. A tree with 200,000 nodes has been executed efficiently (about 0.5 sec per iteration), so it's satisfactory even in a real-time environment. Figure 6 (a), a random drawing, displays

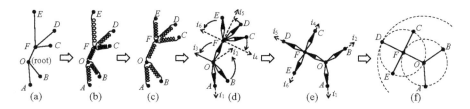

Fig. 5. Illustration of our approach. (a) Initial drawing. (b) Local springs. (c) Spring force equilibrium. (d) Local polar magnetic fields. (e) Magnetic torque equilibrium. (f) Final drawing. Notice that, in fact, (b) and (d) are applied synchronously in our model.

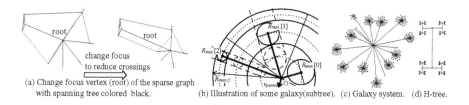

(a) Initial drawing   After 5 iterations.   After 15 iterations.   (b) SNS model   (c) Fractal model

Fig. 6. An incomplete tree with 1,111 nodes, maximum degree 10, and depth 4. (b) runs 99 iterations and costs 0.110 sec. (c) runs 102 iterations and costs 0.109 sec.

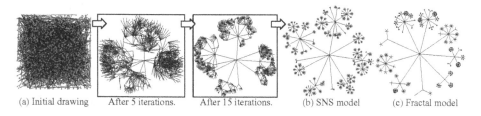

(a) Change focus vertex (root) of the sparse graph with spanning tree colored black.   (b) Illustration of some galaxy(subtree).   (c) Galaxy system.   (d) H-tree.

Fig. 7. Applications

the initial drawing of a tree as the input to our algorithm. As our algorithm progresses, we are able to observe how the evolvement of the drawing preserves the predictability as the frames between (a) and (b) indicate. It's desirable for the dynamic balloon drawing system to offer a capability allowing the user to interact with and/or navigate through the tree effectively. Once we interact with a tree, the corresponding local magnetic spring setting should be modified accordingly, and then the main procedure is performed to yield the new drawing.

In what follows, our balloon drawing algorithm is tailored to cope with three real-world applications. First, to draw a sparse graph with navigation and interaction operations in mind, a good starting point is to find a spanning tree (serving as the skeleton of the sparse graph) to which our balloon drawing algorithm is applied. Following that, the remaining edges are added to the drawing. By interacting with the user, it becomes easier to come up with a nice drawing with fewer edge crossings as shown in Figure 7 (a).

Second, a *galaxy system* involves numerous fixed stars, planets, moons, galaxies, and even huge star clusters. Due to universal gravitation, each of the stars has a revolution around (or related to) some star. Thus stars in a galaxy system form a hierarchical structure, and their revolution orbits are nearly circular and probably concentric circular. The center of the universe can be viewed as the root of the galaxy tree. *Nova*, a new-born star, can be simulated by the operation of adding a node to which a light color is assigned, while a *black hole*, a dying star, is colored dark which disappears after a period of time. When a star or a galaxy dies, the corresponding node (nodes), edge(s), and subtree(s) are deleted. Besides, the behavior that our algorithm propagating the amount of movement and rotation to children in a top-down fashion and making the nodes on lower layer move and rotate faster is similar to the fact that the moon rotates around a planet (earth) faster than around a fixed star (sun). All of the behaviors can therefore be captured by our system, subject to a slight modification. Figure 7 (b) illustrates the modification and (c) is an experimental result.

Finally, for given a binary tree, if we let the R_{min} in Figure 1 (c) and (d) be zero and adjust the polar magnetic fields slightly, then we end up with Figure 7 (d) as the output drawing, which is an example of the so-called H-tree [10].

References

1. J. Carrière and R. Kazman. Reserch report: Interacting with huge hierarchies: Beyond cone trees. In *IV 95*, pages 74–81. IEEE CS Press, 1995.
2. P. Eades. A heuristic for graph drawing. *Congress Numerantium*, 42:149–160, 1984.
3. C.-S. Jeong and A. Pang. Reconfigurable disc trees for visualizing large hierarchical information space. In *InfoVis '98*, pages 19–25. IEEE CS Press, 1998.
4. H. Koike and H. Yoshihara. Fractal approaches for visualizing huge hierarchies. In *VL '93*, pages 55–60. IEEE CS Press, 1993.
5. J. Lamping, R. Rao, and P. Pirolli. A focus+context technique based on hyperbolic geometry for visualizing large hierarchies. In *CHI '95*, pages 401–408. ACM Press, 1995.
6. G. Melançon and I. Herman. Circular drawing of rooted trees. In *Reports of the Centre for Mathematics and Computer Sciences*. Report number INS-9817, available at: http://www.cwi.nl/InfoVis/papers/circular.pdf, 1998.
7. C. H. Papadimitriou and K. Steiglitz. *Combinatorial optimization*. Prentice Hall, Englewood Cliffs, New Jersey, 1982.
8. E. Reingold and J. Tilford. Tidier drawing of trees. *IEEE Trans. Software Eng.*, SE-7(2):223–228, 1981.
9. G. Robertson, J. Mackinlay, and S. Card. Cone trees: Animated 3d visualizations of hierarchical information, human factors in computing systems. In *CHI '91*, pages 189–194. ACM Press, 1991.
10. Y. Shiloach. Arrangements of planar graphs on the planar lattices. Ph D Thesis, Weizmann Instite of Science, Rehovot, Israel, 1976.
11. K. Sugiyama and K. Misue. Graph drawing by the magnetic spring model. *J. Vis. Lang. Comput.*, 6:217–231, 1995.

Convex Drawings of Plane Graphs
of Minimum Outer Apices

Kazuyuki Miura[1], Machiko Azuma[2], and Takao Nishizeki[2]

[1] Faculty of Symbiotic Systems Science,
Fukushima University, Fukushima 960-1296, Japan
miura@sss.fukushima-u.ac.jp
[2] Graduate School of Information Sciences,
Tohoku University, Sendai 980-8579, Japan
azuma@nishizeki.ecei.tohoku.ac.jp, nishi@ecei.tohoku.ac.jp

Abstract. In a convex drawing of a plane graph G, every facial cycle of G is drawn as a convex polygon. A polygon for the outer facial cycle is called an outer convex polygon. A necessary and sufficient condition for a plane graph G to have a convex drawing is known. However, it has not been known how many apices of an outer convex polygon are necessary for G to have a convex drawing. In this paper, we show that the minimum number of apices of an outer convex polygon necessary for G to have a convex drawing is, in effect, equal to the number of leaves in a triconnected component decomposition tree of a new graph constructed from G, and that a convex drawing of G having the minimum number of apices can be found in linear time.

1   Introduction

Recently automatic aesthetic drawing of graphs has created intense interest due to their broad applications, and as a consequence, a number of drawing methods have come out [1, 2, 3, 4, 8]. The most typical drawing of a plane graph G is a *straight line drawing* in which all vertices of G are drawn as points and all edges are drawn as straight line segments without any edge-intersection. A straight line drawing of G is called a *convex drawing* if every facial cycle is drawn as a convex polygon, as illustrated in Fig. 1 [1, 9].

In a convex drawing of a plane graph G, the outer facial cycle $F_o(G)$ of G must be drawn as a convex polygon. A polygonal drawing F_o^* of $F_o(G)$, called an *outer convex polygon*, plays a crucial role in finding a convex drawing of G. The plane graph G in Fig. 2(a) admits a convex drawing if an outer convex polygon F_o^* has four or more apices as illustrated in Fig. 2(b), where apices are drawn as white circles. However, if F_o^* has only three apices, that is, F_o^* is a triangle, then G does not admit a convex drawing as illustrated in Fig. 2(c).

A necessary and sufficient condition for a plane graph G to have a convex drawing is known [1, 9]. A linear-time algorithm is also known for finding a convex drawing of G if G satisfies the condition [1, 7]. We recently give a necessary and sufficient condition for a plane graph G to have a convex drawing such that

P. Healy and N.S. Nikolov (Eds.): GD 2005, LNCS 3843, pp. 297–308, 2005.

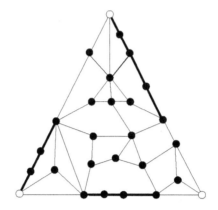

Fig. 1. Convex drawing of a plane graph

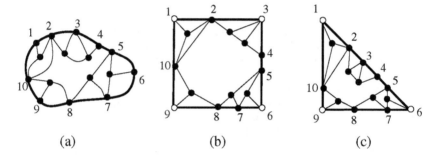

Fig. 2. (a) Plane graph, (b) convex drawing, and (c) non-convex drawings

an outer convex polygon F_o^* has exactly three apices [6]. However, it is not known how many apices of F_o^* are necessary for a plane graph G to have a convex drawing.

In this paper, we show that the minimum number of apices of an outer convex polygon F_o^* necessary for G to have a convex drawing is, in effect, equal to the number of leaves in a triconnected component decomposition tree of a new graph constructed from G, and that a convex drawing of G having the minimum number of apices of F_o^* can be found in linear time.

The remainder of the paper is organized as follows. In Section 2 we give some definitions and two known lemmas. In Section 3 we present our results. Finally we conclude in Section 4.

2   Preliminaries

In this section, we give some definitions and two known lemmas.

We denote by $G = (V, E)$ an undirected connected simple graph with vertex set V and edge set E. An edge joining vertices u and v is denoted by (u, v). The *degree* of a vertex v in G is the number of neighbors of v in G, and is denoted by $d(v)$.

A graph is *planar* if it can be embedded in the plane so that no two edges intersect geometrically except at a vertex to which they are both incident. A *plane graph* is a planar graph with a fixed embedding. A plane graph G divides the plane into connected regions, called *faces*. We denote by $F_o(G)$ the outer face of G. The boundary of $F_o(G)$ is also denoted by $F_o(G)$. A vertex on $F_o(G)$ is called an *outer vertex*, while a vertex not on $F_o(G)$ is called an *inner vertex*. An edge on $F_o(G)$ is called an *outer edge*, while an edge not on $F_o(G)$ is called an *inner edge*.

A polygonal drawing F_o^* of $F_o(G)$ is called an *outer convex polygon* if F_o^* is a convex polygon. A (geometric) vertex of a polygon F_o^* is called an *outer apex*. An outer convex polygon F_o^* is *extendable* if G has a convex drawing in which $F_o(G)$ is drawn as the convex polygon F_o^*. For example, the outer rectangle drawn by thick lines in Fig. 2(b) is extendable, while the outer triangle in Fig. 2(c) is not extendable. We denote by $G - F_o(G)$ the graph obtained from G by deleting all outer vertices.

We call a vertex v of a connected graph G a *cut vertex* if its removal from G results in a disconnected graph. A connected graph G is *biconnected* if G has no cut vertex. If G has a convex drawing, then G is biconnected.

The following necessary and sufficient condition for a plane graph G to have a convex drawing is known.

Lemma 1. [1, 9] Let G be a biconnected plane graph, and let F_o^* be an outer convex polygon of G. Assume that F_o^* is a k-gon, $k \geq 3$, and that $P_1, P_2, \cdots, P_k$ are the k paths in $F_o(G)$, each corresponding to a side of the polygon F_o^*. Then F_o^* is extendable if and only if the following Conditions (a)–(c) hold.

(a) For each inner vertex v with $d(v) \geq 3$, there exist three paths disjoint except v, each joining v and an outer vertex;

(b) The graph $G - F_o(G)$ has no connected component H such that all the outer vertices adjacent to vertices in H lie on a single path P_i, and no two outer vertices in each path P_i are joined by an inner edge (see Fig. 3); and

(c) Every cycle containing no outer edge has at least three vertices of degree ≥ 3.

We call a pair $\{u, v\}$ of vertices in a biconnected graph G a *separation pair* if its removal from G results in a disconnected graph, that is, $G - \{u, v\}$ is not connected. A biconnected graph G is *triconnected* if G has no separation pair. A plane biconnected graph G is *internally triconnected* if, for any separation pair $\{u, v\}$ of G, both u and v are outer vertices and each connected component of $G - \{u, v\}$ contains an outer vertex. In other words, G is internally triconnected if and only if it can be extended to a triconnected graph by adding a vertex in an outer face and joining it to all outer vertices. If a biconnected plane graph G is not internally triconnected, then G has a separation pair $\{u, v\}$ illustrated in Figs. 4(a)–(c) and a "split graph" H contains an inner vertex other than u and v.

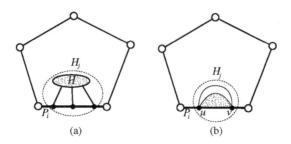

(a)   (b)

Fig. 3. Examples violating Condition (b)

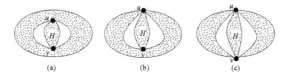

(a)   (b)   (c)

Fig. 4. Biconnected plane graphs which are not internally triconnected

Let $G = (V, E)$ be a biconnected graph, and let $\{u, v\}$ be a separation pair of G. Then, G has two subgraphs $G'_1 = (V_1, E'_1)$ and $G'_2 = (V_2, E'_2)$ such that

(a) $V = V_1 \bigcup V_2$, $V_1 \bigcap V_2 = \{u, v\}$; and
(b) $E = E'_1 \bigcup E'_2$, $E'_1 \bigcap E'_2 = \emptyset$, $|E'_1| \geq 2$, $|E'_2| \geq 2$.

The graph G in Fig. 5(a) has six separation pairs $\{u_1, u_2\}$, $\{u_1, u_3\}$, $\{u_2, u_3\}$, $\{u_2, u_7\}$, $\{u_3, u_6\}$, and $\{u_4, u_5\}$.

For a separation pair $\{u, v\}$ of G, $G_1 = (V_1, E'_1 + (u, v))$ and $G_2 = (V_2, E'_2 + (u, v))$ are called the *split graphs* of G with respect to $\{u, v\}$. The new edges (u, v) added to G_1 and G_2 are called the *virtual edges*. Even if G has no multiple edges, G_1 and G_2 may have. Dividing a graph G into two split graphs G_1 and G_2 are called *splitting*. Reassembling the two split graphs G_1 and G_2 into G is called *merging*. Merging is the inverse of splitting. Suppose that a graph G is split, the split graphs are split, and so on, until no more splits are possible, as illustrated in Fig. 5(b) where virtual edges are drawn by dotted lines. The graphs constructed in this way are called the *split components* of G. The split components are of three types: triple bonds (i.e. a set of three multiple edges), triangles, and triconnected graphs. The *triconnected components* of G are obtained from the split components of G by merging triple bonds into a bond and triangles into a ring, as far as possible, where a *bond* is a set of multiple edges and a *ring* is a cycle. The graph in Fig. 5(a) is decomposed into seven triconnected components $H_1, H_2, \cdots, H_7$ as depicted in Fig. 5(c), where H_1, H_2 and H_6 are triconnected graphs, H_3, H_4 and H_7 are rings, and H_5 is a bond. The split components of G are not necessarily unique, but the triconnected components of G are unique [5].

Let T be a tree in which each node corresponds to a triconnected component H_i and T has an edge (H_i, H_j), $i \neq j$, if and only if H_i and H_j are triconnected components with respect to the same separation pair, as illustrated in Fig. 5(d).

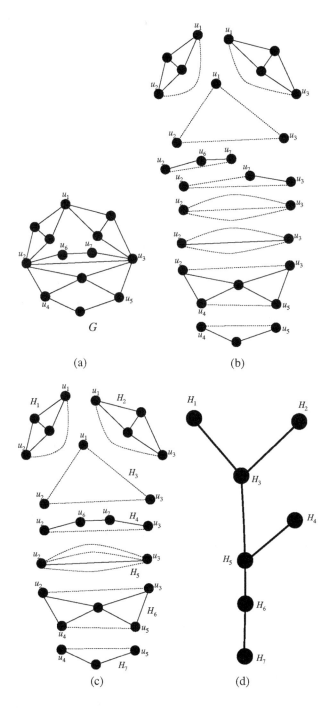

Fig. 5. (a) A biconnected graph G, (b) split components, (c) triconnected components, and (d) a triconnected component decomposition tree T of G

We call T a *triconnected component decomposition tree* of G [5]. Clearly every leaf of T does not correspond to a bond, but corresponds to a triconnected graph or a ring.

The following lemma is known.

Lemma 2. [5] A triconnected component decomposition tree T of a graph G can be found in linear time.

3   Convex Drawing of Minimum Outer Apices

One may obviously assume that a plane graph G is biconnected. One may further assume that every inner vertex of G has degree three or more; if an inner vertex v has degree two in G, then delete v from G, join the neighbors of v, and let G' be the resulting graph; clearly G has a convex drawing if and only if G' has no multiple edges and has a convex drawing. We can then newly formalize a necessary and sufficient condition for G to have a convex drawing, as follows.

Theorem 1. Let G be a plane biconnected graph in which every inner vertex has degree three or more, and let T be a triconnected component decomposition tree of G. Then the following (a)–(c) are equivalent with each other:

(a) G has a convex drawing;

(b) G is internally triconnected; and

(c) both of every separation pair are outer vertices, and every node of T corresponding to a bond has degree two in T.

Proof. We verify (a)⇔(b) and (b)⇔(c), as follows.

(a)⇒(b): Assume that G has a convex drawing. Then G has an extendable outer convex polygon F_o^*, and hence Conditions (a)–(c) in Lemma 1 hold for F_o^*. Suppose for a contradiction that G is not internally triconnected. Then G has a separation pair $\{u, v\}$ illustrated in Figs. 4(a)–(c). In a split graph H with respect to $\{u, v\}$, there is an inner vertex x other than u and v. Vertex x has degree three or more, and hence every path joining x and an outer vertex must pass through either u or v. Therefore there are no three paths disjoint except x, each joining x and an outer vertex. Thus Condition (a) does not hold, a contradiction.

(b)⇒(a): Assume that G is internally triconnected. Let F_o^* be an outer convex polygon in which every outer vertex of G is an apex of F_o^*. We show that F_o^* is extendable and hence G has a convex drawing. Since G is internally triconnected, clearly both Conditions (a) and (c) in Lemma 1 hold. Thus we shall show that Condition (b) holds for F_o^*. Each path P_i in F_o, corresponding to a side of F_o^*, consists of a single edge, and G is a simple graph. Therefore the two outer vertices in P_i are not joined by an inner edge. Since G is internally triconnected, one can easily know that $G - F_o(G)$ has no connected component H such that all the outer vertices adjacent to vertices in H lie on a single path P_i. Thus G satisfies Conditions (a)–(c), and hence by Lemma 1 F_o^* is extendable.

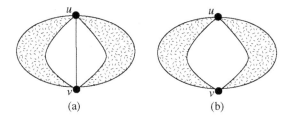

Fig. 6. Plane graphs having separation pairs corresponding to bonds

(b)⇒(c): Assume that G is internally triconnected. Then both of every separation pair of G are outer vertices, and G does not have a separation pair $\{u, v\}$ illustrated in Figs. 4(a)–(c). Therefore, for every separation pair $\{u, v\}$, $G - \{u, v\}$ has exactly two connected components, each containing an outer vertex, as illustrated in Figs. 6(a) and (b); G has an edge (u, v) in Fig. 6(a), while G has no edge (u, v) in Fig. 6(b). Hence one can easily observe that every node of T corresponding to a bond has degree two in T; the bond for Fig. 6(a) is a set of three multiple edges, two of which are virtual ones, and one of which is a real one; the bond for Fig. 6(b) is a pair of virtual multiple edges.

(c)⇒(b): Assume that both of every separation pair are outer vertices, and every node of T corresponding to a bond has degree two in T. We shall show that G does not have a separation pair $\{u, v\}$ illustrated in Figs. 4(a)–(c). Since both of every separation pair are outer vertices, G does not have a separation pair $\{u, v\}$ illustrated in Figs. 4(a) and (b). Assume for a contradiction that G has a separation pair $\{u, v\}$ illustrated in Fig. 4(c). Then a node of T corresponding to a bond containing virtual edges (u, v) has degree three or more, a contradiction. □

By Theorem 1 we may assume that G is internally triconnected and each inner vertex of G has degree three or more, as the graph in Fig. 1. Thus every vertex of degree two must be an outer vertex. Let $P = v_0, v_1, v_2, \cdots, v_{l+1}, l \geq 1$, be a path on $F_o(G)$ such that $d(v_0) \geq 3$, $d(v_1) = d(v_2) = \cdots = d(v_l) = 2$ and $d(v_{l+1}) \geq 3$. Such a path P is called an *outer chain* of G. (The graph in Fig. 1 has four outer chains drawn by thick lines.) We then have the following lemma.

Lemma 3. Let G be an internally triconnected plane graph, and let $P = v_0, v_1, \cdots, v_{l+1}$ be an outer chain of G. Then the following Propositions (a) and (b) hold:

(a) If G has an edge (v_0, v_{l+1}), then G has a convex drawing in which exactly one of the vertices $v_1, v_2, \cdots, v_l$ is an outer apex and G has no convex drawing in which none of them is an outer apex.

(b) If G has no edge (v_0, v_{l+1}), then G has a convex drawing in which none of the vertices $v_1, v_2, \cdots, v_l$ is an outer apex. (See Fig. 7.)

Proof. (a) Suppose that G has an edge (v_0, v_{l+1}). Since G is internally triconnected, by Theorem 1 G has a convex drawing. Let F_o^* be the outer convex

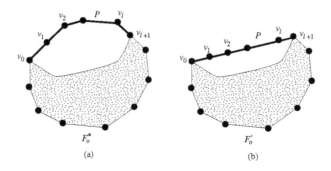

Fig. 7. Outer convex polygons F_o^* and F_o'

polygon of an arbitrary convex drawing of G. Clearly one or more of the vertices $v_1, v_2, \cdots, v_l$ are apices of F_o^*. One can easily modify the drawing to a convex drawing in which exactly one of them is an outer apex.

(b) Suppose that G has no edge (v_0, v_{l+1}). Let F_o^* be the outer convex polygon of an arbitrary convex drawing of G. We may assume that at least one of the vertices $v_1, v_2, \cdots, v_l$ is an apex of F_o^*, as illustrated in Fig. 7(a); otherwise, we have completed a proof for (b). We cut off the region enclosed by the path $P = v_0, v_1, \cdots, v_{l+1}$ and a straight line segment v_{l+1}, v_0 from the polygon F_o^*. Let F_o' be the resulting outer convex polygon, in which $v_0, v_1, \cdots, v_l$ lie on the line segment connecting v_0 and v_{l+1}, as illustrated in Fig. 7(b). We claim that F_o' is extendable and hence G has a convex drawing in which none of the vertices $v_1, v_2, \cdots, v_l$ is an apex of the outer convex polygon. Since G is internally triconnected, both Conditions (a) and (c) in Lemma 1 hold for F_o'. Thus we shall show that Condition (b) holds for F_o'. Since Condition (b) holds for F_o^*, the definition of F_o' implies that Condition (b) holds for all paths of F_o', other than P, each corresponding to a side of the polygon F_o'. Since vertices $v_1, v_2, \cdots, v_l$ have degree two and there is no edge (v_0, v_{l+1}) in G, no two outer vertices in path P are joined by an inner edge. Since G is internally triconnected, $G - F_o(G)$ has no connected component H such that all the outer vertices adjacent to vertices in H lie on the single path P. Thus G satisfies Conditions (a)–(c), and hence F_o' is extendable.   □

By Lemma 3, in order to find a convex drawing of minimum outer apices, we may "contract" an outer chain $P = v_0, v_1, \cdots, v_{l+1}$ of G as follows: if G has an edge (v_0, v_{l+1}), then we replace P in G with an outer chain $P' = v_0, v_i, v_{l+1}$ for an arbitrary index i, $1 \le i \le l$; otherwise, we replace P in G with a single edge (v_0, v_{l+1}). Contract every outer chain of G as above, and let G' be the resulting graph. G' is called a *contracted graph* of G. Figure 8(b) illustrates a contracted graph of the graph G in Fig. 8(a). Since G is internally triconnected, G' is also internally triconnected and hence G' has a convex drawing. One can easily observe that the following lemma holds.

Lemma 4. *Let G be an internally triconnected plane graph, and let G' be a contracted graph of G. If an outer convex polygon F_o^* is extendable for G' (as*

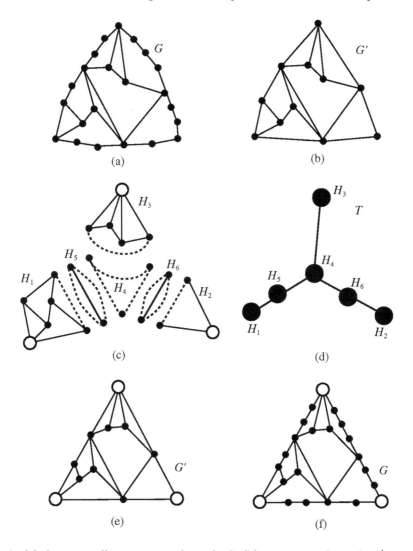

Fig. 8. (a) An internally triconnected graph G, (b) a contracted graph G' of G, (c) triconnected components of G', (d) a triconnected component tree T, (e) a convex drawing of G', and (f) a convex drawing of G

illustrated in Fig. 8(e)), then the same outer convex polygon F_o^* is extendable for G (as illustrated in Fig. 8(f)).

We now have the following lemma.

Lemma 5. Let G be an internally triconnected plane graph, let G' be a contracted graph of G, let T be a triconnected component decomposition tree of G', and let n_l be the number of leaves of T. Then the minimum number of apices of an extendable convex polygon of G' is equal to $\max\{3, n_l\}$.

Proof. If $n_l = 1$, that is, G' is either a cycle or is a triconnected graph, then G' has a convex drawing whose outer facial cycle is drawn as a triangle [6, 8]. Therefore, the minimum number of apices of an extendable convex polygon of G' is $\max\{3, 1\} = 3$. We may thus assume that $n_l \geq 2$.

We first claim that G' has no extendable outer convex polygon such that the number of apices is less than $\max\{3, n_l\}$. If $2 \leq n_l \leq 3$, then $\max\{3, n_l\} = 3$ and hence G' has no extendable outer convex polygon such that the number of outer apices is less than $\max\{3, n_l\}$. We may thus assume that $n_l \geq 4$. Suppose for a contradiction that G' has an extendable outer convex polygon $F_\circ^*$ having less than $\max\{3, n_l\} = n_l$ apices. Then G' has a triconnected component H_j, corresponding to a leaf of T, such that none of the outer vertices in H_j other than the separation pair is an apex of $F_\circ^*$. Hence all the outer vertices in H_j lie on a single side P_i of the polygon $F_\circ^*$. Since H_j corresponds to a leaf of T, H_j is either a triconnected graph or a ring. (H_1 and H_3 in Fig. 8(c) are triconnected graphs, while H_2 is a ring.) Consider first the case where H_j is a triconnected graph. Then $G' - F_\circ(G')$ has a connected component such that all the outer vertices adjacent to vertices in the component lie on the side P_i, as illustrated in Fig. 3(a). Consider next the case where H_j is a ring. Let H_p be the triconnected component corresponding to the node of T which is adjacent in T to the leaf corresponding to H_j. By the definitions of T and G', H_p must be a bond and contain an inner edge joining two outer vertices of G'. Therefore the two outer vertices on the side P_i are joined by the inner edge. Thus, in either case, Condition (b) does not hold for $F_\circ^*$. Hence $F_\circ^*$ is not extendable, a contradiction. We have thus verified the claim.

We then choose a number $k = \max\{3, n_l\}$ of outer vertices as apices of an outer convex polygon $F_\circ^*$ of G', as in the following two cases.

Case 1: $n_l \geq 3$.

In this case, as an apex, we choose an outer vertex v of G' from each triconnected component H_j corresponding to a leaf of T; v must be other than the vertices of the separation pair with respect to H_j. We set these n_l vertices chosen from all leaves of T as the apices of $F_\circ^*$. (In Fig. 8(c), there are three triconnected components H_1, H_2 and H_3 corresponding to leaves of T in Fig. 8(d), and for example the three vertices depicted as white circles are chosen from H_1, H_2 and H_3.)

Case 2: $n_l = 2$.

In this case, as apices, we first choose two outer vertices from the two leaves similarly as in Case 1 above. We then choose an arbitrary outer vertex other than the two vertices as an apex of $F_\circ^*$. We set these three vertices as the apices of $F_\circ^*$. (The triconnected component decomposition tree T of G' for the graph G in Fig. 1 has two leaves, and for example the three vertices depicted as white circles are chosen as outer apices.)

We finally claim that the k-gon $F_\circ^*$ defined above is extendable. We shall show that Conditions (a)–(c) in Lemma 1 hold for $F_\circ^*$. Since G' is internally triconnected, both Conditions (a) and (c) hold. It thus suffices to show that Condition (b) holds for $F_\circ^*$.

Suppose for a contradiction that Condition (b) does not hold for F_o^*. Let P_1, $P_2, \cdots, P_k$ be the k paths in F_o^*, each corresponding to a side of the polygon F_o^*. Since Condition (b) does not hold for F_o^*, one of the following two cases occurs.
Case (i): $G' - F_o(G')$ *has a connected component H such that all the outer vertices adjacent to vertices in H lie on a single path P_i, as illustrated in Fig. 3(a).*
Case (ii): *two outer vertices in a path P_i are joined by an inner edge, as illustrated in Fig. 3(b).*

We first consider Case (i). Since G' is internally triconnected, H is contained in a triconnected component H_j of G' and H_j must be a leaf of T. Then none of the outer vertices in H_j other than the separation pair would be chosen as an outer apices, contrary to the definition of F_o^*.

We then consider Case (ii). Let (u, v) be the innermost one among all the inner edges satisfying Case (ii), and let H_j be the triconnected component containing edge (u, v). Then $\{u, v\}$ is a separation pair of G' and H_j is a leaf of T. None of the outer vertices in H_j other than u and v is an apex of F_o^*, contrary to the definition of F_o^*.   □

Given a plane graph G together with an extendable outer convex polygon F_o^*, one can find in linear time a convex drawing of G which is an extension of F_o^* [1, 7]. Therefore, by Theorem 1 and Lemmas 2, 4 and 5, we have the following theorem.

Theorem 2. Let G be an internally triconnected graph, let T be a triconnected component decomposition tree of a contracted graph G' of G, and let n_l be the number of leaves of T. Then one can find a convex drawing of G having the minimum number of outer apices in linear time, and the minimum number is equal to $\max\{3, n_l\}$.

4   Conclusions

In this paper, we newly formalize a necessary and sufficient condition for a plane graph G to have a convex drawing, and show that a convex drawing having the minimum number of outer apices can be found in linear time.

In a convex grid drawing, all the vertices of G are put on grid points. It is known that if either G is triconnected or a triconnected component decomposition tree T of a contracted graph G' of G has at most three leaves then G has a convex grid drawing of size $(n - 1) \times (n - 1)$ where n is the number of vertices in G [2, 6]. The remaining problem is to obtain a good upper bound on the size of a convex grid drawing of G for which T has four or more leaves.

References

1. N. Chiba, T. Yamanouchi and T. Nishizeki, *Linear algorithms for convex drawings of planar graphs*, in Progress in Graph Theory, J. A. Bondy and U. S. R. Murty (Eds.), Academic Press, pp. 153-173 (1984).
2. M. Chrobak and G. Kant, *Convex grid drawings of 3-connected planar graphs*, International Journal of Computational Geometry and Applications, 7, pp. 211-223 (1997).

3. H. de Fraysseix, J. Pach and R. Pollack, *How to draw a planar graph on a grid*, Combinatorica, 10, pp. 41-51 (1990).
4. G. Di Battista, P. Eades, R. Tamassia and I. G. Tollis, *Graph Drawing*, Prentice Hall, NJ (1999).
5. J. E. Hopcroft and R. E. Tarjan, *Dividing a graph into triconnected components*, SIAM J. Compt., 2, 3, pp. 135-138 (1973).
6. K. Miura, M. Azuma and T. Nishizeki, *Canonical decomposition, realizer, Schnyder labeling and orderly spanning trees of plane graphs*, International Journal of Fundations of Computer Science, 16, 1, pp. 117-141 (2005).
7. T. Nishizeki and Md. S. Rahman, *Planar Graph Drawing*, World Scientific, Singapore (2004).
8. W. Schnyder, *Embedding planar graphs on the grid*, Proc. 1st Annual ACM-SIAM Symp. on Discrete Algorithms, San Francisco, pp. 138-147 (1990).
9. C. Thomassen, *Plane representations of graphs*, in Progress in Graph Theory, J. A. Bondy and U. S. R. Murty (Eds.), Academic Press, pp. 43–69 (1984).

Energy-Based Clustering
of Graphs with Nonuniform Degrees

Andreas Noack

Institute of Computer Science,
Brandenburg University of Technology at Cottbus,
PO Box 10 13 44, 03013 Cottbus, Germany
an@informatik.tu-cottbus.de

Abstract. Widely varying node degrees occur in software dependency graphs, hyperlink structures, social networks, and many other real-world graphs. Finding dense subgraphs in such graphs is of great practical interest, as these clusters may correspond to cohesive software modules, semantically related documents, and groups of friends or collaborators. Many existing clustering criteria and energy models are biased towards clustering together nodes with high degrees. In this paper, we introduce a clustering criterion based on normalizing cuts with edge numbers (instead of node numbers), and a corresponding energy model based on edge repulsion (instead of node repulsion) that reveal clusters without this bias.

1 Introduction

It is increasingly recognized that the degrees of the nodes in many graph models of real-world systems vary widely [1], with examples including dependencies between software artifacts, citations of scientific articles, hyperlink structures (like the World Wide Web, dictionaries, and thesauri), social networks, and neural networks. Dense subgraphs of these graphs are of great scientific and practical interest, because these clusters are candidates for cohesive software modules, research areas, semantically related terms or documents, groups of closely interacting people, and functional units of the nervous system.

The first challenge in the identification of such clusters is to formalize the notion of a cluster. Section 2 shows that several existing cut-based clustering criteria are biased towards certain cluster sizes, and derives two unbiased clustering criteria by appropriately normalizing the cut. There are two unbiased clustering criteria because the two natural measures of cluster size, namely the number of nodes and the number of edges, are equivalent (up to a constant factor) only for graphs with uniform degrees.

The second challenge is the computation and the presentation of the clusters. Section 3 introduces two energy models that reveal the clusters corresponding to the two clustering criteria. This enables the computation of clusters with existing energy minimization algorithms (like the algorithm of Barnes and Hut [3, 19]) that scale to graphs with thousands of nodes. The presentation as graph drawing facilitates the comprehension of the cluster structure, because viewers naturally interpret closely positioned nodes as strongly related [4, 6]. Section 4 presents example drawings of various real-world graphs.

P. Healy and N.S. Nikolov (Eds.): GD 2005, LNCS 3843, pp. 309–320, 2005.

1.1 Basic Definitions

For a set M, let $|M|$ be the number of elements of M, and let $M^{(2)}$ be the set of all subsets of M which have exactly two elements. A *bipartition* of a set M is a pair (M_1, M_2) of sets with $M_1 \cup M_2 = M$, $M_1 \cap M_2 = \emptyset$, $M_1 \neq \emptyset$, and $M_2 \neq \emptyset$.

A *graph* $G = (V, E)$ consists of a finite set V of *nodes* and a finite set E of *edges* with $E \subseteq V^{(2)}$. Because drawings can be computed separately for different components of a graph, we restrict ourselves to connected graphs, i.e. graphs where every pair of nodes is connected by a path.

For a node v, the *degree* $\deg(v)$ is the number $|\{u \mid \{u, v\} \in E\}|$ of nodes adjacent to v. The total degree $\sum_{v \in V_1} \deg(v)$ of all nodes in a set V_1 is denoted by $\deg(V_1)$. For two sets of nodes V_1 and V_2, the number of edges $|\{\{v_1, v_2\} \in E \mid v_1 \in V_1, v_2 \in V_2\}|$ between V_1 and V_2 is called the *cut* between V_1 and V_2 and denoted by $\mathrm{cut}(V_1, V_2)$. We often identify a set of nodes V_1 with the subgraph $(V_1, \{e \in E \mid e \subseteq V_1\})$ it induces.

A *d-dimensional drawing* of the graph G is a vector $p = (p_v)_{v \in V}$ of node positions $p_v \in \mathrm{IR}^d$. For a drawing p and two nodes $u, v \in V$, the length of the difference vector $p_v - p_u$ is called the *distance* of u and v in p and denoted by $\|p_v - p_u\|$.

2 Graph Clustering Criteria

Informally, we denote by a graph cluster a subgraph with many internal edges and few edges to the remaining graph. This can be formalized by defining a measure for the coupling between subgraphs, such that a smaller coupling corresponds to a better clustering. This section discusses such measures, starting with the cut. The main result is that the cut is biased, and has to be normalized with the sizes of the subgraphs. For graphs with uniform degrees, normalizing the cut with the number of nodes of the subgraphs is equivalent to normalizing the cut with the number of edges, but for graphs with nonuniform degrees, these two alternatives lead to considerably different notions of a cluster. For clarity, the discussion is restricted to the coupling between two subgraphs, the generalization to more subgraphs is straightforward.

2.1 The Cut

A simple measure of the coupling between two disjoint sets of nodes V_1 and V_2 of a graph (V, E) is their cut $\mathrm{cut}(V_1, V_2)$. There exist efficient algorithms for finding a bipartition of a given graph with the minimum cut [22].

However, the cut prefers bipartitions that consist of a very small and a very large subgraph, as the following calculation shows. Among the $\frac{1}{2}(|V|^2 - |V|)$ unordered pairs of nodes from V, there are $|V_1| \cdot |V_2|$ pairs of one node from V_1 and one node from V_2. So the expected cut between V_1 and V_2 is $\frac{2|V_1| \cdot |V_2|}{|V|^2 - |V|} |E|$, which is much smaller for bipartitions with $|V_1| \ll |V_2|$ than for bipartitions with $|V_1| = |V_2|$.

2.2 The Node-Normalized Cut

An unbiased measure of the coupling between two disjoint sets of nodes V_1 and V_2 called *node-normalized cut* is obtained by normalizing the cut with the expected cut (and ignoring constant factors for simplicity):

$$\text{nodenormcut}(V_1, V_2) = \frac{\text{cut}(V_1, V_2)}{|V_1| \cdot |V_2|}$$

For a fixed graph (V, E) and all clusters sizes $|V_1|$ and $|V_2|$, the node-normalized cut has the same expected value $\frac{2|E|}{|V|^2 - |V|}$.

This measure is also known as ratio of the cut, and has been used in VLSI design [2] and software engineering [16]. Computing a bipartition with minimum node-normalized cut is NP-complete, but approximable in polynomial time within factor $O(\log(|V|))$ [15].

The node-normalized cut is still biased towards bipartitions with a very small and a very large subgraph if the number of edges is used as measure of subgraph size. Consider two bipartitions of the set of nodes V into two sets V_1 and V_2 of equal cardinality, where $\deg(V_1) = \deg(V_2)$ in the first bipartition, and $\deg(V_1) \ll \deg(V_2)$ in the second bipartition. (Note that such bipartitions only exist in graphs with nonuniform degrees.) Then the expected cut, and therefore the node-normalized cut, is much larger for the first bipartition than for the second.

The following calculation makes this more precise. The $|E|$ edges of a graph (V, E) have $\deg(V) = 2|E|$ end nodes. So there are $\frac{1}{2}\left(\deg(V)^2 - \sum_{v \in V} \deg(v)^2\right)$ unordered pairs of end nodes. (The subtrahend accounts for "pairs" of two equal end nodes.) Among these pairs, there are $\deg(V_1)\deg(V_2)$ pairs of one node from V_1 and one node from V_2. So the expected cut between $|V_1|$ and $|V_2|$ is $\frac{2\deg(V_1)\deg(V_2)}{\deg(V)^2 - \sum_{v \in V}\deg(v)^2}|E|$, which is much smaller for bipartitions with $\deg(V_1) \ll \deg(V_2)$ than for bipartitions with $\deg(V_1) = \deg(V_2)$.

2.3   The Edge-Normalized Cut

Normalizing the cut with the expected cut (without constant factors) results in another measure of coupling called *edge-normalized cut*:

$$\text{edgenormcut}(V_1, V_2) = \frac{\text{cut}(V_1, V_2)}{\deg(V_1)\deg(V_2)}$$

For a fixed graph (V, E) and all clusters sizes $\deg(V_1)$ and $\deg(V_2)$, the edge-normalized cut has the same expected value $\frac{2|E|}{\deg(V)^2 - \sum_{v \in V}\deg(v)^2}$.

A similar measure has been introduced (without a systematic derivation) by Shi and Malik [20] as normalized cut:

$$\text{ncut}(V_1, V_2) = \frac{\text{cut}(V_1, V_2)}{\deg(V_1)} + \frac{\text{cut}(V_1, V_2)}{\deg(V_2)}.$$

Because $(\deg(V_1) + \deg(V_2))\,\text{edgenormcut}(V_1, V_2) = \text{ncut}(V_1, V_2)$, the values of the two measures differ only by the constant factor $\deg(V)$ if $V_1 \cup V_2 = V$. The problem of deciding whether a given graph has a bipartition with an edge-normalized cut smaller than a given constant is NP-complete [20].

2.4   Related Work: Other Measures of Coupling

Other measures of the coupling between two disjoint sets of nodes V_1 and V_2 of a graph (V, E) include the expansion [14]

$$\text{expansion}(V_1, V_2) = \frac{\text{cut}(V_1, V_2)}{\min(|V_1|, |V_2|)}$$

and the conductance [14]

$$\text{conductance}(V_1, V_2) = \frac{\text{cut}(V_1, V_2)}{\min(\deg(V_1), \deg(V_2))}.$$

Computing a bipartition with minimum expansion is NP-complete, but approximable in polynomial time within factor $O(\log(|V|))$ [15].

The expansion is biased towards similarly-sized clusters: For $|V_1| = |V| - 1$ and $|V_2| = 1$, the expected expansion is $\frac{2|E|}{|V|}$, while for $|V_1| = |V_2| = \frac{1}{2}|V|$, the expected expansion is only $\frac{|E|}{|V|-1}$. The conductance has a similar bias when the total degree is used as measure of cluster size.

3   Energy Models for Graph Clustering

One particular way to compute and present the cluster structure of graphs is energy-based graph drawing. That the results are drawings and not partitions of the set of nodes has several benefits: Drawings facilitate the comprehension of the cluster structure, because viewers naturally interpret closely positioned nodes as strongly related [4, 6], and enable the navigation from one cluster to closely related clusters. Drawings show how clearly clusters are separated, and how closely nodes are associated with their cluster.

In an earlier paper [17], we introduced the LinLog energy model for visualizing clusters with respect to the node-normalized cut. The main result of this section is that replacing repulsion between nodes with repulsion between edges adapts the LinLog model to the edge-normalized cut (and thus to graphs with nonuniform degrees).

3.1   The Edge-Repulsion LinLog Energy Model

The *node-repulsion LinLog energy* of a drawing p is defined in [17] as

$$U_{NodeLinLog}(p) = \sum\nolimits_{\{u,v\} \in E} ||p_u - p_v|| - \sum\nolimits_{\{u,v\} \in V^{(2)}} \ln ||p_u - p_v||$$

To avoid infinite energies we assume that different nodes have different positions, which is no serious restriction because we are interested in drawings with low energy. The first term of the difference can be interpreted as attraction between adjacent nodes, the second term as repulsion between different nodes.

In the *edge-repulsion LinLog energy model* the repulsion between nodes is replaced by repulsion between edges. In our formalization, the repulsion does not act between entire edges, but only between their end nodes. So the repulsion between two nodes is weighted by the number of edges of which they are an end node, i.e. by their degrees:

$$U_{EdgeLinLog}(p) = \sum\nolimits_{\{u,v\} \in E} ||p_u - p_v|| - \sum\nolimits_{\{u,v\} \in V^{(2)}} \deg(u) \deg(v) \ln ||p_u - p_v||$$

The beauty of edge repulsion lies in its symmetry: Edges cause both attraction and repulsion. In other words, nodes that attract strongly also repulse strongly. More precisely, each node has consistently – in terms of attraction and repulsion – an influence on the drawing proportional to its degree. (This can be visualized by setting the size of a node to its degree, as in the figures in Sect. 4.) As a beneficial side effect, this symmetry can also facilitate the introduction and weighting of additional forces [18].

In a node-repulsion LinLog drawing of a graph with very nonuniform degrees, the positions of the nodes mainly reflect their degrees: The (strongly attracting) high-degree nodes are mostly placed at the center, and the (weakly attracting, but equally repulsing) low-degree nodes at the borders. This bias is removed in the edge-repulsion LinLog model. For graphs with uniform node degrees, both models have equivalent minima up to scaling.

3.2 Interpretation of Edge-Repulsion LinLog Drawings

The theorems and proofs about the interpretation of node-repulsion LinLog drawings in [17] can be adapted to edge-repulsion LinLog. This subsection only presents a simplified version to illustrate the difference between node repulsion and edge repulsion.

Let $G = (V, E)$ be a graph, and let (V_1, V_2) be a bipartition of the set of nodes V into two cohesive (dense), loosely coupled subgraphs. Let p be a drawing of G with minimum edge-repulsion LinLog energy. How is the distance of V_1 and V_2 in p related to their coupling?

Due to the high cohesion and low coupling, the distances *within* V_1 and *within* V_2 should be much smaller than the distance *between* V_1 and V_2 in p. For our discussion, this situation can be reasonably closely approximated by assuming that all nodes in V_1 have the same position and all nodes in V_2 have the same position in p. Let d be the Euclidean distance of these two positions.

Ignoring the energy between nodes of the same subgraph (which is irrelevant for the distance d between the subgraphs), we obtain the following edge-repulsion LinLog energy of the drawing p:

$$U(d) = \mathrm{cut}(V_1, V_2)\, d - \deg(V_1) \deg(V_2) \ln d$$

Because p is a drawing with minimum energy, this function has a global minimum at d, so $U'(d) = 0$.

$$0 = U'(d) = \mathrm{cut}(V_1, V_2) - \deg(V_1) \deg(V_2)/d$$

$$d = \frac{\deg(V_1) \deg(V_2)}{\mathrm{cut}(V_1, V_2)} = \frac{1}{\mathrm{edgenormcut}(V_1, V_2)}$$

So the distance d between V_1 and V_2 in the drawing with minimum edge-repulsion LinLog energy is the inverse of their edge-normalized cut. For the node-repulsion LinLog energy model, we only need to replace $\deg(V_1) \deg(V_2)$ with $|V_1| \cdot |V_2|$ in all terms, so the distance is the inverse node-normalized cut.

This simple analysis method is not meant to replace a more detailed examination (as done in [17] for node-repulsion LinLog), but it allows a quick approximate assessment of the clustering properties for many energy models.

3.3 Related Work

Energy Models for Clustering. The force and energy models of Eades [7], Fruchterman and Reingold [8], Davidson and Harel [5], and Kamada and Kawai [13] tend to

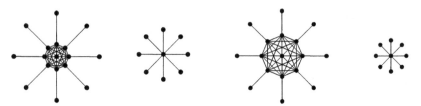

(a) (Node-repulsion) Fruchterman-Reingold   (b) Edge-repulsion Fruchterman-Reingold

Fig. 1. Two small graphs

enforce uniform (or other given) edge lengths, to support tasks like following paths and identifying neighbors. The LinLog energy models reveal clusters, which generally requires some long (between-cluster) and short (within-cluster) edges. So the two goals of clustering and uniform edge lengths are contradictory and cannot be achieved with a single energy model. But classes of energy models like r-PolyLog [17] allow the user to choose any compromise.

Edge Repulsion. In many force and energy models, including those of Eades [7] and Fruchterman and Reingold [8], adjacent nodes attract and all pairs of nodes repulse. Like node-repulsion LinLog, these models tend to draw dense subgraphs too small (because attraction dominates repulsion) and sparse subgraphs too large.

Figure 1a shows examples for the Fruchterman-Reingold model: The complete subgraph of the left graph contains most edges, but uses only a small part of the drawing area. Much area is wasted by the unnecessarily long edges to the eight peripheral nodes. The (sparse) right graph is drawn much larger than the (dense) complete subgraph, although it contains much fewer edges. Further examples are given in Sect. 4.

Like for LinLog, replacing node repulsion with edge repulsion improves the balance between attraction and repulsion, because both are caused by the edges. Figure 1b shows that this leads to a more uniform information density and thus better readability.

A related concept is the repulsive force between edges and nodes proposed by Davidson and Harel [5]. This force was introduced exclusively for improving readability, and not for enabling interpretations with respect to the cluster structure.

Algorithms for Energy Minimization. As usual in force- and energy-based graph drawing (with the exception of Hall's energy model [11]), we have no practical algorithm that finds global minima of the LinLog energy models. In our experiments we use the hierarchical energy minimization algorithm of Barnes and Hut [3], which was introduced to graph drawing by Quigley and Eades [19]. Its runtime is in $O(|E| + |V| \log |V|)$ per iteration. The overall runtime grows somewhat faster because the number of iterations needed for convergence tends to grow with $|V|$. Some other efficient minimization algorithms are not expected to find good energy minima for clustering energy models like LinLog and for graphs with small diameter [9, 12, 21, 10].

4   Examples

This section shows example drawings of the edge-repulsion LinLog energy model, and, for comparison, of the node-repulsion LinLog energy model and the well-known

Fruchterman-Reingold force model [8]. The first subsection illustrates the differences between the models with drawings of a pseudo-random graph. The second subsection shows that drawings of the edge-repulsion LinLog model can provide non-trivial and useful insights into the structure of real-world graphs.

In all figures, the area of each circle that represents a node is proportional to the degree of the node, with the exception that there is a minimum area to ensure visibility. Some drawings were rotated manually. (Rotation does not change the energy.) In most drawings, the edges are omitted to avoid clutter.

An effective visualization of large graphs requires panning and zooming, and inter-active showing and hiding of node labels and edges. Therefore we provide VRML files (offering the first three features) of the drawings on a supplementary web page[1].

4.1 Pseudo-Random Graph

Figure 2 shows a pseudo-random graph with eight cluster of 50 nodes. The probability of an edge $\{u, v\}$ is

- 1 if u and v belong to the same of the first four clusters,
- 0.5 if u and v belong to the same of the second four clusters,
- 0.2 if u and v belong to different of the first four clusters,
- 0.05 if u and v belong to different of the second four clusters, and
- 0.1 if u belongs to one of the first and v belongs to one of the second four clusters.

Both LinLog models reveal the clusters, but their drawings differ because the degrees of the nodes are nonuniform. The node-repulsion LinLog drawing places the first four clusters more closely than the second four clusters, which reflects that node-normalized cuts between the first four clusters are higher than between the second four clusters. In the edge-repulsion LinLog drawing the distances between all clusters are similar, which reflects that the edge-normalized cuts between all pairs of clusters are similar.

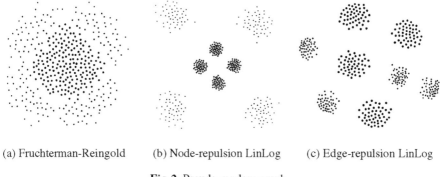

(a) Fruchterman-Reingold   (b) Node-repulsion LinLog   (c) Edge-repulsion LinLog

Fig. 2. Pseudo-random graph

---

[1] http://www-sst.informatik.tu-cottbus.de/GD/erlinlog.html

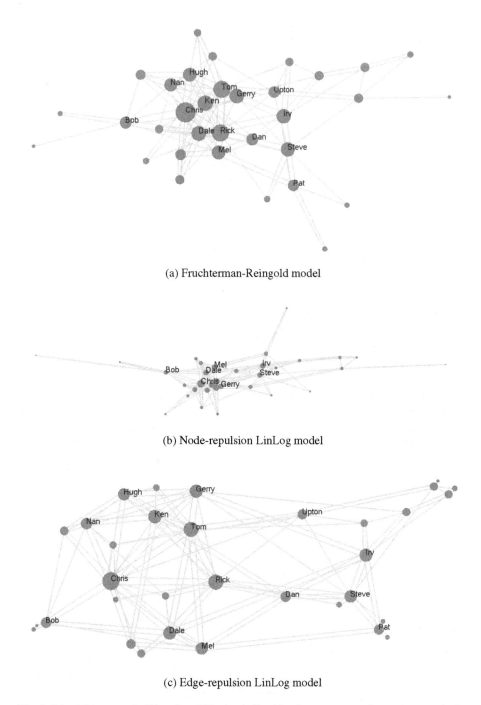

(a) Fruchterman-Reingold model

(b) Node-repulsion LinLog model

(c) Edge-repulsion LinLog model

Fig. 3. Friendship network (33 nodes, 147 edges). Double edges correspond to reciprocated relationships, single edges to non-reciprocated relationships.

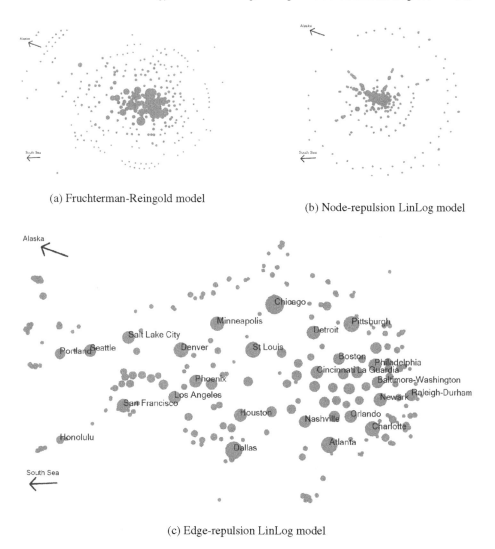

(a) Fruchterman-Reingold model

(b) Node-repulsion LinLog model

(c) Edge-repulsion LinLog model

Fig. 4. Direct flights between US airports (332 nodes, 2126 edges). The airports in Alaska and the South Sea (e.g. Guam) are omitted to improve readability.

4.2 Real-World Graphs

The graphs in Fig. 3 to 5 were obtained from the Pajek project[2]. In the drawings of the Fruchterman-Reingold model (Fig. 3a to 5a) and the node-repulsion LinLog model (Fig. 3b to 5b), nodes with high degree are placed in the center, and nodes with low degree near the borders. So the positions of the nodes mainly reflect their degree.

---

[2] http://vlado.fmf.uni-lj.si/pub/networks/data/

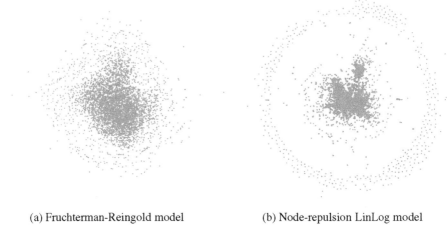

(a) Fruchterman-Reingold model   (b) Node-repulsion LinLog model

(c) Edge-repulsion LinLog model

Fig. 5. Hyperlinks between terms in the Online Dictionary for Library and Information Science ODLIS (2896 nodes, 18238 edges)

In Fig. 3, only the edge-repulsion LinLog drawing clearly reflects that there are two groups of friends – the left group around Chris and Rick and the right group around Steve and Irv – which are mainly connected by Upton and Dan.

Figure 4c shows that the edge-repulsion LinLog model discovers (roughly) the relative geographical locations of the US airports from the airline routing graph. Besides providing insights into the structure of the airline routes, this example impressively shows that the LinLog model can discover non-obvious knowledge in graphs.

The edge-repulsion LinLog drawing of the Online Dictionary for Library and Information Science (ODLIS) is shown in Fig. 5c, but the VRML file on the supplementary web page[3] gives a better impression how well semantically related terms are grouped on all scales. Such grouping is useful e.g. for discovering the global topic areas (like publishing, printing, computer science, etc.), identifying entry points for the exploration of topics, or finding semantically related terms even if they are not explicitly linked.

Drawings of three additional graphs are provided on the supplementary web page[3]. The grouping of papers from the Graph Drawing symposium in a drawing of the citation graph reflects research areas. However, there is some noise in the drawing because many papers have too few citations to be clearly assigned to a group. Again, such drawings have many applications, from getting an overview of the field with its subfields and landmark papers to identifying related papers even if they have no direct citation relationship. A drawing of Roget's thesaurus provides a nice map of (parts of) the English language by grouping semantically related categories, with benefits similar to the ODLIS visualization. The third drawing reflects how often files of a software system changed together in the development process. Because changes should be localized in subsystems, groups of files in this graph help to decompose the system into subsystems or to improve an existing subsystem hierarchy.

5   Conclusion

Cut-based measures for the coupling of subgraphs should be normalized with the size of the subgraphs to avoid biases. For graphs with nonuniform degrees, the number of edges is often a more appropriate measure of the size of subgraphs than the number of nodes. (For uniform degrees, both are equivalent.) Accordingly, energy models should use edge repulsion instead of (or in addition to) node repulsion to avoid dense accumulations of nodes with high degrees. In drawings of one such energy model, called edge-repulsion LinLog, the distance of groups of nodes is approximately inversely proportional to their coupling. Drawings of this energy model can provide deep and useful insights into the structure of real-world graphs from various domains, which are not possible with previous energy models.

References

1. Réka Albert and Albert-László Barabási. Statistical mechanics of complex networks. *Reviews of Modern Physics*, 74(1):47–97, 2002.
2. Charles J. Alpert and Andrew B. Kahng. Recent directions in netlist partitioning: A survey. *Integration, the VLSI Journal*, 19(1-2):1–81, 1995.

---

[3] http://www-sst.informatik.tu-cottbus.de/GD/erlinlog.html

3. Josh Barnes and Piet Hut. A hierarchical O(N log N) force-calculation algorithm. *Nature*, 324:446–449, 1986.
4. Jim Blythe, Cathleen McGrath, and David Krackhardt. The effect of graph layout on inference from social network data. In *Proc. GD 1995*, pages 40–51. Springer-Verlag, 1996.
5. Ron Davidson and David Harel. Drawing graphs nicely using simulated annealing. *ACM Transactions on Graphics*, 15(4):301–331, 1996.
6. Edmund Dengler and William Cowan. Human perception of laid-out graphs. In *Proc. GD 1998*, pages 441–443. Springer-Verlag, 1998.
7. Peter Eades. A heuristic for graph drawing. *Congressus Numerantium*, 42:149–160, 1984.
8. Thomas M. J. Fruchterman and Edward M. Reingold. Graph drawing by force-directed placement. *Software – Practice and Experience*, 21(11):1129–1164, 1991.
9. Pawel Gajer, Michael T. Goodrich, and Stephen G. Kobourov. A multi-dimensional approach to force-directed layouts of large graphs. In *Proc. GD 2000*, pages 211–221. Springer-Verlag, 2001.
10. Stefan Hachul and Michael Jünger. Drawing large graphs with a potential-field-based multi-level algorithm. In *Proc. GD 2004*, pages 285–295. Springer-Verlag, 2004.
11. Kenneth M. Hall. An r-dimensional quadratic placement algorithm. *Management Science*, 17(3):219–229, 1970.
12. David Harel and Yehuda Koren. A fast multi-scale method for drawing large graphs. In *Proc. GD 2000*, pages 183–196. Springer-Verlag, 2001.
13. Tomihisa Kamada and Satoru Kawai. An algorithm for drawing general undirected graphs. *Information Processing Letters*, 31(1):7–15, 1989.
14. Ravi Kannan, Santosh Vempala, and Adrian Vetta. On clusterings: Good, bad and spectral. *Journal of the ACM*, 51(3):497–515, 2004.
15. Tom Leighton and Satish Rao. An approximate max-flow min-cut theorem for uniform multicommodity flow problems with applications to approximation algorithms. In *Proc. 29th Annual Symposium on Foundations of Computer Science (FOCS 1988)*, pages 422–431. IEEE, 1988.
16. S. Mancoridis, B. S. Mitchell, C. Rorres, Y. Chen, and E. R. Gansner. Using automatic clustering to produce high-level system organizations of source code. In *Proc. 6th IEEE International Workshop on Program Comprehension (IWPC 1998)*, pages 45–52. IEEE, 1998.
17. Andreas Noack. An energy model for visual graph clustering. In *Proc. GD 2003*, pages 425–436. Springer-Verlag, 2004.
18. Andreas Noack and Claus Lewerentz. A space of layout styles for hierarchical graph models of software systems. In *Proc. 2nd ACM Symposium on Software Visualization (SoftVis 2005)*, pages 155–164. ACM, 2005.
19. Aaron J. Quigley and Peter Eades. FADE: Graph drawing, clustering, and visual abstraction. In *Proc. GD 2000*, pages 197–210. Springer-Verlag, 2001.
20. Jianbo Shi and Jitendra Malik. Normalized cuts and image segmentation. *IEEE Transaction on Pattern Analysis and Machine Intelligence*, 22(8):888–905, 2000.
21. Chris Walshaw. A multilevel algorithm for force-directed graph drawing. In *Proc. GD 2000*, pages 171–182. Springer-Verlag, 2001.
22. Zhenyu Wu and Richard Leahy. An optimal graph theoretic approach to data clustering: Theory and its application to image segmentation. *IEEE Transaction on Pattern Analysis and Machine Intelligence*, 15(11):1101–1113, 1993.

A Mixed-Integer Program
for Drawing High-Quality Metro Maps*

Martin Nöllenburg and Alexander Wolff

Fakultät für Informatik, Universität Karlsruhe, P.O. Box 6980, D-76128 Karlsruhe
http://i11www.ira.uka.de/algo/group

Abstract. In this paper we investigate the problem of drawing metro maps which is defined as follows. Given a planar graph G of maximum degree 8 with its embedding and vertex locations (e.g. the physical location of the tracks and stations of a metro system) and a set $\mathcal{L}$ of paths or cycles in G (e.g. metro lines), draw G and $\mathcal{L}$ *nicely*. We first specify the niceness of a drawing by listing a number of *hard* and *soft* constraints. Then we present a mixed-integer program (MIP) which always finds a drawing that fulfills all hard constraints (if such a drawing exists) and optimizes a weighted sum of costs corresponding to the soft constraints. We also describe some heuristics that speed up the MIP. We have implemented both the MIP and the heuristics. We compare their output to that of previous algorithms for drawing metro maps and to official metro maps drawn by graphic designers.

1   Introduction

A metro map is a schematic drawing of the underlying geographic network that represents the different stations and metro lines of a metro system. The users of a metro map are the passengers of the public transport system. They want to quickly answer questions like "How do I get from A to B?" or "After how many stops do I have to change trains?". Thus the layout of a metro map must be as clear as possible whereas exact geometry or scale is less important. The problem of drawing maps of metro systems and other means of public transportation is an interesting compromise between schematic road maps [4] where vertex positions are (mostly) fixed and "conventional" graph drawing where vertices can go anywhere. The first approach maximizes maintenance of the user's mental map, the second approach maximizes esthetics. The mother of all modern metro maps is Henry Beck's 1933 map of the London Underground. In the meantime, graphic designers have come up with different layout styles all over the world [8].

After studying a large number of real-world metro maps [8] we formalized the problem of drawing high-quality metro maps as follows. As usual we say that an embedding of a graph G associates to each vertex a list of its adjacent vertices in clockwise order. We say that a set $\mathcal{L}$ of paths and cycles of G is a *line cover* of G if each edge of G belongs to at least one element of $\mathcal{L}$. Now the *metro-map layout*

---

* Work supported by grant WO 758/4-2 of the German Science Foundation (DFG).

P. Healy and N.S. Nikolov (Eds.): GD 2005, LNCS 3843, pp. 321–333, 2005.

problem is the following. Given (a) a planar graph G of maximum degree 8, the *metro graph*, (b) the embedding of G, (c) for each vertex v its location $\pi(v)$ in the plane, and (d) a line cover $\mathcal{L}$ of G, the *metro lines*, find a *nice* drawing μ of G and $\mathcal{L}$. In order to be nice, μ must fulfill a number of *hard* constraints:

(H1) μ must respect the topology of G,
(H2) all edges of $\mu(G)$ must be *octilinear* line segments, i.e. parallel to one of the two coordinate axes or to either of their two bisectors,
(H3) each edge e in $\mu(G)$ has a minimum length ℓ_e, and
(H4) each edge in $\mu(G)$ has a certain minimum distance $d_{\min}$ from each non-incident edge.

Moreover, μ should conform to a number of *soft* constraints as tightly as possible:

(S1) the paths and cycles in $\mu(\mathcal{L})$ should have few bends,
(S2) the total edge length of $\mu(G)$ should be small, and
(S3) for each pair of adjacent vertices (u, v) their *relative position* should be preserved, i.e. the angle $\angle(\mu(u), \mu(v))$ should be similar to the $\angle(\pi(u), \pi(v))$, where $\angle(a, b)$ is the angle between the x-axis in positive direction and the line through a and b directed from a to b.

Note that if the embedding of a metro graph is not planar, this can be achieved by introducing dummy vertices at crossings, of which there are usually not many. We denote the number of vertices (including dummy vertices) and edges of G by n and m, respectively. Let m' be the total number of edges of the paths and cycles in $\mathcal{L}$. We have $m' \geq m$.

While the need for most of the above constraints is immediate, constraint (S3) may need a few explanatory words. The intuition behind requiring the preservation of the relative position is that users of metro system usually have a certain notion of compass directions above ground. Suppose a passenger is in $\pi(u)$ and wants to go to the adjacent metro station $\pi(v)$, which he knows to lie south of $\pi(u)$, then he would be confused if $\mu(u)$ was north of $\mu(v)$. Thus ensuring that the two angles in constraint (S3) do not deviate too much, say by no more than 90 degrees, can be seen as a hard constraint, while it seems to be appropriate to model smaller deviations as a soft constraint, e.g. by charging a cost proportional to the deviation. Our framework reflects this ambivalence, but modeling relative position as a purely soft constraint is also possible, see Sect. 3.1 and 3.6.

Compared to the orthogonal drawing of (embedded) graphs, the introduction of diagonal directions yields drawings that are more similar to the original embedding. In addition, the maximum vertex degree increases from 4 to 8. However, in contrast to the existence of several efficient algorithms for orthogonal drawings [12, 5], the problem becomes NP-complete in the octilinear case as we show in [7]. This partially motivates why we do not follow the topology-shape-metric approach [5] for orthogonal graph drawing: while we could compute a minimum-bend octilinear shape of a metro graph in polynomial time using Tamassia's flow model [12], we cannot efficiently embed the resulting shape without creating crossings even if an octilinear layout exists.

Therefore we decided to model the metro-map layout problem as a MIP, see Sect. 3. This gave us the necessary flexibility to achieve the following. If a layout that conforms to all hard constraints exists (which was the case in all examples we tried), then our MIP finds such a layout. Moreover our MIP optimizes the weighted sum of cost functions each of which corresponds to a soft constraint. Our MIP is the first method that guarantees octilinearity, which is essential for a clear layout of metro maps. Our MIP is also the first method dedicated to drawing metro maps that uses global optimization and thus avoids getting trapped in local minima. This contrasts with methods based on local optimization, see Sect. 2. In [7] we extend our model to combine graph drawing with the placement of non-overlapping station labels. Binucci et al. [2] have used a MIP formulations to combine *orthogonal* graph drawing and label placement.

In order to cope with the running time of MIP solvers, we give several heuristics that speed up our basic MIP, see Sect. 3.7. We have implemented an algorithm based on our MIP formulation. In Sect. 4 we present a metro map that our algorithm drew of a real-world metro system and compare it to the output of previous algorithms and to an official metro map.

We stress that our MIP formulation can be used not only for drawing metro maps, but for any kind of technical drawing with a restricted number of directions. Brandes et al. [3] introduced the concept of a *sketch* of a graph. A sketch can be handmade or the physical embedding of a geometric network like the real position of telephone cables. Brandes et al. compute an *orthogonal* drawing of a sketch in $O(n^2 \log n)$ time. However, their method cannot be extended to more directions or to incorporate the concept of metro lines. In contrast, our framework can be used to draw sketches (possibly dropping constraint (S1)) and can be extended to more than eight directions. Other possible extensions include user interaction (e.g. fixing the direction edges or lines), the drawing of maps in a given format, or the minimization of one dimension of the drawing area (instead of constraint (S2)).

2 Previous Work

To the best of our knowledge the first attempt to automate the drawing of metro maps was made by Barkowsky et al. [1]. They use *discrete curve evolution*, i.e. an algorithm for polygonal line simplification, to treat the lines of the Hamburg subway system. However, their algorithm neither restricts the edge directions nor does it increase station distances in the crowded downtown area. Stations are labeled but no effort is made to avoid label overlap.

Hong et al. [6] give five methods for the metro-map layout problem. The most refined of these methods modifies a topology-maintaining spring embedder such that edge weights are taken into account and such that additional magnetic forces draw the straight-line edges towards the closest octilinear direction. In a preprocessing step the metro graph is simplified by contracting each edge that is incident to a degree-2 vertex. After performing all contractions, the weight of each remaining edge is set to the number of original edges it replaces. After the final layout has been computed, all degree-2 vertices are re-inserted into the corresponding edges in an equidistant manner. The contraction step reduces the running time considerably. Station labels are placed in one out of eight directions. While label–label overlaps are avoided, diagonally placed labels sometimes intersect network edges.

Stott and Rodgers [10] draw metro maps using multi-criteria optimization based on hill climbing. For a given layout they define metrics for evaluating the number of edge intersections, the octilinearity and the length of edges, the angular resolution at vertices and the straightness of metro lines. The quality of a layout is a weighted sum over these five metrics. Their iterative optimization process starts with a layout on the integer grid that is obtained from the original embedding. In each iteration they consider alternative grid positions for each vertex within a certain radius. For each of these grid positions they compute the quality of the modified layout. If any of the positions improves the quality of the layout, they move the current vertex to the best position among those that do not change the topology of the layout. They observed typical problems with local minima during their optimization process and give a heuristic fix that overcomes one of these problems. Stott and Rodgers have experimented with enforcing relative position, but report that it does not really improve the results. They can label stations, but do not check for overlaps other than with the edges incident to the current station. They use the same contraction method as Hong et al. [6] to preprocess the input graph.

The main advantage of our method over its predecessors is that we *guarantee* to keep all hard constraints (among them octilinearity) and that we avoid the problem of local optima.

Interestingly enough the layout principles of metro maps have not only been used in a geographic setting. E.g. Sandvad et al. [9] use the *metro-map metaphor* as a way to visualize abstract information related to the Internet.

3   The Basic MIP Model

A MIP consists of two parts: a set of linear constraints and a linear objective function. In Sect. 3.1 to 3.3 we describe four sets of constraints that model the hard constraints (H1)–(H4). We model the simultaneous optimization of the three soft constraints (S1)–(S3) in Sect. 3.4 to 3.6 using a weighted sum of three individual cost functions:

$$\text{Minimize } \lambda_{\text{length}} \, \text{cost}_{\text{length}} + \lambda_{\text{bends}} \, \text{cost}_{\text{bends}} + \lambda_{\text{dir}} \, \text{cost}_{\text{dir}}, \qquad (1)$$

where the variables λ_i are positive user-defined weights, each of which individually emphasizes a certain esthetic criterion. The total number of constraints and variables in our model is of order $O(n + m' + m^2)$. Note that since G is planar we have $m \leq 3n - 6$ due to Euler's formula.

To be able to treat all four edge directions similarly, we use an (x, y, z_1, z_2)-coordinate system as depicted in Fig. 1, where each axis corresponds to one of the four feasible edge directions in the layout. For each vertex v we define $z_1(v) = x(v) + y(v)$ and $z_2(v) = x(v) - y(v)$.

3.1   Octilinearity and Relative Position

Before modeling the constraints we need some notation to address relative positions between vertices and to denote directions of edges. For each vertex v we define a partition of the plane into eight wedge-shaped sectors, numbered from 0 to 7 counterclockwise starting with the positive x-direction as in Fig. 2. To denote the rough relative position between two vertices u, v in the *original layout* we use the terms $\sec_u(v)$ and $\sec_v(u)$ representing the sector relative to u in which v lies and vice versa. Similarly, for each edge $\{u, v\}$, we define a variable $\mathrm{dir}(u, v)$ to denote the octilinear direction of $\{u, v\}$ in the *new layout*.

As mentioned in the introduction, we partially model the soft constraint (S3) as a hard constraint. As a compromise between conservation of relative positions and flexibility to obtain a nice drawing, we allow that an edge is drawn in three different ways. It can be drawn in the direction corresponding to its original sector relative to either endpoint or it can be drawn in the two neighboring directions. Let $\sec_u^{\mathrm{pred}}(v) = \sec_u(v) - 1 \pmod 8$, $\sec_u^{\mathrm{orig}}(v) = \sec_u(v)$ and $\sec_u^{\mathrm{succ}}(v) = \sec_u(v) + 1 \pmod 8$. We now restrict $\mathrm{dir}(u, v)$, which will be used in Sect. 3.5 and 3.6, to the set $\{\sec_u^{\mathrm{pred}}(v), \sec_u^{\mathrm{orig}}(v), \sec_u^{\mathrm{succ}}(v)\}$. This is expressed by the disjunction

$$\bigvee_{i \in \{\mathrm{pred,orig,succ}\}} (\mathrm{dir}(u, v) = \sec_u^i(v) \ \wedge \ \mathrm{dir}(v, u) = \sec_v^i(u)). \tag{2}$$

To model (2) we introduce binary variables $\alpha_{\mathrm{pred}}, \alpha_{\mathrm{orig}}, \alpha_{\mathrm{succ}}$ and the constraint

$$\alpha_{\mathrm{pred}}(u, v) + \alpha_{\mathrm{orig}}(u, v) + \alpha_{\mathrm{succ}}(u, v) = 1 \qquad \forall \{u, v\} \in E. \tag{3}$$

The variable that takes the value 1 will determine the direction in which edge $\{u, v\}$ is drawn, i.e. the term of disjunction (2) that will evaluate to true.

Now we model the correct assignment of $\mathrm{dir}(u, v)$ and $\mathrm{dir}(v, u)$. For each $i \in \{\mathrm{pred, orig, succ}\}$ we have the following set of constraints

$$\begin{aligned} \mathrm{dir}(u, v) - \sec_u^i(v) &\leq M(1 - \alpha_i(u, v)) \\ -\mathrm{dir}(u, v) + \sec_u^i(v) &\leq M(1 - \alpha_i(u, v)) \\ \mathrm{dir}(v, u) - \sec_v^i(u) &\leq M(1 - \alpha_i(u, v)) \\ -\mathrm{dir}(v, u) + \sec_v^i(u) &\leq M(1 - \alpha_i(u, v)) \end{aligned} \qquad \forall \{u, v\} \in E, \tag{4}$$

where the variables of type $\mathrm{dir}(u, v)$ are integers in the range $\{0, \ldots, 7\}$ and M is a large constant. The use of the large constant M in connection with a set of binary variables as in (3) is a standard trick in MIP modeling for formulating a disjunction of constraints. The constant M must be an upper bound on the left-hand sides of the inequalities. Here, if $\alpha_i(u, v) = 0$, the constraints in (4) are trivially fulfilled and do not influence the left-hand sides. On the other hand, if $\alpha_i(u, v) = 1$, the four inequalities are equivalent to $\mathrm{dir}(u, v) = \sec_u^i(v)$ and $\mathrm{dir}(v, u) = \sec_v^i(u)$ as desired (equality constraints have to be transformed into two inequalities when using this trick). Due to (3), $\alpha_i(u, v) = 1$ for exactly one $i \in \{\mathrm{pred, orig, succ}\}$. Thus, exactly one term of the disjunction (2) must be fulfilled.

Further, depending on the actual values of $\sec_u^i(v)$, we add three more constraints for each $i \in \{\text{pred}, \text{orig}, \text{succ}\}$. For example let $\sec_u^{\text{orig}}(v) = 2$ (meaning v is vertically above u in the original layout). Then the constraints are as follows

$$
\begin{aligned}
x(u) - x(v) &\leq M(1 - \alpha_{\text{orig}}(u, v)) \\
-x(u) + x(v) &\leq M(1 - \alpha_{\text{orig}}(u, v)) & \forall\{u, v\} \in E, \qquad (5) \\
y(u) - y(v) &\leq M(1 - \alpha_{\text{orig}}(u, v)) - \ell_{\{u,v\}}
\end{aligned}
$$

where $\ell_{\{u,v\}} > 0$ is the minimum length of edge $\{u, v\}$. If $\alpha_{\text{orig}}(u, v) = 1$, these constraints force u and v to have the same x-coordinate and to keep a vertical distance of at least $\ell_{\{u,v\}}$. This is exactly what is needed for a vertical upward running edge. The other seven possibilities are formulated similarly by forcing one of the coordinates of both vertices to be equal and the distance along the respective direction to be at least $\ell_{\{u,v\}}$. Overall, this part needs $22m$ constraints and $5m$ variables.

3.2   Conservation of the Embedding

To guarantee conservation of the original embedding it suffices to maintain for each vertex $v \in V$ the circular ordering of all incident edges.

Let $N(v) = \{u_1, u_2, \ldots, u_{\deg(v)}\}$ denote the set of all neighbors of v. The counterclockwise ordering of the edges $\{v, u\} \in E$ incident to v implies an ordering on $N(v)$ by identifying each edge $\{v, u\}$ with the vertex u opposite of v. Assume the ordering is $u_1 < u_2 < \cdots < u_{\deg(v)}$. Then in the metro map layout one of these vertices, say u_i, is assigned the smallest direction number from the set of possible directions $\{0, \ldots, 7\}$. All other vertices in $N(v)$ must follow in the same order as before and must have strictly increasing direction numbers: $\dir(v, u_i) < \dir(v, u_{i+1}) < \cdots < \dir(v, u_{i+\deg(v)-1})$, where in the following all indices greater than $\deg(v)$ are considered modulo $\deg(v)$. In other words, all but one of the inequalities $\dir(v, u_1) < \dir(v, u_2), \ldots, \dir(v, u_{\deg(v)-1}) < \dir(v, u_{\deg(v)}), \dir(v, u_{\deg(v)}) < \dir(v, u_1)$ must hold.

In order to determine the vertex with smallest direction number, we again use binary variables as in Sect. 3.1. But instead of using the standard trick to model a disjunction of $\deg(v)$ many terms with $\deg(v) - 1$ constraints each, we make use of the fact that in each case exactly one of the inequalities may be violated while the rest must hold. This requires about a factor $\deg(v)$ less constraints. They are as follows:

$$
\beta_1(v) + \beta_2(v) + \ldots + \beta_{\deg(v)}(v) = 1 \qquad \forall v \in V, \deg(v) \geq 2, \qquad (6)
$$

with binary variables $\beta_i(v)$, and

$$
\begin{aligned}
\dir(v, u_2) - \dir(v, u_1) &\geq -M\beta_1(v) + 1 \\
\dir(v, u_3) - \dir(v, u_2) &\geq -M\beta_2(v) + 1 \\
&\;\;\vdots & \forall v \in V, \deg(v) \geq 2. \quad (7) \\
\dir(v, u_1) - \dir(v, u_{\deg(v)}) &\geq -M\beta_{\deg(v)}(v) + 1
\end{aligned}
$$

The variable β_i that takes value 1 in constraint (6) determines that vertex u_{i+1} has minimum direction number among $N(v)$ by not enforcing $\text{dir}(v, u_{i+1}) - \text{dir}(v, u_i) \geq 1$ in constraints (7). All other binary variables $\beta_j, (j \neq i)$ are set to 0 and thus $\text{dir}(v, u_{j+1}) - \text{dir}(v, u_j) \geq 1$ holds for all $j \neq i$.

These constraints not only enforce that the embedding is preserved but also that no two edges incident to the same vertex can have the same direction. An upper bound on the number of constraints and variables for this part of the MIP is given by $\sum_{v \in V}(\deg(v) + 1) \in O(m)$.

3.3   Planarity

For preserving planarity we have to ensure that certain pairs of edges do not intersect. This can be done in the octilinear setting by distinguishing eight possible relative positions for a pair $\{e_1, e_2\}$ of edges. We express these relative positions using compass orientations. Fixing an edge e_1 a second, non-intersecting edge e_2 can either be placed north, south, east, west or northeast, northwest, southeast, southwest of e_1. For example northeast means in terms of our coordinate system that both vertices incident to e_1 have strictly smaller z_1-coordinates than both vertices incident to e_2. The other relative positions are defined in a similar way.

Clearly, an octilinear drawing is planar if and only if each pair of non-incident edges is placed according to one of the above relative positions. Indeed, we model this disjunctive constraint for all pairs of edges. The constraint

$$\sum_{i \in \{N,S,E,W,NE,NW,SE,SW\}} \gamma_i(e_1, e_2) \geq 1 \qquad \begin{array}{l} \forall (e_1, e_2) \in \binom{E}{2}, \\ e_1, e_2 \text{ not incident,} \end{array} \qquad (8)$$

introduces the variables $\gamma_N, \ldots, \gamma_{SW}$. As an example we now give the constraints for the condition "e_2 is east of e_1"

$$\begin{array}{l} x(u_1) - x(u_2) \leq M(1 - \gamma_E(e_1, e_2)) - d_{\min} \\ x(u_1) - x(v_2) \leq M(1 - \gamma_E(e_1, e_2)) - d_{\min} \\ x(v_1) - x(u_2) \leq M(1 - \gamma_E(e_1, e_2)) - d_{\min} \\ x(v_1) - x(v_2) \leq M(1 - \gamma_E(e_1, e_2)) - d_{\min} \end{array} \qquad \begin{array}{l} \forall (e_1, e_2) \in \binom{E}{2}, \\ e_1, e_2 \text{ not incident.} \end{array} \qquad (9)$$

Recall that $d_{\min}$ is the minimum distance between non-incident edges as given in (H4). Analogously, each of the other seven relative positions is modeled using four constraints each. This amounts to 33 constraints and 8 variables for each edge pair. The problem is that the number of edge pairs is $O(m^2)$. Therefore, we give several heuristics in Sect. 3.7 to reduce the number of constraints that enforce planarity.

3.4   Minimization of Edge Lengths

For modeling the edge lengths one has to specify the underlying metric. We decided to use the L^∞-metric, which defines the distance of two vertices u and v to be $\max(|x(u) - x(v)|, |y(u) - y(v)|)$. We define new real-valued, non-negative variables $D(u, v)$ for all edges $\{u, v\} \in E$ which serve as upper bounds on the lengths of their respective edges. By setting

$$\text{cost}_{\text{length}} = \sum_{\{u,v\} \in E} D(u, v) \tag{10}$$

and by minimizing $\text{cost}_{\text{length}}$, the variables $D(u, v)$ indeed equal the corresponding edge lengths.

The constraints that bound $D(u, v)$ depend on the respective direction of the edge $\{u, v\}$. Note that the actual direction of this edge is determined according to the constraints in Sect. 3.1. Thus we can reuse the binary variables defined in that section to distinguish the three cases for the edge direction. As an example assume that $\sec_u(v) = 1$. Then the constraints are

$$\begin{aligned}
x(v) - x(u) &\leq M(1 - \alpha_{\text{prev}}(u, v)) + D(u, v) \\
x(v) - x(u) &\leq M(1 - \alpha_{\text{real}}(u, v)) + D(u, v) \qquad \forall \{u, v\} \in E. \\
y(v) - y(u) &\leq M(1 - \alpha_{\text{next}}(u, v)) + D(u, v)
\end{aligned} \tag{11}$$

Note that for an edge $\{u, v\}$ drawn diagonally it holds that $|x(u) - x(v)| = |y(u) - y(v)|$. Hence we can use either of the x- or y-coordinates to determine the length $D(u, v)$. Edge lengths for other values of $\sec_u(v)$ are modeled similarly. In total we use m variables and $3m$ constraints.

3.5   Avoiding Bends Along Lines

Clarity in an octilinear drawing depends crucially on the ability to visually follow the metro lines. This can be partially enhanced by using distinguishable colors, but also by avoiding bends along the lines.

We define the bend cost subject to the actual angle between two adjacent edges on a path. Due to the octilinearity constraints and to the fact that two adjacent edges cannot have the same direction relative to their joint vertex the angles can only equal 180, 135, 90, and 45 degrees. In that order we define the corresponding bend cost to be 0, 1, 2, and 3, such that the cost increases with the acuteness of the angle, see Fig. 3.

In our model we can determine the angle between two adjacent edges $\{u, v\}$ and $\{v, w\}$ by using the values of $\text{dir}(u, v)$ and $\text{dir}(v, w)$. For ease of notation let $\Delta\text{dir}(u, v, w) = \text{dir}(u, v) - \text{dir}(v, w)$. Then, the bend cost can be expressed as

$$\text{bend}(u, v, w) = \begin{cases} |\Delta\text{dir}(u, v, w)| & \text{if } |\Delta\text{dir}(u, v, w)| \leq 4 \\ 8 - |\Delta\text{dir}(u, v, w)| & \text{if } |\Delta\text{dir}(u, v, w)| \geq 5. \end{cases} \tag{12}$$

Now we can set

$$\text{cost}_{\text{bends}} = \sum_{\{u,v\},\{v,w\} \in L, L \in \mathcal{L}} \text{bend}(u, v, w) \tag{13}$$

to minimize the number and acuteness of all bends along lines.

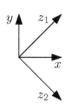

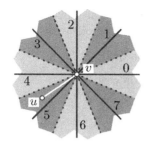

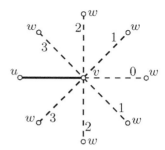

Fig. 1. Octilinear coordinate system

Fig. 2. Sectors relative to v, e.g. $\sec_v(u) = 5$

Fig. 3. Bend cost $b(u, v, w)$ for each value of $\mathrm{dir}_v(w)$

The formulation of bend cost in (12) cannot be transformed directly into a set of linear constraints because it involves absolute values and a case distinction. Here, we solve this problem using instead the following constraints for all lines $L \in \mathcal{L}$ and pairs of incident edges $\{u, v\}, \{v, w\}$ on L. Again, we need some binary variables, namely $\delta_1(u, v, w), \delta_2(u, v, w)$, and $\delta_3(u, v, w)$. The constraint

$$\delta_1(u, v, w) + \delta_2(u, v, w) + \delta_3(u, v, w) = 2 \tag{14}$$

makes sure that exactly one of them takes the value 0. Then, the set of constraints

$$\begin{aligned}
\Delta\mathrm{dir}(u, v, w) &\leq -5 + \delta_1(u, v, w)M \\
\Delta\mathrm{dir}(u, v, w) &\geq 5 - \delta_2(u, v, w)M \\
\Delta\mathrm{dir}(u, v, w) &\leq 4 + \delta_3(u, v, w)M \\
\Delta\mathrm{dir}(u, v, w) &\geq -4 - \delta_3(u, v, w)M
\end{aligned} \tag{15}$$

together with

$$-\mathrm{bend}(u, v, w) \leq \Delta\mathrm{dir}(u, v, w) - 8\delta_1(u, v, w) + 8\delta_2(u, v, w) \leq \mathrm{bend}(u, v, w) \tag{16}$$

assign the bend cost $\mathrm{bend}(u, v, w)$ for the bend between edges $\{u, v\}$ and $\{v, w\}$, where the variable $\mathrm{bend}(u, v, w)$ is integer valued and non-negative. Verify that these constraints in combination with the minimization of (13) indeed model the bend cost as defined in (12). For a detailed explanation we refer to [7].

Minimizing the number of bends thus uses four variables and seven constraints for each pair of incident edges on a path $L \in \mathcal{L}$. Since there are in total at most m' such pairs we are using $4m'$ variables and at most $7m'$ constraints.

3.6 Preservation of Edge Directions

To preserve as much of the overall appearance of the metro system as possible we have already restricted the edge directions to the set of the three directions closest to the original one in Sect. 3.1. Ideally we want to draw an edge $\{u, v\}$ using the closest octilinear approximation, i.e. the direction where $\mathrm{dir}(u, v) = \sec_u(v)$. Hence we introduce a cost in case that the layout does not use this direction. This models (S3).

For each edge $\{u, v\}$ we define as its cost a binary variable $\epsilon(u, v)$ which is 0 if and only if $\operatorname{dir}(u, v) = \sec_u(v)$. This is modeled as follows

$$-M\epsilon(u, v) \leq \operatorname{dir}(u, v) - \sec_u(v) \leq M\epsilon(u, v) \qquad \forall\{u, v\} \in E. \qquad (17)$$

Now we can define the edge-direction cost

$$\operatorname{cost}_{\mathrm{dir}} = \sum_{\{u,v\}\in E} \epsilon(u, v) \qquad (18)$$

which, for each edge, charges 1 when the MIP does not choose the closest octilinear direction. This part of our formulation needs m variables and $2m$ constraints.

3.7 Speed-Up Techniques

A common feature of metro maps is that they tend to have a large number of degree-2 vertices on tracks between two interchange stations. It is useful and common in real metro maps to draw paths between pairs of neighboring interchange or terminal stations as straight as possible. This leads to the idea of replacing chains of degree-2 vertices temporarily by single edges and reinserting the vertices in the final drawing equidistantly on these edge. While this data-reduction trick has been applied before [6, 10], we extend it by keeping two vertices on each chain of degree-2 vertices. The rationale behind this is that it allows for drawing the connection between the corresponding interchange vertices as a polyline with three segments. Our experiments showed that this is a good compromise. Remember that the target function penalizes bends along lines so that in many cases bends at these special degree-2 vertices are in fact avoided.

The only part of our MIP formulation that needs a quadratic number of constraints (and variables) is the one that ensures planarity. This is why we suggest several ways to reduce the number of these constraints. For a planar drawing of an embedded graph it suffices to require that non-incident edges of the same face do not intersect. This already guarantees that no two edges intersect except at common endpoints. So instead of using the constraints in Sect. 3.3 for *all* pairs of non-incident edges we only include them for pairs of non-incident edges of the same face.

In many real-world examples (see Sect. 4) this is still not enough to solve the MIP in an acceptable amount of time. To further reduce the number of constraints we rely on heuristic methods that relax the planarity requirements. These heuristics involve subdividing the external face using the convex hull and considering only pairs of edges where at least one edge is a *pendant* edge, i.e. an edge that leads to a degree-1 vertex. One can also try to skip the planarity constraints completely. In some of these experiments the results were indeed planar in spite of not being enforced in the model. For more details see [7].

4 Experiments

In this section we show how our method performs on the metro system of Sydney because Sydney has been used as a benchmark before [6, 10]. For more examples,

Table 1. Total number of constraints and variables for six different planarity tests

G		n	m	MIP	all pairs	faces	CH	PE	CH & PE	none
Sydney	uncontr.	174	183	constr.	81416	45182	21983	13535	6242	3041
10 lines	contr.	62	71	var.	20329	11545	5921	3873	2105	1329

see [7]. We solved our MIP with the optimizer CPLEX 9.0 running on a Power3-II processor with 375 MHz under the UNIX operating system AIX 5.1, the only system with a CPLEX license accessible to us. We have also experimented with the optimizer XpressMP but found that CPLEX generates better results.

Table 1 shows the size of the uncontracted and contracted network and the number of constraints and variables for the different planarity options in the contracted case. The numbers in the columns *faces* and *none* show that ensuring planarity is in fact responsible for about 90% of the constraints and variables. The other columns show that the convex-hull (CH) and pendant-edge (PE) heuristics as well as their combination effectively reduce the MIP size.

The CityRail system in Sydney (which we restricted to the more interesting suburban part) is a relatively large network and has several multiple edges. The geographic layout is displayed in Fig. 4(a), the official metro map in Fig. 4(f). The weights used in the objective function were $(\lambda_{\text{length}}, \lambda_{\text{bends}}, \lambda_{\text{dir}}) = (1, 5, 5)$. Combining convex-hull and pendant-edge heuristic yielded the planar layout in Fig. 4(e) within 22 minutes. Observe the influence of the soft constraints on the layout: There are no unnecessarily long edges (optimization of (S2)). Moreover, the metro lines only bend where geographically required and pass through interchange stations as straight as possible (optimization of (S1)). And, finally, the simplified edges tend to follow the original directions of Fig. 4(a) (optimization of (S3)). These goals were optimized while guaranteeing octilinearity and preserving the original embedding.

We now compare our layout of the Sydney map to the results of previous algorithms. Figure 4(b) is taken from Hong et al. [6] and shows their layout using a special spring-embedder method. Originally they draw a network that extends slightly further into the periphery but these extensions should not influence the layout of the central part of the network. For ease of comparison we clipped the lines appropriately in Fig. 4(b). Apart from the fact that Hong at al. show station labels, one can observe that edges are not strictly octilinear and that avoiding bends along lines is not a goal of their method. In addition, there is a large variance in the distribution of the edge lengths. Figure 4(c), taken from Stott and Rodgers [10], shows their layout when applying an edge contraction step before actually drawing the network. There are two edges that obviously violate octilinearity, which is an important drawback of this layout. Figure 4(d) displays the result of the same method without prior edge contraction. It again shows an almost octilinear layout, now with the exception of one edge.

Our method overcomes the limitations of the previous results: there are no exceptions to octilinearity and we avoid the problems of local optima in [10]. In

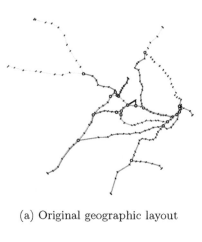

(a) Original geographic layout

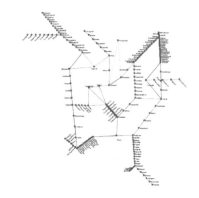

(b) Hong et al. (clipping of Fig. 7(b) in [6])

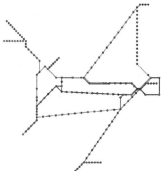

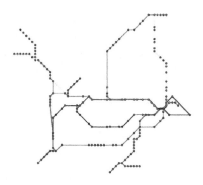

(c) Stott and Rodgers using contracted edges (Fig. 14 in [10])

(d) Stott and Rodgers using uncontracted edges (Fig. 15 in [10])

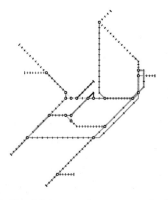

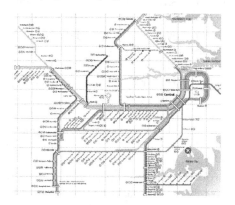

(e) Final layout using our method

(f) Clipping of the official map [11]

Fig. 4. Various drawings of the Sydney CityRail system

contrast to Hong et al. we actively minimize the number of line bends in the layout and maintain the overall geography using the concept of relative position.

The disadvantage of our method is its running time. While we needed 22 minutes to produce our Sydney map, Hong et al. computed the layout in Fig. 4(b) within 7.6 seconds. Stott and Rodgers needed 4 minutes for a Sydney map using a contracted input graph (Fig. 4(c)) and about 28 minutes for the uncontracted graph (Fig. 4(d)). Experiments were carried out on very different machines.

Acknowledgments. We thank Seok-Hee Hong and Herman Haverkort for interesting discussions and Damian Merrick for the Sydney data.

References

1. T. Barkowsky, L. J. Latecki, and K.-F. Richter. Schematizing maps: Simplification of geographic shape by discrete curve evolution. In C. Freksa, W. Brauer, C. Habel, and K. F. Wender, editors, *Proc. Spatial Cognition II*, volume 1849 of *Lecture Notes in Artificial Intelligence*, pages 41–53, 2000.

2. C. Binucci, W. Didimo, G. Liotta, and M. Nonato. Orthogonal drawings of graphs with vertex and edge labels. *Comp. Geometry: Theory & Appl.*, 32(2):71–114, 2005.

3. U. Brandes, M. Eiglsperger, M. Kaufmann, and D. Wagner. Sketch-driven orthogonal graph drawing. In *Proc. 10th Int. Symp. on Graph Drawing (GD'02)*, volume 2528 of *Lecture Notes in Computer Science*, pages 1–11. Springer-Verlag, 2002.

4. S. Cabello, M. d. Berg, S. v. Dijk, M. v. Kreveld, and T. Strijk. Schematization of road networks. In *Proc. 17th Annual Symp. on Computational Geometry (SoCG'01)*, pages 33–39, New York, 2001. ACM Press.

5. G. di Battista, P. Eades, R. Tamassia, and I. G. Tollis. *Graph Drawing: Algorithms for the Visualization of Graphs*. Prentice Hall, 1999.

6. S.-H. Hong, D. Merrick, and H. A. D. d. Nascimento. The metro map layout problem. In J. Pach, editor, *Proc. 12th Int. Symp. on Graph Drawing (GD'04)*, volume 3383 of *Lecture Notes in Computer Science*, pages 482–491. Springer-Verlag, 2005.

7. M. Nöllenburg. Automated drawings of metro maps. Technical Report 2005-25, Universität Karlsruhe, 2005. Available at http://www.ubka.uni-karlsruhe.de/cgi-bin/psview?document=/ira/2005/25.

8. M. Ovenden. *Metro Maps of the World*. Capital Transport Publishing, 2003.

9. E. S. Sandvad, K. Grønbæk, L. Sloth, and J. L. Knudsen. A metro map metaphor for guided tours on the Web: the Webvise Guided Tour System. In *Proc. 10th Int. World Wide Web Conf. (WWW'01)*, p. 326–333, Hong Kong, 2001. ACM Press.

10. J. M. Stott and P. Rodgers. Metro map layout using multicriteria optimization. In *Proc. 8th Int. Conf. on Information Visualisation (IV'04)*, London, pages 355–362. IEEE, 2004.

11. Sydney CityRail. www.cityrail.nsw.gov.au/networkmaps/network_map.pdf.

12. R. Tamassia. On embedding a graph in the grid with the minimum number of bends. *SIAM J. Comput.*, 16(3):421–444, 1987.

Crossing Number of Toroidal Graphs[*]

János Pach[1] and Géza Tóth[2]

[1] City College, CUNY and Courant Institute of Mathematical Sciences,
New York University, New York, NY, USA
pach@cims.nyu.edu

[2] Rényi Institute, Hungarian Academy of Sciences, Budapest, Hungary
geza@renyi.hu

Abstract. It is shown that if a graph of n vertices can be drawn on the torus without edge crossings and the maximum degree of its vertices is at most d, then its planar crossing number cannot exceed cdn, where c is a constant. This bound, conjectured by Brass, cannot be improved, apart from the value of the constant. We strengthen and generalize this result to the case when the graph has a crossing-free drawing on an orientable surface of higher genus and there is no restriction on the degrees of the vertices.

1 Introduction

Let S_g be the compact orientable surface with no boundary, of genus g. Given a simple graph G, a *drawing* of G on S_g is a representation of G such that the vertices of G are represented by points of S_g and the edges are represented by simple (i.e., non-selfintersecting) continuous arcs in S_g, connecting the corresponding point pairs and not passing through any other vertex. The *crossing number* of G on S_g, $\mathrm{cr}_g(G)$, is defined as the minimum number of edge crossings over all drawings of G in S_g. For $\mathrm{cr}_0(G)$, the "usual" planar crossing number, we simply write $\mathrm{cr}(G)$.

Let G be a graph of n vertices and e edges, and suppose that it can be drawn on the torus without crossing, that is, G satisfies $\mathrm{cr}_1(G) = 0$. How large can $\mathrm{cr}(G)$ be? Clearly, we have $\mathrm{cr}(G) < \binom{e}{2}$, and this order of magnitude can be attained, as shown by the following example. Take five vertices and connect any pair of them by $\frac{e}{20}$ vertex-disjoint paths of lengths two. In any drawing of this graph in the plane, every subdivision of K_5 gives rise to a crossing. Therefore, the number of crossings must be at least $\frac{e^2}{400}$.

Peter Brass suggested that this estimate can be substantially improved if we impose an upper bound on the degree of the vertices. More precisely, we have

Theorem 1. *Let G be a graph of n vertices with maximum degree d, and suppose that G has a crossing-free drawing on the torus. Then we have $\mathrm{cr}(G) \leq cdn$, where c is a constant.*

---

[*] János Pach has been supported by NSF Grant CCR-00-98246, and by grants from PSC-CUNY, OTKA, NSA, and BSF. Géza Tóth has been supported by OTKA-T-038397 and T-046246.

For $d \geq 3$, the bound in Theorem 1 cannot be improved, apart from the value of the constant c. Consider the following example. Let $d \geq 4$, $G = C_k \times C_k$, where $k = \sqrt{n/d}$ is a large integer and C_k denotes a cycle of length k. Obviously, this graph can be drawn on the torus without crossings. On the other hand, by a result of Salazar and Ugalde [SU04], its planar crossing number is larger than $(\frac{4}{5} - \varepsilon)k^2$, for any $\varepsilon > 0$, provided that k is large enough. Substitute every edge e of G by $\lfloor \frac{d}{4} \rfloor$ new vertices, each connected to both endpoints of e. The resulting graph G' has at most n vertices, each of degree at most d. It can be drawn on the torus with no crossing, and its planar crossing number is at least

$$\left(\frac{4}{5} - \varepsilon\right) k^2 \times \left\lfloor \frac{d}{4} \right\rfloor^2 > \frac{1}{100} nd.$$

To see this, it is enough to observe that there is an optimal drawing of G' in the plane with the property that any two paths of length two connecting the same pair of vertices cross precisely the same edges. The same construction can be slightly modified to show that $cr(G)$ can also grow linearly in n if the maximum degree d is equal to three.

Theorem 1 can be generalized as follows.

Theorem 2. *Let G be a graph of n vertices of maximum degree d that has a crossing-free drawing on S_g, the orientable surface of genus g. Then we have $cr(G) \leq c_{d,g} n$, where $c_{d,g}$ is a constant depending on d and g.*

We can drop the condition on the maximum degree and obtain an even more general statement.

Theorem 3. *Let G be a graph of n vertices with degrees $d_1, d_2, \ldots, d_n$, and suppose that G has a crossing-free drawing on S_g. Then we have*

$$cr(G) \leq c_g \sum_{i=1}^{n} d_i^2,$$

where c_g is a constant depending on g.

To simplify the presentation and to emphasize the main idea of the proof, in Section 2 first we settle the simplest (planar) case (Theorem 1). In Section 3, we reduce Theorem 3 to a similar upper bound on the crossing number of G in S_{g-1} (Theorem 3.1). This latter result is established in Section 4.

2   The Simplest Case: Proof of Theorem 1

We can assume that $d \geq 3$. It is sufficient to prove that $cr(G) \leq cd(n-1)$ holds for any *two-connected* graph G satisfying the conditions. Indeed, if G is disconnected or has a cut vertex, then it can be obtained as the union of two graphs G_1 and G_2 with n_1 and n_2 vertices that have at most one vertex in common, so that we have $n_1 + n_2 = n$ or $n + 1$. Arguing for G_1 and G_2 separately, we obtain by induction that

$$\mathrm{cr}(G) = \mathrm{cr}(G_1) + \mathrm{cr}(G_2) \leq cd(n_1 - 1) + cd(n_2 - 1) \leq cd(n - 1),$$

as required.

Let G be a two-connected graph with maximum degree d and $\mathrm{cr}_1(G) = 0$. Fix a crossing-free drawing of G on the torus. We can assume that the boundary of each face is connected. Indeed, if one of the faces contains a cycle not contractible within the face, then cutting the torus along this cycle we do not damage any edge of G. Therefore, G is a planar graph and there is nothing to prove.

If our drawing is not a triangulation, then by adding $O(n)$ extra vertices and edges we can turn it into one so that the maximum degree of the vertices increases by at most a factor of three. We have to apply the following easy observation.

Lemma 2.1. *Let G be a two-connected graph with n vertices of degree at most d ($d \geq 3$). Suppose that G has a crossing-free drawing on the orientable surface of genus g such that the boundary of each face is connected. Any such drawing can be extended to a triangulation of the surface with at most $19n + 36(g - 1)$ vertices of maximum degree at most $3d$.*

Proof. First consider a cycle $f = x_1 x_2 \ldots x_{n(f)}$ bounding a single face in the drawing of G. Note that some vertices $x_i \in V(G)$ and even some edges may appear along this cycle several times. Take a simple closed curve $\gamma_0 = p_1 p_2 \ldots p_{n(f)}$ inside the face, running very close to f and passing through the (new) points p_i in this cyclic order. In the ring between f and γ_0, connect each vertex x_i to p_i and p_{i+1} (where $p_{n(f)+1} := p_1$).

Divide γ_0 into $m_0 := \lceil \frac{n(f)}{d-1} \rceil$ connected pieces, each consisting of at most d vertices, such that the last vertex of each piece π_i is the first vertex of π_{i+1}, where $1 \leq i \leq m_0$ and $\pi_{m_0+1} := \pi_1$. Place a simple closed curve $\gamma_1 = q_1 q_2 \ldots q_{m_0}$ in the interior of γ_0. In the ring between γ_0 and γ_1, connect each q_i to all points in π_i. (If $m_0 = 1$ or 2, then γ_1 degenerates into a point or a single edge.) If γ_1 has more than three vertices, repeat the same procedure for γ_1 in the place of γ_0, and continue as long as the interior of the face is not completely triangulated. We added

$$n(f) + m_0 + m_1 + \ldots < n(f) + n(f) + \frac{n(f)}{2} + \frac{n(f)}{4} + \ldots < 3n(f)$$

new vertices, and their maximum degree is at most $d + 4$. The degree of every original vertex of f increased by at most twice the number of times it appeared in f.

If we triangulate every face of G in the above manner, the resulting drawing G' defines a triangulation of the surface with fewer than $n + \sum_f 3n(f) \leq n + 6|E(G)|$ vertices, each of degree at most $d' := 3d$. By Euler's formula, we have $n + 6|E(G)| \leq n + 18(n - 2 + 2g)$, as required.   □

In the sequel, slightly abusing the notation, we write G for the triangulation G' and d for its maximum degree d'.

If G has no *noncontractible* cycle, i.e., no cycle represented on the torus by a closed curve not contractible to a point, then we are done, because G is a planar drawing so that $cr(G) = 0$. Otherwise, choose a noncontractible cycle C with the minimum number of vertices, fix an orientation of C, and let $k := |V(C)|$. Let E_l (and E_r) denote the set of edges not belonging to C that are incident to at least one vertex of C and in a small neighborhood of this vertex lie on the left-hand side (respectively right-hand side) of C. Note that the sets E_l and E_r are disjoint, but this fact is not necessary for the proof.

Replace C by two copies, C_r and C_l, lying on its right-hand side and left-hand side. Connect each edge of E_r (respectively E_l) to the corresponding vertex of C_r (respectively C_l). Cut the torus along C, and attach a disk to each side of the cut.

The resulting spherical (planar) drawing G_1 represents a graph, slightly different from G. To transform it into a drawing of G, we have to remove C_l and (re)connect the edges of E_l to the corresponding vertices of C_r. In what follows, we describe how to do this without creating too many crossings.

Let $\hat{G}_1$ denote the *dual* graph of G_1, that is, place a vertex of $\hat{G}_1$ in each face of G_1, and for any $e \in E(G_1)$ connect the two vertices assigned to the faces meeting at e by an edge $\hat{e} \in E(\hat{G}_1)$. Let r and l denote the vertices of $\hat{G}_1$ lying in the faces bounded by C_r and C_l.

Lemma 2.2. *In $\hat{G}_1$, there are k vertex-disjoint paths between the vertices r and l.*

Proof. By Menger's theorem, the maximum number p of (internally) vertex-disjoint paths connecting r and l in $\hat{G}_1$ is equal to the minimum number of vertices whose deletion separates r from l. Choose p such separating vertices, and denote the corresponding triangular faces of G by $f_1, \ldots, f_p$. The interior of the union of these faces must contain a noncontractible closed curve that does not pass through any vertex of G. Let δ be such a curve whose number of intersection points with the edges of G is minimum. Choose an orientation of δ. Let $e_1, \ldots, e_q$ denote the circular sequence of edges of G intersected by δ. By the minimality of δ, we have $q \leq p$, because the interior of each triangle f_i contains at most one maximal connected piece of δ. Let v_i be the right endpoint of e_i with respect to the orientation of δ. Notice that v_i is adjacent to or identical with v_{i+1}, for every $1 \leq i \leq q$ (where $v_{q+1} := v_1$). Therefore, the circular sequence of vertices $v_1, \ldots, v_q$ induces a cycle in G that can be continuously deformed to δ. Thus, we have a noncontractible cycle of length $q \leq p$ in G, which implies that k, the length of the shortest such cycle, is at most p, as required.   □

By Lemma 2.1, the graph $\hat{G}$ has at most $2|V(G)| \leq 38n$ vertices. According to Lemma 2.2, there is a path connecting r and l in $\hat{G}$ with fewer than $\frac{38n}{k}$ internal vertices. The corresponding faces of G_1 form a "corridor" B between C_r and C_l. Delete now the vertices of C_l from G_1. Pull every edge in E_l through B, and connect each of them to the corresponding vertex of C_r. See Figures 1 and 2. Notice that during this procedure one can avoid creating any crossing between edges belonging to E_l.

We give an upper bound on the number of crossings in the resulting planar drawing of G. Using that $|C| = k$ and $|E_l| \leq dk$, we can conclude that by

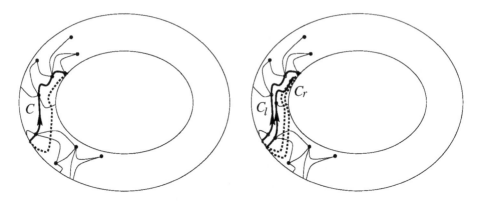

Fig. 1. C is the shortest noncontractible cycle

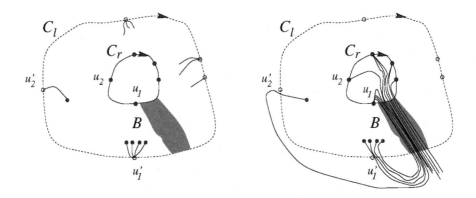

Fig. 2. Pulling the edges in E_l through the corridor B

pulling each edge through the corridor B, we create at most $\frac{38n}{k}$ crossings per edge. Thus, the total number of crossings cannot exceed $dk \cdot \frac{38n}{k} = 38dn$, which completes the proof of Theorem 1.   □

3   Reducing Theorem 3 to Theorem 3.1

Given a graph G, let $n(G)$ and $\sigma(G)$ denote the number of vertices of G and the sum of the squares of their degrees.

Theorem 3 provides an upper bound for the crossing number of a graph G that can be drawn on S_g without crossing. Next we show that this bound can be deduced by repeated application of the following result. In each step, we reduce the genus of the surface by one.

Theorem 3.1. *Let G be a two-connected graph with $\mathrm{cr}_g(G) = 0$. Then we have $\mathrm{cr}_{g-1}(G) \leq c_g^* \sigma(G)$, for some constant $c_g^* \geq 1$.*

Proof of Theorem 3 using Theorem 3.1. As in the proof of Theorem 1, we can assume that G is two-connected. Consider a crossing-free drawing of $G_0 := G$ on S_g. According to Theorem 3.1, G_0 can be drawn on S_{g-1} with at most $c\sigma(G)$ crossings. Place a new vertex at each crossing, and apply Theorem 3.1 to the resulting graph G_1. Proceeding like this, we obtain a series of graphs $G_2, G_3, \ldots, G_g$, drawn on $S_{g-2}, S_{g-3}, \ldots, S_0$, respectively, with no crossing.

We claim that for any i, $0 \le i \le g$,

$$\sigma(G_i) \le (17)^i \left(\prod_{g-i<j\le g} c_j^* \right) \sigma(G)$$

holds. This is obviously true for $i = 0$. Let $0 < i \le g$, and assume that the claim has already been verified for $i - 1$. Notice that, apart from the original vertices of G_{i-1}, every other vertex of G_i has degree four. Thus, applying Theorem 3.1 to the graph G_{i-1} that had a crossing-free drawing on S_{g-i+1}, we obtain

$$\sigma(G_i) \le \sigma(G_{i-1}) + 16\mathrm{cr}_{g-i}(G_{i-1}) \le \sigma(G_{i-1}) + 16c_{g-i+1}^*\sigma(G_{i-1})$$

$$\le (1 + 16c_{g-i+1}^*)(17)^{i-1} \left(\prod_{g-i+1<j\le g} c_j^* \right) \sigma(G) \le (17)^i \left(\prod_{g-i<j\le g} c_j^* \right) \sigma(G),$$

which proves the claim.

It follows from the construction that G_g is a planar graph, and we have

$$n(G_g) - n(G) < \sigma(G_g) \le 17^g \left(\prod_{j=1}^{g} c_j^* \right) \sigma(G).$$

Replacing the $n(G_g) - n(G)$ "new" vertices of G_g by proper crossings, we obtain a drawing of G in the plane with at most $17^g \left(\prod_{j=1}^{g} c_j^* \right) \sigma(G)$ crossings. This completes the proof of Theorem 3. □

4   Reducing the Genus by One: Proof of Theorem 3.1

It remains to prove Theorem 3.1.

All noncrossing closed curves C on S_g belong to one of the following three categories:

1. C is *contractible* (to a point);
2. C is *noncontractible* and *twosided*, i.e., it separates S_g into two connected components;
3. C is *noncontractible* and *onesided*.

Let us cut the surface S_g along C, and attach a disk along each side of the cut. If C is contractible, we obtain two surfaces: one homeomorphic to S_g and the other homeomorphic to the sphere S_0. If C is noncontractible and twosided, then we obtain two surfaces homeomorphic to S_a and S_b, for some $a, b > 0$ with $a + b = g$. Finally, if C is noncontractible and onesided, then we get only *one* surface, S_{g-1} [MT01].

First we need an auxiliary statement, interesting on its own right.

Theorem 4.1. *Let G be a graph with a crossing-free drawing on S_g. If G has no noncontractible onesided cycle, then G is a planar graph.*

Proof. We follow the approach of Cairns and Nikolayevsky [CN00], developed to handle a similar problem on generalized thrackles. Let S be a very small closed neighborhood of the union of all edges of the drawing of G on S_g. Then S is a compact connected surface whose boundary consists of a finite number of closed curves. Attaching a disk to each of these closed curves, we obtain a surface S' with no boundary. We show that S' is a sphere. To verify this claim, consider two closed curves, α' and β', on S'. They can be continuously deformed into closed walks, α_1 and β_1, along the edges of G. Let α and β be the corresponding closed walks along the edges of G in the original drawing on S_g. By the assumption, α divides S_g into two parts, therefore, β crosses α an even number of times. Since the original drawing of G on S_g was crossing-free, every crossing between α and β occurs at a vertex of G. Using the fact that in the new drawing of G on S', the cyclic order of the edges incident to a vertex is the same as the cyclic order of the corresponding edges in the original drawing, we can conclude that α_1 and β_1 cross an even number of times. It is not hard to argue that then the same was true for α' and β'. Thus, S' is a surface with no boundary in which any two closed curves cross an even number of times. This implies that S' is a sphere. Consequently, we have a crossing-free drawing of G on the sphere, that is, G is a planar graph.   □

Proof of Theorem 3.1. As in the previous section, let $\sigma(G)$ denote the the sum of the squared degrees of the vertices of G. A *grid* of size $k \times k$ is the cross product $P_k \times P_k$ of two paths of length k. The vertices of $P_k \times P_k$ with degrees less than four are said to form the *boundary* of the grid. The proof of Theorem 3.1 is based on the same idea as that of Theorem 1, but some important details have to be modified.

Suppose that G is a two-connected graph of n vertices, drawn on S_g without crossing. We can also assume that G has no crossing-free drawing on S_{g-1}, otherwise Theorem 3.1 is trivially true. In particular, it follows that every face of the drawing of G on S_g has a connected boundary.

Replace each vertex v of degree $d(v) > 4$ by a grid of size $d(v) \times d(v)$ and connect the edges incident to v to distinct vertices on the boundary of the grid, preserving their cyclic order. The resulting crossing-free drawing of G' has at most $\sigma(G)$ vertices, each of degree at most four. Every face has a connected boundary, so that we can apply Lemma 2.1 to turn G' into a triangulation G'' with at most $19\sigma(G) + 36(g - 1)$ vertices, each of degree at most twelve. Restricting G' and G'' to any grid substituting for a vertex in G, the only difference between them is that each quadrilateral face in G' is subdivided by one of its diagonals into two triangles in G'''. Color all edges along the boundaries of the grids *blue*, and all other grid and diagonal edges of G''' that lie in the interior of some grid *red*.

If G''' has no noncontractible onesided cycle, then we are done by Theorem 4.1. Otherwise, pick such a cycle C with the smallest number k of vertices. Without increasing its length too much, we can replace all red edges of C by blue edges. Indeed, the first vertex and the last vertex of any maximal red path in C must belong to the boundary of the same grid. Replace each such path by the shortest blue path connecting its first and last vertices along the boundary of the grid containing them. The resulting cycle C' is noncontractible, onesided, and its length is at most $2k$. It has no red edges, and we can assume without loss of generality that it does not intersect itself. Fix an orientation of C'.

Let E_l (and E_r) denote the set of edges not belonging to C' that are incident to at least one vertex of C' and in a small neighborhood of this vertex lie on the left-hand side (respectively right-hand side) of C'.

Replace C' by two copies, C_r and C_l, lying on its right-hand side and left-hand side. Connect each edge of E_r and E_l) to the corresponding vertex of C_r and C_l. Cut S_g along C, and attach a disk to each side of the cut. The resulting surface is S_{g-1}, and it contains a crossing-free drawing G_1 of a graph slightly different from G'''. To obtain a drawing of G''' from G_1, we have to remove C_l and (re)connect the edges of E_l to the corresponding vertices of C_r without creating too many crossings.

Let $\hat{G}_1$ be the *dual* drawing of G_1 on S_{g-1}. Let r (respectively l) be the vertex of $\hat{G}_1$ lying in the face bounded by C_r (respectively C_l). Color *blue* each vertex of $\hat{G}_1$ that corresponds to a face lying inside a grid in G'''.

Repeating the proof of Lemma 2.2, we obtain

Lemma 4.2. *In $\hat{G}_1$, there are k vertex-disjoint paths between the vertices r and l.*

$\square$

The number of cells in G_1 is equal to the number of cells in G''' plus 2. Therefore, by Euler's formula, $\hat{G}_1$ has at most

$$2|V(G''')| + 4(g-1) + 2 \leq 2\left(19\sigma(G) + 36(g-1)\right) + 4(g-1) + 2 < 40(\sigma(G) + 2g)$$

vertices. Thus, by Lemma 4.2, there is a path $P(rl)$ between r and l, of length at most $40(\sigma(G) + 2g)/k$. Replacing all blue vertices of $P(rl)$ by others, we obtain a new path $P'(rl)$, not much longer than $P(rl)$. First observe that r and l, the two endpoints of $P(rl)$, are not blue. Let $uv_1v_2\ldots v_jv$ be an interval along P such that all v_i's are blue ($1 \leq i \leq j$), but u and v are not. Then the faces corresponding to u and v must be adjacent to the boundary of some grid in G_1. These two faces are connected by two chains of faces following the outer boundary of the grid. Replace $v_1, v_2, \ldots, v_j$ by the sequence of vertices corresponding to the shorter of these two chains. Since the degree of every vertex in G_1 is at most twelve, the length of this chain is at most $12j$. Repeating this procedure for each maximal blue interval of $P(rl)$, we obtain a new path $P'(rl)$, whose length is at most $480(\sigma(G) + 2g)/k$.

The corresponding faces of G_1 form a "corridor" B between C_r and C_l. Now delete r, l, and the vertices of C_l. In the same way as in the proof of Theorem 1, "pull" all edges of E_l through B, and connect them to the corresponding vertices of C_r. This step can be carried out without creating any crossing between the edges in E_l.

Now we count the number of crossings in the resulting drawing. Since $|C'| \leq 2k$, $|E_l| \leq 20k$. Pulling them through the corridor B, we create no more than $480(\sigma(G) + 2g)/k$ crossings per edge, that is, at most $X := 9600(\sigma(G) + 2g)$ crossings altogether.

Deleting the extra vertices and edges from G_1 and collapsing each grid into a vertex, we obtain a drawing of G on S_{g-1}, in which the number of crossings cannot exceed X. This concludes the proof of Theorem 3.1.   □

Acknowledgement. We are very grateful to Zoltán Szabó (Princeton) for many valuable suggestions.

References

[CN00]   G. Cairns and Y. Nikolayevsky, Bounds for generalized thrackles, *Discrete Comput. Geom.* **23** (2000), 191–206.

[MT01]   B. Mohar and C. Thomassen: Graphs on surfaces, Johns Hopkins Studies in the Mathematical Sciences. Johns Hopkins University Press, Baltimore, MD, 2001.

[SU04]   G. Salazar and E. Ugalde: An improved bound for the crossing number of $C_m \times C_n$: a self-contained proof using mostly combinatorial arguments, *Graphs Combin.* **20** (2004), 247–253.

Drawing Graphs Using Modular Decomposition

Charis Papadopoulos[1] and Constantinos Voglis[2]

[1] Department of Informatics, University of Bergen, N-5020 Bergen, Norway
charis@ii.uib.no
[2] Department of Computer Science, University of Ioannina, P.O.Box 1186,
GR-45110 Ioannina, Greece
voglis@cs.uoi.gr

Abstract. In this paper we present an algorithm for drawing an undirected graph G which takes advantage of the structure of the modular decomposition tree of G. Specifically, our algorithm works by traversing the modular decomposition tree of the input graph G on n vertices and m edges, in a bottom-up fashion until it reaches the root of the tree, while at the same time intermediate drawings are computed. In order to achieve aesthetically pleasing results, we use grid and circular placement techniques, and utilize an appropriate modification of a well-known spring embedder algorithm. It turns out, that for some classes of graphs, our algorithm runs in $O(n+m)$ time, while in general, the running time is bounded in terms of the processing time of the spring embedder algorithm. The result is a drawing that reveals the structure of the graph G and preserves certain aesthetic criteria.

1 Introduction

The problem of automatically generating a clear and readable layout of complex structures inside a graph is receiving increasing attention in the literature [1]. In this work we present a drawing algorithm which takes advantage of the modular decomposition of a graph. Our goal is to highlight the global structure of the graph and reveal the regular structures within it. The usage of the modular decomposition has been considered by many authors in the past to efficiently solve other algorithmic problems [4].

Our approach, takes advantage of the modular decomposition of the input graph G, which is a recursive tree-like partition that reveals *modules* of G, i.e. sets of vertices having the same neighborhood. By exploiting the properties of these modules and especially the properties of the modular decomposition tree $T(G)$, we are able to draw the modules separately using different techniques for each one. To achieve aesthetically pleasing results, we utilize a grid placement technique, a circular drawing paradigm, and a modification of a spring embedder method, on the appropriate modules. Our algorithm relies on creating intermediate drawings in a systematic fashion by traversing the modular decomposition tree of the input graph from bottom to top, while at the same time certain parameters are appropriately updated. In the end, the drawing of the graph G is

P. Healy and N.S. Nikolov (Eds.): GD 2005, LNCS 3843, pp. 343–354, 2005.
© Springer-Verlag Berlin Heidelberg 2005

obtained by traversing $T(G)$ from the root to the leaves, in order to compute the final coordinates of the vertices in the drawing area, using the parameters computed in the previous traversal of $T(G)$. It turns out that this way of processing $T(G)$, enables us to visualize the graph in various levels of abstraction.

Similar approaches for computing the layout of a graph are based on a specific decomposition of it. Based on this scheme, optimal algorithms have been developed for drawing a series-parallel digraph [1], and for upward planarity testing of a single-source digraph [2]. Also, many techniques for drawing hierarchical clustered graphs, deal with a graph and its tree representation [6, 7, 8]. All these methods address the problem of visualization, by drawing the non-leaf nodes of the tree as simple closed curves. Force directed methods have also been developed to support and show the structure of a clustered graph which is a 2-level decomposition scheme [13, 18].

2   Definitions and Background Results

We consider finite undirected graphs with no loops or multiple edges. For a graph G, we denote by $V(G)$ and $E(G)$ the vertex set and the edge set of G, respectively. Let S be a subset of the vertex set of a graph G. Then, the subgraph of G induced by S is denoted by $G[S]$. A *clique* is a set of pairwise adjacent vertices; a *stable set* is a set of pairwise non-adjacent vertices. The *degree* of a vertex x in the graph G, denoted $d(x)$, is the number of edges incident on x. For a graph G on n vertices and m edges, $D(G) = 2m/n$ is the *average degree* of G. The complement of a graph G is denoted by $\overline{G}$.

Let T be a rooted tree. For convenience, we refer to a vertex of a tree as a node. The parent of a node t of T is denoted by $p(t)$, whereas the node set containing the children of t in T is denoted by $ch(t)$. Let h be the height of the tree T. Then, we denote by L_i the node set containing the nodes of the i-th level of T, for $0 \leq i \leq h$.

2.1   Modular Decomposition

A subset M of vertices of a graph G is said to be a *module* of G, if every vertex outside M is either adjacent to all vertices in M or to none of them. The emptyset, the singletons, and the vertex set $V(G)$ are *trivial* modules and whenever G has only trivial modules it is called a *prime* (or *indecomposable*) *graph*. It is easy to see that the chordless path on four vertices, P_4, is a smallest non-trivial prime graph, since graphs with three vertices are decomposable [4]. A non-trivial module is also called *homogeneous set*. A module M of the graph G is called a *strong module*, if for any module $M' \neq M$ of G, either $M' \cap M = \emptyset$ or one module is included into the other. A module M of a graph G is called *parallel* if $G[M]$ is a disconnected graph, *series* if $\overline{G}[M]$ is a disconnected graph and *neighborhood* if both $G[M]$ and $\overline{G}[M]$ are connected graphs.

The *modular decomposition* of a graph G is a linear-space representation of all the partitions of $V(G)$ where each partition class is a module. The *modular decomposition tree* $T(G)$ of the graph G (or *md-tree* for short) is a unique labelled tree associated with the modular decomposition of G in which the leaves of $T(G)$ are the vertices of G and the set of leaves associated with the subtree rooted at an internal node induces a strong module of G. Thus, the md-tree $T(G)$ represents all the strong modules of G. An internal node is labelled by either P (for parallel module), S (for series module), or N (for neighborhood module). It is shown that for every graph G on n vertices and m edges, the md-tree $T(G)$ is unique up to isomorphism, the number of nodes in $T(G)$ is $O(n)$ and it can be constructed in $O(n + m)$ time [5, 15].

Let t be an internal node of the md-tree $T(G)$ of a graph G. We denote by $M(t)$ the module corresponding to t which consists of the set of vertices of G associated with the subtree of $T(G)$ rooted at node t; note that $M(t)$ is a strong module for every (internal or leaf) node t of $T(G)$. Let $t_1, t_2, \ldots, t_p$ be the children of the node t of md-tree $T(G)$. We denote by $G(t)$ the *representative graph* of node t defined as follows: $V(G(t)) = \{t_1, t_2, \ldots, t_p\}$ and $t_i t_j \in E(G(t))$ if there exists edge $v_k v_\ell \in E(G)$ such that $v_k \in M(t_i)$ and $v_\ell \in M(t_j)$. For the P-, S-, and N-nodes, the following lemma holds (see [4]):

Lemma 1. *Let G be a graph, $T(G)$ its modular decomposition tree, and t an internal node of $T(G)$. Then, $G(t)$ is an edgeless graph if t is a P-node, $G(t)$ is a complete graph if t is an S-node, and $G(t)$ is a prime graph if t is an N-node.*

2.2   Modular Decomposition Based Drawing $\Gamma(G)$

Our drawing algorithm is based on the modular decomposition tree of a given graph G. We deal with box-shaped vertices with a specific size. For every $t \in T(G)$ we define $c(t) = (x(t), y(t)) \in \mathbf{R}^2$ to be the coordinates of the center of node t, and $b(t) = (w(t), h(t)) \in \mathbf{R}^2$ to be the dimensions of the box of node t, where $w(t)$ and $h(t)$ are the width and the height of the box, respectively. In other words, $c(t)$ is the center of the box $b(t)$. We adopt the straight-line drawing convention and we impose the following constraints: (C1) vertices do not overlap; (C2) vertices in every strong module $M(t)$, induced by an internal node t of $T(G)$, are drawn close (in terms of their Euclidean distance) to each other; (C3) vertices in every strong module $M(t)$, induced by an internal node t of $T(G)$, are drawn according to the structure (edgeless or complete or prime) of the representative graph $G(t)$.

Definition 1. *A drawing with the previous constraints is called a modular decomposition based drawing $\Gamma(G)$ of the graph G which is a mapping between the vertices and the Euclidean space $\mathbf{R}^2$: $\Gamma(G) : V(G) \rightarrow \mathbf{R}^2$.*

Definition 2. *A relative drawing $\Gamma'(t, T(G))$ is an md-drawing of the representative graph $G(t)$, relative to $c(t)$.*

3   The Algorithm

Let G be a graph on n vertices $v_1, v_2, \ldots, v_n$ with non-uniform dimensions $b(v_1), b(v_2), \ldots, b(v_n)$, respectively, and m edges. Our algorithm first computes the md-tree $T(G)$ using one of the known linear-time algorithms [5, 15]. In bottom-up fashion, we traverse the md-tree $T(G)$ and calculate the relative drawing $\Gamma'(t, T)$ for every internal node t. In order to apply the new coordinates to the subtree rooted at t, and finally to the graph $G[M(t)]$, we store the displacements from the previous coordinates, $dis(t_i)$ for every t_i. Finally, we traverse the md-tree $T(G)$ in a top-down fashion and for every internal node $t \in T(G)$, we add the displacement $dis(t)$ to the centers of the boxes of every child node $t_i \in ch(t)$. In this way, all the vertices of $G[M(t)]$ obtain the right coordinates relative to the center of their ancestor node t.

We mention that every relative drawing uses a predefined constant k_i as the preferred edge length of the drawing at the level set L_i, $0 \le i \le h - 1$, of the md-tree $T(G)$. The algorithm, called *Module_Drawing*, is given in detail in Algorithm 1.

Algorithm. **Module_Drawing**

Input: A graph G on n vertices and m edges.

Output: An md-drawing $\Gamma(G)$ of the graph G.

---

1. Construct the modular decomposition tree T of the graph G;
2. Initialize the rectangle boxes $b(t)$ and the centers $c(t)$ for every $t \in T$;
3. Compute the node sets $L_0, L_1, \ldots, L_h$ of the levels $0, 1, \ldots, h$ of T, and assign values to the preferred edge lengths k_i;
4. **for** $i = h - 1$ down to 0 **do**   { *bottom-up fashion*}
    for every internal node $t \in L_i$ **do**
      4.1   **if** t is a P-node   **then** $\Gamma'(t, T) \leftarrow Draw\_Edgeless(t, T)$;
      4.2   **else if** t is a S-node   **then** $\Gamma'(t, T) \leftarrow Draw\_Complete(t, T)$;
      4.3   **else**   {*t is a N-node*} $\Gamma'(t, T) \leftarrow Draw\text{-}Prime(t, T)$;
      4.4   Compute the displacement $dis(t_i)$, for each node $t_i \in ch(t)$, with respect to their initial placement;
      4.5   Update the size of the rectangle box $b(t)$, according to the frame boundaries of $\Gamma'(t, T)$;
5. **for** $i = 0$ down to $h - 1$ **do**   { *top-down fashion*}
    for every internal node $t \in L_i$ **do**
      for every child $t_i \in ch(t)$ **do**
      5.1 $c(t_i) \leftarrow c(t_i) + dis(t)$
6. Return the drawing $\Gamma(G) = \Gamma'(r, T)$ computed in the root r of T;

Algorithm 1. *Module_Drawing*

Due to lack of space, the formal description of functions *Draw_Edgeless* and *Draw_Complete* is omitted, whereas the function *Draw-Prime* is described in detail in Sect. 4. All these functions are aware of the preferred edge length, denoted by k, which may be different for each level of $T(G)$. We note here that, one can use different drawing techniques for each relative drawing to fulfill desired aesthetic criteria. Our approach draws edgeless graphs on an underlying grid, complete graphs in a circular way, and prime graphs using a spring embedder method.

Vertices are placed by function Draw_Edgeless, keeping in mind that there are no connecting edges between them. This is achieved by a grid placement of the nodes in an arbitrary order. The Euclidean distance between the boundaries of two nodes placed adjacent on the grid is at least k. For symmetry reasons, we distribute evenly the space between the nodes in each row, so that a complete alignment is achieved. Each row is then processed one by one and it is placed below the previous one, keeping distance of at least k from the bottom boundary of the previous row.

Function Draw_Complete is basically a circular drawing algorithm, even though the representative graph $G(t)$, is a complete graph. We have chosen to draw complete graphs in this way, in order to expose the structure of a series module (see constraint C3). Furthermore, a circular drawing satisfies the aesthetic criterion of symmetry and is the usual way of representing complete graphs in textbooks. The vertices of the series module are placed in an arbitrary order on equal arcs, on the circumference of a cycle centered at $c(t)$. The initial radius is determined by the smallest sized box. Function Draw_Complete process each node $t_i \in ch(t)$ one by one, and calculates its final radius by considering the size of the two adjacent nodes on the cycle. For every node t_i a value $f(t_i)$ is computed that represents the maximum distance from $c(t_i)$ to a point on its boundary $b(t_i)$. Finally, node t_i is positioned on the minimum possible radius, according to $f(t_i)$ and the preferred edge length k, so that any overlapping is avoided. We note that for a complete graph with uniform nodes the drawing is a perfect cycle.

For the time complexity of functions Draw_Edgeless and Draw_Complete, the following holds:

Lemma 2. *Let $T(G)$ be a modular decomposition tree of graph G and let $ch(t)$ be the set of children of a P-node (resp. an S-node) $t \in T(G)$. Function Draw_Edgeless (resp. Draw_Complete) constructs a relative drawing $\Gamma'(t,T)$ in $O(|ch(t)|)$ time.*

4 Modified Spring Embedder

In this section we describe in detail a spring embedder algorithm for the implementation of function Draw_Prime. Recall that this function is applied on a N-node $t \in T(G)$. Since the representative graph $G(t)$ is a prime graph, function Draw_Prime requires the vertex set $V(G(t))$ and the edge set $E(G(t))$.

The main task of Draw_Prime is to combine the aesthetic properties of a spring embedder algorithm, with the constraint that no vertex-to-vertex overlapping occurs. The fact that Draw_Prime is applied on the representative graph $G(t)$ that contains vertices with non-uniform sizes, makes the drawing task more demanding.

The function Draw_Prime falls in the category of force-integration approaches [14, 12, 11]. It is based on the Fruchterman & Reingold (FR) spring embedder algorithm [9] and follows the general guidelines of Harel & Koren [12]. Draw_Prime consists of a main iteration loop, that is repeated until some termination criteria are met. There are three basic steps to each iteration: (i) calculate the effect of the edge-attractive forces (ii) calculate the effect of vertex-to-vertex repulsive forces and (iii) limit the total displacement by a quantity called *temperature* which is decreased over the iterations. The temperature is decreased by a *cooling schedule*, the choice of which greatly affects the quality of the drawing. To summarize, Draw_Prime starts with an initial random placement of the vertices and an initial temperature, and performs the main iteration loop, until the underlying physical system reaches an equilibrium state. As presented in [9], we choose a two phase cooling scheme: the first phase starts with a constant initial temperature and reduces it using an exponential cooling scheme, and the second phase, which starts after a number of iterations, maintains a constant low temperature.

As already mentioned, we must take into account the size of the children t_i of a node t so that vertices of $G(t)$ would not overlap. To achieve this, we have modified the formulas for the attractive and the repulsive forces between the vertices of the graph. The final formulas for the forces will be presented later in the section. We will first describe the heuristics that we use to avoid overlapping. According to [12], the first modification to the original FR algorithm will result the following formulas for the attractive f_a and the repulsive f_r forces:

$$Modified\ FR:\ f_a(r_{MFR}) = \frac{r_{MFR}^2}{k}\quad and\quad f_r(r_{MFR}) = \frac{k^2}{\max(r_{MFR}, \epsilon)},$$

where $r_{MFR} = f(t_i, t_j)$ and $f(t_i, t_j)$ is the shortest distance between the boundaries of the boxes $b(t_i)$ and $b(t_j)$. The variable k is the preferred edge length for the drawing and ϵ is a small positive number.

The next extension is to impose the vertex size constraints gradually. Specifically, at the early iterations of our spring embedder the vertices of the prime graph are considered dimensionless, and thus, we use the forces of the FR algorithm. This policy, combined with a large initial temperature, allows the layout to escape possible local optimum states. In this way a possible cluttered layout is found at early stages of the algorithm, and then, we use the Modified FR repulsive and attractive forces to fully prevent overlaps (see also [12]).

We noticed that the large number of attractive forces, combined with a small value of k, do not allow large vertices to be in a certain distance in order to avoid overlapping. To overcome this problem, we decide to use a factor w in the calculation of the edge attractive forces, inversely proportional to the graph's density. In this manner, we weaken edge attractive forces and allow the algorithm to position vertices without overlaps.

Hereafter we will denote by G the representative graph $G(t)$. To compute the reducing factor w, we use the average degree $D(G)$ that can be thought as a measure for the connectivity of G. To be more precise, we use $D^{-1}(G)$ as the factor in the Modified FR edge attractive force calculation f_a. It follows that the use of $D^{-1}(G)$ as a multiplicative factor weakens the attractive forces between vertices. Note that, since the smallest prime graph is a P_4, for a prime graph G we have: $0 < D^{-1}(G) \le 0.57$.

Using the previous inequality of $D^{-1}(G)$, we set a threshold in the middle of the interval and consider dense the graphs G s.t. $D^{-1}(G) < 0.28$ and sparse the

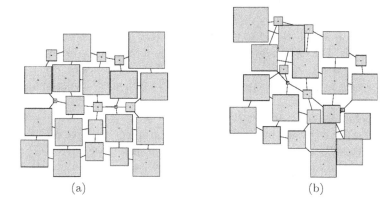

(a)   (b)

Fig. 1. Drawings of a 5×5 grid using (a) $w = D^{-1}(G) = 0.31$ and (b) $w = 1$

graphs s.t. $D^{-1}(G) > 0.28$. If a graph is considered sparse, after a certain point in the algorithm we use $D(G)$ as the multiplicative factor.

In Fig. 1 we show two drawings of a 5×5 grid with random dimensioned vertices. The preferred edge length is set to $k = 60$, which is a small number, with respect to the dimensions of the vertices. In Fig. 1(a) the factor $w = D^{-1}(G) = 0.31$ is used, in the early iterations, for the calculation of the attractive forces. Since the graph is considered sparse, this factor is reversed ($w = D(G)$) at final iterations and so the layout becomes more compact. In Fig. 1(b) the multiplicative factor w is set to one in all iterations.

Having describe the two main features of our spring embedder algorithm, we can present the attractive and repulsive forces of function Draw_Prime (DP) as follows:

$$DP: \quad f_a(r_{DP}) = \frac{w \cdot r_{DP}^2}{k} \quad \text{and} \quad f_r(r_{DP}) = \frac{k^2}{\max(r_{DP}, \epsilon)}$$

where, $\quad r_{DP} = \begin{cases} ||c(t_i) - c(t_j)||, & \text{at early iterations} \\ f(t_i, t_j), & \text{at final iterations} \end{cases}$

and $\quad w = \begin{cases} D^{-1}(G), & \text{at early iterations} \\ & \text{at final iterations, and} \\ D(G), & \text{if } D^{-1}(G) > 0.28. \end{cases}$

We mention that the early and the final iterations coincide with the first and the second part of the cooling schedule, respectively. We denote by ℓ the number of the main iterations needed by our spring embedder algorithm. We conclude with the following lemma.

Lemma 3. *Let $T(G)$ be a modular decomposition tree of graph G and let $ch(t)$ be the set of children of an N-node $t \in T(G)$. Function Draw_Prime constructs a relative drawing $\Gamma'(t, T)$ in $O(\ell \cdot |ch(t)|^2)$ time, where ℓ is the number of main iterations that a spring embedder algorithm performs.*

5   Time Complexity

Next, we introduce the definition of the prime cost of a graph which we will need in our analysis. Let G be a graph and $T(G)$ be its modular decomposition tree. We denote by $\alpha(G) = \{t_1, t_2, \ldots, t_s\}$ the set of the N-nodes of $T(G)$. We define the *prime cost* of G as the value $\phi(G) = \sum\limits_{t \in \alpha(G)} \ell \cdot |ch(t)|^2$, where $ch(t)$ denotes the set of children of node t in $T(G)$.

It is not difficult to see that for any n-vertex graph G, we have $\phi(G) = O(\ell \cdot n^2)$; for an n-vertex P_4-free graph (also known as cograph) G we have $\phi(G) = 0$, since its md-tree (also known as cotree) does not contain any N-node [4]. It follows that in other classes of graphs their prime cost is constant. For example, any N-node of the md-tree of a P_4-reducible graph[1] contains at most five children [4]. Hence for an n-vertex P_4-reducible graph G we have $\phi(G) = O(1)$. We notice that these classes of graphs arise in applications such as examination scheduling problems and semantic clustering of index terms [4].

Theorem 1. *Let G be a graph on n vertices and m edges. Algorithm Module_Drawing constructs an md-drawing $\Gamma(G)$ in $O(n + m + \phi(G))$ time, where $\phi(G)$ is the prime cost of the input graph G.*

6   Implementation and Examples

We have implemented our algorithm in C++. The implementation takes as input an undirected graph G in GraphML format [3]. The vertices are thought of as rectangles with a predefined size, i.e. with a specific height and width. Three files are produced in GraphML format: a file that contains the final drawing of G; a file that contains the md-tree $T(G)$; a file that contains all the relative drawings computed in each level of $T(G)$. For visualization purposes, we use the yEd environment [16].

6.1   An Example of Module_Drawing

In this section, we illustrate how our algorithm produces a final drawing, by showing level-by-level relative drawings, on the md-tree of the input graph. For this purpose we use an input graph from a real life application, which describes a protein interaction network (see [10] for details). More specifically, the input graph, which we will call *Trans* graph, describes a network of proteins that define transcriptional regulator complexes. The md-tree of the Trans graph contains 1 P-node, 6 S-nodes, and 1 N-node. We label the 51 vertices of the graph and assign an additional label, besides P or S or N label, to the 8 internal nodes of the md-tree. In Fig. 2(a) we present the final drawing of Trans graph using Module_Drawing, in Fig. 2(b) we show its modular decomposition tree and in Fig. 2(c) we present level-by-level relative drawings and how they are combined to result the final layout.

---

[1] A P_4-reducible graph is a graph for which no vertex belongs to more than one P_4.

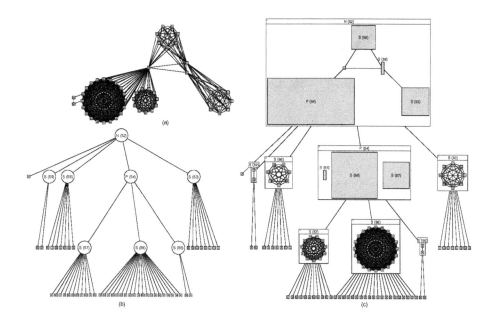

Fig. 2. Illustration of Module_Drawing on *Trans* graph

Starting from level 3 of the tree in Fig. 2(c), we notice three S-nodes. The application of the function Draw_Series results the relative drawings as shown in the corresponding boxes. Their parent, which is a P-node, causes them to be drawn on a 1×3 grid. Finally, the root of the md-tree is an N-node; in particular $G(root)$ is an $\mathcal{A}$-shaped graph, that consists of 1 parallel module, 3 series modules, and 1 simple vertex. The final drawing reveals all modules and gives a useful insight of the structure of the Trans graph. Moreover, function Draw_Prime, which is the most expensive part of our algorithm, in terms of time complexity, is applied on a graph of 5 vertices instead of 51.

6.2   Drawing Examples

In all the examples we choose to draw the vertices of a graph over its edges. The height and width of all the vertices are set to 30 points. As already mentioned in the description of Module_Drawing, we increase the preferred edge length k_i of the i-th level, starting from the level $h-1$ of $T(G)$. Thus, we set k_{h-1} to a constant and $k_i = (h-i) \cdot k_{h-1}$, for $i = h-2, h-3, \ldots, 0$. Obviously, $k_i < k_{i-1}$. We note that an alternative scheme for increasing the preferred edge length between levels is presented in [17].

For each example drawn by our algorithm, we present an additional drawing created by a spring embedder method. For this purpose we apply the Smart Organic Layout (SOL) utility of yEd [16] with desired parameters. We make clear that, there is no reason to compare our method to any spring embedder algorithm, since their drawing goals are different. We use a general purpose

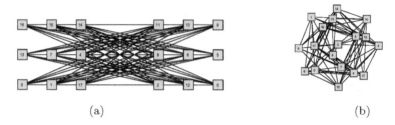

Fig. 3. Drawings of $K_{9,9}$ using (a) Module_Drawing and (b) Smart Organic Layout

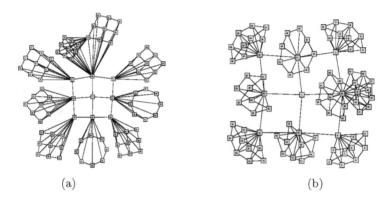

Fig. 4. Drawings of a graph using (a) Module_Drawing and (b) Smart Organic Layout

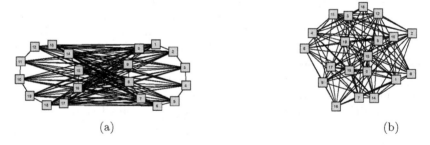

Fig. 5. Drawings of a graph using (a) Module_Drawing and (b) Smart Organic Layout

drawing algorithm, such as spring embedder, to obtain a reference layout of a graph. Note also that we incorporate a spring embedder method in the general framework of our approach.

In Figs. 3–5 the final drawings of our algorithm are shown on the left side whereas the drawings of the same graph using SOL are shown on the right side. Notice that our algorithm manage to expose underlying structures (smaller grids, circles, paths e.t.c) in all the examples. This observation arises from the fact that we apply a spring embedder algorithm without the force impact of the vertices that belong to other modules.

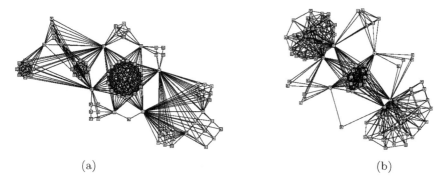

(a)	(b)

Fig. 6. Drawings of a graph using (a) Module_Drawing and (b) Smart Organic Layout

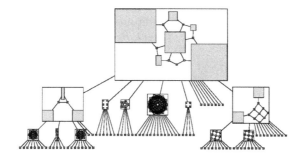

Fig. 7. The md-tree of the graph depicted in Fig. 6

In Fig. 6 we show a graph with an md-tree of 3 levels. Notice that our method reveals three underlying structures: a gear graph[2], an $\mathcal{A}$-shaped graph and a complex of grids. In Fig. 7, we show the md-tree of the graph, in order to illustrate the intermediate steps of our method. It is useful to consider the md-tree representation, as a visualization abstraction of the input graph.

7   Concluding Remarks

In this paper we have presented a divide-and-conquer technique for drawing undirected graphs, based on their modular decomposition tree, where each disjoint induced subgraph (module) is drawn according to its corresponding structure (edgeless, complete or prime). For certain classes of graphs, the structure of their modular decomposition trees ensures that each tree node can be processed in linear time. It turns out that our algorithm, besides its efficiency in terms of time, also exposes the structure of a graph. Revealing the structure of a graph by drawing it, can prove to be helpful in identifying, and thus, recognizing, in which certain class the graph belongs.

---

[2] A gear graph is a wheel graph with a vertex added between each pair of adjacent vertices of the outer cycle.

References

1. G. Di Battista, P. Eades, R. Tamassia, and I. G. Tollis, *Algorithms for the Visualization of Graphs*, Prentice-Hall, 1999.
2. P. Bertolazzi, G. Di Battista, C. Mannino, and R. Tamassia, Optimal upward planarity testing of single-source digraphs, *SIAM J. Comput.* **27** (1998) 132–169.
3. U. Brandes, M. Eiglsperger, I. Herman, M. Himsolt, and M.S. Marshall: GraphML progress report: structural layer proposal, *Proc. 9th Int. Symp. Graph Drawing (GD'01)*, LNCS **2265** (2001) 501–512.
4. A. Brandstädt, V.B. Le, and J.P. Spinrad, *Graph Classes: A Survey*, SIAM Monographs on Discrete Mathematics and Applications, 1999.
5. E. Dahlhaus, J. Gustedt, and R.M. McConnell, Efficient and practical algorithms for sequential modular decomposition, *J. Algorithms* **41** (2001) 360–387.
6. P. Eades and Q.W Feng, Drawing clustered graphs on an orthogonal grid. *Proc. 5th Int. Symp. Graph Drawing (GD'97)*, LNCS **1353** (1997) 146-157.
7. P. Eades, Q.W. Feng, and X Lin, Straight-line drawing algorithms for hierarchical graphs and clustered graphs, *Proc. 4th Int. Symp. Graph Drawing (GD'96)*, LNCS **1190** (1996) 113-128.
8. Q.-W. Feng, R. F. Cohen, and P. Eades, Planarity for clustered graphs. *Proc. 3rd European Symp. Algorithms (ESA'95)*, LNCS **979** (1995) 213-226.
9. T. Fruchterman and E. Reingold, Graph drawing by force-directed placement, *Software-Practice and Experience*, **21** (1991) 1129–1164.
10. J. Gagneur, R. Krause, T. Bouwmeester, and G. Casari, Modular decomposition of protein-protein interaction networks, *Genome Biology* **5**:R57 (2004).
11. E. R. Gansner and S. C. North, Improved force-directed layouts, *Proc. 6th Int. Symp. Graph Drawing (GD'98)*, LNCS **1547** (1998) 364–373.
12. D. Harel and Y. Koren, Drawing graphs with non-uniform vertices, *Proc. of Working Conference on Advanced Visual Interfaces (AVI'02)*, ACM Press 2002, 157–166.
13. M.L. Huang and P. Eades, A fully animated interactive system for clustering and navigating huge graphs, *Proc. 6th Int. Symp. Graph Drawing (GD'98)*, LNCS **1547** (1998) 374-383.
14. W. Li, P. Eades, and N. Nikolov, Using spring algorithms to remove node overlapping, *Proc. Asia Pacific Symp. Information Visualization (APVIS'05)*, 2005.
15. R.M. McConnell and J. Spinrad, Modular decomposition and transitive orientation, *Discrete Math.* **201** (1999) 189–241.
16. yEd - Java Graph Editor, `http://www.yworks.com/en/products_yed_about.htm`.
17. C. Walshaw, A multilevel algorithm for force-directed graph drawing, *J. Graph Algorithms Appl.* **7** (2003) 253–285.
18. X. Wang and I. Miyamoto, Generating customized layouts, *Proc. 3rd Int. Symp. Graph Drawing (GD'95)*, LNCS **1027** (1995) 504–515.

Applications of Parameterized st-Orientations in Graph Drawing Algorithms

Charalampos Papamanthou[1,2] and Ioannis G. Tollis[1,2]

[1] Department of Computer Science, University of Crete, P.O. Box 2208,
Heraklion, Greece
{cpap, tollis}@csd.uoc.gr

[2] Institute of Computer Science, FORTH, Vasilika Vouton, P.O. Box 1385,
Heraklion, GR-71110, Greece
{cpap, tollis}@ics.forth.gr

Abstract. Many graph drawing algorithms use st-numberings (st-orientations or bipolar orientations) as a first step. An st-numbering of a biconnected undirected graph defines a directed graph with no cycles, one single source s and one single sink t. As there exist exponentially many st-numberings that correspond to a certain undirected graph G, using different st-numberings in various graph drawing algorithms can result in aesthetically different drawings with different area bounds. In this paper, we present results concerning new algorithms for parameterized st-orientations, their impact on graph drawing algorithms and especially in visibility representations.

1 Introduction

st-orientations (st-numberings) or bipolar orientations are orientations of undirected graphs that satisfy some certain criteria, i.e., they define no cycles and have exactly one source s and one sink t. Starting with an undirected biconnected graph $G = (V, E)$, many graph drawing algorithms, such as hierarchical drawings [1], visibility representations [2] and orthogonal drawings [3], use an st-orientation of G in order to compute a drawing of G. Therefore, the importance of st-orientations in Graph Drawing is evident.

Given a biconnected undirected graph $G = (V, E)$, with n vertices and m edges, and two nodes s, t, an st-orientation (also known as bipolar orientation or st-numbering) of G is defined as an orientation of its edges such that a directed acyclic graph with exactly one source s and exactly one sink t is produced. An st-orientation of an undirected graph can be easily computed using an st-numbering [4] of the respective graph G and orienting the edges of G from *low* to *high*. An st-numbering of G is a numbering of its vertices such that s receives number 1, t receives number n and every other node except for s, t is adjacent to at least one lower-numbered and at least one higher-numbered node.

st-numberings were first introduced in 1967 in [5], where it is proved (together with an $O(nm)$ time algorithm) that given any edge $\{s, t\}$ of a biconnected undirected graph G, we can define an st-numbering. However, in 1976 Even and

P. Healy and N.S. Nikolov (Eds.): GD 2005, LNCS 3843, pp. 355–367, 2005.
© Springer-Verlag Berlin Heidelberg 2005

Tarjan proposed an algorithm that computes an st-numbering of an undirected biconnected graph in $O(n + m)$ time [4]. Ebert [6] presented a slightly simpler algorithm for the computation of such a numbering, which was further simplified by Tarjan [7]. The planar case has been extensively investigated in [8] where a linear time algorithm is presented which may reach any st-orientation of a planar graph. Finally, in [9] a parallel algorithm is described. An overview of the work concerning bipolar orientations is presented in [10].

However, all developed algorithms compute an st-numbering at random, without expecting any specific properties of the oriented graph. In this paper we present new techniques that produce such orientations with specific properties. Namely, our techniques are able to control the length of the longest path of the resulting directed acyclic graph. This provides significant flexibility to many graph drawing algorithms such as [2, 3]. Actually, st-orientations play a very important role in defining certain aesthetics in the drawings produced by algorithms they use them. The length of the longest path of the final directed graph that is produced is vital in determining the area bounds of the drawing. In this paper, we try to answer these questions by connecting a newly developed algorithm for the computation of st-orientations with graph drawing applications.

The paper is organized as follows. In Section 2 we present the problem, the objectives and some preliminary definitions. In Section 3 we give a brief description of the algorithm and show its implication in defining the longest path length of the final directed graph. A detailed presentation of the algorithm can be found in [11]. In Section 4 we comment on primal and dual st-orientations and Section 5 presents experimental results. Finally, some conclusions are presented in Section 6.

2   Preliminaries

2.1   Motivation and Objectives

Many algorithms in Graph Drawing use st-Orientations as a first step. Additionally, the length of the longest path from s to t of the specific st-orientation determines certain aesthetics of the drawing:

- **Hierarchical Drawings.** One of the most common algorithms in hierarchical drawing is the longest path layering [1]. This algorithm applies to directed acyclic graphs. The height of such a drawing is always equal to the length of the longest path of the directed acyclic graph, l. If we want to visualize an undirected graph G using this algorithm, we must firstly st-orient G. The height of the produced drawing will be equal to the length of the longest path l of the produced st-orientation.
- **Visibility Representations.** In order to compute visibility representations of planar graphs, we must compute an optimal topological numbering of an st-orientation of the input graph [2]. This can be done if we assign unit-weights to the edges of the graph and compute the longest path to each one of its vertices from source s. The y-coordinate of each vertex u in the

visibility representation is equal to the length of the longest path from *s* to *u*. Hence the length of the longest path of the used *st*-orientation is decisive in visibility representations of undirected graphs. Moreover, in the visibility representations, the length of the longest path of the dual graph is also important. How a different primal *st*-orientation impacts on the dual orientation is very crucial for visibility representations.

– **Orthogonal Drawings.** The first step of algorithms that compute orthogonal drawings [3] is to compute an *st*-numbering of the input undirected graph *G*. These algorithms compute some variables (such as the row pairs or the column pairs in [3]) that are functions of the *st*-orientation and which determine the width and the height of the drawing. Applying different *st*-orientations for the orthogonal drawing of a graph *G*, can result in different drawing area bounds.

Figure 1 depicts an undirected graph *G* (Figure 1a) and two different *st*-orientations of it. Figure 2 shows two different longest path and visibility representation layouts for the two different *st*-orientations (1b), (1c) of the same graph (1a). Note that the drawings have different characteristics, which depend on the length of the longest path of the different *st*-orientations.

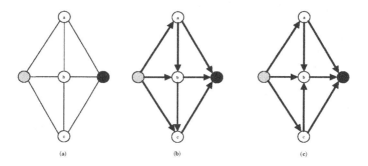

Fig. 1. An undirected graph (a) and two (b), (c) possible *st*-orientations of it

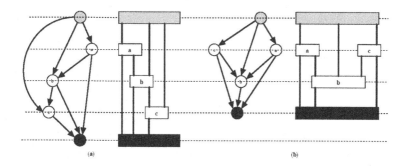

Fig. 2. Longest path layering and visibility representation layouts for the *st*-orientation of Figure 1b (a) and for this of Figure 1c (b)

In order to develop an algorithm for the computation of (longest-path) parameterized st-orientations, there are mainly two things that we should carefully
consider: (1), the correctness of the final st-orientation and (2), the algorithm
should give us the opportunity to control the length of the longest path of the
final directed graph. The idea behind the algorithm is that, beginning with an
undirected biconnected graph G and two nodes of it s, t, we repeatedly remove
a node v_i (different from t), orienting at the same time all its incident edges
from v_i to its neighbors. In this way we build up a directed graph F. The first
node removed is the source s, of the desired st-orientation. Thus, the problem
of computing a correct st-orientation is reduced to this of removing the vertices
of the graph with a correct order $v_1, v_2, \ldots, v_n$ with $v_1 = s$ and $v_n = t$ and
simultaneously maintaining a data structure that will allow us to compute such
a correct order.

2.2   Terminology

In this section, we present some terminology and useful observations. Throughout the paper, $N_G(v)$ denotes the set of neighbors of node v in graph G, s the
source and t the sink of the graph. Additionally, l is the length of the longest
path of the primal graph from s to t, whereas l^* denotes the length of the longest
path of the respective dual graph. Let $G = (V, E)$ be a one-connected undirected
graph, i.e., a graph that contains at least one vertex whose removal causes the
initial graph to disconnect and $T = (B \cup C, U)$ be the respective block-cutpoint
tree [12]. The edges $(i, j) \in U$ of the block-cutpoint tree always connect pairs of
blocks (biconnected components) and cutpoints such that the cutpoint of a tree
edge belongs to the vertex set of the corresponding block (see Figure 3).

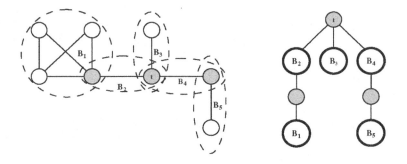

Fig. 3. A one-connected graph and the t-rooted block-cutpoint tree

The block-cutpoint tree is a free tree, i.e., it has no distinct root. In order to
transform this free tree into a rooted tree, we define the t-rooted block-cutpoint
tree with respect to the sink t. Consequently, the root of the block-cutpoint tree
is the block that contains t (see Figure 3).

Finally, we define the leaf-blocks of the t-rooted block-cutpoint tree to be
the blocks, except for the root of the block-cutpoint tree that contain a single

cutpoint. The block-cutpoint tree can be computed in $O(n + m)$ time with an algorithm similar to DFS [12].

Following, we give some results that are necessary for the development of the algorithm.

Lemma 1 *([11]). Let $G = (V, E)$ be an undirected biconnected graph and s, t be two of its nodes. Suppose we remove s and all its incident edges. Then there is at least one neighbor of s lying in a leaf-block of the t-rooted block-cutpoint tree. Moreover, this neighbor is not cutpoint.* □

The main idea of the algorithm is based on the successive removal of nodes and the simultaneous update of the t-rooted block-cutpoint tree. We call each such node a source, because at the time of its removal it is effectively chosen to be a source of the remainder of the graph. We initially remove s, the first source, which is the source of the desired st-orientation and give direction to all its incident edges from s to all its neighbors. After this removal, the graph either remains biconnected or is decomposed into several biconnected components but the number of leaf-blocks remains the same or is decomposed into several biconnected components and the number of leaf-blocks changes.

This procedure continues until all nodes of the graph but one are removed. As it will be clarified in the next sections, at every step of the algorithm there will be a set of potential sources to choose from. Our aim is to establish a connection between the current source choice and the length of the longest path of the produced st-oriented graph.

3   Parameterized st-Orientations

3.1   The Algorithm

Now we describe the procedure in a more formal way. We name this procedure STN. Let $G = (V, E)$ be an undirected biconnected graph and s, t two of its nodes. We will compute an st-orientation of G. Suppose we recursively produce the graphs $G_{i+1} = G_i - \{v_i\}$, where $v_1 = s$ and $G_1 = G$ for all $i = 1, \ldots, n - 1$.

During the procedure we always maintain a t-rooted block-cutpoint tree. Additionally, we maintain a structure Q that plays a major role in the choice of the current source. Q initially contains the desired source for the final orientation, s. Finally we maintain the leaf-blocks of the t-rooted block-cutpoint tree. During every iteration i of the algorithm node v_i is chosen so that

- it is a non-cutpoint node that belongs to Q (1)
- it belongs to a leaf-block of the t-rooted block-cutpoint tree (2)

Note that for $i = 1$ there is a single leaf-block (the initial biconnected graph) and the cutpoint that defines it is the desired sink of the orientation, t. When a source v_i is removed from the graph, we have to update Q in order to be able to choose our next source. Q is then updated by removing v_i and by inserting all of the neighbors of v_i except for t.

By Lemma 1, after the removal of a node v_i, there will always exist at least one node satisfying both (1) and (2). In this way we can reach the final sink of the orientation, t, without disconnecting the graph. Additionally, each time a node v_i is removed we orient all its incident edges from v_i to its neighbors. The procedure continues until Q gets empty. Let $F = (V', E')$ be the directed graph computed by this procedure. We claim [11] that $F = (V', E')$ is an st-oriented graph:

Theorem 2 *([11]). The directed graph $F = (V', E')$ computed by STN is st-oriented.* $\square$

STN is a recursive algorithm for computing an st-orientation of a biconnected undirected graph G. The full pseudocode and an illustrative example can be found in [11]. During the execution of the algorithm we can also compute an st-numbering f of the initial graph. Actually, for each node v_i that is removed from the graph, the subscript i is the final st-number of node v_i. Finally, each node v inserted into Q is associated with a timestamp value $m(v)$ (which will finally determine the longest path length). $m(v)$ is set equal to i, every time that v is discovered by a removed node v_i, i.e., v is a neighbor of v_i. This means that $m(v)$ can be updated many times until the algorithm terminates.

Let us now comment on the execution time of the algorithm. Each time a vertex is removed, we have to update the block-cutpoint tree, which takes time $O(n + m)$ [12]. As all the vertices are removed, the algorithm runs clearly in $O(nm)$ time. However we can use the algorithm for biconnectivity maintenance (which supports edge deletions in $O(\log^5 n)$ time) proposed in [13] and drop the bound to $O(m \log^5 n)$ [11].

3.2   Control of the Length of Longest Path

This section presents methods which can be implemented in order to *control* the length of the longest path of an st-orientation computed with STN. Actually, we take advantage of the timestamps $m(u)$ in order to choose our next source. During iteration j of the algorithm, we have to pick a leaf-block B_j^l of the t-rooted block-cutpoint tree and we always have to make a choice on the structure $Q' = B_j^l \cap Q \sim \{h_j^l\}$, where h_j^l is the cutpoint that defines B_j^l. Our investigation has revealed that if vertices with high timestamp are chosen then long sequences of vertices are formed and thus there is higher probability to obtain a long longest path. We call this way of choosing vertices MAX-STN. Actually, MAX-STN resembles a DFS traversal (it searches the graph at a *maximal* depth). Hence, during MAX-STN, the next source v is arbitrarily chosen from the set

$$\{v \in Q' : m(v) = \max\{m(i) : i \in Q'\}\}.$$

On the contrary, we have observed that if vertices with low timestamp are chosen, then the final st-oriented graph has relatively small longest path. We call this way of choosing vertices MIN-STN, which in turn resembles a BFS traversal. Hence, during MIN-STN, the next source v is arbitrarily chosen from the set

$$\{v \in Q' : m(v) = \min\{m(i) : i \in Q'\}\}.$$

The length of a longest path from s to t computed with MAX-STN is denoted with $\ell(t)$ whereas this computed with MIN-STN is denoted with $\lambda(t)$. As it has already been reported, it would be desirable to be able to compute st-oriented graphs of length of longest path within the interval $[\lambda(t), \ell(t)]$. This is called a parameterized st-orientation. So the question that arises is: Can we insert a parameter into our algorithm, for example a real constant $p \in [0, 1]$ so that our algorithm computes an st-oriented graph of length of longest path that is a function of p?

This is feasible if we modify STN. As the algorithm is executed exactly n times (n vertices are removed from the graph), we can execute the procedure MAX-STN for the first pn iterations and the procedure MIN-STN for the remaining $(1-p)n$ iterations. We call this method PAR-STN(p) and we say that it produces an st-oriented graph with length of longest path from s to t equal to $\Delta(p)$. Note that PAR-STN(0) is equivalent to MIN-STN, thus $\Delta(0) = \lambda(t)$ while PAR-STN(1) is equivalent to MAX-STN and $\Delta(1) = \ell(t)$. PAR-STN has been tested and it seems that when applied to st-Hamiltonian graphs (biconnected graphs that contain at least one path from s to t that contains all the nodes of the graph) there is a high probability that $\Delta(p) \geq p(n-1)$. Actually, $\Delta(p)$ is very close to $p(n-1)$. Additionally, it has been observed that if we switch the order of MAX-STN and MIN-STN execution, i.e., execute MIN-STN for the first pn iterations and MAX-STN for the remaining $(1-p)n$ iterations, there is a high probability that $\Delta(p) \leq p(n-1)$. In this case, $\Delta(p)$ is again very close to $p(n-1)$.

4   Primal and Dual *st*-Orientations

4.1   General

Now we present some results concerning the impact of parameterized st-orientations on st-planar graphs. If we st-orient such a graph, we can define a single orientation for the dual graph G^* which is also an s^*t^*-orientation.

This method is used in the visibility representations algorithms [2], when we have to compute the dual s^*t^*-oriented graph. The length of the longest path of this graph determines the width of the geometric representation. Thus, the questions that arise are natural. What is the impact of the parameter p on the length of the longest path of the dual s^*t^*-oriented graph G^* of an st-planar graph G, which (the graph G) has been st-oriented with PAR-STN(p)? Intuitively, we would expect that l^* (the length of the longest path of the dual graph G^*) will grow inversely proportional to l. As we will see, this is not always the case.

4.2   A Special Class of Planar Graphs

In this section we investigate certain classes of st-planar graphs that can be st-oriented in such a way that certain lengths of primal and dual longest paths can be achieved. This is actually a good reason to justify the fact that different st-orientations are indeed important in many applications.

Definition 3. *We define an n-path planar graph (n ≥ 5) G = (V, E) to be the planar graph that consists of a path P = $v_2, v_3, \ldots, v_{n-1}$ of n − 2 nodes and two other nodes v_1, v_n such that $(v_1, v_i) \in E$, $(v_i, v_n) \in E$ $\forall i = 2, \ldots n - 1$ and $(v_1, v_n) \in E$.*

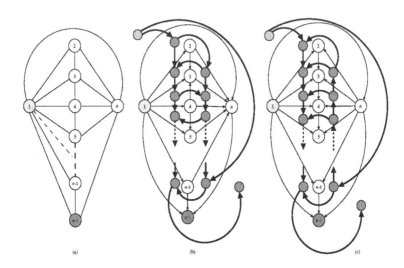

Fig. 4. (a) An n-path planar graph. We define node 1 to be the source of the graph and node $n − 1$ to be the sink of the graph. (b) Primal and Dual st-orientation with $l = 4$ and $l^* = 2n − 4$. (c) Primal and Dual st-orientation with $l = n − 1$ and $l^* = 2n − 4$.

In Figure 4a, one n-path planar graph is depicted. Its source is node 1 whereas its sink is node $n − 1$. Note that an $(n + 1)$-path planar graph G_{n+1} can be obtained from an n-path planar graph G_n if we add a new node and connect it with nodes v_1, v_2 and v_n (nodes v_1 and v_n are the rightmost and leftmost nodes of G_n's embedding in Figure 4). Let now G_n be an n-path planar graph and $\lambda(G_n), \ell(G_n)$ denote the minimum and the maximum longest path length $1(n−1)$-orientations over the set of all the $1(n − 1)$-orientations of G_n respectively. In Figure 4b, the primal orientation of minimum longest path length (together with the respective dual orientation of longest path length $\lambda^*(G_n)$) is depicted while in Figure 4c the orientation of maximum longest path length (together with respective dual orientation of longest path length $\ell^*(G_n)$) is depicted. Inductively, we can prove that for an n-path planar graph the following holds:

Theorem 4. *For all $n ≥ 5$, it is $\lambda(G_n) = 4$, $\ell(G_n) = n − 1$ and $\lambda^*(G_n) = \ell^*(G_n) = 2n − 4$.*

According to Theorem 4, the impact of different st-orientations of an n-path planar graph on the area of their visibility representation is evident. By using the minimum st-orientation, we will need an area equal to

$$\lambda(G_n)\lambda^*(G_n) = 4(2n − 4) = 8n − 16 = O(n)$$

If we use the maximum st-orientation, we will need an area equal to

$$\ell(G_n)\ell^*(G_n) = (n-1)(2n-4) = 2n^2 - 6n + 4 = O(n^2)$$

Note that while $\ell(G_n) + \ell^*(G_n) = 3n - 5 > 2n$, it is $\lambda(G_n) + \lambda^*(G_n) = 2n \leq 2n$. We therefore introduce the following conjecture:

Conjecture 5. *For every n-node planar biconnected graph G, two nodes s, t of its vertex set, there exists at least one st-orientation of G such that $l + l^* \leq 2n + c$, where c is a constant.*

In order to face this conjecture, one should try to devise an algorithm that deterministically st-orients a planar graph in a way that the produced length of the dual longest path grows at most as much as the primal one does.

5 Experimental Results

Following we present our results for different kinds of graphs, st-Hamiltonian graphs (undirected graphs that have at least one Hamilton path from s to t and hence an upper bound for the longest path length equal to $n-1$) and planar graphs. All experiments were run on a Pentium IV machine, 512 MB RAM, 2.8 GH under Windows 2000 professional.

5.1 st-Hamiltonian Graphs

We have implemented the algorithm in Java, using the Java Data Structures Library (www.jdsl.org) [14]. The graphs we have tested are n-node-undirected

Table 1. Results for density 3.5 st-Hamiltonian graphs

n	p=0		p=0.3		p=0.5		p=0.7		p=1	
	l	%(n−1)	l	%(n−1)	l	%(n−1)	l	%(n−1)	l	%(n−1)
100	14.00	0.141	38.90	0.393	59.20	0.598	76.50	0.773	92.20	0.931
200	18.60	0.093	74.10	0.372	113.00	0.568	147.90	0.743	186.60	0.938
300	23.30	0.078	104.80	0.351	165.10	0.552	219.20	0.733	280.70	0.939
400	23.30	0.058	139.10	0.349	213.80	0.536	289.30	0.725	376.30	0.943
500	29.20	0.059	169.40	0.339	267.30	0.536	361.20	0.724	470.70	0.943
600	27.90	0.047	202.10	0.337	318.90	0.532	428.90	0.716	566.60	0.946
800	30.00	0.038	264.90	0.332	415.30	0.520	566.50	0.709	755.60	0.946
900	31.70	0.035	294.30	0.327	469.90	0.523	640.20	0.712	848.10	0.943
1000	36.20	0.036	322.10	0.322	518.20	0.519	709.30	0.710	940.00	0.941
1100	38.90	0.035	353.90	0.322	576.30	0.524	782.90	0.712	1033.40	0.940
1200	34.40	0.029	387.00	0.323	622.10	0.519	845.50	0.705	1127.80	0.941
1300	34.30	0.026	421.10	0.324	674.50	0.519	917.00	0.706	1223.10	0.942
1400	38.90	0.028	448.80	0.321	718.40	0.514	983.90	0.703	1319.90	0.943
1500	38.00	0.025	478.30	0.319	775.70	0.517	1056.40	0.705	1417.10	0.945
1600	39.30	0.025	515.00	0.322	824.30	0.516	1137.20	0.711	1499.10	0.938
1700	38.50	0.023	539.30	0.317	872.00	0.513	1190.40	0.701	1604.00	0.944
1800	41.10	0.023	571.90	0.318	923.60	0.513	1263.80	0.703	1691.30	0.940
1900	41.40	0.022	605.60	0.319	978.60	0.515	1331.80	0.701	1786.30	0.941
2000	44.00	0.022	632.40	0.316	1023.80	0.512	1403.50	0.702	1883.90	0.942

st-Hamiltonian graphs of density d where $n = 100, 200, 300, \ldots, 2000$ and $d = 3.5$. For each pair (n, d) we have tested 10 different randomly generated graphs (and we present the mean of the length of the longest path) in order to get more reliable results. We have similar results for other values of density as well (see Figure 5).

As we can see, the results (Table 1 and Figure 5) are remarkably consistent with the parameter p. The computed longest path length for $p = p_0$ is always very close to $p_0(n-1)$. The computed results are similar for increasing graphs size and density.

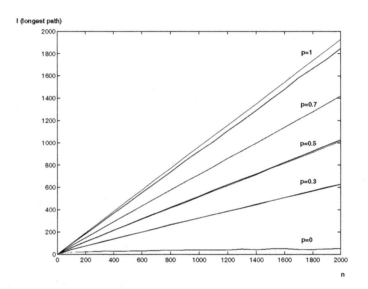

Fig. 5. Longest path length as a function of n, d, p $(d = 2.5, 6.5)$

Table 2. Primal and dual longest path length for triangulated st-planar graphs

n	$2n$	l	l^*	$l+l^*$	l	l^*	$l+l^*$	l	l^*	$l+l^*$	$p=0$	$p=0.5$	$p=1$
		p=0			**p=0.5**			**p=1**			**l × l***		
109	218	31	167	198	75	95	170	100	74	174	5177	7125	7400
310	620	44	503	547	186	319	505	280	163	443	22132	59334	45640
535	1070	98	785	883	240	534	774	402	293	695	76930	128160	117786
763	1526	144	1114	1258	385	780	1165	691	241	932	160416	300300	166531
998	1996	83	1419	1502	425	862	1287	846	340	1186	117777	366350	287640
1302	2604	134	2024	2158	704	1154	1858	1173	451	1624	271216	812416	529023
1501	3002	119	2203	2322	784	1073	1857	1403	224	1627	262157	841232	314272
1719	3438	131	2550	2681	856	1661	2517	1555	515	2070	334050	1421816	800825
1990	3980	208	2339	2547	1013	1581	2594	1773	400	2173	486512	1601553	709200
2159	4318	142	3238	3380	930	1816	2746	1823	445	2268	459796	1688880	811235
2268	4536	148	3136	3284	952	1666	2618	1887	336	2223	464128	1586032	634032
4323	8646	356	5852	6208	2238	3589	5827	3957	841	4798	2083312	8032182	3327837

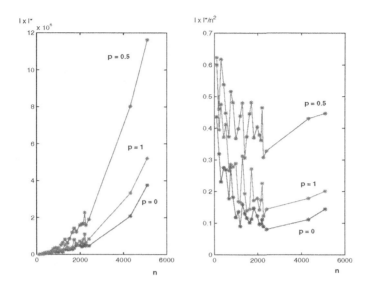

Fig. 6. Absolute (left) and normalized (divided by n^2) (right) results for visibility representation area requirement for different values of the parameter p and triangulated planar graphs. The parameter $p = 0$ (low longest path length *st*-oriented graphs) is clearly preferable.

5.2 Planar Graphs

In this section we present some results for maximum density (triangulated) *st*-planar graphs. We also have similar results for low density planar graphs. We mainly present the impact of the parameter p on the primal and dual longest path length of the planar graphs. From Table 2, it is clear that the primal and the dual longest path length are inversely proportional for various values of the parameter p. We have used the values $p = 0, 0.5, 1$, as the most representative ones. The last three columns of Table 3 show the product $l \times l^*$. This is actually the area that is needed in order to construct a visibility representation of the given graph using the algorithms proposed in [2].

Fig. 7. Visibility Representations of a 21-path planar graph for different *st*-orientations $(p = 0, 0.5, 1)$

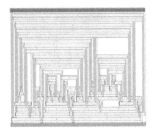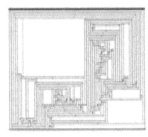

Fig. 8. Visibility Representations of a 85-node triangulated planar graph for different *st*-orientations produced with PAR-STN(*p*) (*p* = 0, 0.5, 1)

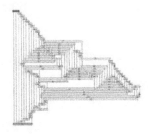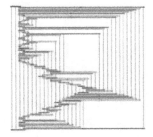

Fig. 9. Visibility Representations of a 10x10 grid graph for different *st*-orientations produced with PAR-STN(*p*) (*p* = 0, 0.25, 1)

Figure 7 shows 3 visibility representation frames of a 21-path planar graph. The difference in the area is evident. Note that the visibility representation that uses the minimum *st*-orientation (*p* = 0) consumes the least area. Figure 8 contains 3 visibility representations frames of a triangulated graph where the value *p* = 0 is preferable. Finally, in Figure 9 we present some visibility representations frames produced by *st*-orienting a grid graph. In this case, the importance of the parameter is clear. Using a parameterized *st*-orientation with *p* = 0.25 is preferable, as it produces a more *compact* drawing.

5.3   Orthogonal Drawings

The impact of the different *st*-orientations is not very clear in orthogonal drawings. However, for the algorithm described in [3], where the area upper bound is roughly $0.76n^2$, we are able to produce *st*-numberings that produce drawings of area upper bound roughly equal to $0.68n^2$ or less, by using the parameterized *st*-orientation algorithm. Due to space limitations, we cannot describe further details.

6   Conclusions

In this paper the application of parameterized *st*-orientations in graph drawing algorithms (mainly in visibility representations) is presented. It seems that there is a way to efficiently control the length of the longest path of an *st*-orientation and keep it "short", "long" or "medium". Experimental results not only on planar graphs but also on non-planar graphs reveal the robustness of the algorithm.

Acknowledgements. The authors would like to thank Hubert de Fraysseix, C.N.R.S. for his help on the visualization frames produced with his software P.I.G.A.L.E..

References

1. G.D. Battista, P. Eades, R. Tamassia, and I.G. Tollis. *Graph Drawing: Algorithms for the Visualization of Graphs.* Prentice Hall, 1999.
2. R. Tamassia and I.G. Tollis. A unified approach to visibility representations of planar graphs. *Disc. and Comp. Geom.*, 1:321–341, 1986.
3. A. Papakostas and I.G. Tollis. Algorithms for area-efficient orthogonal drawings. *Computational Geometry: Theory and Applications*, 9:83–110, 1998.
4. S. Even and R. Tarjan. Computing an st-numbering. *Theoretical Computer Science*, 2:339–344, 1976.
5. A. Lempel, S. Even, and I. Cederbaum. An algorithm for planarity testing of graphs. In *P. Rosestiehl(ed.) Theory of Graphs: International Symposium July 1966*, pages 215–232, 1967.
6. J. Ebert. st-ordering the vertices of biconenected graphs. *Computing*, 30(1):19–33, 1983.
7. R. Tarjan. Two streamlined depth-first search algorithms. *Fundamentae Informatica*, 9:85–94, 1986.
8. P. Rosehnstiehl and R. Tarjan. Rectilinear planar layout and bipolar orientation of planar graphs. *Discrete Comput. Geom.*, 1:343–353, 1986.
9. Y. Maon, B. Schieber, and U. Vishkin. Parallel ear decomposition search (eds) and st-numbering in graphs. *Theoret. Comput. Sci*, 47:277–298, 1986.
10. H.D. Fraysseix, P.O. de Mendez, and P. Rosenstiehl. Bipolar orientations revisited. *Discrete Applied Mathematics*, 56:157–179, 1995.
11. C. Papamanthou and I.G. Tollis. Algorithms for parameterized st-orientations of graphs, *submitted to ESA2005.*
12. J. Hopcroft and R. Tarjan. Efficient algorithms for graph manipulation. *Comm. ACM*, 16:372–378, 1973.
13. J. Holm, K. de Lichtenberg, and M. Thorup. Poly-logarithmic deterministic fully-dynamic algorithms for connectivity, minimum spanning tree, 2-edge and biconnectivity. *J. ACM*, 48(4):723–760, 2001.
14. Michael T. Goodrich and Roberto Tamassia. *Data Structures and Algorithms in Java.* John Wiley & Sons, 2 edition, 2001.

Complexity Results for Three-Dimensional Orthogonal Graph Drawing[*]
(Extended Abstract)

Maurizio Patrignani

Università di Roma Tre, Italy
patrigna@dia.uniroma3.it

Abstract. We introduce the 3SAT reduction framework which can be used to prove the NP-hardness of finding three-dimensional orthogonal drawings with specific constraints. We use it to show that finding a drawing of a graph whose edges have a fixed shape is NP-hard. Also, it is NP-hard finding a drawing of a graph with nodes at prescribed positions when a maximum of two bends per edge is allowed. We comment on the impact of these results on the two open problems of determining whether a graph always admits a 3D orthogonal drawing with at most two bends per edge and of characterizing orthogonal shapes admitting a drawing without intersections.

1 Introduction

Three-dimensional orthogonal graph drawing has attracted a constant research interest throughout the last decade [1, 4, 10, 11, 12, 13, 18, 20]. Nevertheless, some basic questions still lack an answer. It is open, for example, whether a graph of maximum degree six always admits a drawing with at most two bends per edge ([11, 20], and [5], problem #46). In [19] it is shown that such drawings may imply many edges sharing the same axis-perpendicular plane.

Also, a characterization of the orthogonal shapes admitting an orthogonal drawing without intersections (called *simple orthogonal shapes*) is still missing in the general case ([9] and [3], problem 20). Such a characterization would allow the separation of the task of defining the shape of the drawing from the task of computing its coordinates, extending to three-dimensions the well studied and widely adopted two-dimensional approach known as topology-shape-metrics [16, 6].

More formally, we would like to find the solution to the so-called *Simplicity Testing Problem*: Let G be a graph whose edges are directed and labeled with

---

[*] Work partially supported by European Commission – Fet Open project COSIN – COevolution and Self-organisation In dynamical Networks – IST-2001-33555, by European Commission – Fet Open project DELIS – Dynamically Evolving Large Scale Information Systems – Contract no 001907, by MIUR under Project ALGO-NEXT (Algorithms for the Next Generation Internet and Web: Methodologies, Design, and Experiments), and by "The Multichannel Adaptive Information Systems (MAIS) Project", MIUR Fondo per gli Investimenti della Ricerca di Base.

P. Healy and N.S. Nikolov (Eds.): GD 2005, LNCS 3843, pp. 368–379, 2005.

a sequence of labels in the set {x+, x−, y+, y−, z+, z−}. Does a 3D orthogonal drawing of G exist such that each edge has a shape "consistent" with its labeling and no two edges intersect?

While a solution to the 2D counterpart of the Simplicity Testing Problem can be found in the works by Vijaian and Widgerson and by Tamassia [16, 17], only very preliminary results toward the recognition of simple orthogonal 3D shapes are provided in [7, 8] where paths (with further additional constraints) and cycles are considered, respectively. In [9] it is shown that the known characterization for cycles does not immediately extend to even seemingly simple graphs such as theta graphs (simple graphs consisting of three cycles).

In this paper we consider three-dimensional orthogonal drawing of a maximum degree six graph from the computational complexity perspective. We introduce the 3SAT reduction framework which can be used to show that it is NP-hard to decide if an orthogonal 3D drawing of a graph satisfying some constraints exists (Section 3). By using such a framework we show that the *simplicity testing* problem is NP-complete (Section 4) and that the opposite problem of finding a drawing of a graph with nodes at prescribed positions is also NP-complete when a maximum of two bends per edge is allowed (Section 5), while it is polynomial in the general case.

We comment on the impact of these results on the two open problems of determining whether a graph always admits a drawing with at most two bends per edge and of characterizing orthogonal shapes admitting an orthogonal drawing without intersections (Section 6).

2   Background

We assume familiarity with basic graph drawing, graph theory, and computational geometry terminology (see, e.g. [6, 2, 15]).

A *(3D orthogonal) drawing* of a graph is such that nodes are mapped to distinct points of the three dimensional space and edges are chains of axis-parallel segments. A *bend* is a point shared between two subsequent segments of the same edge. An *intersection* in a 3D orthogonal drawing is a pair of edges that overlap in at least one point that does not correspond to a common end-node. A *k-bend drawing* of a graph, where k is a non-negative integer, is a non-intersecting drawing such that each edge has at most k bends.

An *x-plane* (*y-plane*, *z-plane*, respectively) is a plane perpendicular to the x axis (y axis, z axis, respectively). Given a drawing Γ of a graph G and two nodes u and v, we write $u >_x v$ if the x coordinate of u is greater than the x coordinate of v in Γ. Also, we write $u >_x >_y v$ if $u >_x v$ and $u >_y v$.

A *direction label* is a label in the set {x+, x−, y+, y−, z+, z−}. Let G be a graph and Γ be a drawing of G. Let e be an undirected edge of G whose end-nodes are u and v. Select one of the two possible orientations (u, v) and (v, u) of e and call $p_1, p_2, \ldots, p_m$ the end points of the orthogonal segments corresponding to edge e in Γ in the order in which they are encountered while moving along e from u to v. The *shape of e in Γ* is the sequence of the direction labels corresponding to the directions of vectors $\overrightarrow{p_i, p_{i+1}}$, $i = 1, \ldots, m - 1$. For example, consider an edge (u, v) drawn with a single bend b and such that $u <_x b <_y v$. The shape of e consists of the orientation from u to v and the sequence of labels x+, y+. We also write $u \xrightarrow{\text{x+}} \xrightarrow{\text{y+}} v$.

When producing a drawing of a graph one can ask if the positions of the vertices and the shapes of the edges can be computed separately. When computing positions first, one has to solve the following problem.

Problem: **Routing**
Instance: A graph $G(V, E)$ and a mapping between nodes and distinct points of the three-dimensional space.
Question: Does a non-intersecting 3D orthogonal drawing of G exist such that the nodes have the specified coordinates?

We call 2-BEND ROUTING the ROUTING problem when restricted to 2-bend drawings.

Conversely, it can be asked what is the complexity of deciding if a graph admits a drawing such that its edges have a specified shape. We call *shape graph* a graph where a shape (an orientation and a sequence of direction labels) is specified for each one of its edges. A shape graph γ is *simple* if it admits a non-intersecting drawing Γ such that each edge has the specified shape. Formally, the SIMPLICITY TESTING problem is as follows.

Problem: **Simplicity Testing**
Instance: A shape graph γ, that is, a graph $G(V, E)$ and a shape for each edge $e \in E$, consisting of an orientation of e and a sequence of labels in the set $\{$x+, x-, y+, y-, z+, z-$\}$.
Question: Does a non-intersecting drawing of G exist such that each edge has the specified shape?

3 The 3SAT Reduction Framework

The 3SAT reduction framework introduced in this section can be used to show that it is NP-hard finding a 3D drawing of a graph within the orthogonal standard that satisfies some constraints. By using this framework it is shown, in Sections 4 and 5, respectively, the NP-hardness of SIMPLICITY TESTING and of 2-BEND ROUTING. Throughout this section, the target problem is assumed to be as follows:

Problem: **Target problem**
Instance: A graph $G(V, E)$ and a set S of constraints expressed with respect to its nodes and edges.
Question: Does a non-intersecting 3D drawing of G exist such that the constraints in S are satisfied?

The 3SAT problem is as follows:

Problem: **3-Satisfiability (3SAT)**
Instance: A set of clauses $\{c_1, c_2, \ldots, c_m\}$, each containing three literals from a set of boolean variables $\{v_1, v_2, \ldots, v_n\}$.
Question: Can truth values be assigned to the variables so that each clause contains at least one true literal?

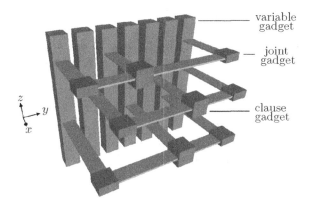

Fig. 1. A representation of the basic blocks of an instance of the target problem built as specified by the 3SAT reduction framework

Given a 3SAT instance ϕ, the 3SAT reduction framework specifies how to build an instance $I_\phi = (G_\phi(V_\phi, E_\phi), S_\phi)$ of the target problem such that ϕ admits a solution if and only if I_ϕ does. $G_\phi(V_\phi, E_\phi)$ is composed of three different types of gadgets connected together. The bounding boxes of the gadgets are depicted in Fig. 1, while the interior components are not shown and depend on the specific target problem.

For each boolean variable v_i of ϕ, instance I_ϕ has a *variable gadget* V_i. Fig. 1 shows the variable gadgets as tall vertical blocks placed in a row along the y axis in such a way that, if $i < j$, variable gadget V_i has lower y coordinates than variable gadget V_j.

For each clause $c_i = l_h \vee l_j \vee l_k$ of ϕ instance I_ϕ has one *clause gadget* C_i. Clause gadgets are represented in Fig. 1 as small cubes. Denoted with v_h, v_j, and v_k the variables of literals l_h, l_j, and l_k, respectively, and assumed that $h < j < k$, clause gadget C_i is placed directly in front of the variable gadget V_j.

For each clause $c_i = l_h \vee l_j \vee l_k$ of ϕ, I_ϕ has two *joint gadgets* $J_{i,h}$ and $J_{i,k}$, depicted in Fig. 1 as flat blocks placed in front of the variable gadgets V_h and V_k, respectively.

In order to use the 3SAT reduction framework for the NP-hardness proof of a specific target problem a complete specification must be provided, where a *specification* for the 3SAT reduction framework is defined as follows.

- Construction rules describing how, starting from an instance ϕ of the 3SAT problem, variable gadgets, joint gadgets, and clause gadgets are built and connected together and an instance $I_\phi = (G_\phi(V_\phi, E_\phi), S_\phi)$ of the target problem is obtained.

- For each variable gadget V_i a bipartition of the non-intersecting drawings of $G_\phi(V_\phi, E_\phi)$ satisfying constraints S_ϕ into two sets, denoted T_{V_i} and F_{V_i}.
- For each joint gadget $J_{i,k}$ a bipartition of the non-intersecting drawings of $G_\phi(V_\phi, E_\phi)$ satisfying constraints S_ϕ into two sets, denoted $T_{J_{i,k}}$ and $F_{J_{i,k}}$.

A specification is said to be *compliant* if, for any 3SAT instance ϕ, the following four statements hold.

Statement 1. *Instance $I_\phi = (G_\phi(V_\phi, E_\phi), S_\phi)$ of the target problem corresponding to instance ϕ of 3SAT can be constructed in polynomial time.*

Statement 2. *If a non-intersecting drawing of $G_\phi(V_\phi, E_\phi)$ satisfying S_ϕ exists, it belongs to $T_{J_{i,h}}$ ($T_{J_{i,k}}$) if and only if it belongs to T_{V_h} (T_{V_k}).*

Statement 3. *For each clause $c_i = l_h \vee l_j \vee l_k$, where l_h (l_j, l_k, respectively) is the positive or the negative literal of variable v_h (v_j, v_k, respectively), and for each non-intersecting drawing Γ of $G_\phi(V_\phi, E_\phi)$ satisfying S_ϕ at least one among the following conditions holds:*

1. *$\Gamma \in T_{J_{i,h}}$ ($\Gamma \in F_{J_{i,h}}$) and l_h is the positive (negative) literal of v_h.*
2. *$\Gamma \in T_{V_j}$ ($\Gamma \in F_{V_j}$) and l_j is the positive (negative) literal of v_j.*
3. *$\Gamma \in T_{J_{i,k}}$ ($\Gamma \in F_{J_{i,k}}$) and l_k is the positive (negative) literal of v_k.*

Statement 4. *Consider a truth assignment to the variables $v_i, \ldots v_n$ satisfying ϕ. The set $\bigcap_{i=0}^n A_i$, where $A_i = T_{V_i}$ if v_i is true and $A_i = F_{V_i}$ if v_i is false, is non-empty.*

Theorem 1. *Given a target problem, whose instance is a graph $G(V, E)$ and a set S of constraints expressed with respect to its nodes and edges, if it admits a compliant specification for the 3SAT reduction framework, then finding a non-intersecting 3D orthogonal drawing of G satisfying the constraints in S is NP-hard.*

Proof sketch. Consider a non-intersecting drawing Γ of $G_\phi(V_\phi, E_\phi)$ satisfying S_ϕ. It is easy to find an assignment of truth values to the boolean variables that satisfies ϕ, by taking $v_i = true$ if $\Gamma \in T_{V_i}$ and $v_i = false$ if $\Gamma \in F_{V_i}$. In fact, because of Statements 2 and 3 we have that each clause $c_i = l_h \vee l_j \vee l_k$ has at least one true literal and thus ϕ is satisfied. Conversely, consider an assignment of truth values to the boolean variables that satisfies ϕ. Statement 4 guarantees the existence of a drawing of $G_\phi(V_\phi, E_\phi)$ satisfying S_ϕ. The proof is completed by showing that instance $I_\phi = (G_\phi(V_\phi, E_\phi), S_\phi)$ can be obtained in polynomial time, which is guaranteed by Statement 1. □

4 Fixing the Shape and Searching for Coordinates

In this section we consider the SIMPLICITY TESTING problem, that is the problem of finding a non-intersecting drawing for a graph whose orthogonal shape is fixed.

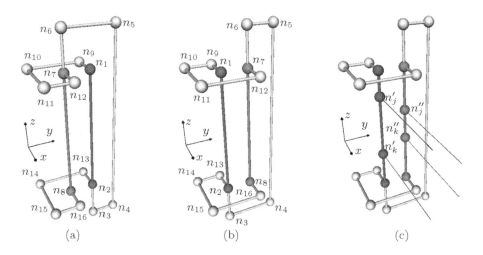

Fig. 2. A drawing of the variable gadget V_i belonging to T_{V_i} (a) and a drawing belonging to F_{V_i} (b). Nodes n'_j, n''_j, n'_k, and n''_k are inserted in order to transmit the geometric constraints to the clause gadgets of clauses C_j and C_k, respectively (c).

It is easy to prove that the problem is in NP (see [14]). In order to prove that SIMPLICITY TESTING is NP-hard we use the framework introduced in Section 3.

The heart of the variable gadget V_i, depicted in Fig. 2, is the path $n_1 \xrightarrow{z-} n_2 \xrightarrow{z-} n_3 \xrightarrow{y+} n_4 \xrightarrow{z+} n_5 \xrightarrow{y-} n_6 \xrightarrow{z-} n_7 \xrightarrow{z-} n_8$, whose nodes lie on the same x-plane. Further, the path $n_1 \xrightarrow{x-} n_9 \xrightarrow{y-} n_{10} \xrightarrow{x+} n_{11} \xrightarrow{y+} n_{12} \xrightarrow{x-} n_7$ constrains nodes n_1 and n_7 to share the same z-plane. Analogously, path $n_2 \xrightarrow{x-} n_{13} \xrightarrow{y-} n_{14} \xrightarrow{x+} n_{15} \xrightarrow{y+} n_{16} \xrightarrow{x-} n_8$ constrains nodes n_2 and n_8 to share the same z-plane. We define T_{V_i} as the set of non-intersecting drawings of $G_\phi(V_\phi, E_\phi)$ satisfying the directions constraints and such that $n_1 >_y n_7$ (as in Fig. 2.a). Analogously, we define F_{V_i} as the set of non-intersecting drawings of $G_\phi(V_\phi, E_\phi)$ satisfying the direction constraints and such that $n_1 <_y n_7$ (as in Fig. 2.b). Observe that T_{V_i} and F_{V_i} form a bipartition of the non-intersecting drawings of $G_\phi(V_\phi, E_\phi)$ satisfying the direction constraints.

For each clause c_j of the 3SAT formula in which the variable participates we insert a node n'_j between nodes n_1 and n_2 and a node n''_j between nodes n_7 and n_8. In any drawing Γ of $G_\phi(V_\phi, E_\phi)$ satisfying the direction constraints, nodes n'_j and n''_j have the same relative position with respect to the y axis as n_1 and n_7, i.e., $n'_j >_y n''_j$ if $\Gamma \in T_{V_i}$ and $n'_j <_y n''_j$ if $\Gamma \in F_{V_i}$. Suitable edges attached to the nodes n'_j and n''_j along the protruding lines shown in Fig. 2.c transmit the above constraints from V_i to the clause gadget C_j (possibly via joint gadget $J_{j,i}$). Note that nodes n'_j and n''_j do not need to lie on the same z-plane.

Given a clause $c_i = l_h \vee l_j \vee l_k$, the joint gadget $J_{i,k}$ is the reflected image with respect to the y axis of the joint gadget $J_{i,h}$. Thus, in the following we will only describe the joint gadget $J_{i,h}$, which is depicted in Fig. 3 and composed of two cycles $\alpha = n_1 \xrightarrow{y+} n_2 \xrightarrow{x-} n_3 \xrightarrow{y-} n_4 \xrightarrow{x+} n_5 \xrightarrow{y-} n_6 \xrightarrow{x+} n_7 \xrightarrow{y+} n_8 \xrightarrow{x-} n_1$, and $\alpha' = n'_1 \xrightarrow{y-} n'_2 \xrightarrow{x+} n'_3 \xrightarrow{y+} n'_4 \xrightarrow{x-} n'_5 \xrightarrow{y+} n'_6 \xrightarrow{x-} n'_7 \xrightarrow{y-} n'_8 \xrightarrow{x+} n'_1$. Nodes n_1 and n'_1

are connected by a path $n_1 \xrightarrow{z-} n_1'' \xrightarrow{z-} n_1'$ while nodes n_5 and n_5' are connected by the path $n_5 \xrightarrow{z-} n_5'' \xrightarrow{z-} n_5'$.

We define $T_{J_{i,h}}$ as the set of non-intersecting drawings of $G_\phi(V_\phi, E_\phi)$ satisfying the directions constraints and such that $n_5'' <_x n_1''$ (as in Fig. 3.a). Analogously, we define F_{V_i} as the set of non-intersecting drawings of $G_\phi(V_\phi, E_\phi)$ satisfying the direction constraints and such that $n_5'' >_x n_1''$ (as in Fig. 3.b). Observe that, since in any non-intersecting drawing of $G_\phi(V_\phi, E_\phi)$ n_1'' and n_5'' have distinct x and y coordinates, $T_{J_{i,h}}$ and $F_{J_{i,h}}$ form a bipartition of the non-intersecting drawings of $G_\phi(V_\phi, E_\phi)$ satisfying the direction constraints.

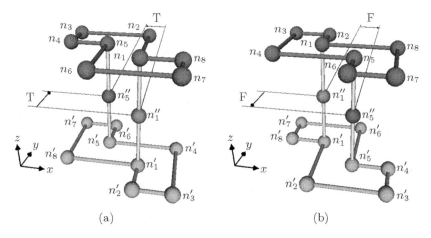

Fig. 3. A drawing of the joint gadget $J_{i,h}$ belonging to $T_{J_{i,h}}$ (a) and a drawing belonging to $F_{J_{i,h}}$ (b)

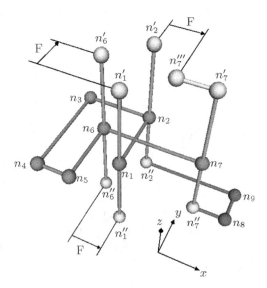

Fig. 4. Clause gadget

The clause gadget is depicted in Fig. 4. Its main component is the path $\alpha = n_1 \xrightarrow{y+} n_2 \xrightarrow{x-} n_3 \xrightarrow{y-} n_4 \xrightarrow{x+} n_5 \xrightarrow{y+} n_6 \xrightarrow{x+} n_7$, whose nodes lie on the same z-plane. Attached to α are the paths $n_1' \xrightarrow{z-} n_1 \xrightarrow{z-} n_1''$, $n_2' \xrightarrow{z-} n_2 \xrightarrow{z-} n_2''$, $n_6' \xrightarrow{z-} n_6 \xrightarrow{z-} n_6''$, and $n_7''' \xrightarrow{x+} n_7' \xrightarrow{z-} n_7 \xrightarrow{z-} n_7'' \xrightarrow{x+} n_8 \xrightarrow{y+} n_9 \xrightarrow{x-} n_2''$.

Joint gadget $J_{i,h}$ is connected to both variable gadget V_h and clause gadget C_i. In particular, n_1'' of $J_{i,h}$ is connected to n_i'' of V_h with the edge $n_i'' \xrightarrow{x+} n_1''$ and n_5'' of $J_{i,h}$ is connected to n_i' of V_h with the edge $n_i' \xrightarrow{x+} n_5''$. Each clause $c_i = l_h \vee l_j \vee l_k$ is connected to joint gadget $J_{i,h}$, variable gadget V_j, and joint gadget $J_{i,k}$. If l_h is the positive (negative) literal of variable v_h, we attach nodes n_1'' and n_5'' of the joint gadget $J_{i,h}$ to nodes n_6'' and n_1'' (n_1'' and n_6''), respectively. If l_j is the positive (negative) literal of variable v_j, we attach nodes n_i' and n_i'' of the variable gadget V_j to n_6' and n_1' (n_1' and n_6'), respectively. If l_k is the positive (negative) literal of variable v_k, we attach nodes n_1 and n_4 of the joint gadget $J_{i,k}$ to nodes n_2' and n_7''' (n_7''' and n_2'), respectively.

It is now easy to prove that the above construction rules are a compliant specification for the 3SAT reduction framework. Hence, we have:

Theorem 2. SIMPLICITY TESTING *is NP-complete.*

5   Fixing the Coordinates and Searching for a Shape

In this section we tackle the reverse problem with respect to the one addressed in Section 4, that is, the problem of finding a routing for the edges when the position of the nodes is fixed. An algorithm to solve ROUTING in $O(|V|log|V|)$ time, where $|V|$ is the number of vertices of the input graph, can be found in [14]. Conversely, we show that the same problem where only two bends per edge are allowed (2-BEND ROUTING) is NP-complete.

In order to show that 2-BEND ROUTING is NP-hard we take advantage of the 3SAT reduction framework introduced in Section 3. The basic gadget shown in Fig. 5 is used as a building block of several parts of the 2-BEND ROUTING instance and is composed of ten nodes. Node n_1 is connected to the three nodes n_2, n_3 and n_4. Analogously, node n_5 is connected to the three nodes n_6, n_7 and n_8. Nodes n_1 and n_5 are connected both with the single edge (n_1, n_5) and with the path of three edges $(n_1, n_{1,5})$, $(n_{1,5}, n_{5,1})$ and $(n_{5,1}, n_5)$.

As for nodes prescribed positions, they are placed in such a way that $n_1 <_x <_y <_z n_2 =_x =_y <_z n_3 =_x =_y <_z n_4$, $n_1 =_x >_y >_z n_{1,5} =_x >_y >_z n_{5,1} =_x >_y >_z n_5$, and $n_5 <_x >_y >_z n_6 =_x =_y >_z n_7 =_x =_y >_z n_8$.

Given a 2-bend drawing of the basic gadget, we call *true* the basic gadget when it is drawn with the bend of edge (n_1, n_5) placed in $p_{t,2}$ (see Fig. 5.a) and *false* the basic gadget when it is drawn with the bend of edge (n_1, n_5) placed in $p_{f,2}$ (see Fig. 5.b). Also, in what follows we use the graphic representation of the basic gadget shown in Fig. 5.c, where the nodes n_1, n_2, n_3, n_4, and $n_{1,5}$ are replaced by their bounding box, and analogously for the nodes n_5, n_6, n_7, n_8, and $n_{5,1}$. In this representation only edge (n_1, n_5) is shown, and it is assumed to have its bend in $p_{t,2}$.

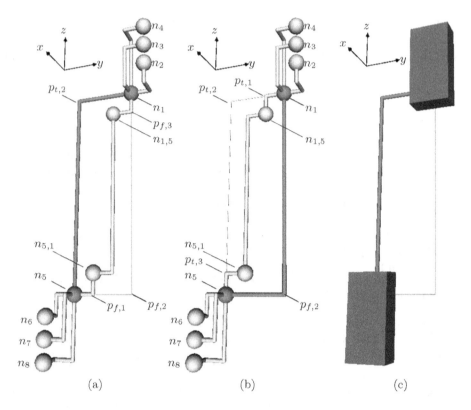

Fig. 5. A true drawing (a) and a false drawing (b) of the basic gadget. In (c) it is shown the schematic representation of the basic gadget that is used in the remaining part of the paper.

The variable gadget V_i is composed of a single basic gadget. Given a variable gadget V_i, we define as T_{V_i} (F_{V_i}) the set of non-intersecting 2-bend drawings of $G_\phi(V_\phi, E_\phi)$ such that the basic gadget is true (false).

The joint gadget $J_{i,h}$, which is depicted in Fig. 6 is built by interleaving four basic gadgets B_1, B_2, B_3, and B_4 as follows. B_1 intersects the variable gadget (not shown in Fig. 6). B_2 is placed on an orthogonal plane as shown in Fig. 6.

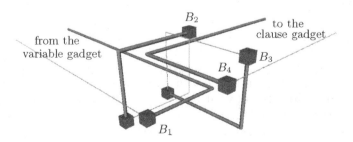

Fig. 6. Joint gadget $J_{i,h}$ is composed of four interleaved basic gadgets

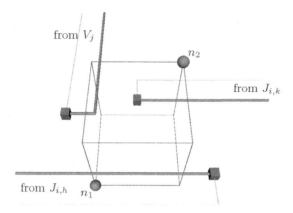

Fig. 7. The clause C_i for clause $c_i = l_h \vee l_j \vee l_k$ in the case in which l_h, l_j and l_k are negative literals: if the three variables v_h, v_h and v_k are true, there is no way of adding edge (n_1, n_2) with at most two bend

B_3 intersects only B_2 and is placed on a plane orthogonal to the first two (see Fig. 6). Finally, B_4 is placed on a plane parallel to the first one and intersects B_3 only as shown in Fig. 6. We define $T_{J_{i,h}}$ $(F_{J_{i,h}})$ as the set of non-intersecting 2-bend drawings of G_ϕ satisfying S_ϕ such that B_4 is true (false).

The clause C_i for clause $c_i = l_h \vee l_j \vee l_k$ is shown in Fig. 7. It is composed of two nodes n_1 and n_2 placed at the opposite vertices of a cube. The two nodes are joined by edge (n_1, n_2) (not shown in Fig. 7). In any 2-bend drawing of the clause gadget edge (n_1, n_2) uses one of the four vertical edges of the cube. The basic gadget B_4 of joint gadgets $J_{i,h}$ and $J_{i,k}$ and the basic gadget coming from V_j suitably intersect the vertical edges of the cube such that only if one literal is true the clause gadget admits a non-intersecting drawing.

It is easy to show that 2-BEND ROUTING is in NP. Since the above described construction rules are a compliant specification for the 3SAT reduction framework, we have:

Theorem 3. 2-BEND ROUTING *is NP-complete.*

6   Discussion and Open Problems

This paper shows that SIMPLICITY TESTING is NP-complete, while the reverse problem, ROUTING, is feasible. This asymmetry may explain why most three-dimensional drawing algorithms in the literature determine edge shapes as a consequence of node relative positions and not vice versa.

With respect to the problem of characterizing simple orthogonal shapes, deciding whether a shape graph is simple is shown here to be NP-complete. Of course, the problem of characterizing simple orthogonal shapes remains open, although we now know that in the general case it implies a heavy computation.

As a consequence of the complexity of the SIMPLICITY TESTING problem in the general case, in any hypothetical 3D drawing process in which the definition of the shape of the drawing is followed by the actual computation of its coordinates, the first step should be very carefully conceived in order for the second step to be efficiently computable. In fact, focusing on peculiar classes of shape graphs seems to be an obliged strategy for practical applications. Are there non trivial families of shape graph for which the simplicity testing is feasible? In particular, is there a "universal" set of shape graphs such that any graph is represented and such that the simplicity testing is guaranteed to be polynomial and to have a positive answer?

With respect to the problem of determining if a graph of degree six always admits a 2-bend drawing, this paper shows the NP-completeness of two problems related with finding such drawings. Namely, it is NP-complete when node positions are fixed (Section 4) and it is NP-complete when edge shapes are fixed (Section 5). Some other 3D drawing problems involving the number of the bends are known to be NP-complete, as, for example, finding a 2-bend drawing when vertices are placed on the diagonal of a cube [21] (provided that the graph admits such a drawing). The number of NP-complete problems related with the computation of a 2-bend drawing raises the following question: What is the complexity of finding a 2-bend drawing of a graph? If finding such a drawing was also NP-hard, then any attempt to prove that such a drawing always exists should produce an algorithm for an intractable problem, which is hard to conceive without resorting to an enumerative approach (which, in turn, assumes the existence of a solution). However both the conception of such an algorithm and the description of a graph not admitting a 2-bend drawing appear to be elusive goals.

References

1. T. C. Biedl. Heuristics for 3D-orthogonal graph drawings. In *Proc. 4th Twente Workshop on Graphs and Combinatorial Optimization*, pages 41–44, 1995.
2. J. A. Bondy and U. S. R. Murty. *Graph Theory with Applications*. Macmillan, London, 1976.
3. F. Brandenburg, D. Eppstein, M. T. Goodrich, S. Kobourov, G. Liotta, and P. Mutzel. Selected open problems in graph drawing. In G. Liotta, editor, *Graph Drawing (Proc. GD 2003)*, volume 2912 of *LNCS*, pages 515–539. Springer-Verlag, 2004.
4. M. Closson, S. Gartshore, J. Johansen, and S. K. Wismath. Fully dynamic 3-dimensional orthogonal graph drawing. *J. of Graph Algorithms and Applications*, 5(2):1–34, 2001.
5. E. D. Demaine, J. S. B. Mitchell, and J. O'Rourke, (eds.). The Open Problems Project. http://cs.smith.edu/~orourke/TOPP/Welcome.html.
6. G. Di Battista, P. Eades, R. Tamassia, and I. G. Tollis. *Graph Drawing*. Prentice Hall, Upper Saddle River, NJ, 1999.
7. G. Di Battista, G. Liotta, A. Lubiw, and S. Whitesides. Orthogonal drawings of cycles in 3d space. In J. Marks, editor, *Graph Drawing (Proc. GD '00)*, volume 1984 of *LNCS*. Springer-Verlag, 2001.

8. G. Di Battista, G. Liotta, A. Lubiw, and S. Whitesides. Embedding problems for paths with direction constrained edges. *Theor. Comp. Sci.*, 289:897–917, 2002.
9. E. Di Giacomo, G. Liotta, and M. Patrignani. A note on 3D orthogonal drawings with direction constrained edges. *Inform. Process. Lett.*, 90:97–101, 2004.
10. P. Eades, C. Stirk, and S. Whitesides. The techniques of Kolmogorov and Bardzin for three dimensional orthogonal graph drawings. *Inform. Process. Lett.*, 60:97–103, 1996.
11. P. Eades, A. Symvonis, and S. Whitesides. Three dimensional orthogonal graph drawing algorithms. *Discrete Applied Math.*, 103(1-3):55–87, 2000.
12. B. Y. S. Lynn, A. Symvonis, and D. R. Wood. Refinement of three-dimensional orthogonal graph drawings. In J. Marks, editor, *Graph Drawing (Proc. GD '00)*, volume 1984 of *LNCS*, pages 308–320. Springer-Verlag, 2001.
13. A. Papakostas and I. G. Tollis. Algorithms for incremental orthogonal graph drawing in three dimensions. *J. of Graph Algorithms and Applications*, 3(4):81–115, 1999.
14. M. Patrignani. Complexity results for three-dimensional orthogonal graph drawing. Tech. Report RT-DIA-94-2005, Dip. Inf. e Automazione, Univ. Roma Tre, 2005. http://dipartimento.dia.uniroma3.it/ricerca/rapporti/rapporti.php.
15. F. P. Preparata and M. I. Shamos. *Computational Geometry: An Introduction.* Springer-Verlag, 3rd edition, Oct. 1990.
16. R. Tamassia. On embedding a graph in the grid with the minimum number of bends. *SIAM J. Comput.*, 16(3):421–444, 1987.
17. G. Vijayan and A. Wigderson. Rectilinear graphs and their embeddings. *SIAM J. Comput.*, 14:355–372, 1985.
18. D. R. Wood. On higher-dimensional orthogonal graph drawing. In J. Harland, editor, *Proc. Computing: the Australasian Theory Symposimum (CATS '97)*, volume 19, pages 3–8. Australian Computer Science Commission, 1997.
19. D. R. Wood. Lower bounds for the number of bends in three-dimensional orthogonal graph drawings. *J. of Graph Algorithms and Applications*, 7:33–77, 2003.
20. D. R. Wood. Optimal three-dimensional orthogonal graph drawing in the general position model. *Theor. Comp. Sci.*, 299:151–178, 2003.
21. D. R. Wood. Minimising the number of bends and volume in 3-dimensional orthogonal graph drawings with a diagonal vertex layout. *Algorithmica*, 39:235–253, 2004.

On Extending a Partial Straight-Line Drawing[*]

Maurizio Patrignani

Roma Tre University
patrigna@dia.uniroma3.it

Abstract. We investigate the computational complexity of the following problem. Given a planar graph in which some vertices have already been placed in the plane, place the remaining vertices to form a planar straight-line drawing of the whole graph. We show that this extensibility problem, proposed in the 2003 "Selected Open Problems in Graph Drawing" [1], is NP-complete.

1 Introduction

A *(simple) graph* $G(V, E)$ consists of a set V of *vertices* and a set E of vertex pairs called *edges*. A *drawing* of G is a mapping of each vertex $v \in V$ to a distinct point of the plane and of each edge $e \in E$ to a Jordan curve connecting its end-vertices. A drawing of G is *planar* if no pair of edges intersect except, possibly, at common end-vertices. A graph G is *planar* if it admits a planar drawing. A *straight-line drawing* of G is a drawing of G where each edge is mapped to a straight segment. Every planar graph admits a straight-line drawing, as independently established by Steinitz and Rademacher [7], Wagner [9], Fary [3], and Stein [6], and such a drawing can be computed in linear time.

In this paper we show that finding a straight-line planar drawing for a graph that is already partially drawn is an NP-complete problem. This extensibility problem was proposed in [1] and thought to be related to the problem of drawing with fixed vertex positions, a problem that was solved by Cabello [2].

Formally, the PARTIAL DRAWING EXTENSIBILITY problem can be stated as follows.

Problem: **Partial Drawing Extensibility (PDE)**
Instance: A planar graph $G(V, E)$ and a mapping between a subset V' of its vertices and a set of distinct points of the plane.
Question: Can coordinates be assigned to the vertices in $V - V'$ such that the resulting straight-line drawing of $G(V, E)$ is planar?

It can be shown that the PDE problem is in NP. In Section 2 we show that it is also NP-hard. Section 3 concludes the paper.

---

[*] Work partially supported by European Commission - Fet Open project DELIS - Dynamically Evolving Large Scale Information Systems - Contract no 001907, by "Project ALGO-NEXT: Algorithms for the Next Generation Internet and Web: Methodologies, Design, and Experiments", MIUR Programmi di Ricerca Scientifica di Rilevante Interesse Nazionale, and by "The Multichannel Adaptive Information Systems (MAIS) Project", MIUR–FIRB.

P. Healy and N.S. Nikolov (Eds.): GD 2005, LNCS 3843, pp. 380–385, 2005.

2   NP-Hardness Proof

In order to show the NP-hardness of the PDE problem we produce a reduction
from the PLANAR 3-SATISFIABILITY (P3SAT) problem, which is strongly NP-
complete [5]. P3SAT is defined as follows:

Problem: **Planar 3-Satisfiability (P3SAT)**

Instance: A set of clauses $C_1, \ldots, C_m$ each one having three literals from
a set of Boolean variables $v_1, \ldots, v_n$. A plane bipartite graph
$G(V_A, V_B, E)$ where nodes in V_A correspond to the variables while
nodes in V_B correspond to the clauses (hence, $|V_A| = n$ and
$|V_B| = m$). Edges connect clauses to the variables of the literals
they contain. Moreover, $G(V_A, V_B, E)$ is drawn without intersections on a rectangular grid of polynomial size in such a way that
nodes in V_A are arranged in a horizontal line that is not crossed
by any edge (see Fig. 1).

Question: Can truth values be assigned to the variables $v_1, \ldots, v_n$ such that
each clause has a true literal?

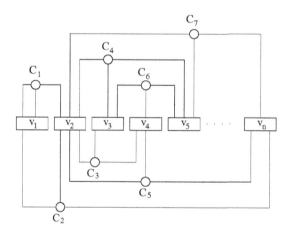

Fig. 1. A planar embedding of graph $G(V_A, V_B, E)$ for a P3SAT instance

Given an instance of the PDE problem, we call *fixed* vertices those in V', i.e.,
those that have assigned coordinates, and we call *free* vertices those in $V - V'$,
whose coordinates have to be found in order to obtain a planar straight-line
drawing of $G(V, E)$. For the construction of the PDE instance we make use of
the *basic gadget* depicted in Fig. 2. The basic gadget only has fixed vertices,
which form the boundary of a chamber. The chamber has two openings on the
bottom side, called *true gate* and *false gate*, respectively, and labeled with a 'T'
and an 'F' in Fig 2. On the top side the chamber has an even number of openings,
that we call *exits*. The vertices and edges near the exits form narrow corridors
pointing towards one of the two gates, and are called *true exits* or *false exits*
depending on which gate they point to. It can be easily checked from Fig. 2,

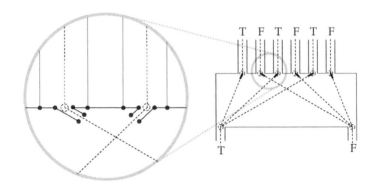

Fig. 2. The basic gadget used to construct the instance of the PDE problem

that only if a path enters a true (resp. false) gate and exits a true (resp. false) exit, it is possible for it to traverse the basic gadget from one gate to one exit leaving only two vertices inside the chamber. In particular, the two vertices of the path must be placed in the spots where dashed circles are drawn in Fig. 2.

Starting from an instance of the P3SAT problem, consisting of the set of clauses $C_1, \ldots, C_m$, each one having three literals from the Boolean variables $v_1, \ldots, v_n$, and a drawing of the graph $G(V_A, V_B, E)$, an instance of the PDE problem can be constructed as follows. For each variable v_i of the P3SAT instance we build a *variable gadget* depicted in Fig. 3. The variable gadget is composed of two basic gadgets, one of which is mirrored with respect to the horizontal axis. The two basic gadgets are glued together in such a way that their true gates and false gates are attached together. The number of the exits of the top (bottom) basic gadget is equal to two times the number of the edges of E that are incident to the node of V_A corresponding to v_i from above (below) in the planar drawing of $G(V_A, V_B, E)$. Also, the small corridors near the exits point alternatively to the true and to the false gate of each gadget.

Consider a clause $C_h = (l_1 \vee l_2 \vee l_3)$, where l_1, l_2, and l_3 are literals of the variables v_{l_1}, v_{l_2}, and v_{l_3}, respectively. (Variables v_{l_1}, v_{l_2}, and v_{l_3} can be assumed to be distinct.) We build a *clause gadget* corresponding to C_h by using three basic gadgets as depicted in Fig. 4. Each basic gadget corresponds to a literal l_i and is attached to a true and a false exit of the variable gadget for v_{l_i} with two "pipes", called the *true* and *false* pipe, respectively, each one bending two times before reaching the variable gadget. Also, the exits of the three basic gadgets point to the same eight points $p_1, \ldots, p_8$, while, internally, the small corridors near the exits of the chambers point to the true gate or the false gate in such a way that each point $p_1, \ldots, p_8$ corresponds to a different combination of the truth values of the basic gadget exits. Further, consider the truth assignment for v_{l_1}, v_{l_2}, and v_{l_3} that does not satisfy the clause and the point p_{false} corresponding to it. The corridors pointing to p_{false} are closed with an edge.

The free vertices of the PDE instance, i.e., those vertices that need to be placed while preserving planarity, are the following. For each variable v_i, we introduce one free vertex $n_{i,\alpha}$ which is adjacent to the fixed vertex of the variable

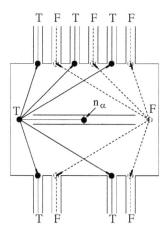

Fig. 3. The variable gadget for a variable which is attached to three clauses from above and to two clauses from below

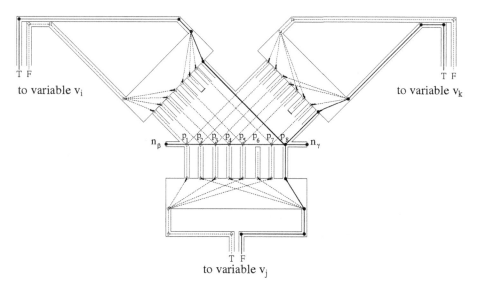

Fig. 4. The clause gadget for a clause $C_h = (\overline{v}_i \vee v_j \vee v_k)$

gadget of v_i labeled n_α in Fig. 3. For each clause C_h, we introduce one free vertex $n_{h,\beta,\gamma}$ which is adjacent to the two fixed vertices of the clause gadget corresponding to C_h labeled n_β and n_γ in Fig. 4. If one literal of variable v_i occurs in clause C_h, vertices $n_{i,\alpha}$ and $n_{h,\beta,\gamma}$ are joined with a path of six edges, that is, containing five other free vertices (see Fig. 5).

Theorem 1. *The* PARTIAL DRAWING EXTENSIBILITY *problem is NP-hard.*

Proof. Suppose that the P3SAT instance admits a truth assignment such that each clause has a true literal. A straight-line drawing of the PDE instance can

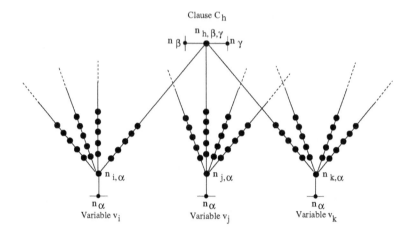

Fig. 5. The free vertices of the PDE instance introduced for a clause C_h with literals of the three variables v_i, v_j, and v_k

be found as follows. Depending on the truth value of variable v_i place vertex $n_{i,\alpha}$ of the variable gadget for v_i on the true gate or false gate, and let each path to a vertex $n_{h,\beta,\gamma}$ exit from the corresponding true or false exit to clause gadget C_h. For each clause C_h with literals l_1, l_2, and l_3, place vertex $n_{h,\beta,\gamma}$ on one point p_{true} different from p_{false}, and let each path to $n_{i,\alpha}$ pass through the (unblocked) corridor of the corresponding basic gadget. The obtained straight-line drawing is planar. In fact, edges between fixed vertices do not intersect, and, if v_i is true (false), for each clause C_h containing a literal of v_i, the five free vertices between each $n_{i,\alpha}$ and $n_{h,\beta,\gamma}$ can be placed inside the true (false) pipe linking the variable gadget for v_i to the clause gadget for C_h.

Suppose now that the free vertices of the PDE instance can be placed in such a way that the resulting straight-line drawing is planar. A truth assignment for the P3SAT instance such that each clause has a true literal can be easily computed as follows. Assign to each variable v_i a true (false) value if the corresponding variable gadget has the vertex $n_{i,\alpha}$ near the true (false) gate. We claim that the truth assignment so computed is such that each clause contains at least a true literal. In fact, consider the clause gadget of clause C_h. Since the paths attached to $n_{h,\beta,\gamma}$ have five internal vertices only, and since each pipe bends two times, the planarity of the drawing implies that $n_{h,\beta,\gamma}$ is placed on a point p_{true} different from p_{false} and that at least one of the three paths joining at $n_{h,\beta,\gamma}$ comes from a variable that has a truth assignment satisfying clause C_h.

Since, starting from a P3SAT instance, the construction of the corresponding PDE instance can be done in polynomial time, the statement follows.

3 Conclusions

We showed that the PARTIAL DRAWING EXTENSIBILITY problem is NP-complete. For simplicity, in the NP-hardness proof we used a reduction from the P3SAT problem producing non-connected PDE instances. We observe that it is not difficult to modify the construction in such a way that the resulting graph is connected. For example, edges can be added to connect each vertex $n_{i,\alpha}$ of the variable gadget for variable v_i to the middle point of each horizontal segment of the same variable gadget. Analogous changes performed on clause gadgets will produce a connected graph.

A similar problem to the one addressed in this paper comes up in mesh generation [4], where the already-placed vertices are usually assumed to form a simple polygon and the graph is assumed to have all interior faces triangles. Do these assumptions simplify the problem?

The drawing method of Tutte [8] may be used to show that the problem becomes tractable when the graph is triconnected and the already-placed vertices form convex faces.

Acknowledgments

We thank the authors of [1] for their help focusing the problem.

References

1. F. Brandenburg, D. Eppstein, M.T. Goodrich, S. Kobourov, G. Liotta, and P. Mutzel. Selected open problems in graph drawing. In G. Liotta, editor, *Graph Drawing (Proc. GD 2003)*, volume 2912 of *Lecture Notes Comput. Sci.*, pages 515–539. Springer-Verlag, 2004.
2. S. Cabello. Planar embeddability of the vertices of a graph using a fixed point set is NP-hard. In *Proc. 20th European Workshop on Computational Geometry*, 2004.
3. I. Fary. On straight lines representation of planar graphs. *Acta Sci. Math. Szeged*, 11:229–233, 1948.
4. L.A. Freitag and P.E. Plassman. Local optimization-based untangling algorithms for quadrilateral meshes. In *Proc. 10th int. Meshing Roundtable*, pages 397–406, 2001.
5. D. Lichtenstein. Planar formulae and their uses. *SIAM J. Comput.*, 11(2):329–343, 1982.
6. S. K. Stein. Convex maps. *Proc. Amer. Math. Soc.*, 2(3):464–466, 1951.
7. E. Steinitz and H. Rademacher. *Vorlesungen über die Theorie der Polyeder*. Julius Springer, Berlin, 1934.
8. W. T. Tutte. How to draw a graph. *Proceedings London Mathematical Society*, 13(52):743–768, 1963.
9. K. Wagner. Bemerkungen zum Vierfarbenproblem. *Jahresbericht der Deutschen Mathematiker-Vereinigung*, 46:26–32, 1936.

Odd Crossing Number Is Not Crossing Number

Michael J. Pelsmajer[1], Marcus Schaefer[2], and Daniel Štefankovič[3,4]

[1] Department of Applied Mathematics, Illinois Institute of Technology,
Chicago, IL 60616
pelsmajer@iit.edu
[2] Department of Computer Science, DePaul University, Chicago, IL 60604
mschaefer@cti.depaul.edu
[3] Department of Computer Science, University of Rochester, Rochester, NY 14627
[4] Department of Computer Science, Comenius University, Bratislava, Slovakia
stefanko@cs.rochester.edu

Abstract. The *crossing number* of a graph is the minimum number of edge intersections in a plane drawing of a graph, where each intersection is counted separately. If instead we count the number of pairs of edges that intersect an odd number of times, we obtain the *odd crossing number*. We show that there is a graph for which these two concepts differ, answering a well-known open question on crossing numbers. To derive the result we study drawings of maps (graphs with rotation systems).

1 A Confusion of Crossing Numbers

Intuitively, the crossing number of a graph is the smallest number of edge crossings in any plane drawing of the graph. As it turns out, this definition leaves room for interpretation, depending on how we answer the questions: what is a drawing, what is a crossing, and how do we count crossings? The papers by Pach and Tóth [7] and Székely [9] discuss the historical development of various interpretations and, often implicit, definitions of the crossing number concept.

A *drawing* D of a graph G is a mapping of the vertices and edges of G to the Euclidean plane, associating a distinct point with each vertex, and a simple plane curve with each edge such that the ends of an edge map to the endpoints of the corresponding curve. For simplicity, we also require that

- a curve does not contain any endpoints of other curves in its interior,
- two curves do not touch (that is, intersect without crossing), and
- no more than two curves intersect in a point (other than at a shared endpoint).

In such a drawing the intersection of the interiors of two curves is called a *crossing*. Note that by the restrictions we placed on a drawing, crossings do not involve endpoints, and at most two curves can intersect in a crossing. We often identify a drawing with the graph it represents. For a drawing D of a graph G in the plane we define

P. Healy and N.S. Nikolov (Eds.): GD 2005, LNCS 3843, pp. 386–396, 2005.

- $\operatorname{cr}(D)$ - the total number of crossings in D;
- $\operatorname{pcr}(D)$ - the number of pairs of edges which cross at least once; and
- $\operatorname{ocr}(D)$ - the number of pairs of edges which cross an odd number of times.

Remark 1. For any drawing D, we have $\operatorname{ocr}(D) \leq \operatorname{pcr}(D) \leq \operatorname{cr}(D)$.

We let $\operatorname{cr}(G) = \min \operatorname{cr}(D)$, where the minimum is taken over all drawings D of G in the plane. We define $\operatorname{ocr}(G)$ and $\operatorname{pcr}(G)$ analogously.

Remark 2. For any graph G, we have $\operatorname{ocr}(G) \leq \operatorname{pcr}(G) \leq \operatorname{cr}(G)$.

The question (first asked by Pach and Tóth [7]) is whether the inequalities are actually equalities.[1] Pach [6] called this "perhaps the most exciting open problem in the area." The only evidence for equality is an old theorem by Chojnacki, which was later rediscovered by Tutte—and the absence of any counterexamples.

Theorem 1 (Chojnacki [4], Tutte [10]). *If* $\operatorname{ocr}(G) = 0$ *then* $\operatorname{cr}(G) = 0$.[2]

In this paper we will construct a simple example of a graph with $\operatorname{ocr}(G) < \operatorname{pcr}(G) = \operatorname{cr}(G)$. We derive this example from studying what we call weighted maps on the annulus. Section 2 introduces the notion of weighted maps on arbitrary surfaces and gives a counterexample to $\operatorname{ocr}(M) = \operatorname{pcr}(M)$ for maps on the annulus. In Section 3 we continue the study of crossing numbers for weighted maps, proving in particular that $\operatorname{cr}(M) \leq c_n \cdot \operatorname{ocr}(M)$ for maps on a plane with n holes. One of the difficulties in dealing with the crossing number is that it is **NP**-complete [2]. In Section 4 we show that the crossing number can be computed in polynomial time for maps on the annulus. Finally, in Section 5 we show how to translate the map counterexample from Section 2 into an infinite family of simple graphs for which $\operatorname{ocr}(G) < \operatorname{pcr}(G)$.

2   Map Crossing Numbers

A *weighted map* M is a 2-manifold S and a set $P = \{(a_1, b_1), \ldots, (a_m, b_m)\}$ of pairs of distinct points on ∂S with positive weights $w_1, \ldots, w_m$. A *realization* R of the map $M = (S, P)$ is a set of m properly embedded arcs $\gamma_1, \ldots, \gamma_m$ in S where γ_i connects a_i and b_i.[3]

Let

$$\operatorname{cr}(R) = \sum_{1 \leq k < \ell \leq m} i(\gamma_k, \gamma_\ell) w_k w_\ell,$$

$$\operatorname{pcr}(R) = \sum_{1 \leq k < \ell \leq m} [i(\gamma_k, \gamma_\ell) > 0] w_k w_\ell,$$

---

[1] Doug West lists the problem on his page of open problems in graph theory [12]. Dan Archdeacon even conjectured that equality holds [1].

[2] In fact they proved something stronger, namely that in any drawing of a non-planar graph there are two non-adjacent edges crossing an odd number of times. Also see [8].

[3] If we take a realization R of a map M, and contract each boundary component to a vertex, we obtain a drawing of a graph with a given rotation system [3]. For our purposes, maps are a more visual way to look at graphs with a rotation system.

$$\mathrm{ocr}(R) = \sum_{1 \leq k < \ell \leq m} [i(\gamma_k, \gamma_\ell) \equiv 1 \ (\mathrm{mod}\ 2)] w_k w_\ell,$$

where $i(\gamma, \gamma')$ is the geometric intersection number of γ and γ' and $[x]$ is 1 if the condition x is true, and 0 otherwise.

We define $\mathrm{cr}(M) = \min \mathrm{cr}(R)$, where the minimum is taken over all realizations R of M. We define $\mathrm{pcr}(M)$ and $\mathrm{ocr}(M)$ analogously.

Remark 3. For every map M, $\mathrm{ocr}(M) \leq \mathrm{pcr}(M) \leq \mathrm{cr}(M)$.

Conjecture 1. For every map M, $\mathrm{cr}(M) = \mathrm{pcr}(M)$.

Lemma 1. *If Conjecture 1 is true then* $\mathrm{cr}(G) = \mathrm{pcr}(G)$ *for every graph* G.

Proof. Let D be a drawing of G with minimal pair crossing number. Drill small holes at the vertices. We obtain a drawing R of a weighted map M. If Conjecture 1 is true, there exists a drawing of M with the same crossing number. Collapse the holes to vertices to obtain a drawing D' of G with $\mathrm{cr}(D') \leq \mathrm{pcr}(G)$.

We can, however, separate the odd crossing number from the crossing number for weighted maps, even in the annulus (a disk with a hole).

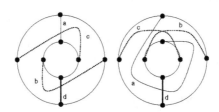

Fig. 1. ocr < pcr

When analyzing crossing numbers of drawings on the annulus, we describe curves with respect to an initial drawing of the curve and a number of *Dehn twists*. Consider, for example, the four curves in the left part of Figure 1. Comparing them to the corresponding curves in the right part, we see that the curves labeled c and d have not changed, but the curves labeled a and b have each undergone a single clockwise twist.

Two curves are *isotopic rel boundary* if they can be obtained from each other by a continuous deformation which does not move the boundary ∂M. Isotopy rel boundary is an equivalence relation, its equivalence classes are called *isotopy classes*. An *isotopy class* on annulus is determined by a properly embedded arc connecting the endpoints, together with the number of twists performed.

Lemma 2. *Let* $a \leq b \leq c \leq d$ *be such that* $a + c \geq d$. *For the weighted map* M *in Figure 1 we have* $\mathrm{cr}(M) = \mathrm{pcr}(M) = ac + bd$ *and* $\mathrm{ocr}(M) = bc + ad$.

Proof. The upper bounds follow from the drawings in Figure 1, the left drawing for crossing and pair crossing number, the right drawing for odd crossing number; it remains to prove the two lower bounds.

First, we claim that

$$\mathrm{pcr}(M) \geq ac + bd.$$

Proof of the claim. Let R be a drawing of M minimizing $\mathrm{pcr}(R)$. We can apply twists so that the thick edge d is drawn as in the left part of Figure 1. Let α, β, γ be the number of clockwise twists that are applied to arcs a, b, c in the left part of Figure 1 to obtain the drawing R. Then,

$$\mathrm{pcr}(R) = cd[\gamma \neq 0] + bd[\beta \neq -1] + ad[\alpha \neq 0] + bc[\beta \neq \gamma] + ab[\alpha \neq \beta] + ac[\alpha \neq \gamma+1]. \tag{1}$$

If $\gamma \neq 0$ then $\mathrm{pcr}(R) \geq cd + ab$ because at least one of the last five conditions in (1) must be true; the last five terms contribute at least ab (since $d \geq c \geq b \geq a$), and the first term contributes cd. Since $d(c - b) \geq a(c - b)$, $cd + ab \geq ac + bd$, and the claim is proved in the case that $\gamma \neq 0$.

Now assume that $\gamma = 0$. Equation (1) becomes

$$\mathrm{pcr}(R) = bd[\beta \neq -1] + bc[\beta \neq 0] + ad[\alpha \neq 0] + ac[\alpha \neq 1] + ab[\alpha \neq \beta]. \tag{2}$$

If $\beta \neq -1$ then $\mathrm{pcr}(R) \geq bd + ac$ because either $\alpha \neq 0$ or $\alpha \neq 1$. Since $bd + ac \geq bc + ad$, the claim is proved in the case that $\beta \neq -1$.

This leaves us with the case that $\beta = -1$. Equation (2) becomes

$$\mathrm{pcr}(R) = bc + ad[\alpha \neq 0] + ac[\alpha \neq 1] + ab[\alpha \neq -1]. \tag{3}$$

The right-hand side of Equation (3) is minimized for $\alpha = 0$. In this case $\mathrm{pcr}(R) = bc + ac + ab \geq ac + bd$ because we assume that $a + c \geq d$. Second, we claim that

$$\mathrm{ocr}(M) \geq bc + ad.$$

Proof of the claim. Let R be a drawing of M minimizing $\mathrm{ocr}(R)$. Let α, β, γ be as in the previous claim. We have

$$\mathrm{ocr}(R) = cd[\gamma]_2 + bd[\beta+1]_2 + ad[\alpha]_2 + bc[\beta+\gamma]_2 + ab[\alpha+\beta]_2 + ac[\alpha+\gamma+1]_2, \tag{4}$$

where $[x]_2$ is 0 if $x \equiv 0 \pmod 2$, and 1 otherwise.

If $\beta \not\equiv \gamma \pmod 2$ then the claim clearly follows unless $\gamma = 0$, $\beta = 1$, and $\alpha = 0$ (all modulo 2). In that case $\mathrm{ocr}(R) \geq bc + ab + ac \geq bc + ad$. Hence, the claim is proved if $\beta \not\equiv \gamma \pmod 2$.

Assume then that $\beta \equiv \gamma \pmod 2$. Equation (4) becomes

$$\mathrm{ocr}(R) = cd[\beta]_2 + bd[\beta + 1]_2 + ad[\alpha]_2 + ab[\alpha + \beta]_2 + ac[\alpha + \beta + 1]_2. \tag{5}$$

If $\alpha \equiv 1 \pmod 2$ then the claim clearly follows because either cd or bd contributes to the ocr. Thus we can assume $\alpha \equiv 0 \pmod 2$. Equation (5) becomes

$$\mathrm{ocr}(R) = (cd + ab)[\beta]_2 + (bd + ac)[\beta + 1]_2. \tag{6}$$

For both $\beta \equiv 0 \pmod 2$ and $\beta \equiv 1 \pmod 2$ we get $\mathrm{ocr}(R) \geq bc + ad$. This finishes the proof of the second claim.   □

We get a separation of pcr and ocr for maps with small integral weights.

Corollary 1. *There is a weighted map M on the annulus with edges of weight $a = 1$, $b = c = 3$, and $d = 4$ for which $\mathrm{cr}(M) = \mathrm{pcr}(M) = 15$ and $\mathrm{ocr}(M) = 13$.*

Optimizing the gap over the reals yields $b = c = 1$, $a = (\sqrt{3} - 1)/2$, and $d = 1 + a$, giving us the following separation of $\mathrm{pcr}(M)$ and $\mathrm{ocr}(M)$.

Corollary 2. *There exists a weighted map M on the annulus with $\mathrm{ocr}(M) \le \sqrt{3}/2\,\mathrm{pcr}(M)$.*

Conjecture 2. For every weighted map M on the annulus, $\mathrm{ocr}(M) \ge \frac{\sqrt{3}}{2}\,\mathrm{pcr}(M)$.

3   Upper Bounds on Crossing Numbers

In Section 5 we will transform the separation of ocr and pcr on maps into a separation on graphs. In particular, we will show that for every $\varepsilon > 0$ there is a graph G such that
$$\mathrm{ocr}(G) < (\sqrt{3}/2 + \varepsilon)\,\mathrm{cr}(G).$$

The gap, however, cannot be arbitrarily large, as Pach and Tóth showed.

Theorem 2 (Pach, Tóth [7]). *Let G be a graph. Then $\mathrm{cr}(G) \le 2(\mathrm{ocr}(G))^2$.* [4]

This result suggests the question whether the linear separation can be improved. We do not believe this to be possible:

Conjecture 3. There is a $c > 0$ such that $\mathrm{cr}(G) < c \cdot \mathrm{ocr}(G)$.

Using a graph redrawing idea from from [8] (which investigates other applications of that idea), we can show something weaker:

Theorem 3. $\mathrm{cr}(M) \le \mathrm{ocr}(M)\binom{n+4}{4}/5$ *for weighted maps M on the plane with n holes, with strict inequality if $n > 1$.*

As a special case of the theorem, we have that if M is a (weighted) map on the annulus ($n = 2$) then $\mathrm{cr}(M) < 3\,\mathrm{ocr}(M)$, which comes reasonably close to the $\sqrt{3}/2$ lower bound from the previous section. The theorem shows that any counterexample to Conjecture 3 cannot be constructed on a plane with a small, fixed number of holes. For reasons of space, we do not include the proof of the theorem.

4   Computing Crossing Numbers on the Annulus

Let M be a map on the annulus. We explained earlier that as far as crossing numbers are concerned we can describe a curve in the realization of M by a properly embedded arc γ_{ab} connecting endpoints a and b on the inner and outer boundary of the annulus, and an integer $k \in \mathbb{Z}$, counting the number of twists

---

[4] In terms of $\mathrm{pcr}(G)$ better upper bounds on $\mathrm{cr}(G)$ are known [11, 5].

applied to the curve γ_{ab}. Our goal is to compute the number of intersections between two arcs after applying a number of twists to each one of them. Since twists can be positive and negative and cancel each other out, we need to count crossings more carefully. Let us orient all arcs from the inner boundary to the outer boundary. Traveling along an arc α, a crossing with β counts as $+1$ if β crosses from right to left, and as -1 if it crosses from left to right. Summing up these numbers over all crossings for two arcs α and β yields $\hat{i}(\alpha, \beta)$, the *algebraic crossing number* of α and β. Tutte [10] introduced the notion

$$\mathrm{acr}(G) = \min_{D} \sum_{\{e,f\}\in\binom{E}{2}} |\hat{i}(\gamma_e, \gamma_f)|,$$

the *algebraic crossing number* of a graph, a notion that apparently has not drawn any attention since.

Let $D^k(\gamma)$ denote the result of adding k twists to the curve γ. For two curves α and β connecting the inner and outer boundary we have:

$$\hat{i}(D^k(\alpha), D^\ell(\beta)) = k - \ell + \hat{i}(\alpha, \beta). \tag{7}$$

Note that $i(\alpha, \beta) = |\hat{i}(\alpha, \beta)|$ for any two curves α, β on the annulus.

Let π be a permutation of $[n]$. A map M_π corresponding to π is constructed as follows. Choose $n+1$ points on each of the two boundaries and number them $0, 1, \ldots, n$ in the clockwise order. Let a_i be the vertex numbered i on the outer boundary and b_i be the vertex numbered π_i on the inner boundary, $i = 1, \ldots, n$. We ask a_i to be connected to b_i in M_π.

We will encode a drawing R of M_π by a sequence of n integers $x_1, \ldots, x_n$ as follows. Fix a curve β connecting the a_0 and b_0 and choose γ_i be such that $i(\beta, \gamma_i) = 0$ (for all i). We will connect a_i, b_i with the arc $D^{x_i}(\gamma_i)$ in R. Note that for $i < j$, $\hat{i}(\gamma_i, \gamma_j) = [\pi_i > \pi_j]$ and hence

$$\hat{i}(D^{x_i}(\gamma_i), D^{x_j}(\gamma_j)) = x_i - x_j + [\pi_i > \pi_j].$$

We have

$$\mathrm{acr}(M_\pi) = \mathrm{cr}(M_\pi) = \min\left\{ \sum_{i<j} |x_i - x_j + [\pi_i > \pi_j]| w_i w_j : x_i \in \mathbb{Z}, i \in [n] \right\}, \tag{8}$$

$$\mathrm{pcr}(M_\pi) = \min\left\{ \sum_{i<j} [x_i - x_j + [\pi_i > \pi_j] \neq 0] w_i w_j : x_i \in \mathbb{Z}, i \in [n] \right\}, \tag{9}$$

$$\mathrm{ocr}(M_\pi) = \min\left\{ \sum_{i<j} [x_i - x_j + [\pi_i > \pi_j] \not\equiv 0 \ (\mathrm{mod}\ 2)] w_i w_j : x_i \in \mathbb{Z}, i \in [n] \right\}. \tag{10}$$

Consider the relaxation of the integer program for $\mathrm{cr}(M_\pi)$:

$$\mathrm{cr}'(M_\pi) = \min\left\{ \sum_{i<j} |x_i - x_j + [\pi_i > \pi_j]| w_i w_j : x_i \in \mathbb{R}, i \in [n] \right\}. \tag{11}$$

Since (11) is a relaxation of (8), we have $\mathrm{cr}'(M_\pi) \leq \mathrm{cr}(M_\pi)$. The following lemma shows that $\mathrm{cr}'(M_\pi) = \mathrm{cr}(M_\pi)$.

Lemma 3. *Let* n *be a positive integer. Let* $b_{ij} \in \mathbb{Z}$ *and let* $a_{ij} \in \mathbb{R}$ *be non-negative,* $1 \leq i < j \leq n$. *Then*

$$\min \left\{ \sum_{i<j} a_{ij} \left| x_i - x_j + b_{ij} \right| \; : \; x_i \in \mathbb{R}, i \in [n] \right\}$$

has an optimal solution with $x_i \in \mathbb{Z}$, $i \in [n]$.

Proof. Let $\bar{x}^*$ be an optimal solution which satisfies the maximum number of $x_i - x_j + b_{ij} = 0$, $1 \leq i < j \leq n$. Without loss of generality, we can assume $x_1^* = 0$. Let G be a graph on vertex set $[n]$ with an edge between vertices i, j if $x_i^* - x_j^* + b_{ij} = 0$. Note that if i, j are connected by an edge and one of x_i^*, x_j^* is an integer then both x_i^* and x_j^* are integers. It is then enough to show that G is connected.

Suppose that G is not connected. There exists non-empty $A \subsetneq V(G)$ such that there are no edges between A and $V(G) - A$. Let χ_A be the characteristic vector of the set A, that is, $(\chi_A)_i = [i \in A]$. Let $f(\lambda)$ be the value of the objective function on $\bar{x} = \bar{x}^* + \lambda \cdot \chi_A$. Let I be the interval on which the signs of the $x_i - x_j + b_{ij}$, $1 \leq i < j \leq n$ are the same as for $\bar{x}^*$. Then I is not the entire line (otherwise G would be connected). Since $f(\lambda)$ is linear on I and an open neighborhood of 0 belongs to I we conclude that f is constant on I. Choosing $x = x^* + \lambda \chi_A$ for λ an endpoint of I gives an optimal solution satisfying more $x_i - x_j + b_{ij} = 0$, $1 \leq i < j \leq n$, a contradiction.

Theorem 4. *The crossing number of maps on the annulus can be computed in polynomial time.*

Proof. Note that $\mathrm{cr}'(M_\pi)$ is computed by the following linear program L_π:

$$\min \sum_{i<j} y_{ij} w_i w_j$$
$$y_{ij} \geq \quad x_i - x_j + [\pi_i > \pi_j], \quad 1 \leq i < j \leq n$$
$$y_{ij} \geq -x_i + x_j - [\pi_i > \pi_j], \quad 1 \leq i < j \leq n.$$

Question 1. Let M be a map on the annulus. Can $\mathrm{ocr}(M)$ be computed in polynomial time?

Conjecture 4. For any map M on the annulus $\mathrm{cr}(M) = \mathrm{pcr}(M)$.

5   Separating Crossing Numbers of Graphs

We modify the map from Lemma 2 to obtain a graph G separating $\mathrm{ocr}(G)$ and $\mathrm{pcr}(G)$. The graph G will have integral weights on edges. From G we can get an unweighted graph G' with $\mathrm{ocr}(G') = \mathrm{ocr}(G)$ and $\mathrm{pcr}(G') = \mathrm{pcr}(G)$ by replacing an edge of weight w by w parallel edges of weight 1 (this does not change any of the crossing numbers). If needed we can get rid of parallel edges by subdividing edges, which does not change any of the crossing numbers.

We start with the map M from Lemma 2 with the following integral weights:

$$a = \left\lfloor \frac{\sqrt{3}-1}{2}m \right\rfloor, \quad b = c = m, \quad d = \left\lfloor \frac{\sqrt{3}+1}{2}m \right\rfloor,$$

where $m \in \mathbb{N}$ will be chosen later.

We replace each pair (a_i, b_i) of M by w_i pairs $(a_{i,1}, b_{i,1}), \ldots, (a_{i,w_i}, b_{i,w_i})$ where the $a_{i,j}$ ($b_{i,j}$) occur on ∂S in clockwise order in a small interval around of a_i (b_i). We can argue that all the curves corresponding to (a_i, b_i) can be routed in parallel in an optimal drawing, and, therefore, the resulting map N with unit weights will have the same crossing numbers as M.

We then replace the boundaries of the annulus by cycles (using one vertex for each $a_{i,j}$ and $b_{i,j}$), obtaining a graph G. We assign weight $W = 1 + \mathrm{cr}(N)$ to the edges in the cycles. This ensures that in a drawing of G minimizing any of the crossing numbers the boundary cycles are embedded without any intersections. This means that a drawing of G minimizing any of the crossing numbers looks very much like the drawing of a map on the annulus. With one subtle difference: one of the boundaries may flip.

Given the map N on the annulus, the *flipped map* N' is obtained by flipping the order of the points on one of the boundaries. In other words, there are essentially two different ways of embedding the two boundary cycles of G on the sphere without intersections depending on the relative orientation of the boundaries. In one of the cases the drawing D of G gives a drawing of N, in the other case it gives a drawing of the flipped map N'. Fortunately, in the flipped case the group of edges corresponding to the weighted edge from a_i to b_i must intersect often with each other (as illustrated in Figure 2).

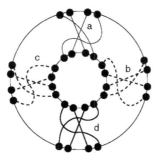

Fig. 2. The inside flipped

Now we know that

$$\mathrm{ocr}(G) \leq \mathrm{ocr}(N) \quad \text{(since every drawing of } N \text{ is a drawing of } G)$$
$$\leq w_1 w_3 + w_2 w_4 \quad \text{(by Lemma 2)}$$
$$\leq \frac{3}{2}m^2 \quad \text{(by the choice of weights).}$$

We will presently prove the following estimate on the flipped map.

Lemma 4. $\mathrm{ocr}(N') \geq 2m^2 - 4m$.

With that estimate and our discussion of flipped maps, we have

$$
\begin{aligned}
\mathrm{cr}(G) &= \min\{\mathrm{cr}(N), \mathrm{cr}(N')\} \\
&\geq \min\{\mathrm{cr}(N), \mathrm{ocr}(N')\} \quad \text{(since } \mathrm{ocr} \leq \mathrm{cr}) \\
&\geq \min\{\sqrt{3}m^2 - 2m, 2m^2 - 4m\} \quad \text{(choice of } w, \text{ and Lemma 4).}
\end{aligned}
$$

By making m sufficiently large, we can make the ratio of $\mathrm{ocr}(G)$ and $\mathrm{cr}(G)$ arbitrarily close to $\sqrt{3}/2$.

Theorem 5. *For any $\varepsilon > 0$ there is a graph G such that*

$$
\mathrm{ocr}(G) < (\sqrt{3}/2 + \varepsilon)\,\mathrm{cr}(G).
$$

The proof of Lemma 4 will require the following estimate.

Lemma 5. *Let $0 \leq a_1 \leq a_2 \leq \cdots \leq a_n$ be such that $a_n \leq a_1 + \cdots + a_{n-1}$. Then*

$$
\max_{|y_i| \leq a_i} \left(\left(\sum_{i=1}^n y_i \right)^2 - 2 \sum_{i=1}^n y_i^2 \right) = \left(\sum_{i=1}^n a_i \right)^2 - 2 \sum_{i=1}^n a_i^2.
$$

Proof of Lemma 4. Let $w_1 = a, w_2 = b, w_3 = d, w_4 = c$ (with a, b, c, d as in the definition of N). In any drawing of N' each group of the edges split into two classes, those with an even number of twists and those with an odd number of twists (two twists make the same contribution to $\mathrm{ocr}(M')$ as no twists). Consequently, we can estimate $\mathrm{ocr}(N')$ as follows.

$$
\begin{aligned}
\mathrm{ocr}(N') &= \min_{0 \leq k_i \leq w_i} \left(\sum_{i=1}^4 \binom{k_i}{2} + \sum_{i=1}^4 \binom{w_i - k_i}{2} + \sum_{i \neq j} k_i(w_j - k_j) \right) \\
&\geq -\frac{1}{2} \sum_{i=1}^4 w_i + \min_{0 \leq x_i \leq w_i} \left(\sum_{i=1}^4 \frac{x_i^2}{2} + \sum_{i=1}^4 \frac{(w_i - x_i)^2}{2} + \sum_{i \neq j} x_i(w_j - x_j) \right) \\
&= -\frac{1}{2} \sum_{i=1}^4 w_i + \frac{1}{4} \left(\sum_{i=1}^4 w_i \right)^2 + \min_{|y_i| \leq w_i/2} \left(2 \sum_{i=1}^4 y_i^2 - \left(\sum_{i=1}^4 y_i \right)^2 \right) \\
&\geq \frac{1}{2} \sum_{i=1}^4 w_i^2 - \frac{1}{2} \sum_{i=1}^4 w_i \quad \text{(using Lemma 5)} \\
&\geq \frac{1}{2} \left(\left(\frac{\sqrt{3}+1}{2}m - 1 \right)^2 + 2m^2 + \left(\frac{\sqrt{3}-1}{2}m - 1 \right)^2 - 4m \right) \\
&\geq 2m^2 - 4m. \tag{12}
\end{aligned}
$$

The equality between the second and third line can be verified by substituting $y_i = x_i - w_i/2$. $\qquad\square$

Proof of Lemma 5. Let $y_1, \ldots, y_n$ achieve the maximum value. Replacing the y_i by $|y_i|$ does not decrease the objective function. Without loss of generality, we can assume $0 \leq y_1 \leq y_2 \leq \cdots \leq y_n$. Note that $y_i < y_j$ then $y_i = a_i$ (otherwise increasing y_i by ε and decreasing y_j by ε increases the objective function for small ε).

Let k be the largest i such that $y_i = a_i$. Let $k = 0$ if no such i exists. We have $y_i = a_i$ for $i \leq k$ and $y_{k+1} = \cdots = y_n$. If $k = n$ we are done. Let

$$f(t) = \left(\sum_{i=1}^{k} a_i + (n-k)t \right)^2 - 2 \left(\sum_{i=1}^{k} a_i^2 + (n-k)t^2 \right).$$

We have

$$f'(t) = 2(n-k) \left(\sum_{i=1}^{k} a_i + (n-k-2)t \right).$$

Note that for $t < a_{k+1}$ we have $f'(t) > 0$ and hence the only optimal choice is $t = a_{k+1}$. Hence $y_{k+1} = a_{k+1}$, a contradiction with our choice of k. $\square$

6   Conclusion

The relationship between the different crossing numbers remains mysterious, and we have already mentioned several open questions and conjectures. Here we want to revive a question first asked by Tutte (in slightly different form). Recall the definition of the algebraic crossing number from Section 4:

$$\mathrm{acr}(G) = \min_{D} \sum_{\{e,f\} \in \binom{E}{2}} |\hat{i}(\gamma_e, \gamma_f)|,$$

where γ_e is a curve representing edge e in a drawing D of G. It is clear that

$$\mathrm{acr}(G) \leq \mathrm{cr}(G).$$

Does equality hold?

References

1. Dan Archdeacon.   Problems in topological graph theory. `http://www.emba.uvm.edu/~archdeac/problems/altcross.html` (accessed April 7th, 2005).
2. Michael R. Garey and David S. Johnson. Crossing number is NP-complete. *SIAM Journal on Algebraic and Discrete Methods*, 4(3):312–316, 1983.
3. Jonathan L. Gross and Thomas W. Tucker. *Topological graph theory*. Dover Publications Inc., Mineola, NY, 2001. Reprint of the 1987 original.
4. Chaim Chojnacki (Haim Hanani). Über wesentlich unplättbare Kurven im dreidimensionalen Raume. *Fundamenta Mathematicae*, 23:135–142, 1934.
5. Petr Kolman and Jiří Matoušek. Crossing number, pair-crossing number, and expansion. *J. Combin. Theory Ser. B*, 92(1):99–113, 2004.

6. János Pach. Crossing numbers. In *Discrete and computational geometry (Tokyo, 1998)*, volume 1763 of *Lecture Notes in Comput. Sci.*, pages 267–273. Springer, Berlin, 2000.

7. János Pach and Géza Tóth. Which crossing number is it anyway? *J. Combin. Theory Ser. B*, 80(2):225–246, 2000.

8. Michael J. Pelsmajer, Marcus Schaefer, and Daniel Štefankovič. Removing even crossings. Manuscript, April 2005.

9. László A. Székely. A successful concept for measuring non-planarity of graphs: the crossing number. *Discrete Math.*, 276(1-3):331–352, 2004. 6th International Conference on Graph Theory.

10. W. T. Tutte. Toward a theory of crossing numbers. *J. Combinatorial Theory*, 8:45–53, 1970.

11. Pavel Valtr. On the pair-crossing number. Manuscript.

12. Douglas West. Open problems - graph theory and combinatorics. http://www.math.uiuc.edu/~west/openp/ (accessed April 7th, 2005).

Minimum Depth Graph Embeddings and Quality of the Drawings: An Experimental Analysis*

Maurizio Pizzonia

Dipartimento di Informatica e Automazione, Università Roma Tre, Italy
pizzonia@dia.uniroma3.it

Abstract. The depth of a planar embedding of a graph is a measure of the topological nesting of the biconnected components of the graph in that embedding. Motivated by the intuition that lower depth values lead to better drawings, previous works proposed efficient algorithms for finding embeddings with minimum depth. We present an experimental study that shows the impact of embedding depth minimization on important aesthetic criteria and relates the effectiveness of this approach with measures of how much the graph resembles a tree or a biconnected graph. In our study, we use a well known test suite of graphs obtained from real-world applications and a randomly generated one with favorable biconnectivity properties. In the experiments we consider orthogonal drawings computed using the topology-shape-metrics approach.

1 Introduction

Well known approaches for drawing graphs compute a planar embedding as an intermediate step [8] and, intuitively, the computed embedding may have a big impact on the quality of the final drawing. This motivated several research efforts to compute a planar embedding of a graph that is optimal with respect to certain cost measures (see for example [3, 11, 5]).

Recently, the concept of *depth* of a planar embedding has been introduced [22]. In a planar embedding, *blocks* (i.e. biconnected components) are inside faces, and faces are inside blocks. The containment relationships between blocks and faces induce a tree rooted at the external face. The depth of the planar embedding is the maximum length of a root-to-leaf path in this tree (see Figure 1, a rigorous definition can be found in Section 2). In [22] it is proved that it is possible to compute a planar embedding with minimum depth in linear time if the embedding of each biconnected component is given and fixed. Gutwenger and Mutzel [14] extended this result by providing an algorithm for computing embeddings with minimum depth among all planar embeddings of a graph. They also provide an algorithm that maximizes the number of vertices of the external face among all the minimum depth embeddings.

---

* A preliminary version of this work appeared in [21]. Work partially supported by European Commission - Fet Open project COSIN - COevolution and Self-organisation In dynamical Networks - IST-2001-33555, by European Commission - Fet Open project DELIS - Dynamically Evolving Large Scale Information Systems - Contract no 001907, by "Project ALGO-NEXT: Algorithms for the Next Generation Internet and Web: Methodologies, Design, and Experiments", MIUR Programmi di Ricerca Scientifica di Rilevante Interesse Nazionale, and by "The Multichannel Adaptive Information Systems (MAIS) Project", MIUR–FIRB.

P. Healy and N.S. Nikolov (Eds.): GD 2005, LNCS 3843, pp. 397–408, 2005.

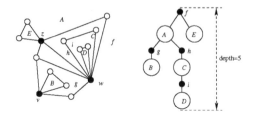

Fig. 1. In this example an embedded graph is shown whose blocks A,B,C,D and E are connected by means of the cutvertices v, w and z. The embedding has cutfaces f,g,h and i. The containment relationship between cutfaces and blocks is represented by a tree of depth 5 rooted at the external face f.

In this paper we quantify the positive effect on the final drawings of the application of embedding depth minimization techniques.

Our experiments are performed with the algorithm described in [22], which is called *MinDepth* throughout this paper and the results are compared with the ones obtained using Algorithm *LargeDepth* (detailed in Section 5) that heuristically computes planar embeddings with large depth.

For our experiments we focus on orthogonal drawings computed using the *topology-shape-metrics* approach [8]. This is a widely used technique for computing orthogonal drawings of general graphs. It has been extensively investigated both theoretically [15, 24, 12, 16, 17, 7, 6, 18] and experimentally [9, 2]. Its practical applicability has been demonstrated by various system prototypes [10, 19] and commercial graph drawing tools [1]. The topology-shape-metrics approach consists of three phases. The first phase takes as input a graph and computes a planar embedding possibly inserting dummy vertices to represent crossings if the graph is not planar. The second phase determines the orthogonal shape (angles) preserving the embedding. The third phase computes the coordinates of the drawing preserving both embedding and shape. With this approach the properties of the embedding computed in the first phase are crucial for the the quality of the final layout.

In our experiments, we apply the topology-shape-metrics approach where we refine the embedding computed by the planarization step using both Algorithm *MinDepth* and Algorithm *LargeDepth*. The quality of the resulting drawings is compared with respect to area, number of bends and total edge length, which are important aesthetic criteria [23]. We relate the effectiveness of embedding depth minimization with two measures, *triviality* and *max-occupancy*, which express how much a graph resembles a tree or a biconnected graph, respectively. Results show that the effectiveness of embedding depth minimization is strongly affected by the values of such measures.

Our experiments are performed on two test suites. The first is a well known set of graphs obtained from real-world applications that allows us to test the effectiveness of embedding depth minimization on realistic instances. The second is randomly generated so that graphs present specific biconnectivity properties which allow the embedding depth to vary over a wide range. This permits us to understand how effective the technique can be in a very favorable setting.

This paper is organized as follows. Section 2 provides basic definitions. In Section 3 we define new biconnectivity-related measures. In Section 4 we analyze the test suites used in this paper for biconnectivity properties. In Section 5 we describe a heuristic algorithm for computing embeddings with large depth. In Section 6 we show the results of our experimental analysis. In Section 7 we draw the conclusions of our work.

2   Basic Definitions

In this section, we review basic concepts about graphs and embeddings, and give definitions that will be used throughout the paper.

Let G be a connected planar graph. For simplicity, we assume that G has no parallel edges or self-loops. A *cutvertex* of G is a vertex whose deletion disconnects G. Graph G is said to be biconnected if it has no cutvertices. A *block* B of G is a maximal subgraph of G such that B is biconnected. A *trivial block* is composed by one edge between two cutvertices with no other path between them. The *block-cutvertex tree* T of G is a tree whose nodes are in one-to-one correspondence with the blocks and the cutvertices of G, and whose edges connect each cutvertex-node to the block-nodes of the blocks containing the cutvertex.

An *embedding* Γ of G is an equivalence class of planar drawings of G with the same circular order of edges around each vertex. Two planar drawings with the same embedding also induce the same circuits of edges bounding corresponding regions in the two drawings. These circuits are called the *faces* of the embedding.

The *dual embedding* Γ' of Γ is the embedded graph induced by the adjacency relations among the faces of Γ through its edges. A *cutface* f of Γ is a face associated with a cutvertex of Γ'. The *block-cutface* tree T^* of Γ is the block-cutvertex tree of Γ'. Since the dual of any biconnected embedding is biconnected, T and T^* contain the same set of block-nodes.

A *planar embedding* is an embedding where a face is chosen as *external face*. We consider the block-cutface tree of a planar embedding either rooted at the external face, if this is a cutface, or rooted at the block that contains the external face.

For a rooted tree T the *depth* of T (depth T) is the length of the longest path from the root of T to one of its leaves. The *diameter* of a tree is the length of the longest path between any two leaves.

Let G be a connected planar graph, and assume that we have a prescribed embedding for each block B of G. We say that Γ is *block-preserving for a cutvertex v* if Γ preserves the embedding of each block B containing v, that is, the circular order of the edges of B incident on v is equal to their circular order in the prescribed embedding of B. We say that Γ is *block-preserving* if it is block-preserving for all cutvertices.

Given a cutvertex v of G and block B containing v, we call the pair (B, v) a *cutpair*. The faces of block B containing v are called the *candidate cutfaces* for the cutpair (B, v) since one or more of them can be cutfaces in block preserving embeddings of G.

3   Biconnectivity-Related Graph Measures

The effectiveness of depth minimization depends on the biconnectivity properties of the graph. It is easy to find families of graphs whose embedding depth cannot be changed by changing the embedding of their cutvertices. Trees and biconnected graphs are examples of such families. Also, graphs that are "almost trees" or "almost biconnected" permit a very small variability of the embedding depth and hence we expect the effectiveness of depth minimization techniques to be small for such instances.

We introduce the following measures for a connected graph. We call *triviality* of a graph the ratio between the number of the trivial blocks and the total number of edges of the graph. For each block of the graph, its *occupancy* is the the ratio between the number of its edges and the total number of edges of the graph. We call *max-occupancy* of a graph the maximum among the occupancies of its blocks.

Intuitively, the triviality is a measure of how much a graph resembles a tree. The triviality of a tree is 1 while the triviality of a biconnected graph is 0. The max-occupancy is a measure of how much a graph resembles a biconnected graph. A tree has max-occupancy near to 0 while a biconnected graph has max-occupancy equal to 1.

We expect depth minimization techniques to be particularly useful on graphs that show both small triviality and small max-occupancy. Small triviality implies that most of the blocks have more than one face, i.e., the blocks contain cycles that may host part of the graph, potentially increasing the embedding depth. On the other hand, small max-occupancy implies that the graph has many blocks of small size making it possible to have many nestings and hence high depth value. Note that, the average of the occupancies of the blocks is not useful for our purposes. Consider a graph that contains one big block and many blocks of only one edge, we have high max-occupancy and low average occupancy. In such a graph, the variability of the embedding introduced by the presence of the trivial blocks have a small effect on the quality of the drawings, since aesthetic measures are largely due to the drawing of the big block.

4   Test Suites Analysis

The experiments described in Section 6 are performed over a test suite of about 3,000 graphs containing primarily graphs that represent real-world data and a test suite of 410 randomly generated graphs with specific characteristics.

Graphs from real-world applications. The graphs that represent real-world data are a subset of the graphs available from the GDToolkit web site [10] (ALF_CU data set) and first used in [9]. We selected the 3724 graphs with no more than 50 nodes from this set. Most of these graphs were non planar. Most of the graphs with more than 50 nodes have one big block and many very small blocks and are not well suited for testing the effectiveness of the depth minimization technique we consider.

In Figure 2(a) the distribution of the diameter of the block-cutvertex tree in the test suite is shown (average 9.21). In Figure 2(b) the distribution of the degree of the cutvertices in the block-cutvertex tree is shown. Most of the cutvertices have degree equal to 2 which is a rather low value. Figures 2(c) and 2(d) show the distribution of the max-occupancy and of triviality respectively. Most of the graphs have large max-occupancy which implies that most have one large block encompassing a large portion of the graph. The above observations show that most of the graphs of this test suite are instances that can be considered hard for depth minimization techniques.

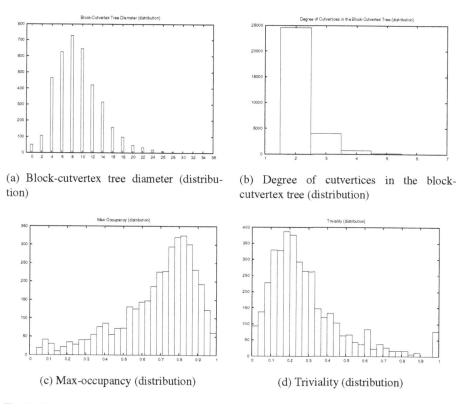

(a) Block-cutvertex tree diameter (distribution)

(b) Degree of cutvertices in the block-cutvertex tree (distribution)

(c) Max-occupancy (distribution)

(d) Triviality (distribution)

Fig. 2. Distributions of biconnectivity-related measurements on the graphs obtained from real-world applications. Statistics are performed on graphs with up to 50 vertices.

Randomly generated graphs. We randomly generated a set of graphs that feature small triviality and small max-occupancy. The purpose of such graphs is to understand the effectiveness of embedding depth minimization in a favorable situation. We generated 410 planar graphs with number of nodes n ranging between 10 and 50. For each value of n we generated 10 graphs. The generation process creates graphs with the number of cutvertices between $n/10$ and $n/5$, for each cutvertex the number of blocks incident to it is between 2 and 5, for each block the number of cutvertices incident to it is no more than 5. The details of the generation algorithm are given in [21].

Some statistics about the randomly generated test suite are shown in Figure 3.

5   Computing Embeddings with Minimum Depth and Large Depth

In this section, we describe in detail Algorithm *LargeDepth* for computing planar embeddings with large depth preserving the embeddings of the blocks, then, to make this paper more self-contained, we briefly sketch Algorithm *MinDepth* whose details are described in [22].

Algorithm *LargeDepth* takes as input a connected planar graph G with a prescribed embedding for each of its blocks, and a block B of G. The output is an embedding Γ

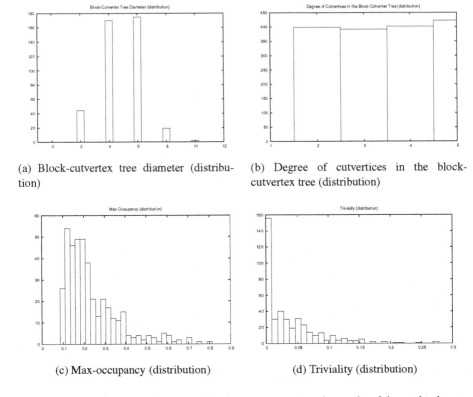

(a) Block-cutvertex tree diameter (distribution)

(b) Degree of cutvertices in the block-cutvertex tree (distribution)

(c) Max-occupancy (distribution)

(d) Triviality (distribution)

Fig. 3. Distributions of biconnectivity-related measurements on the graphs of the randomly generated test suite

of G with large depth which has the external face in B. The algorithm considers the block-cutvertex tree T of G rooted at B and builds the planar embedding Γ by means of a post-order traversal of T.

Given a node x of T (it may be a cutvertex or a block) we denote with $G(x)$ the subgraph of G associated with the subtree of T rooted at x and with $\Gamma(x)$ the planar embedding of $G(x)$ computed by method **embed**(x) of the algorithm.

Method **embed**(x) takes as input graph $G(x)$ and returns a planar embedding $\Gamma(x)$ of $G(x)$ with large depth. Let $y_1, y_2, \ldots, y_m$ be the children of x in T. The embedding $\Gamma(x)$ is computed by assembling the previously computed embeddings $\Gamma(y_i)$ of the children of x. Since we aim at obtaining embeddings whose block-cutface tree shows a high depth, when x is a cutvertex, blocks $y_1, y_2, \ldots, y_m$ are embedded so that the block with the deepest embedding is inside all other blocks. The detailed description of Algorithm *LargeDepth* follows.

Algorithm *LargeDepth*

input. A connected planar graph G with block-cutvertex tree T, a prescribed embedding for the blocks of G, and a block B.

output. A block-preserving planar embedding Γ of G of large depth that has the external face in B.

The algorithm consider T rooted at B. It computes and returns $\Gamma = \mathsf{embed}(B)$.

Method $\mathsf{embed}(B)$

input A block B of T.
output A block-preserving planar embedding $\Gamma(B)$ of large depth that has the external face in B.

    for all children v of B in T **do**
        Let $\Gamma(v) = \mathsf{embed}(v)$.
    end for
    Let $\Gamma(B)$ be equal to the prescribed embedding of B.
    for all children v of B in T **do**
        Modify $\Gamma(B)$ by attaching $\Gamma(v)$ into one of the candidate cutfaces for the cutpair (B, v)
    end for
    if B is the root of the block-cutface tree **then**
        choose the external face of $\Gamma(B)$ such that it is not a cutface, if possible.
    else
        choose an arbitrary external face for $\Gamma(B)$ (since $\mathsf{embed}(v)$ will change it)
    end if

Method $\mathsf{embed}(v)$

input A cutvertex v of T.
output A block-preserving planar embedding $\Gamma(v)$ of large depth that has v on the external face.

    for all children B of v in T **do**
        Let $\Gamma(B) = \mathsf{embed}(B)$.
    end for
    Partition the blocks that are children of v in T into two sets $\mathcal{B}_T$ and $\mathcal{B}_{NT}$:

      - $\mathcal{B}_T$ contains all the trivial blocks (blocks that have only one edge)
      - $\mathcal{B}_{NT}$ contains all the non-trivial blocks

    for all blocks B in $\mathcal{B}_{NT}$ **do**
        For each $B \in \mathcal{B}_{NT}$ select, among the candidate cutfaces for the cutpair (B, v) in $\Gamma(B)$, two distinct candidate cutfaces: $f_{ext}(B)$, $f_{int}(B)$, where possibly $f_{ext}(B)$ does not contain any block.
    end for
    Let $\bar{B}$ be a block in $\mathcal{B}_{NT}$ that shows the maximum of $\mathsf{depth}\,\Gamma(B)$.
    Let Γ' be $\Gamma(\bar{B})$ where all the blocks in $\mathcal{B}_T$ are attached into $f_{int}(\bar{B})$ and the external face of Γ' be $f_{ext}(\bar{B})$.
    for all blocks B in $\mathcal{B}_{NT} - \{\bar{B}\}$ **do**

Modify Γ' such that it is equal to $\Gamma(B)$ where the old Γ' is placed into $f_{int}(B)$ and its external face is $f_{ext}(B)$.

end for

Let $\Gamma(v) = \Gamma'$.

Algorithm *LargeDepth* computes deep embeddings since method **embed**(v) computes an embedding whose depth is the maximum depth shown by the children of v plus 2 times the number of the other non-trivial children of v.

Now we briefly sketch Algorithm *MinDepth* whose details are given in [22]. The algorithm starts from an arbitrary cutvertex v of the block-cutvertex tree and builds an embedding that has v on the external face, by means of a post-order traversal. In choosing how to embed a block and the subgraphs associated with its children, it puts the deepest sub-embeddings into the external face if possible. The result is an embedding with minimum depth among those that have v on the external face. All the subtrees of the rooted block-cutvertex tree turn out to be embedded such that the corresponding block-cutface tree has minimum depth. Starting from this embedding, it is possible to compute a block preserving embedding whose block-cutface tree has minimum diameter by applying a small number of changes. The cutface with minimum eccentricity in the block-cutface tree is chosen to be external which gives a block preserving minimum depth embedding.

6   Experimental Results

In this section we report the results of the experiments we performed on the test suites described in Section 4. Our goal is to show how much the depth of the embeddings affects the area of the drawings, their total edge length, and their number of bends.

The graphs derived from real-world applications may be in general non-planar. According to the topology-shape-metrics approach we planarize them. The planarization heuristic[1] adopted is a well-known one described in [8] and implemented in GDToolkit. This technique does not introduce new cutvertices and hence it does not change the block-cutvertex tree of the graph.

For each graph, after the planarization step, we generated two embeddings: one with minimum depth, by means of Algorithm *MinDepth*, and one with large depth, by means of Algorithm *LargeDepth*. Algorithm *LargeDepth* was run, for each graph, with all possible blocks as root and the deepest among the computed embeddings wasselected.

We draw such two embeddings according to the orthogonal drawing standard introduced in [4] (*simple podevsnef*). The algorithm used to compute the shape minimizes the number of bends within that standard. The compaction technique used is the heuristic[2] presented in [2] which iteratively compacts along the horizontal and the vertical direction until the drawing does not change. It provides good performance in terms of area and total edge length of the drawing.

---

[1] The problem of planarizing a graph introducing the minimum number of crossings is NP-hard [13].

[2] The problems of compacting an orthogonal drawing in order to obtain minimum area or minimum total edge length are NP-hard [20].

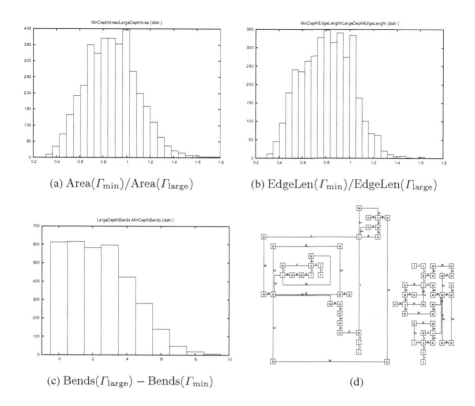

(a) $\mathrm{Area}(\Gamma_{\min})/\mathrm{Area}(\Gamma_{\mathrm{large}})$

(b) $\mathrm{EdgeLen}(\Gamma_{\min})/\mathrm{EdgeLen}(\Gamma_{\mathrm{large}})$

(c) $\mathrm{Bends}(\Gamma_{\mathrm{large}}) - \mathrm{Bends}(\Gamma_{\min})$

(d)

Fig. 4. Measurements performed on the test suite derived from real-world applications. The charts show the distributions of several comparative parameters.

We call $\Gamma_{\min}$ the drawing obtained from the embedding with minimum depth and Γ_{large} the drawing obtained from the embedding with large depth. Figure 4(a) shows the distribution of the ratio between the area of $\Gamma_{\min}$ and the area of Γ_{large}. Figure 4(b) shows the distribution of the ratio between the total edge length of $\Gamma_{\min}$ and the total edge length of Γ_{large}. Figure 4(c) shows the distribution of the difference between the number of bends of Γ_{large} and $\Gamma_{\min}$. Figure 4(d) shows Γ_{large} and $\Gamma_{\min}$ for a graph of 34 vertices (ug31.34).

From Figure 4 we can see that even with the first test suite (the hard one) there is a clear advantage in minimizing the embedding depth. In particular, for 2/3 of the graphs $\Gamma_{\min}$ has better area than Γ_{large} and there are peaks in which the area is decreased to 1/2 of the area of Γ_{large}. A similar behavior may be observed for the total edge length. The average of the ratio between the area of $\Gamma_{\min}$ and Γ_{large} is 0.87. The average of the ratio between the total edge length of $\Gamma_{\min}$ and Γ_{large} is 0.81. For the number of bends the result is not so good. For most of the graphs the number of bends is unchanged. However, in some cases the number decreases up to 9 units.

We repeated the same experiments on the test suite of 410 randomly generated graphs and the corresponding charts are shown in Figures 5(a), 5(b) and 5(c). Figure 5(d) shows Γ_{large} and $\Gamma_{\min}$ for a graph of 26 vertices.

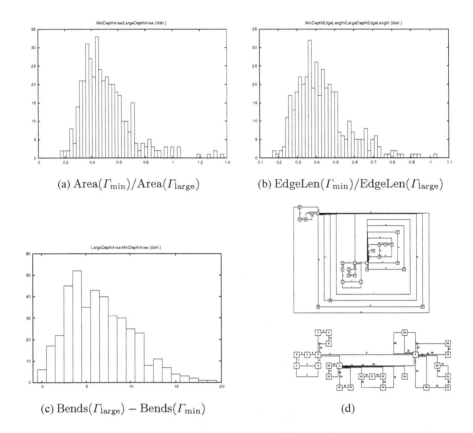

(a) Area($\Gamma_{\min}$)/Area(Γ_{large})

(b) EdgeLen($\Gamma_{\min}$)/EdgeLen(Γ_{large})

(c) Bends(Γ_{large}) − Bends($\Gamma_{\min}$)

(d)

Fig. 5. Measurements performed on the randomly generated test suite. The charts show the distributions of several comparative parameters.

The results in this case are unquestionable. Almost all the considered graphs were drawn with smaller area, smaller total edge length and fewer number of bends. There are $\Gamma_{\min}$ drawings whose area is 1/5 of the area of the corresponding Γ_{large}. The same holds for total edge length. The number of bends shows an average reduction of 6.5 and a peak reduction of 19. The average of the ratio between the area of $\Gamma_{\min}$ and Γ_{large} is 0.51. The average of the ratio between the total edge length of $\Gamma_{\min}$ and Γ_{large} is 0.43.

7   Conclusions

Our experiments prove that the embedding depth minimization technique has a positive effect on the quality of the drawings when adopted as a refinement step after the planarization phase of the topology-shape-metric approach. The effectiveness of the technique was explored using graphs obtained from real-world data, which gives results useful for application, and with graphs randomly generated that show specific biconnectivity properties, which is useful for theoretical investigation. We introduced two new measures to easily characterize the latter family of graphs. Also, we provided a heuristic to compute embeddings with large depth, a problem that may be subject of a more rigorous theoretical investigation.

Other interesting experimental and/or theoretical analysis can be performed in this area. For example, investigate how much better the techniques introduced in [14] perform compared with those used in this paper? Also, suppose that we randomly choose embeddings, how likely is it to obtain embeddings with small (or large) depth?

References

1. D. Alberts, C. Gutwenger, P. Mutzel, and S. Näher. AGD-Library: A library of algorithms for graph drawing. In *Proc. Workshop on Algorithm Engineering*, pages 112–123, 1997.
2. G. Di Battista, W. Didimo, M. Patrignani, and M. Pizzonia. Orthogonal and quasi-upward drawings with vertices of prescribed size. In *Graph Drawing (Proc. GD '99)*, volume 1731 of *Lecture Notes Comput. Sci.* Springer-Verlag, 1999.
3. P. Bertolazzi, G. Di Battista, and W. Didimo. Computing orthogonal drawings with the minimum number of bends. In *Workshop Algorithms Data Struct. (WADS'97)*, volume 1272 of *Lecture Notes Comput. Sci.*, pages 331–344. Springer-Verlag, 1997.
4. P. Bertolazzi, G. Di Battista, and W. Didimo. Computing orthogonal drawings with the minimum number of bends. *IEEETC: IEEE Transactions on Computers*, 49, 2000.
5. D. Bienstock and C. L. Monma. On the complexity of embedding planar graphs to minimize certain distance measures. *Algorithmica*, 5(1):93–109, 1990.
6. U. Brandes and D. Wagner. Dynamic grid embedding with few bends and changes. *Lecture Notes in Computer Science*, 1533, 1998.
7. Ulrik Brandes and Dorothea Wagner. A Bayesian paradigm for dynamic graph layout. In *Graph Drawing (Proc. GD '97)*, volume 1353 of *Lecture Notes Comput. Sci.*, pages 236–247. Springer-Verlag, 1998.
8. G. Di Battista, P. Eades, R. Tamassia, and I. G. Tollis. *Graph Drawing*. Prentice Hall, Upper Saddle River, NJ, 1999.
9. G. Di Battista, A. Garg, G. Liotta, R. Tamassia, E. Tassinari, and F. Vargiu. An experimental comparison of four graph drawing algorithms. *Comput. Geom. Theory Appl.*, 7:303–325, 1997.
10. Giuseppe Di Battista et al. *Graph Drawing Toolkit*. University of Rome III, Italy. http://-www.dia.uniroma3.it/~gdt/.
11. W. Didimo and G. Liotta. Computing orthogonal drawings in a variable embedding setting. In *Algorithms and Computation (Proc. ISAAC '98)*, volume 1533 of *Lecture Notes Comput. Sci.*, pages 79–88. Springer-Verlag, 1998.
12. Ulrich Fößmeier and Michael Kaufmann. Drawing high degree graphs with low bend numbers. In *Graph Drawing (Proc. GD '95)*, volume 1027 of *Lecture Notes Comput. Sci.*, pages 254–266. Springer-Verlag, 1996.
13. M. R. Garey and D. S. Johnson. Crossing number is NP-complete. *SIAM J. Algebraic Discrete Methods*, 4(3):312–316, 1983.
14. C. Gutwenger and P. Mutzel. Graph embedding with minimum depth and maximum external face. In *Graph Drawing (Proc. GD '03)*, volume 2912 of *Lecture Notes Comput. Sci.*, pages 259–272. Springer-Verlag Heidelberg, 2004.
15. J. Hopcroft and R. E. Tarjan. Efficient planarity testing. *J. ACM*, 21(4):549–568, 1974.
16. M. Jünger and P. Mutzel. Maximum planar subgraphs and nice embeddings: Practical layout tools. *Algorithmica*, 16(1):33–59, 1996.

17. Michael Jünger, Sebastian Leipert, and Petra Mutzel. Pitfalls of using PQ-Trees in automatic graph drawing. In *Graph Drawing (Proc. GD '97)*, volume 1353 of *Lecture Notes Comput. Sci.*, pages 193–204. Springer-Verlag, 1997.

18. G. W. Klau and P. Mutzel. Optimal compaction of orthogonal grid drawings. In *IPCO: 7th Integer Programming and Combinatorial Optimization Conference*, volume 1610 of *Lecture Notes Comput. Sci.* Springer-Verlag, 1999.

19. Harald Lauer, Matthias Ettrich, and Klaus Soukup. GraVis - system demonstration. In *Graph Drawing (Proc. GD '97)*, volume 1353 of *Lecture Notes Comput. Sci.*, pages 344–349. Springer-Verlag, 1997.

20. Maurizio Patrignani. On the complexity of orthogonal compaction. *Computational Geometry: Theory and Applications*, 19(1):47–67, 2001.

21. M. Pizzonia. *Engineering of Graph Drawing Algorithms for Applications*. PhD thesis, Dipartimento di Informatica e Sistemistica, University degli Studi "La Sapienza" di Roma, 2001.

22. M. Pizzonia and R. Tamassia. Minimum depth graph embedding. In M. Paterson, editor, *Algorithms – ESA 2000*, volume 1879 of *Lecture Notes Comput. Sci.* Springer-Verlag, 2000.

23. H. C. Purchase, R. F. Cohen, and M. James. Validating graph drawing aesthetics. In F. J. Brandenburg, editor, *Graph Drawing (Proc. GD '95)*, volume 1027 of *Lecture Notes Comput. Sci.*, pages 435–446. Springer-Verlag, 1996.

24. R. Tamassia. On embedding a graph in the grid with the minimum number of bends. *SIAM J. Comput.*, 16(3):421–444, 1987.

No-bend Orthogonal Drawings
of Series-Parallel Graphs
(Extended Abstract)

Md. Saidur Rahman[1], Noritsugu Egi[2], and Takao Nishizeki[2]

[1] Department of Computer Science and Engineering,
Bangladesh University of Engineering and Technology (BUET),
Dhaka 1000, Bangladesh
saidurrahman@cse.buet.ac.bd

[2] Graduate School of Information Sciences, Tohoku University, Aoba-yama 05,
Sendai 980-8579, Japan
egi@nishizeki.ecei.tohoku.ac.jp, nishi@ecei.tohoku.ac.jp

Abstract. In a no-bend orthogonal drawing of a plane graph, each vertex is drawn as a point and each edge is drawn as a single horizontal or vertical line segment. A planar graph is said to have a no-bend orthogonal drawing if at least one of its plane embeddings has a no-bend orthogonal drawing. Every series-parallel graph is planar. In this paper we give a linear-time algorithm to examine whether a series-parallel graph G of the maximum degree three has a no-bend orthogonal drawing and to find one if G has.

Keywords: Planar Graph, Algorithm, Graph Drawing, Orthogonal Drawing, Bend, SPQ tree.

1 Introduction

An *orthogonal drawing* of a planar graph G is a drawing of G such that each vertex is mapped to a point, each edge is drawn as a sequence of alternate horizontal and vertical line segments, and any two edges do not cross except at their common end [NR04, RN02, RNN99, T87]. A *bend* is a point where an edge changes its direction in a drawing. If G has a vertex of degree five or more, then G has no orthogonal drawing. On the other hand, if G has no vertex of degree five or more, that is, the maximum degree Δ of G is at most four, then G has an orthogonal drawing, but may need bends. Minimization of the number of bends in an orthogonal drawing is a challenging problem. A *bend-minimum* orthogonal drawing of a planar graph G has the minimum number of bends among all possible planar orthogonal drawings of G. The problem of finding a bend-minimum orthogonal drawing is one of the most famous problems in the graph drawing literature [BEGKLM04] and has been studied both in the fixed embedding setting [RN02, RNN03, RNN99, T87] and in the variable embedding setting [DLV98, GT01]. Some plane graphs with fixed embeddings have an orthogonal drawing without bends, in which each edge is drawn by a

P. Healy and N.S. Nikolov (Eds.): GD 2005, LNCS 3843, pp. 409–420, 2005.

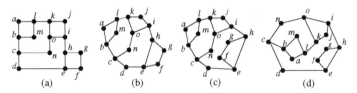

Fig. 1. (a) A no-bend drawing, and (b)–(d) three embeddings of the same planar graph

single horizontal or vertical line segment [RNN03]. We call such a drawing a *no-bend drawing* of a plane graph. Figure 1(a) depicts a no-bend drawing of the plane graph in Fig. 1(b). As a result in the fixed embedding, Rahman *et al.* [RNN03] obtained a necessary and sufficient condition for a plane graph G of $\Delta \leq 3$ to have a no-bend drawing, and gave a linear-time algorithm to find a no-bend drawing if G has.

We say that *a planar graph G has a no-bend drawing* if at least one of the plane embedding of G has a no-bend drawing. Figures 1(b), (c) and (d) depict three of all plane embeddings of the same planar graph G. Among them only the embedding in Fig. 1(b) has a no-bend drawing as illustrated in Fig. 1(a). Thus the *planar* graph G has a no-bend drawing. It is an NP-complete problem to examine whether a planar graph G of $\Delta \leq 4$ has a no-bend drawing in the variable embedding setting [GT01]. However, for a planar graph G of $\Delta \leq 3$, Di Battista *et al.* [DLV98] gave an $O(n^5 \log n)$ time algorithm to find a bend-minimum orthogonal drawing of G. Every series-parallel graph is a planar graph, and their algorithm takes time $O(n^3)$ for a series-parallel graph with $\Delta \leq 3$. Thus, by their algorithm one can examine in time $O(n^3)$ whether a series-parallel graph with $\Delta \leq 3$ has a no-bend drawing. As another result in the variable embedding, Rahman *et al.* [REN05] gave a linear time algorithm to examine whether a subdivision G of a planar triconnected cubic graph has a no-bend drawing, and to find a no-bend drawing of G if G has.

In this paper we study the problem of no-bend orthogonal drawings of series-parallel graphs with $\Delta \leq 3$ in the variable embedding setting, and give a linear algorithm to find a no-bend orthogonal drawing if G has.

The rest of the paper is organized as follows. Section 2 describes some definitions and presents preliminary results. Section 3 presents our algorithm to find a no-bend drawing of a biconnected series-parallel graph G if G has. Finally Section 4 is a conclusion.

2   Preliminaries

In this section we give some definitions and present preliminary results.

Let $G = (V, E)$ be a connected graph with vertex set V and edge set E. The *degree* $d(v)$ of a vertex v is the number of edges incident to v in G. We denote the maximum degree of graph G by $\Delta(G)$ or simply by Δ. The *connectivity* $\kappa(G)$ of a graph G is the minimum number of vertices whose removal results in a disconnected graph or a single-vertex graph K_1. We say that G is *k-connected* if $\kappa(G) \geq k$.

A graph $G = (V, E)$ is called a *series-parallel graph* (with source s and sink t) if either G consist of a pair of vertices connected by a single edge, or there exist two series-parallel graphs $G_i = (V_i, E_i), i = 1, 2$, with source s_i and sink t_i such that $V = V_1 \cup V_2, E = E_1 \cup E_2$, and either $s = s_1, t_1 = s_2$ and $t = t_2$ or $s = s_1 = s_2$ and $t = t_1 = t_2$.

A pair $\{u, v\}$ of vertices of a connected graph G is a *split pair* if there exist two subgraphs $G_1 = (V_1, E_1)$ and $G_2 = (V_2, E_2)$ satisfying the following two conditions: 1. $V = V_1 \cup V_2, V_1 \cap V_2 = \{u, v\}$; and 2. $E = E_1 \cup E_2, E_1 \cap E_2 = \emptyset$, $|E_1| \geq 1, |E_2| \geq 1$. Thus every pair of adjacent vertices is a split pair. A *split component* of a split pair $\{u, v\}$ is either an edge (u, v) or a maximal connected subgraph H of G such that $\{u, v\}$ is not a split pair of H. A split pair $\{u, v\}$ of G is called a *maximal split pair* with respect to a *reference split pair* $\{s, t\}$ if, for any other split pair $\{u', v'\}$, vertices s, t, u and v are in the same split component of $\{u', v'\}$.

Let G be a biconnected series-parallel graph. Let (s, t) be an edge of G. The SPQ-tree T of G with respect to a *reference edge* $e = (s, t)$ describes a recursive decomposition of G induced by its split pairs [GL99]. Tree T is a rooted ordered tree whose nodes are of three types: S, P and Q. Each node x of T corresponds to a subgraph of G, called its *pertinent graph* G_x. Each node x of T has an associated biconnected multigraph, called the *skeleton* of x and denoted by *skeleton(x)*. Tree T is recursively defined as follows.

- *Trivial Case*: In this case, G consists of exactly two parallel edges e and e' joining s and t. T consists of a single Q-node x. The skeleton of x is G itself. The pertinent graph G_x consists of only the edge e'.
- *Parallel Case*: In this case, the split pair $\{s, t\}$ has three or more split components $G_0, G_1, \cdots, G_k, k \geq 2$, and G_0 consists of only a reference edge $e = (s, t)$. The root of T is a P-node x. The *skeleton(x)* consists of $k + 1$ parallel edges $e_0, e_1, \cdots, e_k$ joining s and t. The pertinent graph $G_x = G_1 \cup G_2 \cup \cdots \cup G_k$ is a union of $G_1, G_2, \cdots, G_k$. (The *skeleton* of P-node p_2 in Fig. 2 consists of three parallel edges joining vertices e and g. Figure 2(e) depicts the pertinent graph of p_2.)
- *Series Case*: In this case the split pair $\{s, t\}$ has exactly two split components, and one of them consists of the reference edge e. One may assume that the other split component has cut-vertices $c_1, c_2, \cdots, c_{k-1}, k \geq 2$, that partition the component into its blocks $G_1, G_2, \cdots, G_k$ in this order from s to t. Then the root of T is an S-node x. The skeleton of x is a cycle $e_0, e_1, \cdots, e_k$ where $e_0 = e$, $c_0 = s, c_k = t$, and e_i joins c_{i-1} and $c_i, 1 \leq i \leq k$. The pertinent graph G_x of node x is a union of $G_1, G_2, \cdots, G_k$. (The *skeleton* of S-node s_2 in Fig. 2 is the cycle c, d, e, g, h, a, c. Figure 2(d) depicts the pertinent graph G_{s_2} of s_2.)

In all cases above, we call the edge e the *reference edge* of node x. Except for the trivial case, node x of T has children $x_1, x_2, \cdots, x_k$ in this order; x_i is the root of the SPQ-tree of graph $G_i \cup e_i$ with respect to the reference edge $e_i, 1 \leq i \leq k$. We call edge e_i *the reference edge of node* x_i, and call the endpoints of edge e_i the *poles* of node x_i. The tree obtained so far has a Q-node associated with each

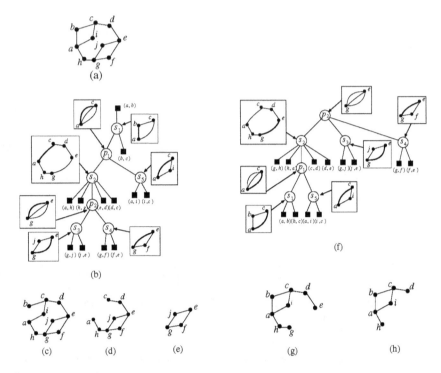

Fig. 2. (a)A biconnected series-parallel graph G with $\Delta = 3$, (b) SPQ-tree $\mathcal{T}$ of G with respect to reference edge (a, b), and skeletons of P- and S-nodes, (c) the pertinent graph G_{s_1} of S-node s_1, (d) the pertinent graph G_{s_2} of S-node s_2, (e) the pertinent graph G_{p_2} of P-node p_2, (f) SPQ-tree $\mathcal{T}$ of G with P-node p_2 as the root, (g) the pertinent graph of S-node s_2, and (h) the core graph of s_2

edge of G, except the reference edge e. We complete the SPQ-tree $\mathcal{T}$ by adding a Q-node, representing the reference edge e, and making it the parent of x so that it becomes the root of $\mathcal{T}$. An example of the SPQ-tree of a biconnected series-parallel graph in Fig. 2(a) is illustrated in Fig. 2(b), where the edge drawn by a thick line in each skeleton is the reference edge of the skeleton.

The SPQ-tree $\mathcal{T}$ defined above is a special case of an "SPQR-tree" [DT96, GL99] where there is no R-node and the root of the tree is a Q-node corresponding to the reference edge e. One can easily modify $\mathcal{T}$ to an SPQ-tree $\mathcal{T}'$ with an arbitrary P-node as the root as illustrated in Fig. 2(f).

In the remainder of this paper, we thus consider a SPQ-tree $\mathcal{T}$ with a P-node as the root. If $\Delta = 2$, then a biconnected series-parallel graph G is a cycle, and a cycle G has a no-bend drawing if and only if G has four or more vertices. One may thus assume that $\Delta \geq 3$, and that the root P-node of $\mathcal{T}$ has three or more children. Then the pertinent graph G_x of each node x is the subgraph of G induced by the edges corresponding to all descendant Q-node of x. The following facts can be easily derived from the fact that each vertex of G has degree at most three and G has no multiple edges.

Fact 1. *Let (s,t) be the reference edge of an S-node x of $\mathcal{T}$, and let $x_1, x_2, \cdots, x_k$ be the children of x in this order from s to t. Then (i) each child x_i of x is either a P-node or a Q-node; (ii) both x_1 and x_k are Q-nodes; and (iii) x_{i-1} and x_{i+1} must be Q-nodes if x_i is a P-node where $2 \leq i \leq k - 1$.*

Fact 2. *Each non-root P-node of $\mathcal{T}$ has exactly two children, and either both of the two children are S-nodes or one of them is an S-node and the other is a Q-node.* ∎

Let x be an S-node of $\mathcal{T}$, and let u and v be the poles of the pertinent graph of x. Let $x_1, x_2, \cdots, x_k$ be the children of x in this order from u to v. From Fact 1, x_1 and x_k are Q-nodes. Thus x_1 and x_k correspond to edges (u, u') and (v', v) of G, respectively. Then the *core graph* for x is a graph obtained from the pertinent graph of x by deleting vertices u and v. (Figure 2(g) illustrates a pertinent graph of S-node s_1 for $\mathcal{T}$ in Fig. 2(f), and Fig. 2(h) illusrates a core graph for s_1.) Vertices u' and v' are called the *poles* of the core graph for x, and edges (u, u') and (v', v) are called *hands* of the core graph for x. (In Figs. 2(g) and (h) the poles of the core graph of S-node s_1 are vertices d and h.) For a P- or Q-node x in $\mathcal{T}$, we define the *core graph* for x as the pertinent graph of x, and the poles of the core graph for x is the same as the poles of the pertinent graph of x. The core graph of a P- or Q-node has no hand.

A drawing of a planar graph G is called an *orthogonal drawing* of G if each vertex is mapped to a point, each edge is drawn as a sequence of alternate horizontal and vertical line segments, and any two edges do not cross except at their common end. We call an orthogonal drawing D of G a *no-bend drawing* if D has no bend, that is, each edge is drawn as a single horizontal or vertical line segment. A *polar drawing* of a series-parallel graph G is a no-bend drawing of G in which the two poles u and v of G are drawn on the outer face F_o of the drawing.

We call a polar drawing D of a series-parallel graph G a *diagonal drawing* if D intersects neither the first quadrant with the origin at pole u nor the third quadrant with the origin at pole v after rotating the drawing and renaming the poles if necessary, as illustrated in Fig. 3(a). Throughout the paper a quadrant is considered to be a closed plane region. Both a drawing of a single vertex as a point and a drawing of a single edge as a straight line-segment are diagonal drawings.

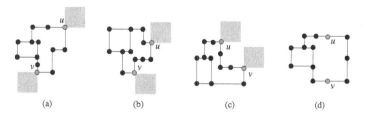

(a)   (b)   (c)   (d)

Fig. 3. Polar drawings of a graph G with poles u and v: (a) a diagonal drawing, (b) a side-on drawing, (c) an L-shape drawing, (d) another polar drawing

We call a polar drawing D of G a *side-on drawing* if D intersects neither the first quadrant with the origin at u nor the fourth quadrant with the origin at v after rotating the drawing and renaming the poles if necessary, as illustrated in Fig. 3(b). A drawing of a single vertex as a point is regarded not to be a side-on drawing, while a drawing of a single edge as a straight line-segment is a side-on drawing.

A polar drawing D is called an *L-shape drawing* if D intersects neither the first quadrant with the origin at u nor the first quadrant with the origin at v after rotating the drawing and renaming the poles if necessary, as illustrated in Fig. 3(c). A drawing of a single vertex as a point is regarded not to be an L-shape drawing. A drawing of a single edge as a straight line-segment is not an L-shape drawing.

We say that a polar drawing is *good* if it is a diagonal, side-on or L-shape drawing. Not every polar drawing D is good. For example, the polar drawing in Fig. 3(d) is not good, because it is not a diagonal, side-on drawing or L-shape drawing.

In the next section we give an algorithm for constructing no-bend drawing of a biconnected series-parallel graph G with $\Delta = 3$.

Our idea is as follows. Let $\mathcal{T}$ be an SPQ-tree of G. The core graph of each leaf-node of $\mathcal{T}$ consists of a single edge. For each leaf-node of $\mathcal{T}$ we first draw the core graph by a line segment as a diagonal or side-on drawing. Then, in bottom up fashion, we find a diagonal drawing, a side-on drawing, and an L-shape drawing of the core graph for each internal node x of $\mathcal{T}$ by merging the drawings corresponding to the children of x if they exist. The drawing of the graph corresponding to the root-node of $\mathcal{T}$ yields a no-bend drawing of G if G has a polar drawing with the split pair, corresponding to the root P-node, as the poles. Our algorithm eventually chooses an appropriate SPQ-tree $\mathcal{T}$ of G such that the drawing of a plane graph corresponding to the root-node of $\mathcal{T}$ yields a no-bend drawing of G if G has. (See Fig. 8 for illustration.)

As we see later, we construct a no-bend drawing of the core graph for a node x in $\mathcal{T}$ by merging the no-bend drawings of the core graphs for the childrens of x; the no-bend drawing of the core graph for each children of x must be a polar drawing with the two poles of the core graph. A side-on drawing is found more suitable for merging than a diagonal drawing, and an L-shape drawing is found more suitable for merging than a side-on drawing. Intuitively, to connect the two poles by a sequence of horizontal and vertical line segments, at least three turns are required for a diagonal drawing, at least two turns are required for a side-on drawing and only one turn is required for an L-shape drawing. A graph may have a diagonal drawing although it has no side-on or L-shape drawing and a graph may have a side-on drawing although it has no L-shape drawing. We call a polar drawing D of a core graph $H(x)$ for a node x in $\mathcal{T}$ a *desirable* drawing if one of the following (a), (b) and (c) holds: (a) D is an L-shape drawing; (b) D is a side-on drawing, and $H(x)$ has no L-shape drawing; (c) D is a diagonal drawing, and $H(x)$ has neither an L-shape drawing nor a side-on drawing. Throughout the paper we denote by $D(x)$ a *desirable* drawing of the core graph $H(x)$ for a node x in $\mathcal{T}$.

3 No-bend Drawings of Biconnected Series-Parallel Graphs

In this section we give an algorithm to construct a no-bend orthogonal drawing of a biconnected series-parallel graph G whenever G has.

If G is a cycle, then it is easy to find a no-bend drawing of G; G has a no-bend drawing if and only if G has four or more vertices. We thus assume that G is not a cycle.

Let $\mathcal{T}$ be an SPQ-tree of G whose root is a P-node x_p having three children. (See Fig. 2(f).) We now have the following lemma.

Lemma 3. *Let G be a series-parallel graph with $\Delta \leq 3$, let $\mathcal{T}$ be an SPQ-tree with a P-node x_p as the root, and let x be a non-root node in $\mathcal{T}$. If the core graph $H(x)$ of x has a no-bend drawing, then the following (a) and (b) hold: (a) $H(x)$ has a side-on or diagonal drawing, and hence $H(x)$ has a desirable drawing $D(x)$; and (b) if a desirable drawing of $H(x)$ is a diagonal drawing, then every no-bend drawing of $H(x)$ is a diagonal drawing for the poles of $H(x)$.*

Proof. We will prove the claim by induction based on $\mathcal{T}$.

We first assume that x is a leaf-node, that is, a Q-node. In this case $H(x)$ consists of a single edge $e = (u, v)$, and u and v are the poles of $H(x)$. We thus draw e as a single vertical line segment, which is a side-on drawing $D(x)$ of $H(x)$. Since $H(x)$ has no L-shape drawing, $D(x)$ is a desirable drawing. Thus (a) and (b) hold.

We next assume that x is an inner node other than the root x_p and that $H(x)$ has a no-bend drawing. Let u and v are the poles of $H(x)$. Let $x_1, x_2, \cdots, x_k$ ($k \geq 2$) be the children of x in this order from u to v. Since $H(x)$ has a no-bend drawing, each $H(x_i)$ has a no-bend drawing. Thus we suppose inductively that (a) and (b) hold for each child of x. We now have two cases to consider.

Case 1: x is an S-node.

Suppose that x has exactly two children. Then $H(x)$ consists of a single vertex. We draw $H(x)$ as a point. Then the diagonal drawing is a desirable drawing $D(x)$. Thus (a) and (b) hold.

We thus assume that x has exactly k children and $k \geq 3$. Then $H(x) = H(x_2) \cup H(x_3) \cup \cdots \cup H(x_{k-1})$, where $H(x_i)$ is the core graph of x_i. The hypothesis implies that, for each i, $2 \leq i \leq k-1$, (a) and (b) hold for the core graph $H(x_i)$. We now have the following four subcases to consider.

Case 1(a): $k = 3$.

In this case $H(x) = H(x_2)$, hence (a) and (b) hold for $H(x)$.

Case 1(b): $k = 4$.

In this case $H(x) = H(x_2) \cup H(x_3)$. Fact 1(iii) implies that either both x_2 and x_3 are Q-nodes or one of them is a P-node and the other one is a Q-node.

If x_2 and x_3 are Q-nodes, then we can construct both an L-shape drawing and a side-on drawing of $H(x)$, as illustrated in Figs. 4(a) and 5(a). Thus a desirable drawing of $H(x)$ is an L-shape drawing, and hence (a) and (b) hold. We thus assume that one of them, say x_2, is a P-node and the other is a Q-node.

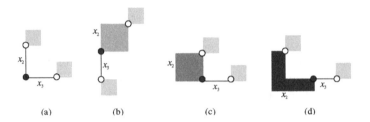

Fig. 4. Desirable drawings of the core graph for S-nodes with four children

We first consider the case where a desirable drawing $D(x_2)$ of $H(x_2)$ is a diagonal drawing. In this case we can construct a side-on drawing $D(x)$ of $H(x)$ as illustrated in Fig. 4(b). Since the desirable drawing of $H(x_2)$ is a diagonal drawing, $H(x_2)$ has neither an L-shape drawing nor a side-on drawing, and hence clearly $H(x)$ has no L-shape drawing. Therefore the side-on drawing $D(x)$ of $H(x)$ is a desirable drawing. Hence (a) and (b) hold.

We next consider the case where the desirable drawing $D(x_2)$ of $H(x_2)$ is a side-on drawing. Then we can construct both an L-shape drawing $D(x)$ and a side-on drawing of $H(x)$ as illustrated in Figs. 4(c) and 5(c). Hence (a) and (b) hold.

We finally consider the case where the desirable drawing $D(x_2)$ of $H(x_2)$ is an L-shape drawing. Then we can construct an L-shape drawing $D(x)$ of $H(x)$ as illustrated in Fig. 4(d). $H(x_2)$ has a side-on or diagonal drawing. From it one can easily construct a side-on drawing of $H(x)$ as illustrated in Figs. 5(b) and (c). Therefore (a) and (b) hold.

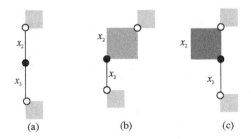

Fig. 5. Side-on drawings of the core graph for S-nodes with four children

Case 1(c): $k = 5$.

In this case, $H = H(x_2) \cup H(x_3) \cup H(x_4)$. Fact 1(iii) implies that at least one of x_2, x_3 and x_4 is a Q-node. In this case we can construct a no-bend drawing of $H(x)$ such that (a) and (b) hold. The details are omitted in this extended abstract.

Case 1(d): $k \geq 6$.

In this case $H = H(x_2) \cup H(x_3) \cup \cdots \cup H(x_{k-1}), k \geq 6$. Fact 1(iii) implies that there are two or more Q-nodes among $x_2, x_3, \cdots x_{k-1}$. Therefore we can easily

construct both an L-shape drawing and a side-on drawing D of $H(x)$, and hence
(a) and (b) hold.

Case 2: x is a P-node.

In this case $k = 2$ and x has exactly two children x_1 and x_2. Then the
hypothesis implies that, for $i = 1, 2$, (a) and (b) hold for $H(x_i)$. By Fact 2 either
both x_1 and x_2 are S-nodes or one of x_1 and x_2 is an S-node and the other is
a Q-node. We first assume that one of x_1 and x_2, say x_1, is a Q-node, then we
have the following two subcases.

Case 2(a): The desirable drawing $D(x_2)$ of $H(x_2)$ is a diagonal drawing.

In this case $H(x_2)$ has neither an L-shape drawing nor a side-on drawing.
Furthermore, every no-bend drawing of $H(x_2)$ is a diagonal drawing by induction
hypothesis. Then $D(x_1), D(x_2)$ and the drawings of hands of $H(x_2)$ cannot be
merged without bends as illustrated in Fig. 6(a). Therefore $H(x)$ does not have a
no-bend drawing, contrary to the assumption that $H(x)$ has a no-bend drawing.
Therefore this case does not occur.

*Case 2(b): The desirable drawing $D(x_2)$ of $H(x_2)$ is a side-on or L-shape
drawing.*

In this case we can construct a no-bend drawing $D(x)$ of $H(x)$ such that (a)
and (b) hold as illustrated in Figs. 6(b)–(i). $\mathcal{Q.E.D.}$

We call the algorithm described in the proof of Lemma 3 for finding a desirable
drawing $D(x)$ of $H(x)$ Algorithm **Desirable-Drawing** whenever $H(x)$ has a no-
bend drawing. Clearly Algorithm **Desirable-Drawing** takes linear-time.

In the rest of the section we give Algorithm **Biconnected-Draw** for finding
a no-bend drawing of G whenever G has. Remember that the root node x_p in $\mathcal{T}$

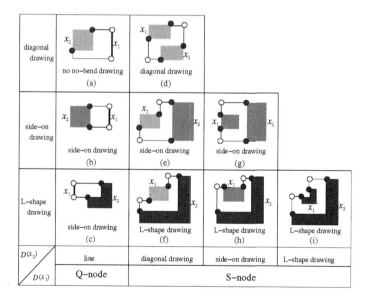

Fig. 6. Drawings of $H(x)$ for a P-node $x \neq x_p$

has three children as depicted in Fig. 2(f). Let x_1, x_2 and x_3 be the three children of x_p in $\mathcal{T}$. If G has a no-bend drawing, then $H(x_i), 1 \leq i \leq 3$, has a no-bend drawing. For $1 \leq i \leq 3$, we find a desirable drawing $D(x_i)$ of $H(x_i)$ by Algorithm **Desirable-Drawing**. If G has a polar drawing for the poles corresponding to x_p, then we now find a no-bend drawing of $G = H(x_p)$ by merging the drawings of $D(x_1), D(x_2), D(x_3)$ and the drawings of their hands. Otherwise, we find appropriate poles for which G has a no-bend polar drawing. Since G is a simple graph, at most one of x_1, x_2 and x_3 is a Q-node. We now have the following two cases to consider.

Case 1: one of them, say x_3, is a Q-node.

In this case only x_3 is a Q-node. If at least one of $D(x_1)$ and $D(x_2)$ is a diagonal drawing, Then G does not have a no-bend drawing as illustrated in Fig. 7(a)-(c). Otherwise, G has a no-bend drawing as illustrated in Fig. 7(d)-(f). The details are omitted.

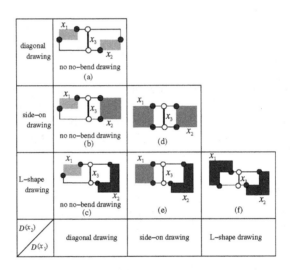

Fig. 7. Illustration for Case 1 of Algorithm **Biconnected-Draw**

Case 2: all of x_1, x_2 and x_3 are S-nodes.

If at most one of $D(x_1), D(x_2)$ and $D(x_3)$ is a diagonal drawing, then we can easily construct a no-bend drawing of G. If all of $D(x_1), D(x_2)$ and $D(x_3)$ are diagonal drawings, then one can easily observe that G does not have a no-bend drawing.

We thus consider the case where exactly two of $D(x_1), D(x_2)$ and $D(x_3)$ are diagonal drawings. If two of $D(x_1), D(x_2)$ and $D(x_3)$ are diagonal drawings and the other is an L-shape drawing, then clearly we can construct a no-bend drawing of G. We may thus assume that two of $D(x_1), D(x_2)$ and $D(x_3)$ are diagonal drawings and the other is a side-on drawing.

We may assume without loss of generality that $D(x_1)$ and $D(x_2)$ are diagonal drawings and $D(x_3)$ is a side-on drawing. By Lemma 3(b) every no-bend drawing of each of $H(x_1)$ and $H(x_2)$ is a diagonal drawing. By merging $D(x_1)$ and $D(x_2)$

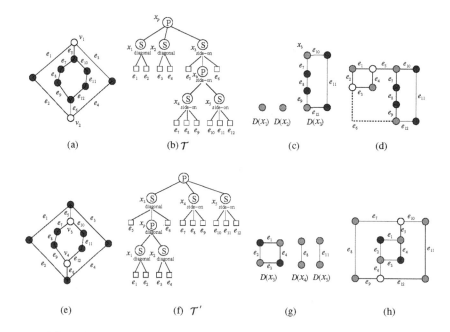

Fig. 8. (a)—(d) A no-bend drawing of G cannot be found using tree $\mathcal{T}$, and (e)—(h) a no-bend drawing of G can be found using tree $\mathcal{T}'$

we can obtain only a diagonal drawing D'. Since $D(x_3)$ is a side-on drawing, D' and $D(x_3)$ cannot be merged to produce a no-bend drawing of G. However, we can construct a no-bend drawing of G if $H(x_3)$ has another appropriate no-bend drawing.

We give an illustrative example in Figure 8 and omit the details of the proof. G has no polar drawing with the poles corresponding to x_p as illustrated in Fig. 8(d). However, G may have a no-bend drawing when one considers some other split pair as poles. We therefore consider an SPQ-tree $\mathcal{T}'$ of G with x_b as the root, as illustrated in Fig. 8(f), where x_3, x_4 and x_5 are the children of x_b. Each of $D(x_4)$ and $D(x_5)$ remains same as one obtained for the SPQ-tree $\mathcal{T}$. Considering $\mathcal{T}'$, $D(x_3)$ is a diagonal drawing D'. We can thus find a no-bend drawing of G by recursively applying Algorithm **Biconnected-Draw** regarding $D(x_3)$, $D(x_4)$ and $D(x_5)$ as new $D(x_1)$, $D(x_2)$ and $D(x_3)$, respectively. (Figure 8(h) shows that G has a no-bend polar drawing with the poles corresponding to root x_b.) If we cannot draw a no-bend orthogonal drawing of G by repeating the operation above, then G does not have a no-bend drawing.

Thus Algorithm **Biconnected-Draw** finds a no-bend drawing of G if G has. One can efficiently implement Algorithm **Biconnected-Draw** so that it takes time $O(n)$. The details are omitted in this extended abstract.

Theorem 1. *Let G be a biconnected series-parallel graph of the maximum degree three. Then Algorithm **Biconnected-Draw** finds a no-bend drawing of G in time $O(n)$ whenever G has, where n is the number of vertices of G.*

4   Conclusions

In this paper, we gave a linear-time algorithm to find a no-bend drawing of a biconnected series-parallel graph G of maximum degree at most three. We also gave an algorithm to find a no-bend drawing of a series-parallel graph G which is not always biconnected. However, the algorithm is omitted in this extended abstract due to page limitation. It is left as a future work to find a bend-minimum drawing of series-parallel graphs and to find a linear-time algorithm for a larger class of planar graphs.

References

[BEGKLM04] F. Brandenburg, D. Eppstein, M. T. Goodrich, S. Kobourov, G. Liotta and P. Mutzel, *Selected open problems in graph drawings*, Proc. of GD '03, Lect, Notes in Computer Science, 1912, pp. 515-539, 2004.

[DLV98] G. Di Battista, G. Liotta and F. Vargiu, *Spirality and optimal orthogonal drawings*, SIAM J. Comput., 27(6), pp. 1764-1811, 1998.

[DT96] G. Di Battista and R. Tamassia, *On-line planarity testing*, SIAM J. Comput., 25(5), pp. 956-997, 1996.

[GL99] A. Garg and G. Liotta, *Almost bend-optimal planar orthogonal drawings of biconnected degree-3 planar graphs in quadratic time*, Proc. of GD '99, Lect. Notes in Computer Science, 1731, pp. 38-48, 1999.

[GT01] A. Garg and R. Tamassia, *On the computational complexity of upward and rectilinear planarity testing*, SIAM J. Comput., 31(2), pp. 601-625, 2001.

[NR04] T. Nishizeki and M. S. Rahman, *Planar Graph Drawing*, World Scientific, Singapore, 2004.

[REN05] M. S. Rahman, N. Egi and T. Nishizeki, *No-bend orthogonal drawings of subdivisions of planar triconnected cubic graphs*, IEICE Trans. Inf. & Syst., E88-D (1), pp.23-30, 2005.

[RN02] M. S. Rahman and T. Nishizeki, *Bend-minimum orthogonal drawings of plane 3-graphs*, Proc. of WG '02, Lect. Notes in Computer Science, 2573, pp. 265-276, 2002.

[RNN03] M. S. Rahman, M. Naznin and T. Nishizeki, *Orthogonal drawings of plane graphs without bends*, Journal of Graph Alg. and Appl., 7(4), pp. 335-362, 2003.

[RNN99] M.S. Rahman, S. Nakano and T. Nishizeki, *A linear algorithm for bend-optimal orthogonal drawings of triconnected cubic plane graphs*, Journal of Graph Alg. and Appl., http://www.cs.brown.edu/publications/jgaa/, 3(4), pp. 31-62, 1999.

[T87] R. Tamassia, *On embedding a graph in the grid with the minimum number of bends*, SIAM J. Comput., 16, pp. 421-444, 1987.

Parallel-Redrawing Mechanisms, Pseudo-Triangulations and Kinetic Planar Graphs

Ileana Streinu

Computer Science Department, Smith College, Northampton, MA 01063
streinu@cs.smith.edu

Abstract. We study *parallel redrawing graphs*: graphs embedded on moving point sets in such a way that edges maintain their slopes all throughout the motion.

The configuration space of such a graph is of an oriented-projective nature, and its combinatorial structure relates to rigidity theoretic parameters of the graph. For an appropriate parametrization the points move with constant speeds on linear trajectories. A special type of kinetic structure emerges, whose events can be analyzed combinatorially. They correspond to collisions of subsets of points, and are in one-to-one correspondence with contractions of the underlying graph on rigid components. We show how to process them algorithmically via a *parallel redrawing sweep.*

Of particular interest are those *planar* graphs which maintain non-crossing edges throughout the motion. Our main result is that they are (essentially) pseudo-triangulation mechanisms: pointed pseudo-triangulations with a convex hull edge removed. These kinetic graph structures have potential applications in morphing of more complex shapes than just simple polygons.

1 Introduction

Consider a straight-line drawing of a graph $G = (V, E)$ in the plane. A *parallel redrawing* of G is another drawing so that for every edge $ij \in E$, the corresponding line segments in the two drawings are parallel (Fig. 1). A parallel redrawing is *trivial* if it is similar to the original drawing (via a rescaling or translation of). See Fig. 1(A-B). Classical results in Rigidity Theory, see [16], establish combinatorial criteria for a graph to admit non-trivial parallel redrawings, generically. In particular, such a graph has a certain number of *degrees of freedom* (for the example in Fig. 1(C-D) this number is one).

A *kinetic point set* is (for the purpose of this paper) a set of points in the plane moving with constant velocities, and a *kinetic graph* is a graph drawn on a kinetic point set. An illustraton is found in the rightmost example of Fig. 2. As the points move, the shape of the embedding changes: edges may shrink to zero-length, points may collide and edges may cross for certain time intervals, and be non-crossing for others. A natural graph drawing problem is whether these phenomena may be

P. Healy and N.S. Nikolov (Eds.): GD 2005, LNCS 3843, pp. 421–433, 2005.

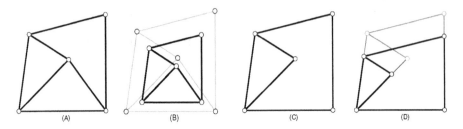

Fig. 1. A direction network (A) with only similar (trivial) parallel redrawings (B), and another one (C) with a non-similar parallel redrawing (D)

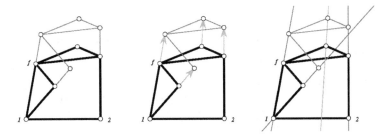

Fig. 2. The kinetic point set underlying a Laman mechanism: a parallel redrawing, the point velocities and the linear trajectories of the vertices

predicted from the law of motion of the points (given by their initial positions and velocities). In this paper, we give a complete answer to this problem under the additional constraint that throughout the motion, the graphs are parallel redrawings of each other. To keep the presentation short, we'll discuss here only the case when the graphs have exactly *one degree of freedom*. Our **main algorithmic result** is an efficient technique for predicting the relevant events via a new process called the *parallel redrawing sweep*, of which Figure 3 is a preview.

Of particular interest is the *planar* case. Given a plane embedding of a planar graph, start moving the points (kinetically): *are there graphs which remain non-crossing throughout the whole duration of the motion?*

The answer to this question is intrinsically related to a special class of planar graphs, the *pointed pseudo-triangulation mechanisms*. A pseudo-triangulation is a plane graph with a convex outer face and with all inner faces embedded as pseudo-triangles: simple polygons with exactly three inner convex vertices (called the *corners* of the pseudo-triangle). In a *pointed pseudo-triangulation*, each vertex is incident to an angle larger than π. Removing a convex hull edge from such a graph produces a *pointed pseudo-triangulation mechanism* (ppt-mechanism), see Fig. 3. Our **main theorem** is that these are (essentially) the only one-degree-of-freedom graphs which *maintain planarity throughout a parallel redrawing motion*. Figure 3 gives a preview.

Historical background. Points moving with constant velocities appear in a popular morphing technique for planar polygonal shapes. It is known that the simplicity of the shape may be violated during the morph. While there are morphing techniques which achieve simplicity, e.g. via compatible planar triangulations as in [5], or by maintaining the slopes of the polygon edges, as in [7], nobody seems to have analyzed theoretically what this simple paradigm for motion has to say about edge crossings.

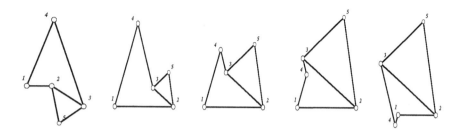

Fig. 3. A parallel redrawing pointed pseudo-triangulation mechanism: snapshots from its motion. The second and third snapshot have the same combinatorial structure, but all the others differ in the convex/reflex angles per face. Note the flips in rigid bodies between consecutive snapshots with different combinatorial structure.

Pseudo-triangulations are relatively new objects, applied in Computational Geometry for problems such as visibility [10] and kinetic data structures [1]. The *pointed* or *minimal* pseudo-triangulations [14, 15], have interesting rigidity-theoretical properties and applications in designing efficient motion planning algorithms for planar robot arms. Recently many papers have investigated their rich combinatorial, rigidity theoretic and polyhedral properties, e.g. [8, 11, 2, 3] and applications [12].

The main result of this paper, as stated above, is in fact a new rigidity-theoretic property of pointed pseudo-triangulation *mechanisms*. Whereas in the fixed-edge-length model of rigidity, these graphs are 1-dof *expansive* mechanisms, the *new result* of this paper is that in the *fixed-edge-direction* model, they capture the essence of kinetic parallel-redrawing *planar graphs*. The property is surprisingly simple, has an elementary proof and is entirely new: I am not aware of anybody even asking such a question before. I am presenting it to the Graph Drawing community with the expectation that it may find applications beyond those that originally motivated my investigation.

How to read this extended abstract. To formally state the results (theorems and algorithms), I have no choice but to plunge into a fair amount of definitions: it will take a few pages, since there is no standard reference where I can send the reader to gather them all (but please skip directly to page 426 if this is familiar material). To help the reader less fluent with all these concepts get faster to the ideas, and especially since many of them have a kinetic nature which can hardly be conveyed only with static printed images, I have assembled a web site with animations and interactive applets illustrating them: http://cs.smith.edu/~streinu/Research/KineticPT/

2   Preliminaries

A (planar) *kinetic point set* $p(t) = \{p_1(t), \cdots, p_n(t)\}$ is a finite collection of time-dependent points $p_i(t) \in R^2$ in the Euclidean plane, each one moving on a *linear trajectory* at constant *speed*. The continuous *time* parameter t runs over the real line R, but it will become apparent during our proofs and analysis that it is very useful to think of it sometimes as running over the *projective line*: R extended with a *point at infinity*.

A *graph drawing* or *embedding* $G(p)$ is a mapping of the vertices V of a graph $G = (V, E)$ to a set $p = \{p_1, \cdots, p_n\}$ of points in the plane, $i \mapsto p_i$. A *planar graph embedding* (or a *plane graph*) has no crossing edges. A *topological plane graph* is the (planar) graph with the additional information regarding its faces and their incidences, given for instance by the *rotations* of edges around vertices (but not necessarily by a concrete embedding).

A *kinetic graph* $G(p(t))_{t \in I}$ is a graph G embedded on a kinetic point set $p(t)$. Of particular interest are the kinetic graphs which remain *planar* throughout the entire motion $t \in R$. To formally state our results on planar kinetic graphs, we give now the necessary definitions regarding Parallel Drawings, Rigidity Theory and Pseudo-triangulations.

A *direction network* (G, D) is a graph together with a set of directions (slopes) d_{ij} associated to its edges $ij \in E$. A *drawing* [1] or *realization* $G(p)$ of (G, D) is an embedding $G(p)$ of G on a set of points p such that for each edge $ij \in E$, the direction of the line through p_i and p_j is d_{ij}.

If we denote the coordinates of the unknown points as $p_i = \{x_i, y_i\}$ and the known directions as $d_{ij} = [a_{ij} : b_{ij}]$ (as projective points, expressed with homogeneous coordinates), we obtain each realization as a solution of the homogeneous linear system:

$$\langle p_i - p_j, d_{ij}^{\perp} \rangle = 0, \quad \forall ij \in E \tag{1}$$

where $d_{ij}^{\perp} = [b_{ij} : -a_{ij}]$ is the vector orthogonal to d_{ij}. Since every translation of a realization yields another realization, we can factor them out by pinning down a vertex at the origin, e.g. as $x_1 = y_1 = 0$. This shows that the set of all the realizations (modulo translations) is a linear space (of some dimension k), and contains the *trivial realization*, where all the points coincide (with the origin).

A *parallel redrawing* of a graph embedding $G(p)$ is another embedding $G(p')$ of the same graph, such that the corresponding edges have parallel directions in the two drawings. A parallel redrawing is *trivial* if it is a translation $G(p + p_0)$ or a rescaling $G(\alpha p)$ of $G(p)$. In this case the two figures are *similar*. See Fig. 1. The *configuration space* of a direction network is the space of all possible realizations, modulo the trivial parallel redrawings.

A direction network is *consistent* if it has a realization with not all the points coinciding (i.e. its configuration space is not trivial). It is *generic* if small per-

---

[1] It is worth alerting the reader at this point that although our graph drawings are taking place in the Euclidean plane, the resulting space of *all possible drawings* will turn out to be a *real projective space*.

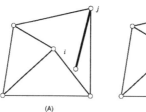
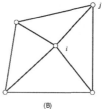

Fig. 4. (A) An inconsistent direction network: only the trivial realization matches for the edge ij the direction of the thick segment. (B) Changing the direction of the inconsistent edge to a well defined value turns it into a consistent network.

turbations of the directions do not change the dimension of the configuration space. It is *(generically) tight* if it is generic and the configuration space contains exactly one non-trivial embedding (modulo parallel redrawings). Otherwise, it is called a (generically) *loose* graph. A graph is *tight* if for some choice of directions, it becomes a generic tight network, and it is *minimally tight* if it is tight, but the removal of any edge makes it loose. See Fig. 4 for some examples.

A point set is *trivial* if all the points coincide, $p_i = p_j$, $\forall i, j \in [n]$, otherwise it is non-trivial. A point set is *degenerate* if some of its points coincide. Otherwise, if *all the points* are distinct, it is called *non-degenerate* or *sharp*.

What is the point of all these definitions? The realizations of direction networks are solutions to linear systems, which may or may not have solutions, or have too many (if they are *under-determined*). The interesting thing is that the nature of the linear solution space is controlled by the combinatorial structure of the graph, generically. For almost all possible choices of slopes, we can read off whether a network is consistent, tight, etc. using a simple counting criterion (this is well known from rigidity theory). And where I am trying to get to, eventually (and this is one of the results of the paper), is that even questions such as whether there are realizations where some points *coincide*, or whether *all* the realizations of a direction network are non-crossing (planar) can be answered combinatorially. To get there, we need two more concepts: Laman graphs and pseudo-triangulations.

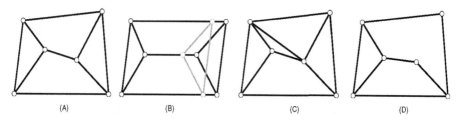

Fig. 5. A generically minimally tight graph (Laman graph) in a generic (A) and non-generic (B) embedding. A parallel redrawing is shown for (B), which is loose in this non-generic embedding. (C) A generically tight graph which is not minimally tight (Laman-plus-one), and (D) a generically loose graph (a 1dof Laman mechanism).

In a graph G, a subset of some k vertices *satisfies the Laman count* if it spans at most $2k - 3$ edges. G is a *Laman graph* if $m = 2n - 3$ and every subset satisfies the Laman count. See Figure 5(A).

Laman graphs are known in the Rigidity Theory literature as *generic minimal infinitesimally rigid graphs*, due to their rigidity theoretic properties (on which we won't elaborate here, since they will not be used in this paper). It is a well known fact in Rigidity Theory (see [16]) that *minimally tight graphs* coincide with the class of Laman graphs. Their configuration space is zero-dimensional, for generic choices of directions. Removing any subset of k edges creates a *(redrawing) mechanism*: a direction network with a k-dimensional configuration space.

In a Laman graph, a subset for which the Laman count is satisfied with equality is called a *block*. An *independent set of edges*[2] is a Laman graph with some edges (possibly none) removed. If at least one edge is missing, it is called a *Laman mechanism*. Of special interest is the *one-degree-of-freedom (1 dof) Laman mechanism*, with only one missing edge. It has $2n - 4$ edges and every subset satisfies the Laman count. See Fig. 5 (D). A *maximal block* of a Laman mechanism is called a *rigid component* (shortly, an *r*-component).

A *pointed planar graph* is a non-crossing (planar) graph embedding where each vertex is incident to an angle larger than π. A *pointed pseudo-triangulation* is a pointed planar graph where each face is a pseudo-triangle. In [14] it was shown that pointed pseudo-triangulations are *maximal* pointed plane graph embeddings (maximal with respect to edge set inclusion), and that they are planar Laman graphs (and hence generically minimally rigid). When a convex hull edge is removed, the resulting *pseudo-triangulation mechanism* has several rigid components, which are themselves pointed pseudo-triangulations if they contain more than two vertices.

3   Overview of the Results

We show that *generic* parallel redrawings of 1-dof Laman mechanisms induce *kinetic point sets* and *kinetic graphs* with a rich inner structure. Although the definition of configuration space is static and seems to be purely algebraic, it inherently - and naturally - has *motion* in it! Our results include a complete characterization of all the collision situations in such kinetic point sets by relating them to the *rigid components* in the associated kinetic graph. In particular, we prove that, generically, at most $2n - 4$ collision events may occur and that this bound is attained. We also describe how to efficiently predict combinatorially which clusters of points will collide, and when.

For this purpose, we introduce an algorithmic tool, the *parallel redrawing sweep*, which follows the time parameter of the kinetic point set and predicts combinatorially all the collision events in its configuration space.

---

[2] This terminology relates to the matroidal point of view of rigidity in dimension two, see [6].

Finally, we give a characterization of the kinetic parallel redrawing planar 1-dof Laman mechanism graphs which distort linearly [3] for *any* interval of time without the occurrence of crossings, except at collision events, which take place in clusters corresponding to rigid components. Surprisingly, they turn out to be exactly [4] the **pointed pseudo-triangulation mechanisms**.

Let us emphasize that the results described in this paper hold *generically*. One can always find non-generic situations, resulting in more collisions and other special types of behavior not addressed in this paper. The set of non-generic direction networks is described as an algebraic condition (technically, the rank of the parallel redrawing matrix is not maximum). The complement of this algebraic set, consisting in all the generic cases, is an open set characterizing *most of* the situations one would encounter in a random parallel redrawing kinetic point set.

The following theorems relate the configuration spaces of arbitrary *generic* and pseudo-triangulation direction networks of Laman mechanisms to certain types of kinetic point sets. The inner structure of the configuration space, viewed as an oriented-projective (defined in the spirit of [13]), rather than as an affine or projective real space) is captured in Theorem 1. A *(collision) cluster* is a maximal set of colliding points.

Theorem 1. Parallel-Redrawing Laman Mechanisms. *The oriented-projective configuration space of a 1-dof Laman mechanism is a (topological) circle. It contains exactly r antipodal pairs of isolated points (configurations), each pair corresponding to a colliding cluster of points induced by one of the r rigid components of the mechanism. The regions obtained by removing the isolated points are one-dimensional segments, corresponding to classes of collision free configurations.*

Figure 6 illustrates this theorem (and also Theorems 3 and 4) projectively (rather than oriented-projectively, which would just wrap around the sequence of events once more and show them rotated around by π). The reader is advised to use the interactive applet from my web page to get a better sense of what it actually means for *moving points*.

The theorem can be extended to k-dof parallel redrawing Laman mechanisms. We focus now on the 1-dof case. An affine part of the configuration space, obtained by removing one *point at infinity*, can be described as follows. Pin down to the origin an arbitrary vertex (called the *grounded vertex*) of the direction network. This eliminates translations. Then pin down a whole incident edge. This eliminates *rescalings*. By varying the length of another edge incident to the grounded vertex (i.e. choosing one of its coordinates to act as *time-parameter*), we *sweep* an affine part of the configuration space. The point at infinity corresponds to the *scale edge* shrinking to zero-length. We show that the underlying

---

[3] A technical detail, not addressed in this extended abstract, is that the kinetic point set *may* obey other absolute laws of motion (not necessarily linear). It is their *relative* motion that must be *linear*.

[4] Modulo a technical detail. The precise concept is that of a collapsed pseudo-triangulation mechanism, as in [11].

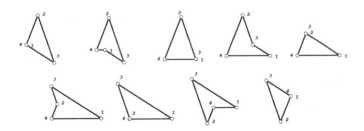

Fig. 6. The (projective) configuration space of a 4-bar parallel redrawing Laman mechanism. The sweep events and sweep segments are illustrated by a representative embedding. The first image represents the *event at infinity*: the collapse of the *scale edge* 14.

point set of a *grounded 1-dof parallel redrawing Laman mechanism*, parametrized by time, behaves like a kinetic point set, and call it the *parallel redrawing kinetic point set*. Alternate linear parametrizations, or circular versions sweeping the entire oriented projective trajectory may also be considered.

Theorem 2. Kinetic Parallel Redrawing Point Sets. *The points of a grounded parallel redrawing Laman mechanism with 1-dof move with constant velocities. As the time parameter changes, all the configurations (modulo translations and scalings) are encountered, except the one corresponding to the scale edge being reduced to zero length.*

See Fig. 2. The next Theorem 3 refines the previous result for pseudo-triangulation mechanisms.

Theorem 3. Parallel Redrawing Pseudo-Triangulation Mechanisms.
All the realizations with fixed directions of a pseudo-triangulation mechanism maintain the pseudo-triangulation mechanism property.
All these pseudo-triangulations have the same plane graph structure, but the combinatorial pseudo-triangulation structure (defined in [8]) varies.

From the algorithmic point of view, we investigated a continuous process, called the *parallel redrawing sweep*, which generates (sweeps) the configuration space of such a one-degree-of-freedom mechanism. This can be seen as an animation of the kinetic graph in time, with particular attention being paid to the *events*, when *the combinatorial structure of the graph embedding changes*. See Fig. 3 for an illustration.

Theorem 4. Parallel Redrawing Sweep. *As the time parameter sweeps the real line, the events encountered correspond each to the collapse of a rigid component to a single point (and thus capture a collision of an entire cluster in the kinetic point set).*
Each event is characterized combinatorially by a Laman graph, obtained as the contraction of the Laman mechanism on a rigid component.

The events and their sequence can be predicted algorithmically as follows:

1. **Prediction of all the events.** *The events can be computed a priori in $O(n^2)$ time using an algorithm for computing rigid components in a Laman mechanism (see [9] and the references given there).*

2. **Prediction of the next event.** *Given a time t and the direction of the sweep, the time of the next collision event can be computed in linear time.*

3. **Computing the combinatorial event.** *Given a collision time t, the colliding cluster can be computed combinatorially in $O(n^2)$ time, or algebraically in the time needed to solve a linear system.*

Next section defines the concepts needed to understand the concept of *shape similarity* used during the parallel redrawing sweep. The rest of the abstract is a high level sketch of the proofs.

4   Shape Similarity of Embedded Graphs

Let $G(p)$ be a graph embeddeding. We define and collect several types of combinatorial information from the embedding, depending on both p and G. See the example in Fig. 7.

Signed circular hyperlines. for each vertex index i, the ith hyperline ($1 \leq i \leq n$) is the signed sequence of indices $j \neq i$, in the circular order in which a line rotating through the point p_i encounters all the other points p_j. See [4].
Unsigned circular hyperlines. same as before, but we ignore the signs.
Signed linear hyperlines. same as the circular ones, but recorded starting from the vertical direction.
Unsigned linear hyperlines. the unsigned version of the previous ones.

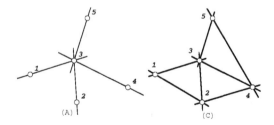

Fig. 7. (a) The signed linear hyperline is $3 : 24\overline{1}5$. (b) A graph embedding, with the set of signed linear hyperlines $1 : 23$, $2 : \overline{3}\overline{1}4$, $3 : 24\overline{1}5$, $4 : \overline{5}32$, $5 : 4\overline{3}$.

Let G be a graph and let $G(p)$, $G(p')$ be two non-degenerate embeddings. We say that the two embeddings are *combinatorially equivalent* if they have the same set of signed circular hyperlines. For a fixed graph G, a *combinatorial class* of embeddings $[G(p)]$ contains all the embeddings which are combinatorially equivalent to $G(p)$, and is identified by their common set of signed circular hyperlines.

This concept suggests the use of the partial signed hyperlines as a combinatorial criterion for discerning *shape similarity* of two embeddings of the same graph. This measure is invariant under *translations, rotations and scalings.* For pseudo-triangulations, roughly the same information is contained in the *combinatorial pseudo-triangulations* defined in [8].

5   The Parallel Redrawing Sweep

Let G be a Laman graph with some edges (possibly none) added or removed. Let $m = 2n - 3 - k$ be the number of edges, $k < 0, = 0$ or $k > 0$. Let (G, D) be a generic direction network on G.

Lemma 1. (The Dimension Lemma). *The projective configuration space of* (G, D) *is trivial for extra added edges* ($k \leq 0$). *Otherwise, it is a projective subspace of dimension equal to the number* $0 < k \leq 2n - 3$ *of removed edges, embedded in the projective* $(2n - 3)$-*dimensional space* P^{2n-3}.

We can visualize, via a continuous linear motion, an affine part of the projective configuration space of a 1dof Laman direction network. This is the **Parallel redrawing sweep.**

A *parallel redrawing mechanism* $G(p(t))_{t \in R}$ is a continuously deforming family of embeddings of a 1dof Laman mechanism direction network (G, D). Out of the many possible parametrizations, here's one. An edge $gs \in E$ is pinned down, to eliminate translations and rescalings. This eliminates the point at infinity of the projective configuration space. The affine part can now be swept through by the continuous motion of a *free vertex* f on a linear trajectory by varying one of its coordinates, the *time parameter.* All the vertices of the mechanism (except those that were pinned down) move with constant speed along linear trajectories. If we lift the arrangement to the third dimension as *time,* it becomes a line arrangement in space, which is swept by a plane orthogonal to the time dimension. Combinatorially, we prove that the sweep events (the vertices of the arrangement) correspond to a *reorientation* of the local hyperline sequences at certain vertices and are in one-to-one correspondence with the rigid components of the mechanism.

The following lemmas clarify the occurrence of degenerate embeddings in Laman networks and contribute to the proof of Theorem 1. The proofs are elementary and make use of basic combinatorial properties of Laman graphs.

Lemma 2. *Let* (G, D) *be a generic Laman direction network. Then all its non-trivial realizations are non-degenerate.*

Lemma 3. *Let* (G, D) *be a generic Laman direction network.*

1. *For every non-zero scaling factor* $s \neq 0$ *there exists a unique embedding* $G(p)$. *For opposite scale factors* s *and* $-s$, *the two embeddings are one the rotation by* π *of the other.*
2. *All the non-trivial realizations of a Laman direction network have the same signed circular hyperlines. The realizations fall into two classes of signed linear hyperlines: one for positive and one for negative scale factors, which are reversal equivalent.*

We introduce now the operation of *contracting a rigid component* and relate it the different signed linear hyperlines of realizations $G(p)$ and to degenerate embeddings of (G, D).

Let G be an arbitrary graph, $e = ij \in G$ an edge. The *contraction G/e of G on e* is a new graph (possibly not simple) obtained by *collapsing* the endpoints i and j into a single vertex i'. In the contraction, every vertex incident to i or j becomes instead incident to i' and the edge ij is discarded. The contraction operation is extended naturally to subsets E' of edges, which are contracted one by one in an arbitrary order. The contracted graph is denoted by G/E'.

Lemma 4. *Let G be a Laman mechanism and $R \subset V$ a subset of vertices spanning a rigid component (R, E_R). Then the contraction G/E_R of a $1dof$ Laman mechanism on a rigid component is a Laman graph. The contraction G/E_R of a connected k-dof Laman mechanism on a rigid component is a $(k-1)$-dof Laman graph.*

Let (G, D) be a direction network, $E' \in E$ a subset of edges, V' the set of endpoints of the edges in E' and G/E' the contraction of G on E'. As a result of the contraction on E', all the vertices in V' have been collapsed into one. We denote this new vertex by $[V']$, when we want to remind the set of vertices that have been collapsed into one, or shortly by i'.

We define the *contracted direction network* $(G/E', D')$ obtained from (G, D) by keeping the directions of the edges not touched by the contraction operation, discarding the deleted edges and giving the direction $d_{i'j} = d_{ij}$ to a new edge $i'j$ resulting from an old edge ij, $i \in E'$, $j \notin E'$. We are interested in characterizing the configuration space of the contracted network in terms of the configuration space of the original network, for Laman graphs and mechanisms. We show first that the contraction can be realized algebraically by the addition of a single linear equation.

Lemma 5. *Let (G, D) be a direction network. Let $e = ij \in E$. If d_{ij} is not vertical, add the constraint $x_i = x_j$ to its direction network system, otherwise add the constraint $y_i = y_j$. This is equivalent to asking for an embedding where this edge has length zero, hence its endpoints are collapsed. Then, the resulting system has the same realization space as the direction network on the contraction G/e.*

The proof follows from elementary linear algebra. We apply now this fact to Laman graphs and mechanisms, under the same assumptions as in Lemma 5.

Lemma 6. *Adding the constraint $x_i = x_j$ to the direction network system of a Laman graph produces only trivial realizations.*

Lemma 7. *Adding the constraint $x_i = x_j$ to the direction network system of a Laman mechanism produces a unique realization (up to translations and scalings). This realization is non-trivial, and contains a unique trivial part corresponding to the collapsed vertices of the rigid component to which the edge ij belongs.*

Lemma 8. *The only degenerate embeddings of a $1dof$ Laman mechanism are those obtained by the collapsing of rigid components.*

The Parallel Redrawing Sweep. The inner structure of the configuration space of a generic 1dof Laman mechanism consists of:

Events. If G has r rigid components, the (oriented projective) configuration space contains exactly r pairs of *special* degenerate embeddings. Each pair corresponds to the two *antipodal* realizations of a contraction of (G, D) on one of the r rigid components.

Sweep segments. The removal of the points corresponding to the collapsed embeddings from the circular configuration space leaves r pairs of connected components (antipodal circle, or *sweep segments*, in the oriented projective space), each one corresponding to a pair of reversal-equivalent combinatorial classes of embeddings. By Lemma 8, these realizations are non-degenerate.

Theorem 5. *All the embeddings within a sweep segment have the same signed linear hyperlines. Embeddings in antipodal segments differ by a complete reversal of signs in their signed linear hyperlines.*

6   Parallel Redrawing Pseudo-Triangulation Mechanisms

A *parallel redrawing pseudo-triangulation mechanism* inherits all the properties proved in Section 5. Theorem 3 is a consequence of the following lemmas.

Lemma 9. *Let $G(p)$ be a generic embedding of a Laman mechanism and let $G(p')$ be a parallel redrawing of $G(p)$. Viewing now $G(p)$ as a mechanism with fixed edge lengths, let $\mathcal{V}$ be the linear space of all the infinitesimal motions of $G(p)$. Let $v = \{v_1, \cdots, v_n\} \in \mathcal{V}$ be a non-trivial infinitesimal motion and define $\sigma_{ij}(p, v) := \langle p_i - p_j, v_i - v_j \rangle, \forall i, j \in [n]$. Then $\sigma_{ij}(p, v) = \sigma_{ij}(p', v), \ \forall i, j, \forall v \in \mathcal{V}$.*

Under the same assumptions as in Lemma 9, the following two properties hold:

Corollary 6. *When viewed as fixed-edge lengths mechanisms, $G(p)$ and $G(p')$ have the same linear space of infinitesimal motions, in particular they have the same pattern of infinitesimal expansion.*

Indeed, the pattern of infinitesimal expansion is given by the signs of $\sigma_{ij}(p, v)$.

Corollary 7. *If $G(p)$ is a pseudo-triangulation, then $G(p')$ is a pseudo-triangulation. We pin down an edge and parameterize by the position of a third vertex k. Then all the resulting parallel redrawings are pseudo-triangulations, and the plane graph structure doesn't change.*

This completes the high-level sketch of the results and proofs.

References

1. Pankaj Agarwal, Julien Basch, Leonidas Guibas, John Hershberger, and Li Zhang. Deformable free space tilings for kinetic collision detection. *International Journal of Robotics Research*, 21:179–197, March 2003. Preliminary version appeared in Proc. 4th International Workshop on Algorithmic Foundations of Robotics (WAFR), 2000.

2. Oswin Aichholzer, Günter Rote, Bettina Speckmann, and Ileana Streinu. The zig-zag path of a pseudo-triangulation. In *Proc. 8th International Workshop on Algorithms and Data Structures (WADS)*, Lecture Notes in Computer Science 2748, pages 377–388, Ottawa, Canada, 2003. Springer Verlag.

3. Sergei Bespamyatnikh. Enumerating pseudo-triangulations in the plane. *Comput. Geom. Theory Appl.*, 30(3):207–222, 2005.

4. Jürgen Bokowski, Susanne Mock, and Ileana Streinu. The folkman-lawrence topological representation theorem for oriented matroids - an elementary proof in rank 3. *European Journal of Combinatorics*, 22:601–615, July 2001.

5. Craig Gotsman and Vitaly Surazhsky. Guaranteed intersection-free polygon morphing. *Computers and Graphics*, 25(1):67–75, 2001.

6. Jack Graver, Brigitte Servatius, and Herman Servatius. *Combinatorial Rigidity*. Graduate Studies in Mathematics vol. 2. American Mathematical Society, 1993.

7. Leonidas Guibas, John Hershberger, and Subash Suri. Morphing simple polygons. *Discrete and Computational Geometry*, 24:1–34, 2000.

8. Ruth Haas, David Orden, Günter Rote, Francisco Santos, Brigitte Servatius, Herman Servatius, Diane Souvaine, Ileana Streinu, and Walter Whiteley. Planar minimally rigid graphs and pseudo-triangulations. *Computational Geometry: Theory and Applications*, pages 31–61, May 2005.

9. Audrey Lee, Ileana Streinu, and Louis Theran. Finding and maintaining rigid components. In *Proc. Canad. Conf. Comp. Geom.*, Windsor, Canada, August 2005.

10. Michel Pocchiola and Gert Vegter. Topologically sweeping visibility complexes via pseudo-triangulations. *Discrete & Computational Geometry*, 16(4):419–453, Dec. 1996.

11. Günter Rote, Francisco Santos, and Ileana Streinu. Expansive motions and the polytope of pointed pseudo-triangulations. In Janos Pach Boris Aronov, Saugata Basu and Micha Sharir, editors, *Discrete and Computational Geometry - The Goodman-Pollack Festschrift*, Algorithms and Combinatorics, pages 699–736. Springer Verlag, Berlin, 2003.

12. Bettina Speckmann and Csaba Tóth. Allocating vertex π-guards in simple polygons via pseudo-triangulations. In *Proc. ACM-SIAM Symp. Discrete Algorithms (SODA)*, pages 109–118, 2003. To appear in *Discrete and Computational Geometry*, 2004.

13. Jorge Stolfi. *Oriented Projective Geometry: A Framework for Geometric Computations*. Academic Press, New York, NY, 1991.

14. Ileana Streinu. A combinatorial approach to planar non-colliding robot arm motion planning. In *IEEE Symposium on Foundations of Computer Science*, pages 443–453, 2000.

15. Ileana Streinu. Pseudo-triangulations, rigidity and motion planning. *Discrete and Computational Geometry*, to appear, 2005. A preliminary version appeared in [14].

16. Walter Whiteley. Some matroids from discrete applied geometry. In J. Oxley J. Bonin and B. Servatius, editors, *Matroid Theory*, volume 197 of *Contemporary Mathematics*, pages 171–311. American Mathematical Society, 1996.

Proper and Planar Drawings of Graphs on Three Layers

Matthew Suderman

School of Computer Science, McGill University
msuder@cs.mcgill.ca

Abstract. A graph is *proper k-layer planar*, for an integer $k \geq 0$, if it admits a planar drawing in which the vertices are drawn on k horizontal lines called layers and each edge is drawn as a straight-line segment between end-vertices on adjacent layers. In this paper, we point out errors in an algorithm of Fößmeier and Kaufmann (CIAC, 1997) for recognizing proper 3-layer planar graphs, and then present a new characterization of this set of graphs that is partially based on their algorithm. Using the characterization, we then derive corresponding linear-time algorithms for recognizing and drawing proper 3-layer planar graphs. On the basis of our results, we predict that the approach of Fößmeier and Kaufmann will not easily generalize for drawings on four or more layers and suggest another possible approach along with some of the reasons why it may be more successful.

Layered graph drawings [16] have applications in visualization [2, 9], DNA mapping [17], and VLSI layout [10]. In a layered drawing, the vertices are drawn on a set of horizontal lines called layers, and edges are drawn as straight-line segments between their end-vertices. Depending on the purpose of the drawing, it may also satisfy additional constraints. Common constraints include bounds on the number of layers in the drawing, and restrictions on the edges that may intersect one another.

In this paper, we consider layered drawings that are proper and planar; that is, we consider drawings in which the end-vertices of each edge lie on adjacent layers, and edges intersect only at common end-vertices. Heath and Rosenberg [8] show that the problem of recognizing graphs with proper and planar drawings on layers is $\mathcal{NP}$-complete. In a more restricted version of this problem, the input is not only a graph but also a number $k \geq 0$, and the problem asks whether or not the graph has a proper and planar drawing on k layers. Though the $\mathcal{NP}$-completeness of the original problem implies that this problem is also $\mathcal{NP}$-complete, Dujmović *et al.* [4] show that it can be solved in polynomial time when k is bounded by a constant. Unfortunately, the constants in the running time are impractically large even for $k = 3$.

The difficulty of this problem seems to increase as the number of layers increases. Consequently, this motivates a study of proper and planar drawings on a very small number of layers in hopes of obtaining insights for drawings on a

P. Healy and N.S. Nikolov (Eds.): GD 2005, LNCS 3843, pp. 434–445, 2005.
© Springer-Verlag Berlin Heidelberg 2005

larger number of layers. Interestingly, this approach has had some limited success for planar layered drawings that are not proper. In particular, Cornelsen, Schank and Wagner [3] show that a graph G has a planar layered drawing on three layers if and only if a certain transformation of G has a drawing on two layers. Using this result, they obtain a linear-time recognition and graph drawing algorithm for three layers.

Proper planar drawings on up to three layers have also been studied. For one and two layers, the drawings are quite simple and it is easy to determine in linear-time whether or not a graph admits such a drawing [5, 7, 15]. For three layers, Fößmeier and Kaufmann [6] also claim to have a linear-time recognition algorithm; however, we will show that, even though their algorithm seems plausible, it contains significant errors.

Following a few preliminary definitions and results in Section 1, we will briefly describe their algorithm in Section 2, and then discuss its flaws in Section 3. We will then describe a new characterization of graphs that have proper and planar drawings on three layers, and derive a corresponding linear-time recognition and drawing algorithm in Section 4.

The overall purpose of our work is not to correct an error, but rather to obtain efficient algorithms for layered graph drawing. Therefore, we would like know if the approach of Fößmeier and Kaufmann [6] can be generalized to obtain recognition and drawing algorithms for four or more layers. The simplicity of their algorithm seems to suggest a positive answer; however, based on the complexity of our attempt to correctly handle all cases, such a generalization would probably be very long and tedious. In Section 5, then, we will conclude by suggesting another approach that may lead to efficient algorithms for layered graph drawing.

1 Preliminaries

In this paper, each graph $G = (V, E)$ is simple, undirected and connected. A graph $G = (V, E)$ is *bipartite* if its vertices can be partitioned into two disjoint sets V_0 and V_1 such that each edge in E has one end-vertex in V_0 and the other in V_1. We call V_0 and V_1 the *bipartition classes* of G and write $G = (V_0, V_1; E)$.

A *leaf* vertex in a graph is a vertex with exactly one neighbor. Any graph that can be transformed into a path $v_1, v_2, \ldots, v_p$ by removing all of its leaves is called a *caterpillar*, and the path $v_1, v_2, \ldots, v_p$ is called the *spine* of the caterpillar. The *2-claw* is the smallest tree that is not a caterpillar. It consists of a vertex called the *root* that has three neighbors, and each neighbor is additionally adjacent to a leaf. See Figure 1 for a drawing of a caterpillar and a 2-claw.

A *cut-vertex* in a graph is a vertex whose removal disconnects the graph.

A *planar drawing* of a graph is a two-dimensional drawing in which each pair of edges may intersect only at a common end-vertex. A *planar embedding* of a graph defines a clockwise order of the neighbors of each vertex that corresponds to a planar drawing of the graph. A graph is *k-layer planar* if it has a planar drawing in which each edge is drawn as a straight-line between its end-vertices

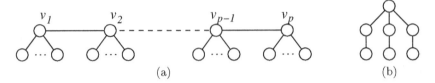

Fig. 1. (a) Caterpillar, (b) 2-Claw

and the vertices lie on k horizontal lines called *layers*. Such a drawing is called a *k-layer planar drawing*. A graph is said to be *proper k-layer planar* if it has a k-layer planar drawing in which the end-vertices of each edge lie on adjacent layers. In other words, the edges in such drawings intersect layers only at points that coincide with the drawings of their end-vertices. Such a drawing is called a *proper k-layer planar drawing*.

As mentioned in the introduction, there is a simple characterization of proper 2-layer planar graphs:

Lemma 1 ([5, 7, 15]). *Let G be a graph. The following are equivalent:*

1. *G is proper 2-layer planar;*
2. *G is a forest of caterpillars;*
3. *G is acyclic and contains no 2-claw; and*
4. *The graph obtained from G by deleting all leaves is a forest and contains no vertex with degree three or greater.*

In this paper, we will prove a similar though much more complicated characterization theorem for proper 3-layer planar graphs.

2   A Recursive Approach for Recognizing Proper 3-Layer Planar Graphs

Rather than present the entire algorithm of Fößmeier and Kaufmann [6], we will describe only the basic approach of the algorithm and then, in the following section, describe in detail only those steps that contain significant flaws.

The algorithm depends on a few simplifying assumptions. First of all, the algorithm assumes that the input graph is bipartite. This is because every proper 3-layer planar graph is bipartite (the vertices on the top and bottom layers in a proper 3-layer planar drawing belong to one bipartition class and the remaining vertices belong to the other), and it is easy to test whether or not a graph is bipartite. The algorithm also assumes that the vertices of some given bipartition class must be drawn on the top and bottom layers. Thus, to recognize all proper 3-layer planar graphs, the algorithm would need to be applied two times, each time with a different bipartition class selected as the one that must be drawn on the top and bottom layers. In the algorithm, we will denote the bipartition classes as V_0 and V_1 and assume that the vertices of V_0 must be drawn on the top and bottom layers.

Given these assumptions, the algorithm divides the recognition problem into several cases and then handles nearly each case recursively. For example, in one case, the input graph G contains a cut-vertex $v \in V_1$ with at least four non-leaf neighbors. If v has more than four non-leaf neighbors, then it is not too difficult to see that G does not have a proper 3-layer planar drawing with v on the middle layer; therefore, the algorithm returns false. Likewise, if v has exactly four non-leaf neighbors and $G - v$ contains three connected components that are each not caterpillars (i.e. not proper 2-layer planar), then G is also not proper 3-layer planar so the algorithm returns false. If G and v pass these two tests, then v has exactly four non-leaf neighbors and $G - v$ contains at most two non-caterpillar components. The algorithm then returns true if and only if each non-caterpillar component (plus v) has a proper 3-layer planar drawing in which v has the smallest or largest x-coordinate of any vertex in the drawing.

We observe that the previous case is handled recursively but with an additional constraint on the position of v. In fact, the input to the algorithm consists not only of a graph but also of a set of vertices called *borders*. This set may contain up to two vertices and the algorithm returns true if and only if the graph has a proper 3-layer planar drawing with one of the vertices in *borders*, if $|borders| > 0$, has the smallest x-coordinate of any vertex in the drawing, and the other vertex in *borders*, if $|borders| > 1$, has the largest x-coordinate of any vertex in the drawing.

Even without describing the remainder of the algorithm, one can see that this basic approach appears to be promising and seems to suggest a very simple algorithm for recognizing proper 3-layer planar graphs. Unfortunately, as we will show in the next section, the algorithm contains significant flaws in the way it handles some of the other cases.

3   Shortcomings of the Algorithm

In this section, we describe some of the errors in the algorithm of Fößmeier and Kaufmann [6] and provide examples of graphs that it either incorrectly accepts or rejects as a result of these errors.

The algorithm divides the problem into four main cases numbered (a)-(d), and each main case is divided into two or more subcases numbered 1, 2, and so on. The cases are handled in alpha-numeric order; thus, for example, the algorithm handles cases (b)-(d) only if case (a) does not apply to the input. The algorithm is called **test** and, as described in the previous section, the input to the algorithm is a bipartite graph $G = (V_0, V_1; E)$ and a set of vertices *borders*.

- Case (a2) states:

> **if** there is a small vertex[1] $v \in V_1$ **then**
> **if** $v \in borders$ **then** insert v's neighbor into *borders* **fi**;
> call test($G \setminus \{v\}$, *borders* $\setminus \{v\}$);

---

[1] A small vertex is a leaf.

Consider the following graph $G = (V_0, V_1; E)$ where the vertices of V_0 are darkened:

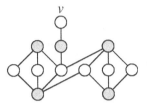

The algorithm would first remove v from the graph as described in (a2). The remaining graph $G - v$ is clearly proper 3-layer planar so the recursive call to the algorithm should return true. However, by our characterization theorem in Section 4, this graph is not proper 3-layer planar because the main biconnected component is not safe (see Definition 4).

– Case (c2) does not contain an error, however, we mention it because it is never fully described anywhere in the literature. This case assumes that V_1 contains no vertices with four non-leaf neighbors, contains no vertices that have three non-leaf neighbors and are cut-vertices, but does contain at least one vertex with three non-leaf neighbors. Case (c2) states:

> we need a more special case analysis involving separator edges[2]which is omitted;

Unfortunately, by definition, the separator edge referred to belongs to a cycle. Based on the fact that the bulk of the complexity in our characterization theorem in Section 4 relates to cycles, we would be very surprised if there is a straightforward way to handle this case.

– Case (d3) states (where G is biconnected and the vertices of V_1 have degree equal to 2):

> let L be the graph obtained by replacing all $v \in V_1$ by edges between its neighbors. **if** L is a ladder graph (an outerplanar graph with completely nested shortcut edges[3]) **then** return true **else** return false;

We assume that the authors mean that, not only must L be a ladder graph, but it must have a drawing in which one vertex of *border*, if $|border| > 0$, has the smallest x-coordinate in the drawing and the other vertex, if $|border| > 1$, has the largest x-coordinate in the drawing.

If this is the case, then the algorithm would reject the following graph even though it is proper 3-layer planar:

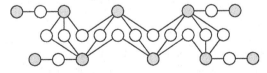

---

[2] A separator edge is an edge (u, v) such that removing u and v from the graph disconnects the graph.

[3] A shortcut edge in a biconnected outerplanar graph does not lie on the external face.

Since the graph contains four separator vertices in V_0, the algorithm would return false because any attempt to recursively draw the main biconnected component in the graph would require the parameter *borders* to contain all four separator vertices. As mentioned above, *borders* may contain at most two vertices.

If this is not the case (i.e. the algorithm ignores *borders* in case (d3)), then the algorithm would return true for the following graph even though it is not proper 3-layer planar:

We address this difficulty in our characterization by defining safe biconnected components (see Definition 4).

4   Characterizing Proper, 3-Layer Planarity

Our characterization of proper 3-layer planar graphs is based on many of the observations contained in the algorithm of Fößmeier and Kaufmann [6]; however, as we mentioned earlier, correct handling of all cases can be very tedious.

Our characterization consists of constraints on vertices and biconnected components. For the restricted case where the input graph G is a tree, our characterization is very similar to their algorithm: roughly it says that G is 3-layer planar if and only if, for each vertex v in G, at most two connected components of $G - v$ are not proper 2-layer planar. The reason is that, if there are three components that are not proper 2-layer planar, then each component in the drawing of G occupies all three layers. Consequently, one component must be drawn between the other two components. The problem, however, arises when we want to add v to the drawing. Since it is adjacent to all three components, its edge incident on the leftmost component or its edge incident on the rightmost component will cross an edge of the middle component.

In the more general setting, G may not be a tree because even-length cycles are proper 3-layer planar. Consequently, to characterize proper 3-layer planar graphs, we must handle biconnected components containing more than two vertices. As we will show, we can handle biconnected components by generalizing the way we handle vertices in trees. For example, it is not difficult to see that if C is a biconnected component in a proper 3-layer planar graph G, then $G - C$ contains at most two connected components that are not proper 2-layer planar. Unfortunately, this in itself is not sufficient because not all biconnected components are proper 3-layer planar. For an example of one, see Figure 2. Consequently, our characterization must contain additional constraints for biconnected components.

In the following, we describe the constraints on vertices and biconnected components more formally and completely. A vertex or biconnected component that

Fig. 2. A biconnected graph that is not 3-layer planar

satisfies these constraints will be called *safe*. We recall that proper 3-layer planar graphs are bipartite so our definitions will apply to bipartite graphs and be given with respect to a given bipartition class of the graph.

Definition 1. *Let V_0 be a bipartition class of a bipartite graph G. A vertex v in G is* safe *with respect to V_0 if:*

1. *$v \in V_0$ and $G - v$ contains at most two components that are not caterpillars (e.g see Figure 3(b)); or*
2. *$v \notin V_0$ and v has at most four non-leaf neighbors, and:*
   (a) *$G - v$ contains at most two components H_1 and H_2 such that $G(V(H_1) + v)$ and $G(V(H_2) + v)$ are not caterpillars.*
   (b) *if v has four non-leaf neighbors or v belongs to a cycle, then $G - v$ contains at most two components H_1 and H_2 such that $G(V(H_1) + v)$ plus a leaf attached to v and $G(V(H_2) + v)$ plus a leaf attached to v are not caterpillars (e.g. see Figure 3(c)).*

It is not too difficult to see that if a bipartite graph $G = (V_0, V_1; E)$ contains a vertex that is not safe with respect to V_0, then G does not admit a proper 3-layer planar drawing in which the vertices of V_0 lie on the top and bottom layers.

A vertex v may be safe but only just "barely safe" because the connected components H_1 and H_2 of $G - v$ mentioned above are in fact not caterpillars. As a result, in every proper 3-layer planar drawing of G, one of these components must be drawn to the left of v and the other component must be drawn to the

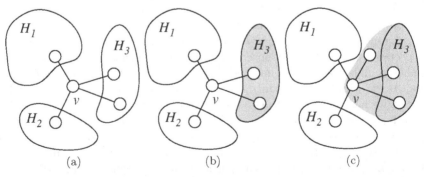

(a)   (b)   (c)

Fig. 3. (a) Suppose that, for some vertex v in $G = (V_0, V_1; E)$, $G - v$ has three connected components H_1, H_2 and H_3. (b) If H_3 is a caterpillar and $v \in V_0$, then v is safe with respect to V_0. (c) If $v \notin V_0$ and $G(V(H_3) + v)$ plus a leaf attached to v is a caterpillar, then v is safe with respect to V_0.

right. Thus, we call v a *connecting vertex* because it "connects" the left part of the drawing with the right part of the drawing.

Definition 2. *A vertex v is a* connecting vertex *with respect to V_0 if:*

1. $v \notin V_0$ *and v has four non-leaf neighbors in G; or*
2. v *does not belong to a cycle and $G - v$ contains two components H_1 and H_2 such that $G(V(H_1) + v)$ and $G(V(H_2) + v)$ are not caterpillars; or*
3. v *belongs to a cycle and $G - v$ contains two components H_1 and H_2 such that the graph containing $G(V(H_1) + v)$ plus a leaf attached to v and the graph containing $G(V(H_2) + v)$ plus a leaf attached to v are not caterpillars.*

To describe the properties of a *safe* biconnected component, it is necessary to know how the biconnected component is connected to the remainder of the graph:

Definition 3. *Let C be a biconnected component of a bipartite graph $G = (V_0, V_1; E)$. The* extension *of C with respect to vertices v_1 and v_2 in C and V_0, is a graph obtained from C by attaching leaves and pendant 2-paths to certain vertices in C. More specifically, if a vertex of C is adjacent to a leaf in G, then, in the extension of C, this vertex is adjacent to a leaf. If a vertex v in C has $d \geq 1$ non-leaf neighbors in G that do not belong to C, then:*

1. *If $v \neq v_1, v_2$ or $v \notin V_0$, then we attach d pendant 2-paths to v in the extension of C.*
2. *If $v = v_1 = v_2 \in V_0$ and $d \geq 2$, then we attach 2 pendant 2-paths to v in the extension of C.*
3. *Otherwise, we attach exactly one pendant 2-path to v in the extension of C.*

This definition is illustrated in Figure 4.

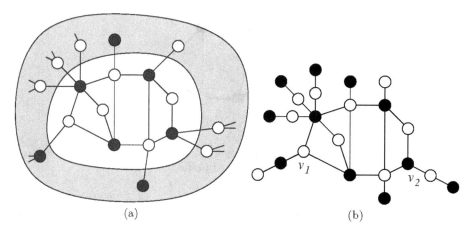

(a)   (b)

Fig. 4. (a) A biconnected component C in a bipartite graph $G = (V_0, V_1; E)$ where the darkened vertices belong to V_0, and (b) the extension of C with respect to v_1, v_2 and V_0

Definition 4. *Let C be a biconnected component containing at least three vertices in a bipartite graph $G = (V_0, V_1; E)$. Biconnected component C is* safe *with respect to V_0 if there exists a* safety certificate *for C with respect to V_0, namely at tuple $\langle v_1, v_2, P_1, P_2, \Psi_C \rangle$ consisting of two vertices v_1 and v_2 in C, two simple paths P_1 and P_2 in C_e, the extension of C with respect to v_1, v_2 and V_0, and a planar embedding Ψ_C of C, such that:*

1. *The vertices of P_1 and P_2 in C lie on the external face of Ψ_C;*
2. *P_1 and P_2 each contain a vertex of V_0 on each face of Ψ_C;*
3. *If a vertex of C belongs to V_0, then it belongs to P_1 or P_2 but not both;*
4. *If a vertex v in C has a neighbor outside C, then v belongs to P_1 or P_2;*
5. *If a vertex in C is a connecting vertex, then it is equal to v_1 or v_2;*
6. *Both v_1 and v_2 are end-vertices of the subpaths of P_1 or P_2 in C such that each path from v_1 to v_2 on the external face cycle of C in Ψ_C contains the subpath of P_1 in C or the subpath of P_2 in C;*
7. *If, for some vertex v in C, $G - v$ contains two components H_1 and H_2 vertex-disjoint with C that are not caterpillars, then $v = v_1 = v_2$; and*
8. *Each pendant 2-path in C_e belongs to P_1 or P_2.*

A safety certificate $\langle v_1, v_2, P_1, P_2, \Psi_C \rangle$ is said to be *tied* if $v_1 = v_2$. We note that the extension of Figure 4(b) does not have a safety certificate because one of the vertices has three pendant 2-paths; as a result, there are no paths P_1 and P_2 that contain all pendant 2-paths and each contain a vertex of V_0 on each face. Figure 5, however, shows that the biconnected component of Figure 4(a) is safe. The safety certificate showing this consists of the vertices labelled v_1 and v_2, the two highlighted paths P_1 and P_2, and the embedding of the component shown in the drawing.

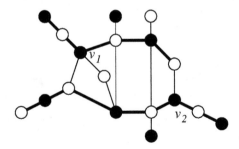

Fig. 5. The biconnected component of Figure 4(a) has a safety certificate $\langle v_1, v_2, P_1, P_2, \Psi \rangle$ where P_1 and P_2 are the highlighted paths

Based on these definitions, we state our characterization theorem:

Theorem 1. *A graph G is proper 3-layer planar if and only if G is bipartite and each vertex and each biconnected component of G is safe with respect to some bipartition class of G.*

We prove this theorem in two parts: we first prove the necessity of the conditions, and then we prove their sufficiency. The full proof is several pages long so we include only a sketch here.

To prove the necessity of the conditions, we consider a proper 3-layer planar drawing of a graph G. From the drawing, it is easy to see that G is bipartite. It is also easy though tedious to show that each vertex and each biconnected component is safe with respect to the bipartition class corresponding to the vertices drawn on the top and bottom layers. For example, to prove that a biconnected component C is safe, we must obtain the necessary safety certificate $\langle v_1, v_2, P_1, P_2, \Psi_C \rangle$. This is done first by selecting any simple path P from a leftmost to a rightmost vertex in the drawing. Letting the embedding Ψ_C of C be the embedding of C in the given drawing, we let v_1 be the first vertex of P in C, and v_2 be the last vertex of P in C. We obtain P_1 and P_2 by starting with two subpaths of the external face cycle of Ψ_C between v_1 and v_2, and then extending them as necessary to contain any pendant 2-paths in the extension of C. It is then straightforward though tedious to prove that $\langle v_1, v_2, P_1, P_2, \Psi_C \rangle$ is a safety certificate for C.

To prove the sufficiency of the conditions, we consider a graph that is bipartite, and for some bipartition class, each vertex is safe and each biconnected component has a safety certificate. Using this information, we show how to construct a proper 3-layer planar drawing of the graph. We first construct a special path P in the graph that contains each connecting vertex and, for each biconnected component C_i with safety certificate $\langle v_1^i, v_2^i, P_1^i, P_2^i, \Psi_{C_i} \rangle$, P also contains v_1^i and v_2^i. We then obtain the drawing as follows:

(a) We draw each biconnected component C_i according to the embedding Ψ_{C_i} in its safety certificate and insert each of these drawings into the main drawing in the order that they appear in P.

(b) We draw the remaining subpaths of P that connect consecutive biconnected components.

(c) Finally, we insert drawings of the pendant trees that are attached to vertices already drawn.

This completes our sketch of the proof of Theorem 1.

Testing whether or not a given graph is proper 3-layer planar is simply an application of the characterization given in Theorem 1. Many of the definitions of safety, both for vertices and biconnected components, depend on knowing whether or not various subgraphs of the input graph G are caterpillars. This can be computed in linear time by first computing a block-cut tree of G [14] and then applying a dynamic programming algorithm to the block-cut tree. A *block-cut tree* of a graph is a tree whose vertices correspond to cut-vertices and biconnected components in the graph. Edges in the block-cut tree connect biconnected components with the cut-vertices that they contain in the graph. We call the resulting algorithm IsPROPER3LAYERPLANAR:

Theorem 2. *Algorithm* IsProper3LayerPlanar *determines whether or not a graph is proper 3-layer planar in linear time.*

We can transform this recognition algorithm into a drawing algorithm by returning a proper 3-layer planar drawing anytime the recognition algorithm returns TRUE. The drawing is constructed as described in the sufficiency proof of Theorem 1. To do this, we require the set of connecting vertices and a safety certificate for each biconnected component. Fortunately, as described above, the recognition algorithm already computes these things.

5   Conclusions

We have shown how to determine whether or not a graph is proper 3-layer planar in linear time, and, if it is, we show how to obtain a proper 3-layer planar drawing in the same asymptotic running time. It is not too difficult to see that our basic approach is identical to that of Fößmeier and Kaufmann [6]. For example, we observe that the path mentioned in our algorithm actually contains all vertices that trigger recursive calls in their algorithm. Unfortunately, as can be seen from the length of our characterization statement, the effort required to obtain correct algorithms for three layers is much greater than for two layers. Therefore, we believe that a new approach will be required to obtain efficient algorithms for four or more layers.

One possible approach involves applying graph operations that reduce the graph to an empty graph if and only if the graph has a planar drawing on a given number of layers. Arnborg and Proskurowski [1] use this type of approach to recognize graphs of treewidth three, and Matousek and Thomas [11] modify their set of reductions to obtain an efficient quadratic-time recognition algorithm. Generalizing these results, Sanders [12] shows that this approach can be used to efficiently recognize graphs with treewidth four. Our hope is to similarly find a set reductions for proper 3-layer planar graphs and likewise use them to obtain reductions for proper 4-layer planar graphs. We believe that this approach might be successful because Dujmović *et al.* [4] use pathwidth, a restricted version of treewidth, to obtain algorithms for recognizing k-layer planar graphs (inefficient algorithms though they may be). In addition to this, we show in [13] how to use pathwidth to obtain planar drawings of trees on an minimum number of layers. These results show that pathwidth and hence treewidth are closely related to the number of layers in planar layered graph drawings.

Acknowledgements

I thank my supervisor Sue Whitesides for introducing me to this problem at one of her workshops, for double-checking the correctness of my results, and for many helpful suggestions for improving the presentation of this paper. I also thank Vida Dujmović for productive discussions about the algorithm of Fößmeier and Kaufmann.

References

1. Arnborg, S., Proskurowski, A.: Characterization and recognition of partial 3-trees. SIAM Journal of Algebraic and Discrete Methods **7** (1986) 305–314
2. Battista, G.D., Eades, P., Tamassia, R., Tollis, I.G.: Graph Drawing: Algorithms for the Visualization of Graphs. Prentice-Hall (1999)
3. Cornelsen, S., Schank, T., Wagner, D.: Drawing graphs on two and three lines. In Goodrich, M., Kobourov, S., eds.: Graph Drawing, 10th International Symposium (GD 2002). Volume 2528 of Lecture Notes in Computer Science., Springer-Verlag (2002) 31–41
4. Dujmović, V., Fellows, M.R., Hallett, M.T., Kitching, M., Liotta, G., McCartin, C., Nishimura, N., Ragde, P., Rosamond, F.A., Suderman, M., Whitesides, S., Wood, D.R.: On the parameterized complexity of layered graph drawing. In auf der Heide, F.M., ed.: Algorithms, 9th European Symposium (ESA 2001). Volume 2161 of Lecture Notes in Computer Science., Springer-Verlag (2001) 488–499
5. Eades, P., McKay, B., Wormald, N.: On an edge crossing problem. In: Proceedings of the 9th Australian Computer Science Conference, Australian National University (1986) 327–334
6. Fößmeier, U., Kaufmann, M.: Nice drawings for planar bipartite graphs. In Bongiovanni, G.C., Bovet, D.P., Battista, G.D., eds.: Proceedings of the 3rd Italian Conference on Algorithms and Complexity (CIAC 1997). Volume 1203 of Lecture Notes in Computer Science., Springer-Verlag (1997) 122–134
7. Harary, F., Schwenk, A.: A new crossing number for bipartite graphs. Utilitas Mathematica **1** (1972) 203–209
8. Heath, L.S., Rosenberg, A.L.: Laying out graphs using queues. SIAM Journal on Computing **21** (1992) 927–958
9. Kaufmann, M., Wagner, D.: Drawing Graphs: Methods and Models. Volume 2025 of Lecture Notes in Computer Science. Springer-Verlag (2001)
10. Lengauer, T.: Combinatorial Algorithms for Integrated Circuit Layout. Wiley (1990)
11. Matousek, J., Thomas, R.: Algorithms finding tree-decompositions of graphs. Journal of Algorithms **12** (1991) 1–22
12. Sanders, D.P.: On linear recognition of tree-width at most four. SIAM Journal on Discrete Mathematics **9** (1995) 101–117
13. Suderman, M.: Pathwidth and layered drawings of trees. International Journal of Computational Geometry & Applications **14** (2004) 203–225
14. Tarjan, R.: Depth-first search and linear graph algorithms. SIAM Journal on Computing **1** (1972) 146–160
15. Tomii, N., Kambayashi, Y., Yajima, S.: On planarization algorithms of 2-level graphs. Technical Report EC77-38, Institute of Electronic and Communication Engineers of Japan (IECEJ) (1977)
16. Warfield, J.N.: Crossing theory and hierarchy mapping. IEEE Transactions on Systems, Man, and Cybernetics **7** (1977) 502–523
17. Waterman, M.S., Griggs, J.R.: Interval graphs and maps of DNA. Bulletin of Mathematical Biology **48** (1986) 189–195

Incremental Connector Routing

Michael Wybrow[1], Kim Marriott[1], and Peter J. Stuckey[2]

[1] Clayton School of Information Technology,
Monash University, Clayton, Victoria 3800, Australia
{mwybrow, marriott}@csse.monash.edu.au
[2] NICTA Victoria Laboratory, Dept. of Comp. Science & Soft. Eng.,
University of Melbourne, Victoria 3010, Australia
pjs@cs.mu.oz.au

Abstract. Most diagram editors and graph construction tools provide some form of automatic connector routing, typically providing orthogonal or poly-line connectors. Usually the editor provides an initial automatic route when the connector is created and then modifies this when the connector end-points are moved. None that we know of ensure that the route is of minimal length while avoiding other objects in the diagram. We study the problem of incrementally computing minimal length object-avoiding poly-line connector routings. Our algorithms are surprisingly fast and allow us to recalculate optimal connector routings fast enough to reroute connectors even during direct manipulation of an object's position, thus giving instant feedback to the diagram author.

1 Introduction

Most diagram editors and graph construction tools provide some form of automatic connector routing. They typically provide orthogonal and some form of poly-line or curved connectors. Usually the editor provides an initial automatic route when the connector is created and again each time the connector end-points (or attached shapes) are moved. The automatic routing is usually chosen by an ad hoc heuristic.

In more detail the graphic editors OmniGraffle [1] and Dia [2] provide connector routing when attached objects are moved, though these routes may overlap other objects in the diagram. Both Microsoft Visio [3] and ConceptDraw [4] provide object-avoiding connector routing. In both applications the routes are updated only after object movement has been completed, rather than as the action is happening. In the case of ConceptDraw, its orthogonal object-avoiding connectors are updated as attached objects are dragged, though not if an object is moved or dropped onto an existing connector's path. The method used for routing does not use a predictable heuristic and often creates surprising paths. Visio offers orthogonal connectors, as well as curved connectors that follow roughly orthogonal routes. Visio's connectors are updated when the attached shapes are moved or when objects are placed over the connector paths, but only in response to either of these events. Again, connector routing does not use a predictable

P. Healy and N.S. Nikolov (Eds.): GD 2005, LNCS 3843, pp. 446–457, 2005.
© Springer-Verlag Berlin Heidelberg 2005

heuristic, such as minimizing distance or number of segments. Visio does update these connectors dynamically as objects are resized or rotated, though if there are too many objects for this to be responsive Visio reverts to calculating paths only when the operation finishes. The Graph layout library yFiles [5] and demonstration editor yEd offers both orthogonal and "organic" edge routing—a curved force directed layout where nodes repel edges. Both of these are layout options that can be applied to a diagram, but are not maintained throughout further editing. We know of no editor which ensures that the connectors are optimally routed in any meaningful sense.

Automatic connector routing in diagram editors is, of course, essentially the same problem as edge routing in graph layout, especially when edge routing is a separate phase in graph layout performed after nodes have been positioned. Like connector routing it is the problem of routing a poly-line, orthogonal poly-line or spline between two nodes and finding a route which does not overlap other nodes and which is aesthetically pleasing, i.e., short, with few bends and minimal edge crossings. The main difference is that edge routing is typically performed for a once-off static layout of graphs while automatic connector routing is dynamic and needs to be performed whenever the diagram is modified.

One well-known library for edge routing in graph layout is the Spline-o-matic library developed for GraphViz [6]. This supports poly-line and Bezier curve edge routing and has two stages. The first stage is to compute a *visibility graph* for the diagram. The visibility graph contains a node for each vertex of each object in the diagram. There is an edge between two nodes iff they are mutually visible, i.e., there is no intervening object. In the second stage connectors are routed using Dijkstra's shortest path algorithm to compute the minimal length paths in the visibility graph between two points. A third stage, actually the responsibility of the diagram editor, is to compute the visual representation of the connector This might include adding rounded corners, ensuring connectors don't overlap unnecessarily when going around the same object vertex, etc.

Here we describe how this three stage approach to edge routing can be modified to support incremental shortest path poly-line connector routing in diagram editors. We support the following user interactions:

- *Object addition:* This makes existing connector routes invalid if they overlap the new object and requires the visibility graph to be updated.
- *Connector addition:* This simply requires routing the new connector.
- *Object removal:* This makes existing connector routes sub-optimal if there is a better route through the region previously occupied by the deleted object. It also requires the visibility graph to be updated.
- *Connector removal:* This is simple–just delete the connector.
- *Direct manipulation of object placement:* This is the most difficult since it is essentially object removal followed by addition.

To be useful in a diagram or graph editor we need these operations to be fast enough for reasonable size diagrams or graphs with up to say 100 nodes. The performance requirement for direct manipulation is especially stringent if we wish to update the connector routing during direct manipulation, i.e., to reroute the connectors during the movement of the object, rather than re-routing only after the final placement. This requires visibility graph updating and connector re-routing to be performed in milliseconds.

Somewhat surprisingly our incremental algorithms are fast enough to support this. Two key innovations allowing this are: an "invisibility graph" which associates each object with a set of visibility edges that it obscures (this speeds up visibility graph update for object removal and direct manipulation); and a simple pruning function which significantly reduces the number of connectors that must be considered for re-routing after object removal. In addition we investigate the use of an $A^\star$ algorithm rather than Dijkstra's shortest path algorithm to compute optimal paths.

There has been considerable work on finding shortest poly-line paths and shortest orthogonal poly-line paths. Most of this has focused on finding paths given a fixed object layout and has not considered the problem of dynamically changing objects and the need to incrementally update an underlying visibility structure. The most closely related work is that of Miriyala $et.$ $al.$ [7] who give an efficient $A^\star$ algorithm for computing orthogonal connector paths. Like us they are interested in doing this incrementally and rely on a $rectangulation$ of the graph and previously drawn edges which is essentially a visibility graph. The main difference to this paper is that they only consider orthogonal paths. Other differences are that their algorithm is heuristic and routes are not guaranteed to be optimal even if minimizing edge crossings is ignored (see e.g. Figure 9 of [7]). They do not discuss object removal and how to maintain optimality of connectors.

There are several well known algorithms for constructing visibility graphs that run in less than the naive $O(n^3)$ approach. In [8] Lee provided an $O(n^2 \log n)$ solution with a rotational sweep. Later, Welzl presented an $O(n^2)$ duality-based arrangements approach [9]. Asano, et. al., presented two more arrangement based solutions in [10] both running in $O(n^2)$ time. Another $O(n^2)$ approach was given by Overmars and Welzl using rotational trees in [11]. Ghosh and Mount showed an output sensitive solution to the problem in [12] which uses plane sweep triangulation and funnel splits. It runs in $O(m + n \log n)$ time, where m is the number of visibility edges. Only Lee's algorithm and Asano's algorithm support incremental update of a visibility graph.

Given a visibility graph with m edges and n nodes the standard implementation of Dijkstra's shortest path algorithm is $O(n^2)$ or $O(m \log n)$ if a binary heap based priority queue representation is used. Fredman and Tarjan [13] give an $O(m + n \log n)$ implementation using Fibonacci heaps. The $A^\star$ algorithm has similar worst case complexity but in practice we have found it to be faster. There are techniques based on the continuous Dijkstra method to compute a single shortest path in $O(n \log n)$ time which do not require computation of a visibility graph [14]. These methods are more complex and so we chose to use a visibility graph-based approach. In practice we conjecture that they will lead to similar complexity assuming $O(n^2)$ connectors.

2   Algorithms

We assume that we have objects which have an associated list of vertices and connector points. For simplicity we restrict ourselves to convex objects and for the purposes of complexity analysis we assume the number of vertices is a fixed constant, i.e. say four, since the bounding box can be used for many objects. The algorithms themselves work for arbitrary convex polygons, so circles can be approximated by a dodecagon for example.

We also have connectors, these are connected to a particular connection point on an object and have a current routing which consists of a list of edges in the visibility graph. Of course, connectors are not always connected to objects, and may have end-points which are neither object vertices or connection points. In this case an extra node is added to the visibility and invisibility graphs for this end-point.

The most important data structure is the visibility graph. Edges in this graph are stored in a distributed sparse fashion. Each object has a list of vertices, and each of these vertices has a list of vertices that are visible from it. We treat the connection points on objects as if they are vertices. They behave like standard vertices in the visibility graph, but, of course, must occur at the start or end of a path, not in the middle. They have connectors associated with them.

Actually, not all visibility edges are placed in the visibility graph. As noted in [15] in any shortest path the path will bend around vertices, making an angle of less than 180° around each object. This means that we do not need to include edges to vertices which are visible in the sector opposite to the object. Consider the vertex v with incoming visibility edge shown in Figure 1. Clearly in any shortest path the outgoing visibility edge must be to a vertex in the region indicated. And so, in general, we need not include edges in the visibility graph that are between v and vertices in either of the two "hidden" regions. Note that this generalizes straightforwardly for any convex object.

The other important data structure is the *"invisibility graph."* This is a new data structure we have introduced to support incremental recomputation of the visibility graph when an object is deleted. It associates each non-visible edge with a blocking object. This should not be confused with the invisibility graph of [16] which simply represents the visibility graph negatively.

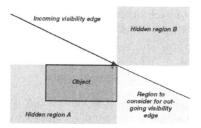

Fig. 1. Hidden regions which can be ignored when constructing the visibility graph and when finding the shortest path

The invisibility graph consists of all edges between object vertices which could be in the visibility graph except that there is an object intersecting the edge and so obscuring visibility. Edges in the invisibility graph are associated with the obscuring object and with the objects they connect. Thus each object has a list of visibility edges that it obscures. If the edge intersects more than one object the edge is associated with only one of the intersecting objects.

2.1   Connector Addition and Deletion

Connector deletion is simple, all we need to do is to remove the connector and references to the connector from its component edges in the visibility graph.

Connector addition requires us to determine the optimal route for the connector. The simplest approach is use a shortest path algorithm such as Dijkstra's [17]. Dijkstra's method has $O(n^2)$ worst case complexity while a priority queue based approach has worst case complexity $O(m \log n)$ where m is the number of edges in the visibility graph and n the number of objects in the diagram.

A (hopefully) better approach is to use an $A^\star$ algorithm which uses the Euclidean distance between the current vertex on the path as a lower bound on the total remaining cost [14]. The idea is to modify the priority queue based approach so that the priority for each frontier node x is the cost of reaching x plus $\|(x, v)\|$ where v is the endpoint of the connector. In practice we would hope that this is faster than Dijkstra's shortest path algorithm since the search is directed towards the goal vertex v rather than exploring all directions from the start vertex in a breadth-first fashion.

2.2   Object Addition

When we add an object we must first incrementally update the visibility and invisibility graphs, then recompute the route for any connectors whose current route has been invalidated by the addition. The precise steps when an object o is added are

1. Find the set of edges E_o in the visibility graph that intersect o.
2. Find the set of connectors C_o that use an edge from E_o.
3. Remove the edges in E_o from the visibility graph and place them in the invisibility graph, associating them with o.
4. For each vertex (and connection point) v of o and for each vertex (and connection point) u of each other object in the diagram o' determine if there is another object o'' which intersects the segment (v, u). If there is add (v, u) to the invisibility graph and associate it with o''. If not add (v, u) to the visibility graph.
5. For each connector $c \in C_o$ find its new route.

The two steps with greatest expected complexity are Step 1, computing the visibility edges E_o obscured by o, and Step 4, computing the visible and obscured vertices for each vertex v of o.

The simplest implementation of Step 1 has $O(m)$ complexity since we must examine all edges in the visibility graph to see if they intersect o. We could reduce this to an average case $O(\log m)$ using a spatial data structure such as a PMR quad-tree [18].

Naive computation of the visible and obscured vertices from a single vertex has $O(n^2)$ complexity. However more sophisticated algorithms for computation of the visibility graph have been developed. One reasonably simple approach which appears to work well in practice Lee's rotational sweep algorithm [8] in which the vertices of all objects are sorted w.r.t. the angle they make with the original vertex v of o and then these are processed in sorted order. It has $O(n \log n)$ complexity.

2.3   Object Deletion

Perhaps surprisingly, object deletion is potentially considerably more expensive than object creation. The first stage is to incrementally update the visibility graph.

Assume initially that we do not have an invisibility graph. We first need to remove all edges in the visibility graph that are connected to the object being deleted, o. Then when need to add edges to the visibility graph that were previously obscured by o. For each vertex (and connection point) v of each object and for each vertex (and connection point) u of some other object in the diagram we must check that (u, v) is not in the visibility graph and that it intersects o. If so we need to check whether there is any other object which intersects the segment (u, v). If there is not then it must be added to the visibility graph.

Identifying these previously obscured edges is potentially very expensive: $O(n^2)$ to compute the candidate new edges and then an $O(n)$ test for non-overlap for each edge of which there may be $O(n^2)$. Thus the worst case complexity of this method is $O(n^3)$.[1]

In order to reduce the expected (but not the worst case) cost we have introduced the invisibility graph. By recording the reason for not including an edge between two vertices in the visibility graph we know almost exactly which edges we need to retest for visibility. More exactly when we remove o we take the set of edges I_o associated with o in the invisibility graph and then test for each of these whether they intersect any other objects. Note that I_o can be expected to be considerably smaller than the candidate edges identified above since an edge (u, v) is only in I_o if it intersects o *and* o was the object first discovered to intersect (u, v).

Thus, although the invisibility graph does not reduce the worst case cost, we can expect it to substantially reduce the average cost of updating the visibility graph. Furthermore, construction of the invisibility graph does not introduce substantial overhead in any of the other operations, the only overhead is the space required to store the edges. Note that when we remove an object we also need to remove edges to it that are in the invisibility graph.

The second stage in object deletion is to recompute the connector routes. This is potentially very expensive since removing an object means that we could have

---

[1] Based on this one might consider recomputing the entire visibility graph using the Sweep Algorithm since this has $O(n^2 \log n)$ complexity.

to recompute the best route for all connectors since there may be a better route that was previously blocked by the object just removed.

However, we can use a simple strategy to limit the number of connectors reconsidered. Let A be the region of the object removed and let u and v be the two ends of the connector C. We need only reconsider the current route for C if $\exists y \in A$ s.t. $||(u, y)|| + ||(y, v)||$ is less than the cost of the current route since otherwise any route going through A will be at least as expensive as the current one.

Thus we need to compute $\min_{y \in A} ||(x, y)|| + ||(y, v)||$. Assuming A is convex we can compute this relatively easily. If the line segment (u, v) intersects A then the lower bound is $||(u, v)||$ and we must reroute C. Otherwise we find for each line segment (s, t) on the boundary of A the closest point y to A on that segment. The closest point in A is simply the closest of these. Now consider the line segment (s, t). We first compute the closest point x on the line $\overline{st}$. If x is in the segment (s, t) it is the closest point, otherwise we set y to s or t whichever is closest to x. W.lo.g. we can assume that (s, t) is horizontal. Let b and c be the vertical distance from $\overline{st}$ to u and v respectively and a the horizontal distance between u and v, as shown in Figure 2(a). We are finding the value for x which minimizes $\sqrt{x^2 + b^2} + \sqrt{(x - a)^2 + c^2}$. There are two solutions: $x = \frac{ab}{b+c}$ when $b \cdot c \geq 0$ and $x = \frac{ab}{b-c}$ when $b \cdot c \leq 0$. In the case $b = c = 0$, x is any value in $[0, a]$.

Now consider the case when we have determined that there may be a better path for the connector because of the removal. Instead of investigating all possible paths for the connector we need only investigate those that pass through the deleted object. Let A be the region of the object removed and let u and v be the two ends of the connector C and assume that the current length of the connector is c. Requiring the path to go through A means that we can use the above idea to provide a better lower-bound when computing a lower bound on the remaining length of the connector. The priority for each frontier node x is the cost of reaching x plus $\min_{y \in A} ||(x, y)|| + ||(y, v)||$ if the path has not yet

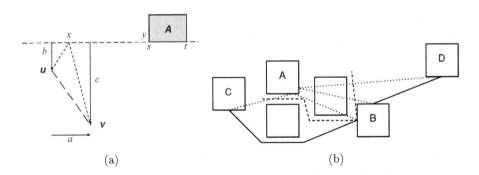

(a)   (b)

Fig. 2. (a) Computing the closest point $y \in A$ to the segment (u, v). (b) Example recomputation of connectors after deleting object A. Connectors are shown as solid lines and lower-bound connector paths through A are shown as dotted lines. The re-exploration to try and find a better path from B to C is shown as dashed lines.

gone through A. Furthermore we can remove any node whose priority is $\geq c$ since this cannot lead to a better route than the current one.

For example consider deleting object A from the diagram in Figure 2(b).The connector from B to D does not need to be reconsidered since the shortest path from the connection points (dotted) is clearly longer than the current path. But the connector from B to C needs to be reconsidered (even though in this case it will not move). The A^* algorithm will compute the dashed paths whose endpoints fail the lower bound test.

2.4   Direct Manipulation of Object Placement

The standard approach in diagram and graph editors for updating connectors during direct manipulation is to only reroute the connectors once the object has been moved to its final position. The obvious disadvantage is that the user does not have any visual feedback about what the new routes will be and may well be (unpleasantly) surprised by the result. One of the main ideas behind direct manipulation [19] is that the user should be given visual feedback about the consequences of the manipulation as they perform it rather than waiting for the manipulation to be completed. Thus it seems better for diagram and graph editors to reroute connectors during direct manipulation.

We have identified two possible approaches. In the *complete feedback* approach all connectors are rerouted at each mouse move during the direct manipulation. The advantage is that the user knows exactly what would happen if they left the object at its current position. The disadvantage is that this is very expensive. Another possible disadvantage is that it might be distracting to reroute connectors under the object being moved during the direct manipulation—for positions between the first and final position of the object the user knows that the object will not be placed there and so it is distracting to see the effect on connectors that are only temporarily under the object being manipulated. For these reasons we have also investigated a *partial feedback* approach in which for intermediate positions we only update the routes for connectors attached to the object being manipulated and leave other connectors alone.

The simplest way of implementing complete connector-routing feedback is to regard each move as an object deletion followed by object addition. Assume that we move object o from region R_{old} to R_{new}. Then we

1. Find the set of edges I_o associated with o in the invisibility graph which do not intersect R_{new} and remove them from the invisibility graph.
2. For each edge $(u, v) \in I_o$ determine if there is another object $o' \in O$ which intersects the segment (u, v). If there is add (v, u) to the invisibility graph and associate it with o'. If not add (v, u) to the visibility graph.
3. Find the set of edges E_o in the visibility graph that intersect $R_{new} \setminus R_{old}$ but are not from o.
4. Find the set of connectors C_o that use an edge from E_o.
5. Remove the edges in E_o from the visibility graph and place them in the invisibility graph, associating them with o.

6. For each vertex (and connection point) u of o and edge (u, v) in the invisibility graph check that the object o' associated with the edge still intersects it. If it does not, temporarily add the edge to the visibility graph.

7. For each vertex (and connection point) u of o and edge (u, v) in the visibility graph check if there is another object $o' \in O$ which intersects the segment (u, v). If there is add (u, v) to the invisibility graph and associate it with o'. If not, keep (u, v) in the visibility graph.

8. For each connector $c \in C_o$ find its new route.

9. For every connector not in C_o determine if there is a better route through $R_{old} \setminus R_{new}$.

Note that in the above we can conservatively approximate the regions $R_{new} \setminus R_{old}$ or $R_{old} \setminus R_{new}$ by any enclosing region such as their bounding box.

The simplest way of implementing partial connector-routing feedback is perform object deletion once the object o has moved and then at each step

1. Compute the vertex and connector points which are visible from a vertex of o and temporarily add these to the visibility graph

2. Recompute shortest routes for all connectors to/from o

Once the move has finished we perform object addition for o. Clearly this is substantially less work than required for complete feedback.

3   Evaluation

We have implemented our incremental connector algorithms in the Dunnart diagram editor and have conducted an experiment to evaluate our algorithms.[2] Dunnart is written in C++ and compiled with **gcc** 3.2.2 at **-O3**. We ran Dunnart on a Linux machine (glibc 2.3.3) with 512MB memory and Pentium 4, 2.4GHz processor.

In our experiment we compared the Spline-o-matic (SoM) connector routing library of Graph Viz (which is non-incremental) with a static version (Static) of our algorithm in which the visibility graph and connector routes are recomputed from scratch after each editor action, and the incremental algorithm (Inc) given here with various options.

The experiment used various sized grid arrangement of boxes, where each outside box is connected to the diagonally opposite box by a connector and each box except those on the right and bottom edge is connected to the box directly down and right. Figure 3(a) show an example layout for a 6x6 grid. For an $n \times n$ grid we have n^2 objects and $2(n-1) + (n-1)^2$ connectors. We also used a smallish but more realistic diagram **bayes** from [20] (a Bayesian network with 35 objects and 61 connectors) shown in Figure 3(b). The experiments were: for each object to delete it from the grid and the add it back

---

[2] Dunnart including this feature is downloadable from http://www.csse.monash. edu.au/~mwybrow/dunnart/

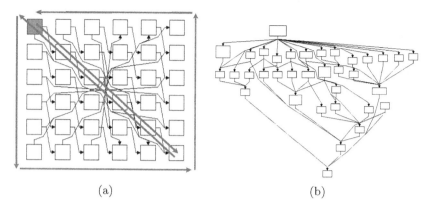

(a)   (b)

Fig. 3. (a) 6x6 Grid layout, showing the path taken through grid for **Move** experiment, and (b) Layout of the **bayes** diagram

in. We measured the time for each deletion and each addition giving the average under the **Add** and **Delete** rows. We have separated the time into that for manipulating the visibility graph (*Vis*) and that for performing all connector (re)routing (*Paths*). We also measured the average time taken to compute the new layout for each mouse position when moving the marked corner box through and around the grid as shown in Figure 3(a). The move of **bayes** is similar using the top leftmost box. This results are is given in the **Move** rows.

Both our static (Static) and incremental (Inc) version use the $A^{\star}$ algorithm to compute connector paths and give complete feedback. The incremental version computes the invisibility graph while the static one does not since this is not needed. We also give times for versions of the incremental algorithm which do not construct the invisibility graph (Inc-noInv). and use Dijkstra's shortest path algorithm rather than the $A^{\star}$ algorithm (Inc-noA*). For the **Move** sub-experiment we also give times for an incremental version providing partial feedback (Inc-par) rather than complete feedback.

The results are shown in Table 1, for grids of size 6, 8, 10 and 12, and **bayes**. A "—" indicates that the approach failed to complete the total experiment in three hours.

We can see from the table that the static version of our algorithm is considerably faster than Spline-O-Matic. The incremental versions are orders of magnitude faster than static algorithms. While the incremental version is usable for direct manipulation with complete feedback at least until **grid10** (and with difficulty at **grid12**), the static algorithms become unusable at **grid08**. The results show how important incremental recomputation is. The importance of the invisibility graph is clearly illustrated by the difference between Inc-noInv and Inc, improving visibility graph recomputation by orders of magnitude for **Delete** and **Move**. The $A^{\star}$ algorithm gives around 50% improvement in path re-routing. Partial feedback reduces the overhead of movement considerably particularly as the number of connectors grows.

Table 1. Average visibility graph and connector routing times (in msec.)

		grid06		grid08		grid10		grid12		bayes	
Op	Algorithm	Vis	Paths	Vis	Paths	Vis	Paths	Vis	Paths	Vis	Paths
	SoM	152	198	752	881	2449	2831	6669	1166	122	284
	Static	40	67	154	313	475	1024	1209	1064	29	87
Add	Inc-nolnv	0	13	8	90	15	304	14	761	0	1
	Inc-noA*	0	11	9	61	18	195	16	347	0	7
	Inc	0	1	9	20	18	58	16	138	0	0
	SoM	146	185	724	853	2385	2779	6542	1149	110	269
	Static	40	64	153	296	463	1003	1186	1042	29	80
Delete	Inc-nolnv	53	16	266	87	1006	298	1146	749	77	3
	Inc-noA*	2	38	17	223	49	734	55	1331	7	9
	Inc	2	8	18	58	48	204	55	504	8	0
	SoM	149	188	742	863	—	—	—	—	114	282
	Static	31	69	156	310	461	1004	1214	1026	29	79
Move	Inc-nolnv	43	15	230	80	950	289	1167	708	80	0
	Inc-noA*	0	25	12	102	34	261	37	449	7	8
	Inc	0	10	12	28	34	77	37	213	7	0
	Inc-par	0	3	10	7	26	20	28	11	0	3

4   Conclusion

Most diagram editors and graph construction tools provide some form of automatic connector routing, usually providing orthogonal or poly-line connectors. However the routes are typically computed using ad hoc techniques and updated in an ad hoc manner with no feedback during direct manipulation.

We have investigated the problem of incrementally computing minimal length object-avoiding poly-line connector routings. Our algorithms are surprisingly fast and allow us to recalculate optimal connector routings fast enough to reroute connectors even during direct manipulation of an object's position, thus giving instant feedback to the diagram author.

References

1. The Omni Group: OmniGraffle Product Page. Web Page (2002) http://www.omnigroup.com/omnigraffle/
2. Larsson, A.: Dia Home Page. Web Page (2002) http://www.gnome.org/projects/dia/
3. Microsoft Corporation: Microsoft Visio Home Page. Web Page (2002) http://office.microsoft.com/visio/
4. Computer Systems Odessa: ConceptDraw Home Page. Web Page (2002) http://www.conceptdraw.com/
5. yWorks: yFiles - Java Graph Layout and Visualization Library. Web Page (2005) www.yworks.com/ products/yfiles/
6. AT&T Research: Spline-o-matic library. Web Page (1999) http://www.graphviz.org/Misc/spline-o-matic/

7. Miriyala, K., Hornick, S.W., Tamassia, R.: An incremental approach to aesthetic graph layout. In: Proceedings of the Sixth International Workshop on Computer-Aided Software Engineering, IEEE Computer Society (1993) 297–308

8. Lee, D.T.: Proximity and reachability in the plane. PhD thesis, Department of Electrical Engineering, University of Illinois, Urbana, IL (1978)

9. Welzl, E.: Constructing the visibility graph for n line segments in $O(n^2)$ time. Information Processing Letters **20** (1985) 167–171

10. Asano, T., Asano, T., Guibas, L., Hershberger, J., Imai, H.: Visibility of disjoint polygons. Algorithmica **1** (1986) 49–63

11. Overmars, M.H., Welzl, E.: New methods for computing visibility graphs. In: Proceedings of the fourth annual symposium on Computational geometry, ACM Press (1988) 164–171

12. Ghosh, S.K., Mount, D.M.: An output-sensitive algorithm for computing visibility. SIAM Journal on Computing **20** (1991) 888–910

13. Fredman, M.L., Tarjan, R.E.: Fibonacci heaps and their uses in improved network optimization algorithms. J. ACM **34** (1987) 596–615

14. Mitchell, J.S.: Geometric shortest paths and network optimization. In Sack, J.R., Urrutia, J., eds.: Handbook of Computational Geometry. Elsevier Science Publishers B.V., Amsterdam (2000) 633–701

15. Rohnert, H.: Shortest paths in the plane with convex polygonal obstacles. Information Processing Letters **23** (1986) 71–76

16. Ben-Moshe, B., Hall-Holt, O., Katz, M.J., Mitchell, J.S.B.: Computing the visibility graph of points within a polygon. In: SCG '04: Proceedings of the twentieth annual symposium on Computational geometry, New York, NY, USA, ACM Press (2004) 27–35

17. Dijkstra, E.W.: A note on two problems in connection with graphs. Numerische Mathematik (1959) 269–271

18. Nelson, R.C., Samet, H.: A consistent hierarchical representation for vector data. In: SIGGRAPH '86: Proceedings of the 13th annual conference on Computer graphics and interactive techniques, New York, NY, USA, ACM Press (1986) 197–206

19. Shneiderman, B.: Direct manipulation: A step beyond programming languages. IEEE Computer **16** (1983) 57–69

20. Woodberry, O.J.: Knowledge engineering a Bayesian network for an ecological risk assessment. Honours thesis, Monash University, CSSE, Australia (2003) http://www.csse.monash.edu.au/hons/projects/2003/Owen.Woodberry/.

An Application of Well-Orderly Trees
in Graph Drawing

Huaming Zhang[1] and Xin He[2,*]

[1] Department of Computer Science, University of Alabama in Huntsville,
Huntsville, AL, 35899, USA
hzhang@cs.uah.edu

[2] Department of Computer Science and Engineering, SUNY at Buffalo,
Buffalo, NY, 14260, USA
xinhe@cse.buffalo.edu

Abstract. *Well-orderly trees* seem to have the potential of becoming a powerful technique capable of deriving new results in graph encoding, graph enumeration and graph generation [3, 4]. In this paper, we reduce the height of the visibility representation of plane graphs from $5n/6$ to $(4n-1)/5$, by using well-orderly trees.

1 Introduction

Graph drawing has emerged as an exciting and fast growing area of research in the computer science community in recent years [1]. Among various techniques for drawing planar graphs, the *canonical orderings and canonical ordering trees* of 3-connected plane graphs have served as a fundamental step upon which drawing algorithms are built [7, 8, 9, 12]. The work by de Frayseix, Pach and Pollack [9] is considered to be the first using the canonical orderings to produce straight-line drawings with polynomial sizes for planar graphs. The technique of canonical orderings has subsequently been applied to drawing graphs with respect to a variety of aesthetic constraints, including straight-line, convexity, orthogonality, visibility representation, 2-visibility, floor-planning, and others.

Later on, Chiang et. al. introduced the concept of *orderly spanning tree* [6], which generalizes canonical ordering tree and leads to several improvements in various styles of graph drawings [6, 5, 16]. In [3], Bonichon, Gavoille and Hanusse introduced *well-orderly trees*, which are canonical ordering trees with some special properties. These special properties have been successfully used in graph encoding, graph enumeration, and graph generation [3, 4]. More importantly, well-orderly trees are closely related to the concept of *Schnyder's realizers* [20, 21], which has also been widely used in graph drawing. We believe, well-orderly trees will be a promising technique of unifying known results as well as deriving new results in various styles in graph drawings. In this paper, we are going to derive an application of well-orderly trees in graph drawing.

A *visibility representation* (VR for short) of a plane graph G is a representation, where the vertices of G are represented by non-overlapping horizontal line

---

* Research supported in part by NSF Grant CCR-0309953.

segments (called *vertex segment*), and each edge of G is represented by a vertical line segment touching the vertex segments of its end vertices. The problem of computing a compact VR is important not only in algorithmic graph theory, but also in practical applications such as VLSI layout. A simple linear time VR algorithm was given in [19, 22] for a 2-connected plane graph G. It only uses an *st-orientation* of G and the corresponding *st*-orientation of its dual G^* to construct a VR of G.

One of the main concerns afterwards for VR is the size of the representation, i.e., the height and width of VR. Some work has been done to reduce the size of the VR by carefully choosing a special *st*-orientation of G. We summarize related previous results in the following table:

References	Plane graph G	4-Connected plane graph G
[19, 22]	Width of VR $\leq (2n - 5)$	Height of VR $\leq (n - 1)$
[13]	Width of VR $\leq \lfloor \frac{3n-6}{2} \rfloor$	
[17]	Width of VR $\leq \lfloor \frac{22n-42}{15} \rfloor$	
[14]		Width of VR $\leq (n - 1)$
[25]	Height of VR $\leq \lfloor \frac{5n}{6} \rfloor$	
[24, 26]	Width of VR $\leq \lfloor \frac{13n-24}{9} \rfloor$	Height of VR $\leq \lceil \frac{3n}{4} \rceil$

In this paper, we prove that every plane graph G has a VR with height at most $\frac{4n-1}{5}$, and it can be obtained in linear time.

The present paper is organized as follows. Section 2 introduces preliminaries. Section 3 presents the construction of a VR with height bounded by $\frac{4n-1}{5}$.

2   Preliminaries

In this section, we give definitions and preliminary results. Definitions not mentioned here are standard.

G is called a *directed graph* (digraph for short) if each edge of G is assigned a direction. We abbreviate the words "counterclockwise" and "clockwise" as **ccw** and **cw** respectively.

An *orientation* of a graph G is a digraph obtained from G by assigning a direction to each edge of G. We will use G to denote both the resulting digraph and the underlying undirected graph unless otherwise specified. (Its meaning will be clear from the context.) For a 2-connected plane graph G and an exterior edge (s, t), an orientation of G is called an *st-orientation* if the resulting digraph is acyclic with s as the only source and t as the only sink. For more information on *st*-orientation, we refer readers to [18].

Let G be a 2-connected plane graph and (s, t) an exterior edge. An *st-numbering* of G is a one-to-one mapping $\xi : V \rightarrow \{1, 2, \cdots, n\}$, such that $\xi(s) = 1$, $\xi(t) = n$, and each vertex $v \neq s, t$ has two neighbors u, w with $\xi(u) < \xi(v) < \xi(w)$, where u (w, resp.) is called a *smaller neighbor (bigger neighbor, resp.)* of v. Given an *st*-numbering ξ of G, we can orient G by directing each edge in E from its lower numbered end vertex to its higher numbered end vertex. The resulting orientation is called the *orientation derived from ξ* which, obviously, is an *st*-orientation of G. On the other hand, if $G = (V, E)$ has an *st*-orientation $\mathcal{O}$, we can define an 1-1 mapping $\xi : V \rightarrow \{1, \cdots, n\}$ by topological sort. It is easy to see that ξ is an *st*-numbering and the orientation derived from ξ is $\mathcal{O}$. From now on, we will interchangeably use the term an *st*-numbering of G and the term an *st*-orientation of G, where each edge of G is directed accordingly.

Lempel et. al. [15] showed that for every 2-connected plane graph G and an exterior edge (s, t), there exists an *st*-numbering. The following lemma was given in [19, 22]:

Lemma 1. *Let G be a 2-connected plane graph. Let $\mathcal{O}$ be an st-orientation of G. A VR of G can be obtained from $\mathcal{O}$ in linear time. The height of the VR is the length of the longest directed path in $\mathcal{O}$.*

Let T be a rooted spanning tree of a plane graph G. Two nodes are *unrelated* if neither of them is an ancestor of the other in T. An edge of G is unrelated if its endpoints are unrelated.

Bonichon et. al. introduced *well-orderly trees* [3], a special case of *orderly spanning trees* defined by Chiang, Lin and Lu in [6], referred as simply *orderly trees* afterwards. Let $v_1, v_2, \cdots, v_n$ be the ccw preordering of the nodes in T. A node v_i is *orderly* in T with respect to T if the incident edges of v_i in T form the following four blocks (possibly empty) in ccw order around v_i:

- $B_p(v_i)$: the edge incident to the parent of v_i;
- $B_<(v_i)$: unrelated edges incident to nodes v_j with $j < i$;
- $B_C(v_i)$: edges incident to the children of v_i; and
- $B_>(v_i)$: unrelated edges incident to nodes v_j with $j > i$.

A node v_i is *well-orderly* in G with respect to T if it is orderly, and if:

- the first ccw edge $(v_i, v_j) \in B_>(v_i)$, if it exists, verifies that the parent of v_j is an ancestor of v_i.

T is a *well-orderly tree* of G is all the nodes of T are well-orderly in G, and if the root of T belongs to the boundary of the exterior face of G (similarly for simply orderly tree). Note that an orderly tree (simply orderly or well-orderly) is necessarily a spanning tree.

A plane triangulation is a plane graph where every face is a triangle (including the exterior face). Let G be a plane triangulation of n vertices with three exterior vertices v_1, v_2, v_n in ccw order. A *realizer* $\mathcal{R} = \{T_1, T_2, T_n\}$ of G is a partition of its interior edges into three sets T_1, T_2, T_n of directed edges such that the following holds:

- for each $i \in \{1, 2, n\}$, the interior edges incident to v_i are in T_i and directed toward v_i.
- For each interior vertex of G, v has exactly one edge leaving v in each of T_1, T_2, T_n. The ccw order of the edges incident to v is: leaving in T_1, entering in T_n, leaving in $T2$, entering in T_1, and entering in T_2 (See Fig. 1). Each entering block could be empty.

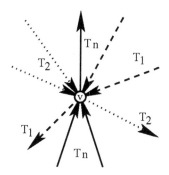

Fig. 1. Edge directions around an interior vertex v

Normally, realizers of a plane triangulation G are not unique. Among all the realizer of G, there is an unique realizer $\mathcal{R}_0$ of G, where according to the edge directions in $\mathcal{R}_0$, there are no ccw-triangles. This realizer of G will be called the minimum realizer of G. For example, in Fig. 2, the three trees of the realizer are drawn in solid lines, dashed lines and dotted lines respectively. There are three cw cyclic faces (marked by empty circles) but no ccw cyclic triangles, so it is the minimum realizer of G.

Schnyder showed in [20] that each set T_i of a realizer is a tree rooted at the exterior vertex v_i. For each tree T_i of a realizer, we denote by $\bar{T}_i$ the tree composed of T_i augmented with the two edges of the exterior face incident to the root of T_i, i.e. the vertex v_i. For example, in Fig. 2, $\bar{T}_i$ is T_n (the tree in thick solid lines) augmented with edges (v_n, v_1) and (v_n, v_2).

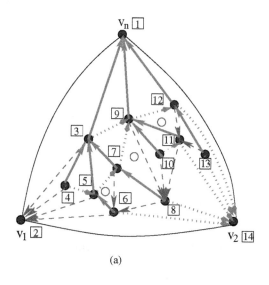

(a)

Fig. 2. A plane triangulation G and the minimum realizer $\mathcal{R}_0$ of G

We summarize related results in the following lemma [3, 6, 20, 21]:

Lemma 2. *Let G be a plane triangulation of n vertices with three exterior vertices v_1, v_2, v_n in ccw order. Let $\mathcal{R} = \{T_1, T_2, T_n\}$ be any realizer of G. Then,*

1. *Each $\bar{T}_i$, $i \in \{1, 2, n\}$ is a simply orderly tree. In addition, if $\mathcal{R}$ is the minimum realizer $\mathcal{R}_0$, then each $\bar{T}_i$, $i \in \{1, 2, n\}$ is a well-orderly tree.*
2. *Given the tree $\bar{T}_1$ ($\bar{T}_2$, $\bar{T}_n$ resp.), all the first ccw edge $(u, v_j) \in B_>(u)$ for each node u with respect to $\bar{T}_1$ ($\bar{T}_2$, $\bar{T}_n$ resp.) form the tree $\bar{T}_n$. ($\bar{T}_1$, $\bar{T}_2$ resp.)*
3. *The minimum realizer can be computed in linear time.*

For example, in Fig. 2, $\bar{T}_n$ is a well-orderly tree for G. And the first ccw edge $(9, 12)$ in $B_>(9)$ for the node 9 is in $\bar{T}_2$.

Let $v_1, v_2, \cdots, v_n$ be the ccw preordering of the nodes of a tree T. The subsequence $v_i, \cdots, v_j$ is a *branch* of T if it is a chain (i.e., v_t is the parent of v_{t+1} for every $i \le t < j$), and if $j - i$ is maximal. Branches partition the nodes of T, and each branch contains exactly one leaf.

Bonichon et. al. proved the following [3]: The well-orderly tree $\bar{T}_n$ of a minimum realizer $\mathcal{R}_0 = \{T_1, T_2, T_n\}$ has the *branch property*: All nodes of a given branch of $\bar{T}_n$ must have the same parent in $\bar{T}_1$ (except the root of $\bar{T}_n$). (Similar results hold for $\bar{T}_1$ and $\bar{T}_2$.) For example, in Fig. 2, nodes $3, 4$ form a branch, they have the same parent in $\bar{T}_1$.

3   More Compact VR of Plane Graphs

Let T be a tree drawn in the plane. Let $t_1, t_2, \cdots, t_n$ be the cw postordering of the nodes of T. A node of T is a *glue node* of T if it is right before a leaf node in the ordering $t_1, t_2, \cdots, t_n$. For example, considering $\bar{T}_n$ in Fig. 2, nodes $14, 12, 11, 9, 7, 5, 3$ are the glue nodes. Note that, the set of the first node of all branches of T except the root is the set of glue nodes. Also observe that the number of glue nodes of T is the number of leaves of T minus 1.

Next, let's explore another property of a well-orderly tree of a plane triangulation.

Lemma 3. *Let $\mathcal{R}_0 = \{T_1, T_2, T_n\}$ be the minimum realizer of a plane triangulation G with n vertices. Let ξ_1, ξ_2, ξ_n be the number of internal nodes (i.e, non-leaf node) of $\bar{T}_1, \bar{T}_2, \bar{T}_n$, l_1, l_2, l_n be the number of the leaves of $\bar{T}_1, \bar{T}_2, \bar{T}_n$ respectively. Then,*

1. *The internal nodes of $\bar{T}_2$ ($\bar{T}_n, \bar{T}_1$ resp.) must be the glue nodes of $\bar{T}_n$ ($\bar{T}_1, \bar{T}_2$ resp.).*
2. *$l_n - 1 \ge \xi_2$, $l_1 - 1 \ge \xi_n$, $l_2 - 1 \ge \xi_1$.*

Proof. According to Lemma 2, each $\bar{T}_i$ is a well-orderly tree of G. We only prove the case of $\bar{T}_2$. The other two cases are similar.

1. Let w be an internal node in $\bar{T}_2$. Therefore, there is an edge (u, w) in $\bar{T}_2$ such that w is the parent of u in $\bar{T}_2$. Applying Lemma 2 2, for the node u in $\bar{T}_n$, (u, w) is the first ccw edge in $B_>(u)$ with respect to $\bar{T}_n$. Since $\bar{T}_n$ is a well-orderly tree, the parent of w must be the ancestor of u in $\bar{T}_n$. So w must be a glue node of $\bar{T}_n$.

2. Applying to the observation that the number of glue nodes of T is the number of leaves of T minus 1, we have $l_n - 1 \geq \xi_2$.

For example, in Fig. 2, the internal nodes of $\bar{T}_2$ are $14, 12, 9, 7, 5, 11$. All of them are glue nodes of $\bar{T}_n$.

Next we use the three well-orderly trees from the minimum realizer to obtain more compact VR of a plane triangulation G.

Let $\mathcal{R}_0 = \{T_1, T_2, T_n\}$ be the minimum realizer of a plane triangulation G with n vertices.

Let's construct an st-numbering of G using $\bar{T}_n$ step by step. (The cases of using $\bar{T}_1, \bar{T}_2$ are similar.)

Each step begins from a leaf of $\bar{T}_n$. Suppose the leftmost unassigned leaf is u_1, the second leftmost unassigned leaf is q_1. The rightmost unassigned leaf if w_1, the second rightmost unassigned leaf if w_1'. The ordering of vertices of G by **ccw postordering**, starting from u_1 with respect to $\bar{T}_n$ is $u_1, u_2, \cdots, u_a, q_1, \cdots, q_b$. And q_b is the last vertex before the third leaf in this **ccw postordering**. The branch of $\bar{T}_n$ containing q_1 contains $q_{b+c}, \cdots, q_{b+1}, q_b, \cdots, q_1$ (which will be needed later.). The ordering of vertices of G by **cw postordering**, starting from w_1 with respect to $\bar{T}_n$ is $w_1, w_2, \cdots, w_d, w_1', \cdots, w_e'$. And w_e' is the last vertex before the third leaf in this **cw postordering**.

See Fig. 3 for an illustration. Only part of the graph is drawn. Edges and paths of $\bar{T}_n$ are drawn in solid lines. Note that q_{b+1} must have a child on the right of q_b.

Each step is classified into one of the following two cases:

Case 1: If there is no edge between u_a and w_1, then we first assign numbers to $u_1, u_2, \cdots, u_a$ by **ccw postordering** with respect to $\bar{T}_n$, then continue to assign numbers to $w_1, \cdots, w_d$ by **cw postordering** with respect to $\bar{T}_n$.

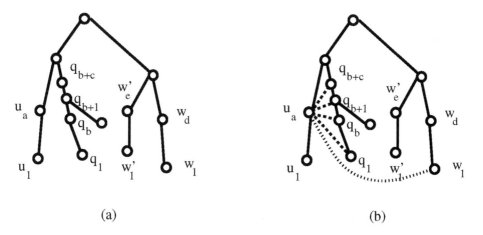

(a)   (b)

Fig. 3. (a) There is no edge between u_a and w_1. (b) There is an edge between u_a and w_1. (u_a, w_1) could be in $\bar{T}_1$ or $\bar{T}_2$. Then there must be edges between $q_1, \cdots, q_b, q_{b+1}, \cdots, q_{b+c}$ with u_a and they must be in $\bar{T}_1$.

Case 2: (u_a, w_1) is an edge in G. Note that q_1 is a leaf in $\bar{T}_n$, and u_a is the only vertex of G in $B_<(q_1)$. Therefore (u_a, q_1) must be an edge of G and it is in T_1. According to the branch property for $\bar{T}_n$, all the edges (u_a, q_i), $i = 1, \cdots, b, (b+1), \cdots, (b+c)$ must also be in $\bar{T}_1$. For the vertex q_{b+c}, u_a is the only vertex of G in $B_<(q_{b+c})$, and (q_{b+c}, u_a) is in $\bar{T}_1$. Hence, q_{b+c} cannot be an internal node in the tree $\bar{T}_2$. Also observe that q_{b+c} is a glue node of $\bar{T}_n$. In this case, we first assign numbers to $w_1, \cdots, w_d, w'_1, \cdots, w'_e$ by **cw postordering** with respect to $\bar{T}_n$. Then we assign numbers to $u_1, u_2, \cdots, u_a$ by **ccw postordering** with respect to $\bar{T}_n$.

Continue to next step if there are leaves left unassigned.

Note: If there are only 1 or 2 leaves left in the end, then we assign the remaining numbers to them either using **ccw postordering** or using **cw postordering** until we finish at the root of T. We do not count this as a step. Note that, for each node, either it is assigned a number in a **cw postordering** setting, or it is assigned a number in a **ccw postordering** setting.

We have the following two key observations:

Observation 1: For each step, at most three leaves are assigned numbers.

Observation 2: If Case 2 is applied k_n times, then k_n glue nodes of $\bar{T}_n$ cannot be internal nodes of $\bar{T}_2$. Therefore, according to Lemma 3 (1), $l_n - k_n - 1 \geq \xi_2$.

Lemma 4. *Let G be a plane triangulation, $\mathcal{R}_0 = \{T_1, T_2, T_n\}$ be the minimum realizer of G. Then, using $\bar{T}_i$, $i = 1, 2, n$,*

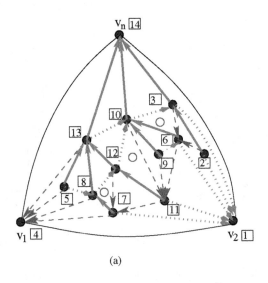

(a)

Fig. 4. An st-numbering of G in Fig. 2, obtained from $\bar{T}_n$ by using our numbering scheme

1. *The numbering of the vertices of G constructed by the above numbering scheme is an st-numbering of G.*
2. *If Case 2 is applied k_i times for $\bar{T}_i$, then any directed path in the resulting st-orientation is at most $n - \frac{l_i - k_i}{2}$.*
3. *Any directed path in the resulting st-orientation is at most $n - \frac{l_i}{3}, i = 1, 2, n$.*

Proof. We only prove the case $i = n$. The other two cases are similar.

1. First observe that, for any node other than the root of $\bar{T}_n$, its parent is assigned a bigger number. And the root of $\bar{T}_n$ is assigned n.

   For any internal node of $\bar{T}_n$, their children are assigned smaller numbers. For a leaf $u \neq v_1, v_2$ of $\bar{T}_n$, either it is assigned a number in a **ccw** setting, then the non-empty block $B_<(u)$ contains its smaller neighbors; or it is assigned a number in a **cw** setting, then the non-empty block $B_>(u)$ contains its smaller neighbors. For v_1, v_2, one of them is assigned 1, and it becomes a smaller neighbor for the other. Therefore, this numbering is an st-numbering for G.
2. Observe that, if Case 1 is applied to a step, then one of u_a and w_1 has to be bypassed by any directed path. On the other hand, if Case 2 is applied to a step, then one of w'_e and u_1 must be bypassed by any directed path. Therefore, from the nodes assigned numbers within the same step, at least one node has to be bypassed by any directed path.

   Suppose Case 2 is applied k_n times, then the total number of steps is at least $\frac{l_n - 3k_n}{2} - 1 + k_n$. (The subtraction of 1 comes from the last 1 or 2 leaves which do not form a step.) Therefore, any directed path has to bypass at least $\frac{l_n - 3k_n}{2} - 1 + k_n$ vertices. Therefore, its length is at most $n - \left(\frac{l_n - 3k_n}{2} - 1 + k_n\right) - 1 = n - \frac{l_n - k_n}{2}$.
3. In the worst scenario, each step assigns numbers to three leaves, then we have $\lfloor \frac{l_n}{3} \rfloor$ steps. So any directed path must bypass at least $\lfloor \frac{l_n}{3} \rfloor$ vertices, so it length is at most $n - \lfloor \frac{l_n}{3} \rfloor - 1 \leq n - \frac{l_n}{3}$.

For example, Fig. 4 shows an st-numbering of G, using our numbering scheme to $\bar{T}_n$. The first step numbers $1, 2, 3$ by **cw postordering**, then it numbers 4 by **ccw postordering**. The second step numbers 5 by **ccw postordering**, then it numbers 6 by **cw postordering**.

Next we present our main theorem:

Theorem 1. *Let G be a plane triangulation with n vertices, then there is a VR of G whose height is at most $\frac{4n-1}{5}$. And it can be constructed in linear time.*

Proof. Let $\mathcal{R}_0 = \{T_1, T_2, T_n\}$ be the minimum realizer of G. Apply our st-numbering scheme, suppose for $\bar{T}_1, \bar{T}_2, \bar{T}_n$, the number of their Case 2 steps are k_1, k_2, k_n respectively. Then we have $\xi_2 \leq l_n - k_n - 1$. Symmetrically, we have $\xi_1 \leq l_2 - k_2 - 1$, and $\xi_n \leq l_1 - k_1 - 1$. Summing them up and moving 3 to the left side, we have:

$$\xi_1 + \xi_2 + \xi_n + 3 \leq (l_1 + l_2 + l_n) - (k_1 + k_2 + k_n). \tag{1}$$

Pick a longest directed path for each st-orientation. By Lemma 4 (2), the sum of their lengths is at most:

$$(n - \frac{l_n - k_n}{2}) + (n - \frac{l_2 - k_2}{2}) + (n - \frac{l_1 - k_1}{2})$$
$$= 3n - \frac{l_1 + l2 + l_n}{2} + \frac{k_1 + k_2 + k_n}{2}$$
$$= 3n - \frac{(l_1 + l_2 + l_n) - (k_1 + k_2 + k_n)}{2}$$
$$\leq 3n - \frac{\xi_1 + \xi_2 + \xi_n + 3}{2} \tag{2}$$

The last inequality comes from Equation (1).

By Lemma 4 (3), the sum of their length is at most:

$$n - \frac{l_1}{3} + n - \frac{l_2}{3} + n - \frac{l_n}{3}$$
$$= 3n - \frac{l_1 + l_2 + l_n}{3} \tag{3}$$

Multiply Equation (2) by 2 and multiply Equation (3) by 3. Adding them up, we have that 5 times the sum of the lengths of the three longest directed paths is at most:

$$6n - (\xi_1 + \xi_2 + \xi_n + 3) + 9n - (l_1 + l_2 + l_n)$$
$$= 15n - (\xi_1 + \xi_2 + \xi_n + l_1 + l_2 + l_n) - 3$$
$$= 15n - 3n - 3$$
$$= 12n - 3. \tag{4}$$

Therefore, one of the longest directed path from these three paths must be at most $\frac{12n-3}{15} \leq \frac{4n-1}{5}$. Applying Lemma 1, G admits a VR whose height is at most $\frac{4n-1}{5}$, and it can be constructed in linear time.

References

1. G. di Battista, P. Eades, R. Tammassia, and I. Tollis, Graph Drawing: Algorithms for the Visualization of Graphs, Princeton Hall, 1998
2. N. Bonichon, B. Le Saëc and M. Mosbah, Wagner's theorem on realizers, in: Proc. ICALP'02, Lecture Notes in Computer Science, Vol. 2380, (Springer, Berlin, 2002) 1043-1053.
3. N. Bonichon, C. Gavoille, and N. Hanusse, An information-theoretic upper bound of planar graphs using triangulation, in *Proc. STACS'03*, pp 499-510, Lectures Notes in Computer Science, Vol. 2607, Springer-Verlag, 2003.
4. N. Bonichon, C. Gavoille, and N. Hanusse, Canonical decomposition of outerplanar maps and application to enumeration, coding and generation, In *Proc. WG'2003*, pp. 81-92, Lecture Notes in Computer Science, Vol. 2880, Springer-Verlag, 2003
5. H.-L. Chen, C.-C. Liao, H.-I. Lu and H.-C. Yen, Some applications of orderly spanning trees in graph drawing, in *Proc. Graph Drawing'02*, pp. 332-343, Lecture Notes in Computer Science, Vol. 2528, Springer-Verlag, Berlin, 2002.
6. Y.-T. Chiang, C.-C. Lin and H.-I. Lu, Orderly spanning trees with applications to graph encoding and graph drawing, in *Proc. of the 12th Annual ACM-SIAM SODA*, pp. 506-515, ACM Press, New York, 2001.

7. M. Chrobak and G. Kant, Convex grid drawings of 3-connected planar graphs, Technical Report RUU-CS-93-45, Department of Computer Science, Utrecht University, Holland, 1993.
8. U. Fößmeier, G. Kant and M. Kaufmann, 2-Visibility drawings of planar graphs, in *Proc. 4th International Symposium on Graph Drawing*, pp. 155-168, Lecture Notes in Computer Science, Vol. 1190, Springer-Verlag, Berlin, 1996.
9. H. de Fraysseix, J. Pach and R. Pollack, How to draw a planar graph on a grid. *Combinatorica* 10 (1990), 41-51.
10. G. Kant, Drawing planar graphs using the lmc-ordering, in *Proc. 33rd Symposium on Foundations of Computer Science*, pp.101-110, IEEE, Pittsburgh, 1992.
11. G. Kant, Algorithms for drawing planar graphs, Ph.D. Dissertation, Department of Computer Science, University of Utrecht, Holland, 1993.
12. G. Kant, Drawing planar graphs using the canonical ordering, *Algorithmica* 16 (1996), 4-32.
13. G. Kant, A more compact visibility representation. *International Journal of Computational Geometry and Applications* 7 (1997), 197-210.
14. G. Kant and X. He, Regular edge labeling of 4-connected plane graphs and its applications in graph drawing problems. *Theoretical Computer Science* 172 (1997), 175-193.
15. A. Lempel, S. Even and I. Cederbaum, An algorithm for planarity testing of graphs, in *Theory of Graphs (Proc. of an International Symposium, Rome, July 1966)*, pp. 215-232, Rome, 1967.
16. C.-C. Liao, H.-I. Lu and H.-C. Yen, Floor-planning via orderly spanning trees, in *Proc. 9th International Symposium on Graph Drawing*, pp. 367-377, Lecture Notes in Computer Science, Vol. 2265, Springer-Verlag, Berlin, 2002.
17. C.-C. Lin, H.-I. Lu and I-F. Sun, Improved compact visibility representation of planar graph via Schnyder's realizer, in: *SIAM Journal on Discrete Mathematics*, 18 (2004), 19-29.
18. P. Ossona de Mendez, Orientations bipolaires. PhD thesis, Ecole des Hautes Etudes en Sciences Sociales, Paris, 1994.
19. P. Rosenstiehl and R. E. Tarjan, Rectilinear planar layouts and bipolar orientations of planar graphs. *Discrete Comput. Geom.* 1 (1986), 343-353.
20. W. Schnyder, Planar graphs and poset dimension. *Order* 5 (1989), 323-343.
21. W. Schnyder, Embedding planar graphs on the grid, in *Proc. of the First Annual ACM-SIAM Symposium on Discrete Algorithms*, pp. 138-148, SIAM, Philadelphia, 1990.
22. R. Tamassia and I.G.Tollis, An unified approach to visibility representations of planar graphs. *Discrete Comput. Geom.* 1 (1986), 321-341.
23. H. Zhang and X. He, Compact Visibility Representation and Straight-Line Grid Embedding of Plane Graphs, in: Proc. WADS'03, Lecture Notes in Computer Science, Vol. 2748, (Springer-Verlag Heidelberg, 2003) 493-504.
24. H. Zhang and X. He, On Visibility Representation of Plane Graphs, in: Proc. STACS'04, Lecture Notes in Computer Science, Vol. 2996, (Springer-Verlag Heidelberg, 2004) 477-488.
25. H. Zhang and X. He, *New Theoretical Bounds of Visibility Representation of Plane Graphs*, in: Proc. GD'2004, Lecture Notes in Computer Science, Vol. 3383, (Springer-Verlag, 2005) pp. 425-430.
26. H. Zhang and X. He, Improved Visibility Representation of Plane Graphs, *Discrete Comput. Geom.* 30 (2005), 29-29.

GEOMI: GEOmetry for Maximum Insight

Adel Ahmed[1,2], Tim Dwyer[3], Michael Forster[1], Xiaoyan Fu[1], Joshua Ho[2],
Seok-Hee Hong[1,2], Dirk Koschützki[5], Colin Murray[1,2], Nikola S. Nikolov[1,4],
Ronnie Taib[1], Alexandre Tarassov[1,4], and Kai Xu[1]

[1] IMAGEN Program, National ICT Australia, Sydney, Australia
{adel.ahmed, michael.forster, xiaoyan.fu, seokhee.hong, nikola.nikolov,
ronnie.taib, alexander.tarasov, kai.xu}@nicta.com.au
[2] School of IT, The University of Sydney, Sydney, Australia
joho4868@mail.usyd.edu.au
[3] Monash University, Melbourne, Australia
Tim.Dwyer@infotech.monash.edu.au
[4] Department of CSIS, University of Limerick, Limerick, Republic of Ireland
[5] Institute of Plant Genetics and Crop Plant Research, Gatersleben, Germany
koschuet@ipk-gatersleben.de

Abstract. This paper describes the GEOMI system, a visual analysis tool for the visualisation and analysis of large and complex networks. GEOMI provides a collection of network analysis methods, graph layout algorithms and several graph navigation and interaction methods. GEOMI is part of a new generation of visual analysis tools combining graph visualisation techniques with network analysis methods. GEOMI is available from http://www.cs.usyd.edu.au/~visual/valacon/geomi/.

1   Introduction

The GEOMI system is a visual analysis tool for the visualisation and analysis of large and complex networks such as web-graphs, social networks, biological networks, sensor networks and transportation networks. Such visual analysis tools take advantage of the graphics capabilities of computers to support the analysis of network structure. Using GEOMI, one can visually explore networks and discover patterns and trends that can provide critical insights. GEOMI is being developed by VALACON (Visualisation and Analysis of Large and Complex Networks) project team members in the National ICT Australia (NICTA) IMAGEN program.

Figure 1 briefly describes the architecture of GEOMI. Its core consists of three main components: network analysis, graph layout and interaction, using an extended version of WilmaScope [1] as its graph visualisation library. GEOMI can easily be extended by various types of plug-ins. It integrates the JUNG library [2] as a plug-in to utilise its many network analysis algorithms. We also have added many new plug-ins for network analysis, graph layout as well as interaction methods. In summary, GEOMI currently provides the following functionalities:

P. Healy and N.S. Nikolov (Eds.): GD 2005, LNCS 3843, pp. 468–479, 2005.

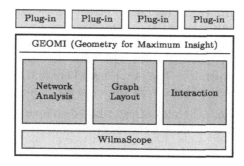

Fig. 1. The system architecture of GEOMI

1. Graph generators: clustered graph, clustered general graph, clustered tree, GML graph reader, grid, Erd.-Ren. random graph, scale-free graph, stratified graph, tree, CGF graph reader, random clustered graph converter.
2. Network analysis: shortest path/random walk betweenness centrality, closeness centrality, degree centrality, eccentricity centrality, eigenvector centrality, uniqueness centrality, 3D parallel coordinates, blockmodel, hierarchical centrality comparison, orbital centrality comparison, k-means clustering.
3. Graph algorithms: bi-connected components, biggest component, directed cycle removal, edge weight filter, longest path layering, parallel edge filter.
4. Graph layout: circular, clustered circular, clustered clone tree, clustered force directed, clustered free tree, clustered rod tree, column, cone tree, force directed, free tree, hierarchical, high-dimensional embedding, multiscale, random, rod tree, simulated annealing, spectral, stratified.
5. Interaction: HTML graph generator, head gesture interaction.

This paper is organised as follows: GEOMI's plug-ins for network analysis, graph layout, and interaction are presented in Sect. 2, 3, and 4, respectively. Section 5 concludes the paper.

2   Network Analysis Plug-Ins

2.1   Centrality Analysis

Centrality analysis is an effective tool to study graph nodes. The importance of nodes is measured by their degree, their neighbouring nodes, or other node/graph properties. Centrality analysis has found many applications in social and biological networks. The centralities implemented in GEOMI are listed in Sect. 1. GEOMI can map the results of centrality analysis to node visual attributes.

2.2   Centrality Comparison

Comparing the relative importance of a node in different centrality measures provides an overview of the node from different perspectives. Besides mapping centralities to various visual node attributes, three methods designed specifically for centrality comparison are implemented in GEOMI. See [3] for details.

3D Parallel Coordinates. This method treats every node with multiple centrality values as a multivariate data point. To demonstrate nodes sharing the same value, they are displayed using the third dimension. In Fig. 2, each axis represents a centrality measure and nodes with the same centrality value are shown in the third dimension. The graph used here is the "Krackhardt's High-tech managers" dataset, which is well-known in social network analysis.

Orbital Comparison. The idea of orbit-based comparison of centralities can be summarised as follows: copies of the analysed network are stacked, every copy

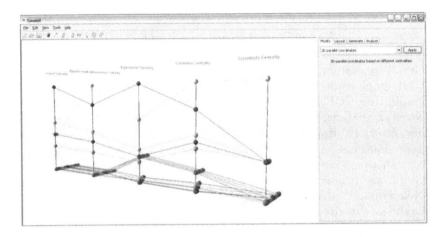

Fig. 2. 3D parallel coordinates

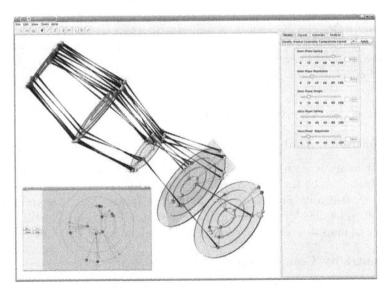

Fig. 3. The created stack of networks with centralities used for orbital placement of the nodes. On the bottom left the detail view shows the highlighted plane and on the right side the layout modification plug-in controls are shown.

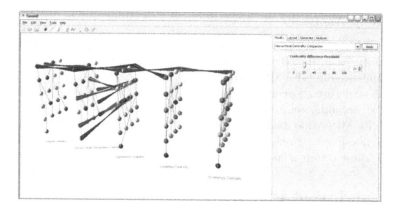

Fig. 4. Hierarchical comparison

is used to visualise one centrality measure and within a copy the nodes are placed on concentric circles depending on the centrality value of the vertex. Figure 3 shows an example. The dataset used here is the same as the one in Fig. 2.

Hierarchical Comparison. The hierarchal comparison is similar to orbital comparison, but uses a hierarchical layout instead. For each graph the layering is based on centrality values, i.e., nodes in the upper layer have larger centrality values than those in the lower layer. Within each layer nodes are ordered to reduce edge crossings. Figure 4 shows the result of hierarchical comparison. The user has the option to show nodes whose centrality value changes significantly between measures. In this example, edges are shown only if the centrality of the connected nodes differs by more than 20% in two consecutive measures.

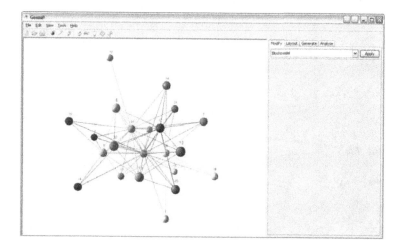

Fig. 5. Blockmodel analysis

2.3 Group Analysis

While centrality analysis focuses on the properties of individual nodes, group analysis focuses on the overall graph structure. One of the group analysis methods implemented in GEOMI is blockmodel [4], which groups nodes according to the graph structure associated with them. After blockmodel analysis, the nodes that are *structurally equivalent* are put into the same cluster. The implementation in GEOMI is that two nodes are structurally equivalent if they have the same neighbour set, which is the original definition for structural equivalence. In Fig. 5, same colour is assigned to nodes that are structurally equivalent.

3 Graph Layout Plug-Ins

3.1 Hierarchical Layout

The Hierarchical Layout plug-in implements the algorithm for drawing directed graphs in three dimensions [5], a 3D extension to the Sugiyama method which includes an additional step after the layering step. It further partition the layer into a set of $k > 1$ subsets, called *walls*. For details, see [5].

Currently the following algorithms are available for each step of the algorithm:

- Layer-assignment algorithms: longest-path, longest-path followed by node promotion, network simplex, minwidth. [6,7]
- Wall-assignment algorithms: balanced min-cut, zig-zag, dominating wall, k-wall min-span, k-wall balanced. [8,5]
- Node-ordering algorithms: layer-by-layer sweep with barycenter heuristic for two-layer crossing reduction.
- Horizontal assignment: Brandes-Köpf algorithm [9].

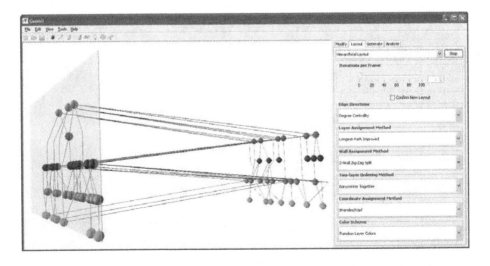

Fig. 6. HLP applied to one of the Rome graphs

An example of one of the Rome graphs with 62 nodes and 79 edges is illustrated in Figure 6. The algorithms used for each step of the extended Sugiyama method are listed in the dialogue box in right-hand side of the screen. The user can choose a method for each step of the algorithm. Further, the user can also choose a colour scheme for the hierarchical layout.

3.2   3D Tree Layout Plug-Ins

GEOMI supports three linear time 3D tree drawing algorithms that can support nodes with different sizes - namely *cone tree*, *rod tree* and *free tree* (see Fig. 7). This layout is also used for drawing clustered graphs in three dimensions. For details, see [10].

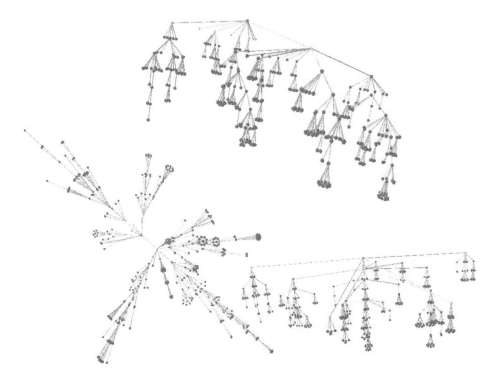

Fig. 7. Tree layout styles in GEOMI

If the given graph is not a tree, the plug-ins can automatically compute a spanning tree for the graph and compute the layout of the spanning tree. All non-tree edges are then added back to the final drawing.

3.3   Circular Layout Plug-In

A circular layout plug-in is implemented for visualisation of social networks using the two pass crossing reduction algorithm [11].

3.4   Clustered Graph Layout Plug-Ins

Clustered Graph Generator. A clustered graph in GEOMI can be generated in three ways. Firstly, it can be randomly generated. Two random clustered graph generators have been implemented to generate clustered trees and clustered general graphs. In both generators, the user can control the graph generation process by changing a number of parameters. Secondly, a clustered graph can be generated by reading from the clustered graph data file. Thirdly, a normal GEOMI graph can be converted into a clustered graph by applying a clustering algorithm.

Clustered Graph Layout. A series of six clustered graph layout plug-ins has been created by combinations of three 3D tree layouts and two 2D cluster layouts. It implements the four step method for drawing clustered graphs in 3D. For details, see [10].

The plug-in provides a control panel by using the super graph layout control panel as a sub-panel. This also consists of a general clustered graph layout control sub-panel that allow users to control the general clustered graph layout process. See Fig. 8 for an example.

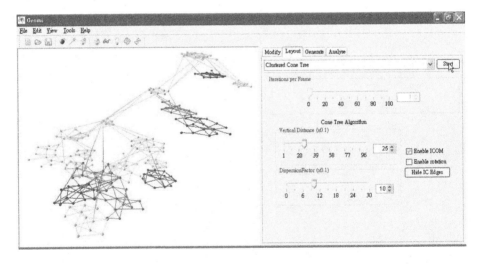

Fig. 8. A clustered graph layout plug-in. This example shows a layout that draws each cluster using circular layout and combined the whole clustered graph as cone tree.

3.5   Scale-Free Network Plug-In

The scale-free layout plug-in is an implementation of the FADE fast force directed algorithm allowing for interactive modification of the force parameters. It also allows for layering based on degree centrality which is particularly useful for visualisation of scale-free networks, see [12].

The edge force controls allow the user to turn edge forces on or off. It also allows for the resilience of the edges to be increased or decreased. Other options

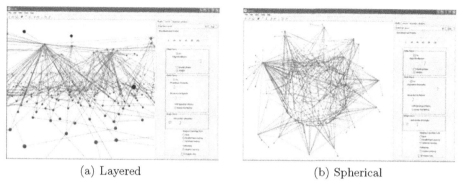

(a) Layered   (b) Spherical

Fig. 9. Scale-free plug-in

include the ability to have edges repel instead of attract and to use the weights associated with an edge in calculating the edge force.

The node force controls allow the user to turn node forces on or off. It also allows the user to modify the extent of the repulsion force as well as the accuracy parameter of the FADE algorithm, which allows for an accuracy vs speed trade-off. An attractive force towards the origin can also be turned on or off and modified.

The degree layering controls allow the user to restrict nodes to layers based on degree. The nodes can be restricted to either *parallel planes* (see Fig. 9(a)) or *concentric spheres* (see Fig. 9(b)). The nodes are partitioned into layers such that nodes with degree greater than or equal to 10 are at the highest layer and nodes with degree less than 5 are on the lowest layer. Single degree nodes are placed on their neighbour's level. The user can also select to colour the different partitions differently and can select to have only incoming edges count towards the degree total.

3.6   Temporal Network Plug-In

Temporal networks, which describe graph changes over time, attract growing research interests for their analysis. The $2\frac{1}{2}$D method is one of the solutions to represent temporal network data. In such a method, a graph snapshot at a particular time is placed on a 2D plane, in which a layout algorithm can be applied; a series of such planes are stacked together following time order to show the changes. In order to identify a particular node in different time plane, same nodes in different planes are connected by edges. Combined with navigation tools in GEOMI, users can trace the change of each individual node's relationship to others and also can evaluate the evolution of the whole network in general.

This method is implemented in GEOMI as a generator plug-in. It can convert a series of data files, with one file containing information of one time frame only, into GEOMI data structure, so that layout methods (force-directed layout by default) can be used. Potentially, all layout algorithms and analysis methods in GEOMI can be applied, even with different layout and in different plane.

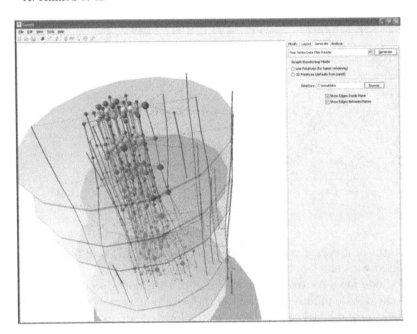

Fig. 10. Email connections of a research group represented in time series

As an example, Fig. 10 shows the email connections of a certain research group. Each plane represents one month while each node is one person. The edges between nodes in same plate shows the email traffic between person. In addition, degree centrality is mapped to node size while node colour represents betweenness centrality.

4   Interaction Plug-Ins

4.1   HTML Graph Generator

The HTML graph generator plug-in generates an undirected web-graph where web pages are represented by graph nodes and hyper-links are represented by edges between two nodes. Two nodes are connected by an edge if one is referenced by a hyper-link in the other. Given a URL and the depth of parsing, a web crawler is employed starting at the specified URL. A fetched web page is parsed for hyper-links and their respective web pages are acquired recursively. A graph node is added for every acquired page and edges are added accordingly. As the graph is being generated on the fly the force directed layout algorithm is started by default to calculate the layout of the dynamically growing graph.

The generated graph in conjunction with different centrality analysis tools may be used to extract website structural information. Various page properties may be defined and identified. In the initial web-graph, the high degree of the node indicates the most referencing pages, pages that have the most

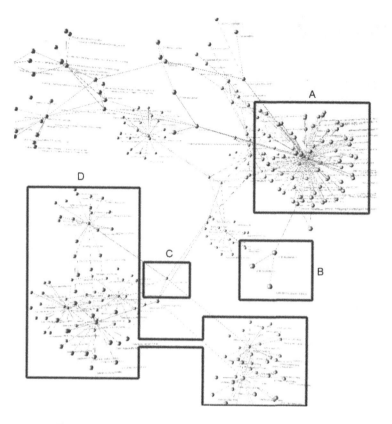

Fig. 11. Anatomy of the http://www.cnn.com page

hyper-links to other pages as shown in Fig. 11A. On the other hand, low degree nodes or pendent nodes, shown in Fig. 11B, may express remote pages access to them may not be easy or trivial from the root of the web-graph, or the start URL. The web-graph may be used to categorize pages according to some predefined conventions that identify the importance of a page. For example the page shown in Fig. 11C is considered to be an important one for if a user misses this page while navigating the website, all the information in subgraph in Fig. 11D will be kept hidden from the user, in other words, the user will not be aware of them. A demo movie is available from http://www.cs.usyd.edu.au/~visual/valacon/geomi/movies/.

4.2   Head Gesture Plug-In

The head gesture plug-in allows immersive navigation of the data using 3D head gestures instead of the classical mouse input. The plug-in relies on two gesture recognition modules, receiving inputs from two low-cost web cameras located orthogonally, one in front of the user, one on her/his side. See Fig. 12. A demo movie is available from http://www.cs.usyd.edu.au/~visual/valacon/geomi/movies/.

 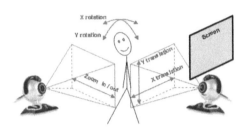

(a) Nodding down triggers a view from above

(b) 3D Gesture detection and corresponding actions

Fig. 12. GEOMI gesture plug-in

The user can literally walk into the network, move closer to nodes or clusters by simply aiming in their direction. Nodding and tilting the head rotate the entire network along the X and Y axis respectively. The command mapping is as follows:

HEAD MOVEMENT	NAVIGATION
Horizontal sideways	Mirrored X translation
Vertical up/down	Mirrored Y translation
Horizontal back/forth (towards screen)	Zoom out/in
Nodding up/down	X rotation
Tilting left/right	Y rotation

5   Conclusion

GEOMI is a generic visual analysis tool and can be extended to a special tool for visual analysis of biological networks or social networks. We will add more analysis methods, graph algorithms and layout methods including interaction methods. The analysis methods include social network analysis and graph mining methods. Layouts methods include implementation of algorithms for various graph models such as planar graphs and general graphs, and network models such as evolution networks and dynamic networks. Interaction methods include navigation methods and various user interactions.

References

1. Dwyer, T.: Extending the WilmaScope 3D graph visuzlisation system - software demonstration. In: In Proceeding of Asia-Pacific Symposium on Information Visualisation 2005 (APVIS2005). (2005) 35–42
2. Madadhain, J.O., Fisher, D., Smyth, P., White, S., Boey, Y.B.: Analysis and visualization of network data using JUNG. http://jung.sourceforge.net (2005)
3. Dwyer, T., Hong, S.H., Koschützki, D., Schreiber, F., Xu, K.: Visual analysis of network centralities. In: Planed submission for APvis 2006. (2005)

4. Wasserman, S., Faust, K.: Social Network Analysis: Methods and Applications. Cambridge Univ. Press (1994)

5. Hong, S.H., Nikolov, N.S.: Hierarchical layouts of directed graphs in three dimensions. Technical report, National ICT Australia Ltd. (NICTA) (2005)

6. Nikolov, N., Tarassov, A.: Graph layering by promotion of nodes. Special issue of Discrete Applied Mathematics associated with the IV ALIO/EURO Workshop on Applied Combinatorial Optimization (to appear)

7. Tarassov, A., Nikolov, N.S., Branke, J.: A heuristic for minimum-width of graph layering with consideration of dummy nodes. In: Experimental and Efficient Algorithms, Third International Workshop. Lecture Notes in Computer Science, Springer-Verlag (2004) 570–583

8. Hong, S.H., Nikolov, N.S.: Layered drawings of directed graphs in three dimensions. In: Information Visualisation 2005: Asia-Pacific Symposium on Information Visualisation (APVIS2005), CRPIT (2005) 69–74

9. Brandes, U., Köpf, B.: Fast and simple horizontal coordinate assignment. In Mutzel, P., Jünger, M., S., L., eds.: Graph Drawing: Proceedings of 9th International Symposium, GD 2001. Volume 2265 of Lecture Notes in Computer Science., Springer-Verlag (2002) 31–44

10. Ho, J., Hong, S.H.: Drawing clustered graphs in three dimensions. to appear in the proceedings of 13th International Symposium on Graph Drawing (GD2005) (2005)

11. Baur, M., Brandes, U.: Crossing reduction in circular layouts. In: Proceedings on 30th International Workshop Graph-Theorectic Concepts in Computer Sciecne. (2004) 332–343

12. Ahmed, A., Dwyer, T., Hong, S.H., Murray, C., Song, L., Wu, Y.X.: Visualisation and analysis of large and complex scale-free networks. In: in Proceedings Eurographics / IEEE VGTC Symposium on Visualization (EuroVis 2005). (2005) to appear

WhatsOnWeb: Using Graph Drawing to Search the Web⋆

Emilio Di Giacomo, Walter Didimo, Luca Grilli, and Giuseppe Liotta

Università di Perugia, Italy
{digiacomo, didimo, grilli, liotta}@diei.unipg.it

Abstract. One of the most challenging issues in mining information from the World Wide Web is the design of systems that can present the data to the end user by clustering them into meaningful semantic categories. We envision that the analysis of the results of a Web search can significantly take advantage of advanced graph drawing techniques. In this paper we strengthen our point by describing the visual functionalities of WhatsOnWeb, a meta search clustering engine explicitly designed to make it possible for the user to browse the Web by means of drawings of graphs whose nodes represent clusters of coherent data and whose edges describe semantic relationships between pairs of clusters. A prototype of WhatsOnWeb is available at http://whatsonweb.diei.unipg.it/.

1 Introduction

Increasing attention has been recently devoted to the development of new systems that support users in searching the Web. Although classical search engines perform well in many circumstances, there are some practical situations in which the data that they return as a reply to a user's query are not structured enough to easily convey the information that the user may be looking for (see also [6, 10]). Indeed, the output of a classical search engine consists of an ordered list of document links (URLs), where each link comes together with a brief summary (called snippet) of its associated document. The links are selected and ranked by the search engine according to some criteria that depend on the user's query, on the documents content, and (in some cases, like Google) on the popularity of the links in the World Wide Web. The list of URLs returned by a search engine is often very long and the consequence is that users may omit to check URLs that can be relevant for them, just because these links do not appear in the first positions of the list. This problem is even more evident when the user's query presents polisemy, that is words with different meanings. For example, suppose that the user submits the query "Jaguar"; is she interested in the "car" or in the "animal"?

A *Web meta-search clustering engine* is a system conceived to cope with the above described limitations of classical search engines. As shown in Figure 1, a

⋆ This work is partially supported by the MIUR Project ALGO-NEXT: Algorithms for the Next Generation Internet and Web: Methodologies, Design and Applications.

P. Healy and N.S. Nikolov (Eds.): GD 2005, LNCS 3843, pp. 480–491, 2005.

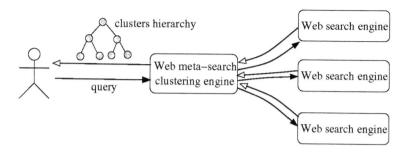

Fig. 1. Schema of the working model of a Web meta-search clustering engine

Web meta-search clustering engine provides a visual interface to the user who submits a query; it forwards the query to (one or more) traditional search engines, and it displays a set of clusters, also called *categories*, which are typically organized in a hierarchy. Each category contains URLs of documents that are semantically related with each other and is labelled with a string that describes its content. With this approach, the user has a global view of the different semantic areas involved by her query, and can explore those areas in which she is mostly interested.

Despite the graphical user interface of a Web meta-search clustering engines plays a fundamental role for efficiently retrieving the wanted information, the vast majority of Web meta-search clustering engines (see, e.g., Vivisimo, iBoogie, SnakeT [6] [1]) have a GUI in which the set of clusters is represented simply as a tree of directories and subdirectories. We find this type of representation unsatisfactory for the following reasons.

– A tree of clusters represents inclusion relationships between a cluster and its sub-clusters. However, nodes that are not in an ancestor/descendant relation can also have strong semantic connections, which may be underestimated if not completely lost by following the tree structure. Suppose for example that the user's query is "Armstrong" and that the tree of clusters is as the one depicted in Figure 2(a). Is the category "Biography" related to "Louis" or to "Lance" or to both (or to neither of them but to the astronaut Neil Armstrong?). If there were one edge as in Figure 2(b) or if there were two edges with different weights as in Figure 2(c), the user could decide whether or not the category "Biography" is of her interest.
– Viewing the clusters as the vertices of a graph can also help the user to detect communities of clusters which identify a common topic and/or to highlight isolated clusters which may be pruned if their labels are not meaningful for the user's query. This type of analysis is clearly unfeasible if all clusters are displayed in a tree structure.
– Finally, the visualization paradigm adopted by systems like Vivisimo often uses basic drawing techniques: The tree is displayed with poor aspect ratio on the left-hand side of the interaction window. The visual exploration is

---

[1] http://vivisimo.com/; http://www.iboogie.com/;http://snaket.di.unipi.it/

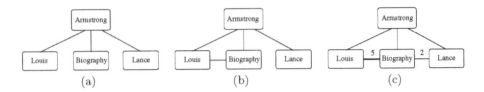

Fig. 2. (a) A portion of a tree of categories for the query "Armstrong". (b)-(c) The same tree, plus edges that highlight cluster relationships. In (c), each edge is labelled with a number that represents the strength of the edge.

particularly uncomfortable when the degree of the tree is high: Expanding a few nodes at the same level can make the rest of the tree no longer visible on the screen.

We envision that the analysis of the results of a Web meta search clustering engine can significantly take advantage of advanced graph drawing techniques. In this paper we strengthen our point by describing the user interface of WhatsOnWeb[2]. WhatsOnWeb, a meta search clustering engine explicitly designed to make it possible for the user to browse the Web by means of drawings of graphs whose nodes represent clusters of coherent data and whose edges describe semantic relationships between pairs of clusters. The produced diagrams are in general structurally more complex than trees and the user can interact with them in a number of different ways. The main visual functionalities and tools of the GUI of WhatsOnWeb are shortly summarized below.

– WhatsOnWeb adopts a visual paradigm that makes it possible to dynamically interact with the graph of categories by expanding and/or contracting its nodes while preserving the user's mental-map. Particular attention is given to the area required by the representation, since the graph of the clusters may become relatively large after the expansion of a few categories.
– WhatsOnWeb emphasizes the relevance of a category with respect to other categories by using a suitable score function. The relevance (score) of each category is conveyed in the drawing by using different colors for the vertices of the graph. It is also possible to visualize the "strength" of a semantic relationship between two categories expressed as another suitable score function. Furthermore, the user can analyze the relationship between two categories c_1 and c_2 by visualizing a bipartite graph with all URLs that give rise to the relationship between c_1 and c_2.
– The user is provided with several automatic or semiautomatic tools to simplify and/or to refine the clustered graph returned by the system. For example, the user can ask the system to prune those categories whose relevance is below a given threshold and that are not connected to any other relevant categories.

---

[2] A prototype of WhatsOnWeb is available at http://whatsonweb.diei.unipg.it/

We remark that, to the best of our knowledge, the only other Web meta-search clustering engine that supports the user with a graph visualization approach is Kartoo [3]. Besides being very appealing from a graphic view point, the graph visualization offered by Kartoo is however not much more expressive than the tree of clusters of Vivisimo. The nodes in the graph of Kartoo are the individual URLs and the clusters are represented as regions of a virtual map. By pointing the page with the mouse, the system highlights all categories in which the Web page is contained; by pointing a category, the system highlights all URLs that form the specified category. Also, every map of Kartoo contains only a limited number of URLs (about twelve) and a global view of the relationships between all categories (and thus an enhanced visual analysis) is not possible.

2   Principles of the System WhatsOnWeb

The system WhatsOnWeb computes the clusters hierarchy based on the *topology-driven approach* [5], which allows us to define a new visualization paradigm for representing and browsing the clusters hierarchy. The topology-driven approach relies on the basic concept of *snippet graph*. Intuitively, a snippet graph represents the URLs returned by a search engine in response to a user query, along with their relationships. The relationships between pairs of URLs are computed by analyzing the *snippet* that a search engine returns along with each URL; the snippet of a URL is a text that represents a brief summary of the document associated with the URL. Each vertex of the snippet graph is a URL and there is an edge between two URLs when the associated snippets share some terms. In the following, we first give a formal definition of snippet graph, which is refined with respect to that introduced in [5]; then, we recall the principles of a visualization paradigm related to the definition of a snippet graph.

Let U be a set of URLs returned by a search-engine in response to a user query. If $u \in U$, we call *text of* u the concatenation of the title and the snippet associated with u. In order to construct the snippet graph of U we perform a sequence of preliminary steps:

Cleaning step. Stop-words and HTML symbols are removed in the text of each element of U.

Stemming step. A stemming algorithm is applied on the cleaned texts, and for every computed stem s we associate a word z_s chosen among those whose stem is equal to s; words z_s is used later to label the clusters. After this step, the text of each URL u is viewed as a sequence of stems, denoted by S_u.

Scoring step. We assign a score w_s to each stem s; w_s measures the "relevance" of s in all URL texts containing s. Several criteria can be used to determine w_s. We adopt a standard function called *tf-idf* (*term frequency-inverse document frequency*), introduced by Salton in 1989 [9]. Namely, let D be a set of documents, let s be a term in a document $d \in D$, and let $D_s \subseteq D$ be the subset of documents containing s. The tf-idf of s with respect to document

---

[3] http://www.kartoo.com/

d is defined as: $t(s, d) = f(s, d) \log(|D|/|D_s|)$, where $f(s, d)$ is the frequency of s in d. In our case, assuming D as the set of URL texts, we define w_s as the average of the tf-idf of s over all documents of D_s, i.e., $w_s = \frac{\sum_{d \in D_s} t(s,d)}{|D_s|}$.

The *snippet graph* G of U is a labeled weighted graph defined as follows:

- G has a vertex v_u associated with each element $u \in U$. The label of v_u can be either the title of u or its description as URL.
- G has an edge $e = (v_{u_1}, v_{u_2})$ (where $u_1, u_2 \in U$) if the texts of u_1 and u_2 share some stems, i.e. if $S_{u_1} \cap S_{u_2}$ is not empty. To determine the weight and the label of e, the following procedure is applied. Compute the set $\mathcal{S}_e$ of all maximal sub-sequences of consecutive stems (each sub-sequence with single multiplicity) shared by S_{u_1} and S_{u_2}; each element of $\mathcal{S}_e$ is also called a *sentence*. The *score of a sentence* $\sigma \in \mathcal{S}_e$ is defined as $w_\sigma = \sum_{s \in \sigma} w_s$, and the weight of e is $w_e = \sum_{\sigma \in \mathcal{S}_e} w_\sigma$. Finally, the label of e is set equal to $\mathcal{S}_e$.

In the snippet graph, the weight of an edge measures the strength of the semantic relationship between their end-vertices. This strength depends both on the number of sub-sequences of stems shared by their texts and on the relevance (scores) of these sub-sequences. As also pointed out by other authors (see e.g. [6, 11]), considering sub-sequences of terms (i.e., sentences) shared by two texts is in general more informative than considering unordered sets of terms, and it allows better estimation of the strength of the connections between the two texts. Also, sentences can be used to compute more effective labels for clusters.

In order to determine clusters in the snippet graph we compute *communities of vertices*, i.e. sets of vertices that are strongly connected from a topological point of view. We adopt a recursive divisive strategy based on graph connectivity (see, e.g. [2, 7, 8]): The clusters hierarchy is determined by recursively cutting some edges that disconnect the graph; the algorithms based on this strategy mainly differ for the criteria used to choose the next edge to be removed. Our algorithm is a variation of an elegant technique recently proposed by Brinkmeier [2] (see [5] for details). We remark here that for any pair of categories A and B in the computed hierarchy, we have that either A and B are disjoint or one of them includes the other. In fact, in order to simplify the information returned to the user, we aim at assigning each URL to its most representative category, and possibly relationships between disjoint categories will be highlighted using the new visualization paradigm.

To label a category μ of our hierarchy tree, we apply a procedure that takes advantage of the new definition of the snippet graph. Namely, let $G_\mu = (V_\mu, E_\mu)$ be the subgraph of G induced by the vertices of G contained in μ, and let $\mathcal{S}_\mu$ be the set of all labels of the edges of G_μ, that is $\mathcal{S}_\mu = \cup_{e \in E_\mu} \mathcal{S}_e$. If $\sigma \in \mathcal{S}_\mu$ occurs in the label of k edges ($k > 0$) of G_μ, then $k w_\sigma$ is called the *total score of* σ in G_μ. Denoted by $\sigma = s_1, s_2, \ldots, s_h$ ($h > 0$) any sentence of S_μ with maximum total score, the label of μ is defined as the concatenation of the (spaced) words $z_{s_1}, z_{s_2}, \ldots, z_{s_h}$.

Let G be the snippet graph of a set of URLs, and $H(G)$ its category tree. For any internal node μ of $H(G)$, we denote by G_μ the subgraph of G induced by the leaves of the subtree rooted at μ. The drawing approach works in three steps.

Edge sparsification. A preprocessing step is applied on G to remove those edges that do not provide useful information to the structure of the clusters. For each internal node μ of $H(G)$, we remove from G_μ every edge whose label only consists of sentences that, in their unstemmed version, are completely contained in the label of μ. Denote by G' the graph obtained from G by applying this cleaning operation over all sub-graphs G_μ. We maintain on G' the same clusters hierarchy as G, so that $H(G') = H(G)$ in the following.

Cluster relationships computation. We transform graph G' into a new *super clustered graph* G'' such that each cluster μ is explicitly represented as a kind of *super node*, which we denote by C_μ. Super nodes can be connected by *super edges* to emphasize the relationships among clusters. Super edges can be seen as extra arcs connecting nodes of $H(G')$. More formally (see also Figures 3(a) and 3(b)), let μ_1 and μ_2 be any two internal nodes of $H(G')$ that are children of the same node μ, and let $E(\mu_1, \mu_2)$ be the subset of edges of G' that connect vertices of G'_{μ_1} to vertices of G'_{μ_2}. If $E(\mu_1, \mu_2)$ is not empty, we replace this set of edges by a new super edge e connecting the super nodes C_{μ_1} and C_{μ_2}. The label of e is the union of all labels of the

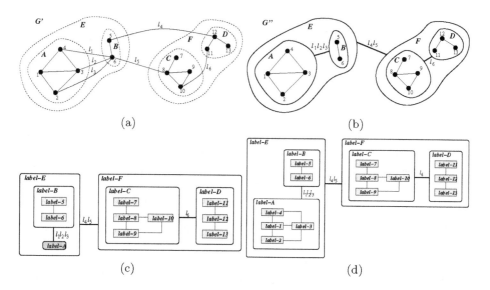

Fig. 3. (a) A clustered graph G' with no redundant edges; the edges connecting vertices in distinct clusters are labelled. (b) The graph G'' obtained from G' by representing clusters with super nodes. (c) A drawing of G''; the string *label-x* denotes the label of object (cluster or vertex) x; cluster A is contracted. (d) A drawing of G'' where all clusters are expanded.

edges in $E(\mu_1, \mu_2)$, and the weight of e is the sum of all weights of the edges in $E(\mu_1, \mu_2)$.

Maps Visualization. Let μ be a cluster of G''. The graph consisting of the children of μ and their relationships is represented as an orthogonal drawing with box-vertices; each vertex is drawn as a box with the size required to host its label or its subclusters, depending on the fact that it is contracted or expanded. Figures 3(c) and 3(d) show two different maps of the clustered graph G'' in Figure 3(b). Details about the algorithms used to compute the drawings are given in Section 4.

3   User Interface and Graph Visualization Functionalities

The user interacts with the system WhatsOnWeb by means of a Web browser and a Java applet. The graphic environment consists of two frames (see, e.g., Figure 4). In the left frame the clusters hierarchy is presented to the user as a classical directories tree. In the right frame, the user interacts with a map that gives a graphical view of the clusters hierarchy at the desired level of abstraction. Initially, the map shows only the first level categories and their relationships, according to the principles given in the previous section. The drawing is computed using an orthogonal drawing style, where each category is represented as a box with prescribed size. The size of each box is chosen as the minimum required to host the label of the category and/or a drawing of its subclasters. The user can decide to expand any of these categories, by simply clicking on the corresponding box, and so on recursively. Expanded categories can also be contracted by the user successively. Figure 4 shows a snapshot of the interface, where the results for the query "Armstrong" are presented; in the figure, the category "Louis Armstrong" has been expanded by the user. During the browsing, the system automatically keeps consistent the map and the tree, i.e., if a category is expanded in the map, it also appears expanded in the tree, and vice-versa. Also, in order to preserve the user mental map, WhatsOnWeb preserves the orthogonal shape of the drawing during any expansion or contraction operation. For example, Figure 5(a) shows the map obtained by expanding the categories "Jazz", "School", and "Louis Armstrong Stamp" in the map of Figure 4.

Besides the browsing functionalities above described, the interface is equipped with several facilities that increase the expressiveness of the output and the interaction between the system and the user. They are listed and described below:

– **Ranking of categories.** WhatsOnWeb ranks the semantic categories at each level, by assigning to each category a *relevance score*. This score is proportional to these three main parameters: (a) the size of the category, (b) the maximum rank (returned by Google) of a document in the category, and (c) the deviation of the score for the label of the category from the average score of all sentences in the same category. We use this third parameter as an estimate of the reliability of the sentence used to label the category. In the tree

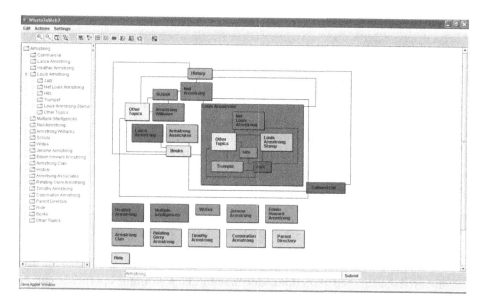

Fig. 4. A map for the query "Armstrong"; in the map the category "Louis Armstrong" has been expanded by the user

representation, the categories are ordered from top to bottom according to their decreasing relevance score. In the map, the relevance score is conveyed to the user adopting different colors in the red scale; highly red categories are the most relevant ones, while yellow categories are the least relevant. For example, in Figure 4, the expanded category "Louis Armstrong" is supposed to be one of the most relevant categories at the first level. Conversely, the category "Ride" has a low relevance score for the system, and its label does

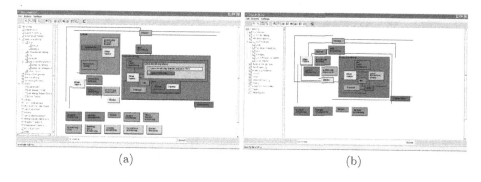

(a)   (b)

Fig. 5. (a) A map obtained from the one of Figure 4 by expanding the categories "Jazz", "School", and "Louis Armstrong Stamp"; this last category contains two URLs, labeled with their titles. (b) Another map where isolated categories of little relevance have been automatically hidden.

not appear meaningful; this category is even completely unrelated with every other category.

– **Automatic filtering.** As the categories have a relevance score, also each relationship between two categories has an assigned weight that measures the strength of that relationship (see Section 2). The user can customize a visibility threshold both for the categories and for their relationships. The system automatically hides those edges whose weight is less than the given threshold, and it hides all categories that are isolated and whose score is less than the specified threshold. For example, in Figure 5(b) several isolated categories of the map in Figure 4 have been automatically hidden by the system, after that the user has increased the visibility threshold for the categories.

– **Pruning of categories.** The user can select a desired subset of categories in which she is no longer interested. After this selection, the system can be forced to only recompute the drawing, without changing the content of the remaining clusters, or it can be forced to also recompute the whole clusters hierarchy using the remaining documents. This pruning operation can be used to progressively reduce the amount of information that the user handles, and to refine the remaining information if necessary. For example, Figure 6 shows the result of a pruning of the categories "Neil Armstrong", "School", and "Armstrong Williams", starting from the map of Figure 5(b). After the pruning only the drawing is recomputed, in order to optimize the occupied area. In the map of Figure 6, three further categories are selected by the

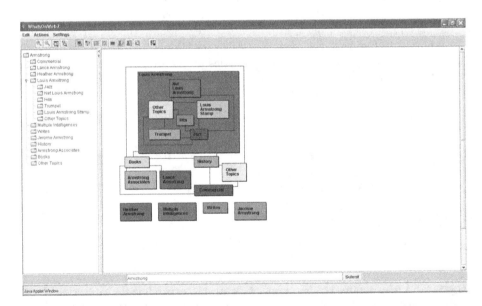

Fig. 6. The categories "Neil Armstrong", "School", and "Armstrong Williams" have been pruned from the map of Figure 5(b), and only the drawing has been recomputed

user to be pruned; in this case the user requires to completely recompute the clusters hierarchy after the pruning operation. The result is depicted in Figure 7; observe how the set of clusters is changed.

- **Edge exploration.** The user can explore the information associated with the edges of the map. If the mouse is positioned on an edge (u, v), a tool-tip is displayed that shows both the weight and the complete list of strings that form the labels of the relationship between u and v. Clicking on the edge,

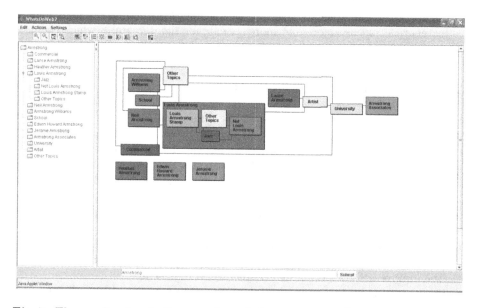

Fig. 7. The result of a further pruning of the categories "Hits", "Trumpet" and "Writes", where also the clustering is recomputed

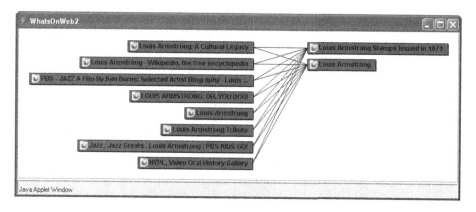

Fig. 8. A 2-layered drawing that shows the relationships between the URLs of "Louis Armstrong Stamp" and those of "Jazz"

a 2-layered drawing is shown to the user, that displays the URLs in u, the URLs in v, and the relationships between these two sets of URLs, as they appear in the snippet graph (see Figure 8). This feature makes it possible to isolate all those documents that give rise to a specific relationship and can be also used as a tool to evaluate and tune the mechanism adopted by the system for creating cluster relationships.

– **Query refinement.** The user can select one or more categories and can ask the system to repeat the query by adding to the original query string the labels of the selected categories. In this way it is possible to obtain semantic categories that are more and more specific for the domain in which the user is interested.

4   Implementation and Graph Drawing Engine

The GUI environment and the client/server communication layer of `WhatsOnWeb` have been implemented using Java technology. From the client side we defined a Java Applet which interacts with the user and forwards the requests to the server. From the server side we defined several Java Servlets; one of these Servlets interacts with a graph drawing engine, which implements engineered versions of orthogonal graph drawing algorithms and which uses the facilities of the GDToolkit library [4]. During the browsing of the user, the map displayed on the screen changes dynamically, depending on the expansion or contraction operations required by the user. `WhatsOnWeb` has two main goals each time a new drawing must be computed: (i) The preservation of the user mental map, and (ii) the minimization of the area occupied by the drawing.

The first goal is achieved by computing an orthogonal shape for each subgraph of the super clustered graph G'' (see Subsection)and by always using the same shape each time a drawing of the subgraph must be displayed. To compute this shape the system first planarizes the subgraph and then uses the *simple Kandisky* model and the flow-based algorithm described in [1].

The second goal is achieved by determining the size of each super node in the map, and by applying the effective compaction algorithm for orthogonal drawings with vertices of prescribed size described in [3]. The compaction algorithm is recursively applied to all those orthogonal shapes (computed in the previous step) that must be visualized. More in detail, during the browsing of the user each cluster μ maintains a state that informs if it is expanded or contracted. If μ is contracted, it is represented as a "small" rectangle containing its label; all its subclusters are hidden. If μ is expanded, its super node C_μ is drawn as a rectangular region r_μ that contains the drawings of its subclusters and the label of μ. The dimensions of r_μ depend on the state of the subclusters of μ, which can be expanded or not. To determine the dimensions of r_μ we apply a procedure that recursively constructs a drawing Γ of the subclusters of μ and then sets the height and the width of r_μ as the height and the width of the bounding box of Γ, plus a small area needed to place the label of μ.

---

[4] http://www.dia.uniroma3.it/~gdt

Finally, we implemented a standard Sugyiama algorithm (see, e.g., [4]) in order to construct the 2-layered drawings that show the relationships between the URLs of two distinct clusters.

References

1. P. Bertolazzi, G. D. Battista, and W. Didimo. Computing orthogonal drawings with the minimum number of bends. *IEEE Trans. on Comp.*, 49(8):826–840, 2000.
2. M. Brinkmeier. Communities in graphs. In *Innovative Internet Community Systems (IICS'03)*, volume 2877 of *LNCS*, 2003.
3. G. Di Battista, W. Didimo, M. Patrignani, and M. Pizzonia. Orthogonal and quasi-upward drawings with vertices of prescribed sizes. In *Graph Drawing (GD '99)*, volume 1731 of *LNCS.*, pages 297–310, 1999.
4. G. Di Battista, P. Eades, R. Tamassia, and I. G. Tollis. *Graph Drawing*. Prentice Hall, Upper Saddle River, NJ, 1999.
5. E. Di Giacomo, W. Didimo, L. Grilli, and G. Liotta. A topology-driven approach to the design of web meta-search clustering engines. In *SOFSEM '05*, volume 3381 of *LNCS.*, pages 106–116, 2005.
6. P. Ferragina and A. Gulli. The anatomy of a clustering engine for web-page snippet. In *The Fourth IEEE International Conference on Data Mining (ICDM'04)*, 2004.
7. E. Hartuv and R. Shamir. A clustering algorithm based on graph connectivity. *Information Processing Letters*, 76:175–181, 2000.
8. M. E. J. Newman and M. Girvan. Finding and evaluating community structure in networks. *Phys. Rev. E 69*, 2004.
9. G. Salton. *Automatic Text Processing. The Transformation, Analysis, and Retrieval of Information by Computer*. Addison-Wesley, 1989.
10. O. Zamir and O. Etzioni. Web document clustering: A feasibility demonstration. In *Research and Development in Information Retrieval*, pages 46–54, 1998.
11. O. Zamir and O. Etzioni. Grouper: a dynamic clustering interface to web search results. *Computer Networks*, 31(11-16):1361–1374, 1999.

Drawing Clustered Graphs in Three Dimensions

Joshua Ho[1,2] and Seok-Hee Hong[1,2]

[1] IMAGEN Program, NICTA (National ICT Australia)
[2] School of IT, University of Sydney, NSW, Australia
joshua.ho@student.usyd.edu.au, seokhee.hong@nicta.com.au

Abstract. Clustered graph is a very useful model for drawing large and complex networks. This paper presents a new method for drawing clustered graphs in three dimensions. The method uses a *divide and conquer* approach. More specifically, it draws each cluster in a 2D plane to minimise occlusion and ease navigation. Then a 3D drawing of the whole graph is constructed by combining these 2D drawings.

Our main contribution is to develop three linear time weighted tree drawing algorithms in three dimensions for clustered graph layout. Further, we have implemented a series of six different layouts for clustered graphs by combining three 3D tree layouts and two 2D graph layouts. The experimental results with metabolic pathways show that our method can produce a nice drawing of a clustered graph which clearly shows visual separation of the clusters, as well as highlighting the relationships between the clusters. Sample drawings are available from http://www.cs.usyd.edu.au/~visual/valacon/gallery/C3D/

1 Introduction

Recent advances in technology have led to many large and complex network models in many domains such as webgraphs, social networks and biological networks. Visualisation can be an effective analysis tool for such networks. Scalability, however, is the most challenging issue, as they may have millions of nodes.

Graph clustering is one of the most efficient approaches to solve the scalability problem. Good clustering methods can identify clusters and the relationships between clusters. Further, many real-world networks have an underlying clustered graph topology.

Recent technological advances in computer graphics hardware made high quality 3D graphics affordable. Further, HCI researchers have established that 3D visualisation can be helpful for giving new insights into abstract data by amplifying human cognition [13].

Clustered graphs have been introduced to both the graph drawing and information visualisation communities. Methods for visualising clustered graphs in two and three dimensions have been developed [6, 7]. Good drawings of clustered graphs should visually separate the clusters effectively as well as reveal the inter-cluster relationships. However, it seems that existing methods fail to satisfy at least one of the following criteria for drawing clustered graphs in 3D:

P. Healy and N.S. Nikolov (Eds.): GD 2005, LNCS 3843, pp. 492–502, 2005.

- minimum number of edge crossings between intra-cluster edges
- minimum volume of the drawing
- minimum sum of inter-cluster edge length
- minimum occlusion views of the drawing
- no overlap between the drawing area of each cluster
- easy navigation

In this paper, we present a new method for drawing clustered graphs in three dimensions. Our work concentrates on layout of flat clustered graphs [7], so we only have to consider one level of clustering. Although this clustered graph model is less general than the common clustered graph model [6], it appears to be a useful model in some real-world applications, like visualization of biological networks. Our proposed method draws each cluster in a 2D plane using an existing 2D drawing algorithm to minimise occlusion and facilitate ease of navigation. Then a 3D drawing of the whole clustered graph is constructed by combining these 2D drawings. For this purpose, we designed three linear time weighted tree drawing algorithms in three dimensions.

Further, we have implemented a series of six different layouts for clustered graphs by combining three 3D tree layouts and two 2D graph layouts. The experimental results with metabolic pathways suggest that it is useful for the visual analysis of large and complex networks.

This separation of dimensionality can help to achieve some of the criteria for clustered graph drawing. In our divide and conquer algorithm, the problem of drawing each cluster in 2D and the problem of arranging each cluster in 3D are addressed separately. However, in order to ensure the overall aesthetics of the drawing of the clustered graph, a post-combination step, called inter-cluster occlusion minimisation, is devised.

Our method also follows Ware's guideline, a $2\frac{1}{2}$ design attitude that uses 3D depth selectively and pays special attention to 2D layout can provide the best match with the limited 3D capabilities of the human visual system" [14].

This paper is organized as follows: the main results of the paper are presented in Section 2. Here we describe a new method for drawing clustered graphs in three dimensions. In particular, three 3D tree drawing algorithms are presented for clustered graph layout. Section 3 presents experimental results and Section 4 concludes.

2   Algorithms for Drawing Clustered Graphs in 3D

First we define our clustered graph model and terminology. We use a flat clustered graph model. That is, we have a set of clusters, $G_1, G_2, \ldots, G_k$ with $G_i = (V_i, E_i)$. Further, we define a supergraph GG such that each G_i is a node v_i in GG, and if there is an edge between a node in G_i and a node in G_j, then there is an edge between v_i and v_j. Note that we can define a weight to each node and edge in GG depending on the size of the cluster and the number of edges between the clusters.

Our algorithm draws a clustered graph using the following four steps:

Algorithm 3D_Clustered_Graph_Drawing
1. Draw each cluster in 2D using a 2D drawing algorithm.
2. Draw the weighted supergraph in 3D.
3. Merge the drawings from Step 1 and 2 to construct a 3D drawing of the given clustered graph.
4. Apply inter-cluster occlusion minimisation procedure.

The first step of the algorithm is to draw each cluster in a 2D plane using a 2D graph drawing algorithm. Each node in the cluster is assigned a coordinate in this step. Then the size of each cluster, i.e. the drawing area of the cluster, is computed and assigned as a weight to the corresponding supernode.

In the second step, the layout of the weighted supergraph is computed. As the supernodes have different sizes, a weighted graph layout algorithm is required for this step.

The third step combines the drawings from step 1 and step 2 to construct a 3D drawing of the whole graph. This step is to transform the coordinates of each node to its final position using the coordinates of the corresponding supernode.

The last step of the algorithm is called the inter-cluster occlusion minimisation (ICOM) step. This step addresses the occlusion problem caused by the insertion of inter-cluster edges. We now describe the details of each step.

2.1   Drawing Each Cluster in 2D

The first step is to draw each cluster using any 2D drawing algorithm. One can choose a method depending on the application [5].

Once each cluster is drawn, its size is calculated. More specifically, the size of a cluster is defined by the radius of an enclosing circle of the drawing area.

2.2   Drawing the Supergraph in 3D

The second step is to compute the layout of the supergraph in 3D. In general, any layout algorithm that can draw graphs with different node sizes can be used. In this paper, we mainly concentrate on drawing the supergraph in 3D, where it has a tree structure.

There are two main reasons for focusing on 3D tree drawing methods. First, in general tree layout algorithms are simple and run in linear time. Second, many real life networks resemble tree-like structures, and sometimes visualising the spanning tree of a general graph can be desirable.

In order to accomodate nodes with different sizes, we have designed three linear time 3D tree drawing algorithms, based on the cone tree algorithm [12]: (a) weighted cone tree drawing algorithm (b) weighted rod tree drawing algorithm, and (c) weighted free tree drawing algorithm. We now describe the details of the algorithms.

Weighted Cone Tree Layout. A cone tree [12] is normally computed by a two-pass algorithm: the first pass computes the cone radius by a post-order tree

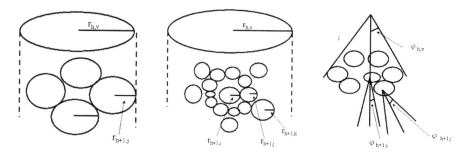

Fig. 1. (a) cone tree (b) rod tree (c) free tree

traversal. In the second pass, a pre-order traversal is used to assign coordinates to each node. For our purpose of drawing weighted trees, we mainly modified the first step.

In a cone tree, each non-leaf node is considered as an apex of a circular cone with cone radius. Let $r_{h,v}$ be the cone radius of the node v at height h. In general, $r_{h,v}$ can be computed from the cone radius $r_{h+1,j}$ of its child nodes j at height $h+1$ using the method described in [4]. However, for our purpose, we need another variable $rad_{h,v}$ to represent the radius (that is weight or size) of the node. In our weighted version of the modified cone tree algorithm, the $r_{h,v}$ of each node is computed as follows (see Figure 1(a)):

$$r_{h,v} = max\{\frac{\sum_j r_{h+1,j}}{\pi} + max_j(r_{h+1,j}), rad_{h,v}\} \qquad (1)$$

Weighted Rod Tree Layout. The cone tree algorithm draws all child nodes on the boundary of the circular base of a cone. There is a drawback of this model: a drawing of a highly unbalanced tree may require a large volume. Further, in real world applications, some subtrees may have different levels of importance and it may be desirable to emphasize those subtrees.

In this section, we describe a variation of the cone tree algorithm, called *rod tree*, which chooses one subtree and places it along the z-axis. A similar idea has been explored in the context of visualisation of state transition graphs [9] or symmetric tree drawing [11].

In our rod tree algorithm, we choose a child subtree of maximum height as the center node and then place it on the z-axis. Then, its zero-height siblings form a inner circle to surround the center node. Finally, all the other non-zero height siblings are placed in the outer circle. Therefore, each cone contains at least one child subtree on the axis of the cone, surrounded by at most two concentric circles.

In a rod tree, the child nodes of node v at level h can be divided into three groups: the center node i with the maximum height, a set of nodes j with zero-height, and a set of nodes k with non-zero height (see Figure 1(b)). The cone radius $r_{h,v}$ of node v can be computed as:

$$r_{h,v} = max\{base\_radius(h,v), rad_{h,v}\} \qquad (2)$$

The *base_radius* function returns the cone radius. It is computed by first considering the radius of the node i of maximum height, as the inner circle and the circle formed by the zero-height nodes j as outer circle. The radius of the inner circle rad_{inner} is defined as $r_{h+1,i}$. The radius of the outer circle rad_{outer} can be calculated using the normal cone tree method.

Note that the inner and outer circle overlap if $rad_{inner} > rad_{outer} - 2max$ $(r_{h+1,j})$. In this case, we can remove overlapping by increasing the radius of the outer circle. Therefore, rad_{outer} is assigned as $rad_{inner} + 2max(r_{h+1,j})$. The nodes k are then then packed around nodes j. The positions of nodes k are calculated in the same manner as positioning nodes j around node i.

The coordinate assignment step in the rod tree algorithm is similar to the one in the cone tree algorithm.

Weighted Free Tree Layout. Both cone tree and rod tree aim to draw a rooted tree where the hierarchical relationship is important. However, not all clustered graphs have a hierarchical relationship. To cover this case, we now present a 3D free tree drawing algorithm that considers nodes of different sizes.

The main idea is to divide the tree into two subtrees, then draw each subtree in a hemisphere. For each subtree, the layout is computed by placing the child node with maximum height along the axis of the parent node, surrounded by all the other siblings on a spherical surface.

The method for computing the size of a cone is quite different from the one for cone tree and rod tree. In a free tree, each non-leaf node v at level h is an apex of the *spherical* cone. The size of the cone is computed by the angle $\varphi_{h,v}$, an angle between its main axis and the conic surface (see Figure 1(c)).

Let i represent the center node and j represent all other child nodes. Let l be the edge length between adjacent node. The angle $\varphi_{h,v}$ of a leaf node is defined as:

$$\varphi_{h,v} = tan(\frac{2rad_{h,v}}{l}) \tag{3}$$

For all non-leaf nodes, $\varphi_{h,v}$ can be computed as:

$$\varphi_{h,v} = max\{contribution(h+1, i), 2max_j[contribution(h+1, j)]\} \tag{4}$$

where the function contribution() returns a value that contributes to the computation of $\varphi_{h,v}$:

$$contribution(a, b) = max\{arctan\frac{rad_{a,b}}{l}, arctan\frac{tan\,\varphi_{a,b}}{2}\} \tag{5}$$

Note that the previous steps pack the spherical cones as tightly as possible. In general, a good layout uses the space evenly, therefore a *scaling* procedure is applied. Let $\varphi_{h,available}$ be the total amount of angle available to draw the tree at level h. At each level, $\varphi_{h,v}$ is scaled according to the function:

$$\varphi_{h,v} = \varphi_{h,available} \times \frac{\varphi_{h,v}}{\sum_v \varphi_{h,v}} \tag{6}$$

At the next level $\varphi_{h+1,available} = \varphi_{h,v}$.

2.3   Merging the 2D Drawings into a 3D Drawing

This step combines all drawings of the clusters together according to the supergraph layout. The drawings are transformed to the position specified by the corresponding supernode's position. We used the following two simple methods.

The first method is a simple translation. Since all clusters are initially drawn on the xy plane, all the planes are placed in parallel along the z-axis in 3D.

The second method is a combination of translation and rotation. Each drawing is first translated to the corresponding position, and then rotated towards the center of the drawing. This method places each plane parallel to each other along a concentric sphere.

2.4   The Inter-cluster Occlusion Minimisation Step

The last step of our algorithm considers the placement of inter-cluster edges in order to minimise occlusion and the sum of inter-cluster edge length. A method for Inter-Cluster Occlusion Minimisation (ICOM) is designed for drawing a clustered graph using a 2D spring algorithm for drawing each cluster and one of the 3D tree drawing algorithms. The main idea is to place the nodes connecting an inter-cluster edge close to the boundary of the cluster, as close to the adjacent cluster as possible.

More specifically, our method is based on the spring algorithm with specialized forces. These forces include spring force, repulsion, planar force, ICOM force, and boundary force. The ICOM force can be seen as the spring force for inter-cluster edges. It only pulls nodes that connect to inter-cluster edges along the 2D plane towards the adjacent cluster.

The ICOM force for each node $u \in V$ is defined as follows:

$$f_{icom}(u) = I \sum_{(v \in N)} [\|p_v - p_u\| + refl(\|p_v - p_u\|)] \tag{7}$$

where I is the ICOM force constant and N is the set of nodes connected to u via an inter-cluster edge. The function $refl()$ is a reflection function that returns a vector, a reflection about the cluster plane. Therefore, $\|p_v - p_u\| + refl(\|p_v - p_u\|)$ is a vector parallel to the cluster plane that points to the direction of a neighbour node in the adjacent cluster.

In order to prevent this ICOM force from indefinitely enlarging the cluster, a boundary force is added. The boundary force sets a circular boundary for the cluster. Inside the boundary, a node can move freely. If a node goes over the boundary, a strong force is applied to pull the node back. More specifically, the boundary force for each node $u \in V$ is defined as follows. If $\|p_u\| > bound$, then:

$$f_{boundary}(u) = -Bp_u \tag{8}$$

where p_u is defined as the position of node u. If u is out of the circular boundary of radius bound, a force of $-Bp_u$ is applied to pull the node back inside the boundary. B is a positive boundary force constant.

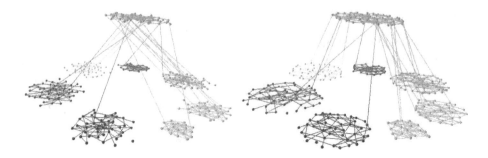

Fig. 2. Comparison of clustered graph layout (a) without ICOM (b) with ICOM

For example, see Figure 2. A randomly generated clustered graph of 329 nodes, 674 edges and 8 clusters is drawn with the weighted cone tree and spring algorithm. ICOM allows a node that is connected to another node in an adjacent cluster to be drawn closer along its own 2D plane. This reduces the visual complexity of the whole drawing by less occlusion and shorter inter-cluster edges. Since the boundary force is applied to restrict the size of the cluster, all resulting clusters are circular in shape. The normal spring, repulsion and planar force allow each cluster to optimise the aesthetic criteria.

2.5   Extension to the General Case

Not all clustered graphs have a supergraph of tree structure. In this case, the maximum spanning tree of the supergraph can be used. Since the weight of a superedge is the actual number of edges between the two clusteres it connect to, the use of a maximum spanning tree ensures most of the inter-cluster edges are drawn between nodes on different tree levels.

3   Implementation and Experimental Results

Our algorithm has been implemented as a set of plug-in modules for GEOMI, a visual analysis tool for large and complex networks [1]. More specifically, we have implemented a series of six different layouts for clustered graphs by combining three 3D tree layouts (weighted cone tree, weighted rod tree and weighted free tree) and two 2D graph layouts (force directed layout and circular layout).

For data sets, we have used both randomly generated data sets and real world data sets. For this purpose, we implement a generator to randomly generate clustered trees, see [1].

Figure 3, 4, 5 shows three different layouts of clustered graphs. From the drawing, it is easy to identify each cluster separately. Further, it is easy to understand the relationship between clusters.

For real world networks, we used metabolic networks. A metabolic network is a collection of chemical reactions in cells. Visualisation of these networks can help identify key features and possible malfunctions of cells. The whole metabolic

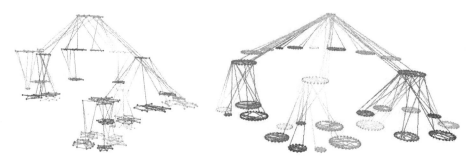

Fig. 3. Cone tree layout (a) a tree with 341 vertices, 1269 edges and 30 clusters drawn with spring layout (b) a tree with 616 vertices, 1636 edges and 31 clusters drawn with circular layout

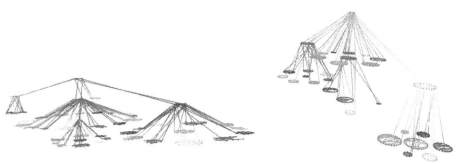

Fig. 4. Rod tree layout (a) a tree with 620 vertices, 1769 edges and 47 clusters drawn with spring layout (b) a tree with 616 vertices, 1636 edges and 47 clusters drawn with circular layout

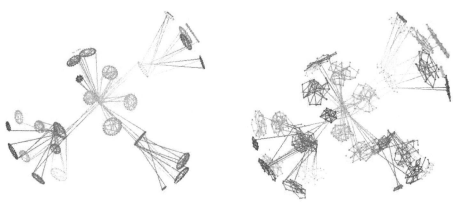

Fig. 5. Free tree layout a tree with 616 vertices, 1636 edges and 31 clusters drawn with (a) circular layout (b) spring layout

network can be divided into many functional units called pathways. Different pathways are connected by sharing the same chemicals. Existing approaches either focus on only visualising individual pathways or the entire network [2, 3, 8].

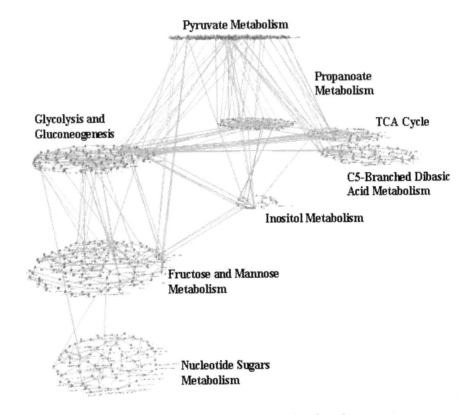

Fig. 6. Metabolic Network of eight related pathways

Clearly, these approaches are not sufficient in conveying the biological functionality of the network. A visualisation method that can effectively display the whole network as well as highlighting the functional independence of the individual pathway is needed.

In order to address this problem, we use the cluster graph model to visualise the metabolic network. Chemicals are represented by nodes and reactions are represented by edges. Each pathway is a cluster, and the whole metabolic network forms a clustered graph. Sharing of chemicals between pathways is represented by inter-cluster edges.

In our study, we use metabolic network data retrieved from the KEGG database (http://www.genome.ad.jp/kegg/). GML [10] files containing individual pathways are read from the Clustered Graph Reader in GEOMI, then the whole network is constructed by adding inter-cluster edges between the same chemicals on different pathways. Since the supergraph of the metabolic network is not a tree, a maximum spanning tree of the original supergraph is used. The supernode with the highest degree centrality is chosen to be the root of the spanning tree.

Figure 6 shows a metabolic network of eight related carbohydrate metabolic pathways. The graph contains 600 nodes and 745 edges. Overall, our approach in visualising this metabolic network has a number of advantages. Firstly, the functional independence of each pathway is emphasized. Since each pathway is drawn on a different 2D plane, individual pathways can be readily identified, while retaining the overview of how the pathways are connected in the entire network in 3D.

Next, the *central* pathway in the network is identified. For example, both pyruvate metabolism and the glycolysis pathways are connected to most of the other pathways in the network. This means that they may have a significant biological role as the removal of these pathways would essentially disconnect the network. On the other hand, the Nucleotide Sugars Metabolism pathway is quite isolated from the rest of the network.

Thirdly, the relative size of the pathway is easily comparable. For example, the Inosital metabolism pathway is relatively small compared to the other seven similar-sized pathways.

Finally, the connectivity between adjacent pathways is effectively visualised. For example, glycolysis and gluconeogenesis share a lot of chemicals with the TCA cycle as shown by the large number of inter-cluster edges between them. This infers that these two pathways are probably biologically closely related.

4   Conclusion and Future Work

A new method for drawing clustered graph in 3D is presented. The divide-and-conquer algorithm draws each cluster separately in a 2D plane and then combines each drawing of a cluster to construct a drawing of the whole clustered graph in 3D. For this, we designed three linear time weighted tree drawing algorithms in 3D.

We have implemented a series of six different layouts for clustered graphs by combining three 3D tree layouts and two 2D graph layouts. The experimental results show that the resulting drawing can clearly display the structure of the cluster and the relationships between the clusters. Further the use of 2D plane ideas reduces the occlusion problem in 3D and the resulting drawing is easy to navigate.

In the future, different combination of cluster/supergraph layout algorithms can be explored. Application of this clustered graph drawing method can be further explored in the context of other network visualisation applications, for example social networks, communication networks and web-graphs. Further one can design efficient navigation methods for the clustered graph layout for user interaction.

Acknowledgement

The work was supported by a NICTA summer vacation scholarship and an Australian Research Council (ARC) grant. We also thank Dr. Falk Schreiber for providing the metabolic network dataset in GML format.

References

1. A. Ahmed, T. Dwyer, M. Forster, X. Fu, J. Ho, S. Hong, D. Koschützki, C. Murray, N. Nikolov, A. Tarassov, R. Taib and K. Xu, GEOMI: GEOmetry for MAximum Insight, Proceedings of Graph Drawing 2005, 2005.
2. F. Brandenburg, M. Forster, A. Pick, M. Raitner and F. Schreiber, BioPath, Proceedings of Graph Drawing 2001, pp. 455-456, 2001.
3. U. Brandes, T. Dwyer, and F. Schreiber, Visualization of Related Metabolic Pathways in Two and a Half Dimensions, Proceedings of Graph Drawing 2003, pp.111-121, 2003.
4. J. Carriere and R. Kazman, Visualization of Huge Hierarchies: Beyond Cone Trees, Proceedings of IEEE Symposium on Information Visualization 1995, pp. 74-81, 1995.
5. G. Di Battista, P. Eades, R. Tamassia and I. G. Tollis, *Graph Drawing: Algorithms for the Visualization of Graphs*, Prentice-Hall, 1998.
6. P. Eades and Q. Feng, Multilevel Visualization of Clustered Graphs, Proceedings of Graph Drawing 1996, pp. 101-112, 1996.
7. Y. Frishman and A. Tal, Dynamic Drawing of Clustered Graphs, Proceedings of IEEE Symposium on Information Visualization 2004, pp. 191-198, 2004.
8. B. Genc and U. Dogrusoz, A Constrained, Force-Directed Layout Algorithm for Biological Pathways, Proceedings of Graph Drawing 2003, pp. 314-319, 2003.
9. F. van Ham, H. van de Wetering and J. can Wijk, Visualization of State Transition Graph, Proceedings of IEEE Symposium on Information Visualization 2001, pp. 59-63, 2003.
10. M. Himsolt, GML: Graph Modelling Language. Report, University of Passau, Germany, December 1996. http://www.uni-passau.de/
11. S. Hong and P. Eades, Drawing Trees Symmetrically in Three Dimensions, Algorithmica, 36(2), pp. 153-178, 2003.
12. G. Robertson, J. Mackinkay, and S. Card, Cone Trees: Animated 3D Visualizations of Hierarchical Information, Proceedings of CHI'91, pp. 189-194, 1991.
13. C. Ware and G. Franck, Viewing a Graph in a Virtual Reality Display is Three Times as Good as a 2-D Diagram, IEEE Conference on Visual Languages, pp. 182183, 1994.
14. C. Ware, Designing with a 2 1/2D Attitude, Information Design Journal 10 (3), pp. 171-182, 2001.

BLer: A Boundary Labeller for Technical Drawings*

Michael A. Bekos and Antonios Symvonis

National Technical University of Athens,
School of Applied Mathematical & Physical Sciences,
15780 Zografou, Athens, Greece
{mikebekos, symvonis}@math.ntua.gr

Abstract. BLer is a prototype tool aiming to automate the boundary labelling process [1]. It targets the area of technical and medical drawings, where it is often common to explain certain features of the drawing by blocks of text that are arranged on its boundary.

1 Introduction

In technical drawings and medical atlases, it is quite common to place the labels on the boundary of the drawing and to connect them to the features they describe by non-crossing poly-lines (*leaders*). To the best of our knowledge, no drawing software includes support for automated placement of labels on (or near) the boundary of the drawing. BLer is a prototype tool that supports the boundary labelling model and facilitates the annotation of drawings with text labels. It is suitable for the production of medical atlases and technical drawings, where the basic requirement is large labels. It is entirely written in Java based on the yFiles class library (http://www.yworks.com).

2 The Labelling Process

BLer enables the user to quickly generate boundary labellings from scratch. Its environment (see Figure 1) supports multiple views of the labelling and directs the user through the steps of the labelling process (*diagram loading, definition of enclosing rectangle, definition of point features, production of boundary labelling*).

The labellings produced by the current version are based on algorithms developed by Bekos et. al. [1]. They minimize the total leader length and contain no crossings. The resulting drawings are simple, in terms of readability, ambiguity and legibility.

BLer supports several labelling models. These include: the sides of the enclosing rectangle where labels can be placed, the type of the leaders (*opo, po*, or *s* [1]), the type of the *ports*.

---

* This work has been partially founded by the program "Pythagoras" which is co-funded by the European Social Fund (75%) and Greek National Resources (25%).

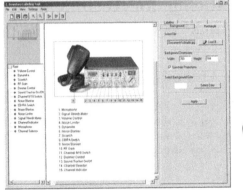

Fig. 1. A snapshot of Bler

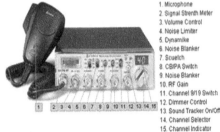

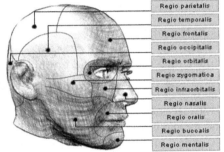

Fig. 2. A technical drawing of a cb radio

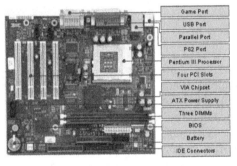

Fig. 3. A technical drawing of a mother-board

Fig. 4. A medical map of the regions of a human head

BLer can also be instructed to use a *legend* (see Figure 2), which is useful in cases where the labels (due to their size and number) do not all fit on the boundary of the enclosing rectangle. When a legend is used, labels of appropriate uniform size are used, each containing a number which refers to a particular line of the legend. The legend is treated as a floating object and can be manually placed anywhere around the resulting labelling.

As expected from a labelling tool, it supports storing, reloading and post-processing of labellings. It provides advanced graphic functionality, including popup menus, printing capabilities, custom-zoom, fit-in window, selection, dragging and resizing of objects.

Figures 2, 3 and 4 depict some characteristic output examples of BLer.

Reference

1. M. Bekos, M. Kaufmann, A. Symvonis, and A. Wolff. Boundary labeling: Models and efficient algorithms for rectangular maps. In Janos Pach, editor, *Proc. 12th Int. Symposium on Graph Drawing (GD'04)*, LNCS 3383, pp. 49-59, 2004.

D-Dupe: An Interactive Tool for Entity Resolution in Social Networks

Mustafa Bilgic, Louis Licamele, Lise Getoor, and Ben Shneiderman

Computer Science Department, University of Maryland,
College Park, MD 20742, USA
{mbilgic, licamele, getoor, ben}@cs.umd.edu

Abstract. Graphs describing real world data often contain duplicate entries for names, cities, or other entities. This paper presents D-Dupe, an interactive visualization tool designed to help users to discover and resolve duplicate nodes in a social network. Users can resolve the ambiguity by merging nodes, or by specifying that the nodes are in fact distinct. The entity resolution process is iterative; as pairs of nodes are merged, additional duplicates may become apparent.

1 Introduction

The typical assumption in network visualization is that the underlying data is clean and the nodes refer to distinct entities while edges represent unique relationships. However, this presumption is rarely true. Networks are extracted from databases which may contain errors and inconsistencies. As data collection increases, and the databases themselves are being extended through automatic extraction techniques, or through the combination of multiple sources, the duplicate entries become more common place. This is especially true when the duplicate entries are not identical but slight variations of one another like abbreviations.

Duplicates may lead to inappropriate conclusions when the underlying data is visualized. Consider the example of a citation graph in which nodes correspond to authors and the node sizes are drawn in proportional to the number of publications of each author. If an author's name had multiple spellings and each of them were treated as distinct, the true entity will be represented not as one large node but as many unrelated small nodes. With such data (and representation), our conclusions about which author is most prolific will be incorrect.

In many cases, using the underlying structure of the network helps in resolving duplicates. For instance, suppose a bibliographic dataset that consists of four author references: "James Smith," "John Smith," "J. Smith," and "Mary Ann." We are interested in determining if "J. Smith" refers to "James Smith," to "John Smith" or it is a distinct author. Given no other information, we have no choice but to guess. But, if we know that "Mary Ann" collaborates with "John Smith" and "J. Smith" we are more likely to believe that "J. Smith" and "John Smith" refer to the same author. D-Dupe utilizes this idea by making such underlying structures apparent to users.

P. Healy and N.S. Nikolov (Eds.): GD 2005, LNCS 3843, pp. 505–507, 2005.

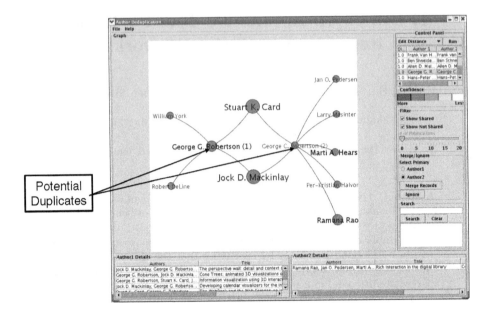

Fig. 1. The D-Dupe Interface

2   Overview

D-Dupe consists of three coordinated windows (Fig. 1). The left panel is the context collaboration graph, main controls are on the right panel, and the bottom panel displays the details on demand for the nodes.

Context Collaboration Graph (CCG): One novelty of D-Dupe is that the network visualization is tuned to the deduplication task. The CCG is the relevant subgraph of the whole network, where one duplicate pair and their immediate neighbors are displayed at a time. D-Dupe simplifies the graph further by showing only the edges between the possible duplicates and their neighbors. The potential duplicate pairs are colored in a shade of red in the tool; dark red pairs are more likely to be duplicates.

Control Panel: D-Dupe allows integration of a variety of machine learning algorithms for finding possible duplicates. After users select one of the algorithms, a table of potential duplicates is populated for users to inspect. Clicking on a potential duplicate pair will show the corresponding collaboration graph in the CCG. After inspecting the CCG, users can decide to either merge these duplicate records or disambiguate them by marking them as distinct entities. D-Dupe provides further graph filtering options such as filtering the authors based on number of publications.

Details on Demand: This panel displays descriptive information about the potential duplicate pairs; for instance in bibliographic domain, this panel displays the publications (the title, the date, the source, other authors, etc.) of the potential duplicate authors under inspection.

Acknowledgments. The work of Mustafa Bilgic, Louis Licamele, and Lise Getoor has been supported by the National Science Foundation, the National Geospatial Agency, and the UMD Joint Institute for Knowledge Discovery.

A New Method for Efficiently Generating Planar Graph Visibility Representations

John M. Boyer

IBM Victoria Software Lab, Victoria, BC Canada
boyerj@ca.ibm.com, jboyer@acm.org

1   Introduction

A planar graph *visibility representation* maps each vertex to a horizontal segment at a vertical position and each edge to a vertical segment at a horizontal position such that each edge segment terminates at the vertical positions of its endpoint vertices and intersects no other horizontal vertex segments. The first $O(n)$ algorithms for producing visibility representations were presented in [4, 5]. These were based on pre-processing to compute both an st-numbering and the dual of the planar graph, which were then used with the combinatorial planar embedding to produce a visibility representation. Greater efficiency is obtained in [3] by eliminating the need for the planar graph dual and by re-using the pre-computed st-numbering in the PQ-tree [1] algorithm.

Recently, the Boyer-Myrvold *edge addition* planarity method was presented [2]. The benefits relative to many prior methods, including simpler proof of correctness and $O(n)$ implementation, are due in part to eliminating the PQ-tree's st-numbering. Hence, a new approach was required in order to extend the efficiency and simplicity of edge addition planarity into the realm of generating visibility representations.

2   Computing Vertical Positions of Vertices

During the execution of the edge addition planarity algorithm, the vertices are assigned a relative position of 'between' or 'beyond' the depth first search (DFS) parent relative to some selected DFS ancestor. Each time a back edge is embedded, if its endpoints are in separate biconnected components of the partial planar embedding, then all components that become biconnected by the new edge are merged. Each edge along the external face that is incident to a merge point is marked so that when the edge is moved off of the external face (by embedding another back edge around it), the relative position of the merge point and one of its DFS children can be assigned. Figure 1 shows how the edge marks are made, and Figure 2 depicts how they are resolved into relative vertex placements.

The relative vertex placements assigned during planar embedding are converted in a post-processing step into a vertical *vertex order* using pre-order DFS tree traversal. When a vertex is visited, its ancestors have already been

P. Healy and N.S. Nikolov (Eds.): GD 2005, LNCS 3843, pp. 508–511, 2005.

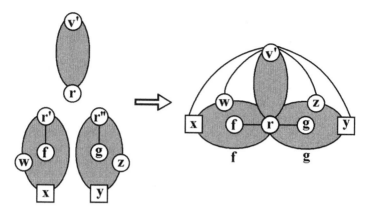

Fig. 1. When merging biconnected components, the external face edges incident to the merge points are marked with the identity of a DFS child. The children f and g are 'tied' with parent r in vertical placement until these marks are resolved.

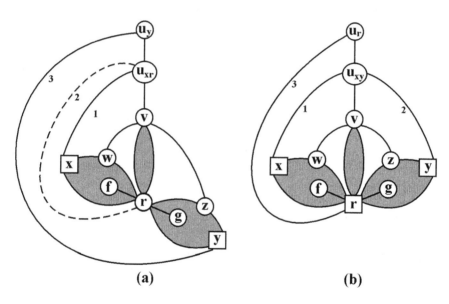

(a)   **(b)**

Fig. 2. (a) In a step u_x of the embedding, traversing from the descendant x of f to the parent r of f means that f is placed between the parent r and the ancestor u_x. Traversing from the parent r of g to the descendant y of g means that g is placed beyond parent r relative to ancestor u_x. (b) External activity at r can result in both children f and g being placed between r and some ancestor.

added to the vertex order. The localized information includes a marking of 'between' the DFS parent and a given ancestor or 'beyond' the DFS parent relative to the given ancestor. This is converted to be 'above' or 'below' the DFS parent, then the vertex is inserted immediately above or below its parent in the vertex order. More information is required to perform this conversion without resorting to non-linear time techniques like dynamic topological sorting. Each vertex v is positioned relative to its DFS parent p and an ancestor a, and both are added to the vertex order beforehand, but the relative positions of a and p in the vertex order are needed. Fortunately, we already store the placement of each vertex relative to its parent, and the placement of a vertex relative to its parent controls the placement of the entire DFS subtree rooted by that vertex relative to the DFS parent. Hence, the child c of a that roots the subtree containing p and v is stored during planar embedding when the relationship between v, p and a is made. Then, during this post-processing step, the relative order of p and a is obtained by query of the relative order of c and a.

3   Computing Horizontal Positions of Edges

A sweep algorithm is performed on the combinatorial planar embedding, using the vertical positions of the vertices to advance a horizontal sweep line, a data structure in which the **edge order** is developed. Also, each vertex keeps track of its **generator edge** in the edge order, which is just the first edge incident to the vertex that is added to the edge order. The generator edge provides an insertion point along the horizontal sweep line for the edges emanating from the vertex to the vertices that are below it (which have a greater vertex position number).

For starters, each edge e incident to the DFS tree root is added to the edge order according to the cyclic order in the embedding, and the generator edge of the child endpoint is set to e. For each vertex v below the DFS tree root in vertex order, we obtain the generator edge e as the starting point of the cyclic traversal of the adjacency list. The subset of edges emanating from v to vertices with greater vertex positions (i.e. below v) are added in cyclic order immediately after e. Also, for each such edge (v, w) that is added, if w has no generator edge, then (v, w) becomes the generator edge of w.

4   Conclusion

This research has yielded a new method for generating planar graph visibility representations. A linear-time reference implementation is available from the author based on the edge addition reference implementation that accompanies [2]. A pre-computed st-numbering was found to not be necessary, though the vertex ordering method produces an st-numbering as an output. It would be of theoretical interest to determine whether the notions of visibility representation and st-numbering could be completely decoupled, but first appearances suggest there would be little practical benefit as the sweep algorithm for edge ordering appears to become much more complicated with multiple source vertices. Future work would more easily find ways to compact the drawings via refinement of the algorithm presented in this paper.

References

1. K. S. Booth and G. S. Lueker. Testing for the consecutive ones property, interval graphs, and graph planarity using PQ–tree algorithms. *Journal of Computer and Systems Sciences*, 13:335–379, 1976.
2. J. Boyer and W. Myrvold. On the cutting edge: Simplified $O(n)$ planarity by edge addition. *Journal of Graph Algorithms and Applications*, 8(3):241–273, 2004.
3. R. Jayakumar, K. Thulasiraman, and M. N. S. Swamy. Planar embedding: Linear-time algorithms for vertex placement and edge ordering. *IEEE Transactions on Circuits and Systems*, 35(3):334–344, 1988.
4. P. Rosenstiehl and R. Tarjan. Rectilinear planar layouts and bipolar orientations of planar graphs. *Discrete and Computational Geometry*, 1(4):343–353, 1986.
5. R. Tamassia and I. G. Tollis. A unified approach to visibility representations of planar graphs. *Discrete and Computational Geometry*, 1(4):321–341, 1986.

SDE: Graph Drawing Using Spectral Distance Embedding*

Ali Civril, Malik Magdon-Ismail, and Eli Bocek-Rivele

Computer Science Department, RPI, 110 8th Street, Troy, NY 12180
{civria, magdon, boceke}@cs.rpi.edu

Abstract. We present a novel graph drawing algorithm which uses a spectral decomposition of the distance matrix to approximate the graph theoretical distances. The algorithm preserves symmetry and node densities, i.e., the drawings are aesthetically pleasing. The runtime for typical 20,000 node graphs ranges from 100 to 150 seconds.

1   Introduction

The graph drawing problem is to compute an aesthetically pleasing layout of vertices and edges so that it is easy to grasp visually the inherent structure of the graph. A general survey can be found in [4]. We consider only straight-line edge drawings, which reduces to embedding the vertices in two dimensions.

Spectral graph drawing, which has become popular recently [2, 3], produces a layout using the spectral decomposition of some matrix related to the vertex and edge sets. We present a spectral graph drawing algorithm SDE (Spectral Distance Embedding), in which we use the spectral decomposition of the graph theoretical distance matrix. We present the results of our algorithm through several examples, including run-times. Compared to similar techniques, we observe that our results achieve superior drawings, while at the same time not significantly sacrificing on computation time. The details of the algorithms, implementations and performance analysis can be found in an accompanying technical report [1].

The two stages in the algorithm are: (i) computing all-pairs shortest path (APSP) lengths; (ii) finding a rank-2 approximation to the matrix of squared distances $\mathbf{L}$. Step (i) involves a BFS for each node. For (ii), we use a standard procedure referred to as the power iteration to compute the eigenvalues and eigenvectors of $\mathbf{M} = -\frac{1}{2}\gamma\mathbf{L}\gamma$, where $\gamma = \mathbf{I}_n - \frac{1}{n}\mathbf{1}_n\mathbf{1}_n^T$. The complexity of the algorithm is $O(|V||E|)$, using $O(|V|^2)$ space (to store all the pair-wise distances). The algorithm is summarized below (PowerIteration returns the final coordinates specified by the top 2 eigenvectors of the input matrix, to a precision specified by ϵ). Let $\mathbf{D}$ be the matrix of distances.

SDE(G) 1: Use an APSP algorithm to compute $\mathbf{L}$, where $L_{ij} = D_{ij}^2$.

2: **return** $\mathbf{Y} = $ PowerIteration$(-\frac{1}{2}\gamma\mathbf{L}\gamma, \epsilon)$ % epsilon is a tolerance.

---

* This material is based upon work partially supported by the National Science Foundation under Grant Nos. 0323324, 0324947.

P. Healy and N.S. Nikolov (Eds.): GD 2005, LNCS 3843, pp. 512–513, 2005.

2 Results and Conclusion

The table below shows that SDE is reasonably fast for graphs up to 20,000 nodes. As can be seen from the following figures, it also produces aesthetically pleasing drawings of graphs varying in size, node density and degree of symmetry. Our algorithm has the advantages of exact (as opposed to iterative) computation of spectral graph drawing techniques and the quality of slower force-directed methods.

Graph	$|V|$	$|E|$	APSP time	Power Iteration time	Total time
jagmesh1	936	2664	0.10	0.11	0.21
Grid 50x50	2500	4900	0.65	0.54	1.19
3elt	4720	13722	4.67	3.80	8.47
whitaker3	9800	28989	25.24	8.18	33.42
sphere	16386	49152	106.96	29.73	136.69

The justification for SDE stems from the following (heuristically stated) theorem,

Theorem 1 (Theorem 3, [1]). *When the distance matrix is nearly embedable and satisfies some regularity conditions, SDE recovers (up to rotation) a close approximation to the optimal embedding.*

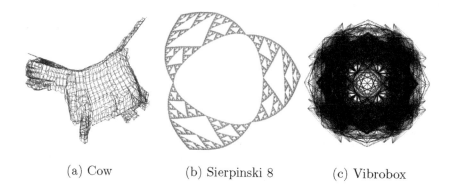

(a) Cow   (b) Sierpinski 8   (c) Vibrobox

References

1. A. Civril, M. Magdon-Ismail, and E. B. Rivele. SDE: Graph drawing using spectral distance embedding. Technical report, Rensselaer Polytechnic Institute, 2005.
2. D. Harel and Y. Koren. Graph drawing by high-dimensional embedding. In *GD02*, LNCS. Springer-Verlag, 2002.
3. Y. Koren, D. Harel, and L. Carmel. Drawing graphs by algebraic multigrid opti- mization. *Multiscale Modeling and Simulation*, 1(4):645–673, 2003. SIAM.
4. I. G. Tollis, G. D. Battista, P. Eades, and R. Tamassia. *Graph Drawing: Algorithms for the Visualization of Graphs*. Prentice Hall, 1999.

MultiPlane: A New Framework for Drawing Graphs in Three Dimensions*

Seok-Hee Hong

School of Information Technologies, University of Sydney, Australia
NICTA (National ICT Australia)
shhong@it.usyd.edu.au

This poster presents a new framework for drawing graphs in three dimensions, which can be used effectively to visualise large and complex real world networks.

The new framework uses a divide and conquer approach. More specifically, the framework divides a graph into a set of smaller subgraphs, and then draws each subgraph in a 2D plane using well-known 2D drawing algorithms. Finally, a 3D drawing of the whole graph is constructed by combining each drawing in a plane, satisfying defined criteria.

The framework is very flexible. Algorithms that follow this framework vary in computational complexity, depending on the type of graph and the optimisation criteria that are used. Specific instantiations of the framework require solutions to optimisation problems. A simple example of the framework is illustrated in Figure 1.

Our framework generalises some existing methods. For example, PolyPlane methods draw trees in 3D [6]. Another method is to use two and a half dimensional methods to visualise related networks in parallel planes [3, 4, 9].

Further, the design principle behind the framework also confirms Ware's guideline, a $2\frac{1}{2}$ design attitude that uses 3D depth selectively and pays special attention to 2D layout may provide the best match with the limited 3D capabilities of the human visual system [9].

As examined with the PolyPlane methods, the resulting drawing can reduce visual complexity and occlusion, and ease navigation. While rotating the drawing, some of the planes can be displayed as lines; this both reduce visual complexity and occlusion and allow the user concentrate on their plane of interest.

Preliminary results suggest that the framework can be useful for visual analysis and insight into large and complex networks such as hierarchical graphs and clustered graphs arising in social networks and biological networks domains. For details, see [2] for scale-free networks, [7, 8] for directed graphs and [5] for clustered graphs. Further, these methods are implemented in GEOMI, a visual analysis tool for large and complex networks [1].

Current work is to further develop the framework for various graph models and application domains.

---

* This research has been partially supported by a grant from the Australian Research Council.

P. Healy and N.S. Nikolov (Eds.): GD 2005, LNCS 3843, pp. 514–515, 2005.

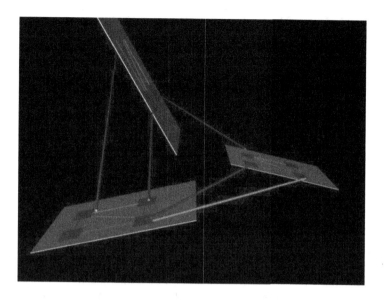

Fig. 1. The MultiPlane framework

References

1. A. Ahmed, T. Dwyer, M. Forster, X. Fu, J. Ho, S. Hong, D. Koschützki, C. Murray, N. Nikolov, A. Tarassov, R. Taib and K. Xu, GEOMI: GEOmetry for Maximum Insight, Proceedings of Graph Drawing 2005.
2. A. Ahmed, T. Dwyer, S.Hong, C. Murray, L. Song and Y. Wu, Visualisation and Analysis of Large and Complex Scale-free Networks, Proceedings of EuroVis 2005.
3. M. Baur, U. Brandes, M. Gaertler and D. Wagner, Drawing the AS Graph in 2.5 Dimensions, Proceedings of Graph Drawing 2004, Lecture Notes in Computer Science 3383, pp. 43-48, 2004.
4. U. Brandes, T. Dwyer and F. Schreiber, Visualizing Related Metabolic Pathways in Two and a Half Dimensions, Proceedings of Graph Drawing 2003, pp. 111-122, 2003.
5. J. Ho and S. Hong, Drawings Clustered Graphs in Three Dimensions, Proceedings of Graph Drawing 2005.
6. S. Hong and T. Murtagh, Visualization of Large and Complex Networks Using PolyPlane, Proceedings of Graph Drawing 2004, Lecture Notes in Computer Science 3383, pp. 471-482, Springer Verlag, 2004.
7. S. Hong and N. Nikolov, Layered Drawings of Directed Graphs in Three Dimensions, Proceedings of APVIS 2005: Asia Pacific Symposium on Information Visualisation, 2005, CPRIT 45, pp. 69-74, 2005.
8. S. Hong and N. Nikolov, Hierarchical Layouts of Directed Graphs in Three Dimensions, Proceedings of Graph Drawing 2005.
9. C. Ware, Designing with a 2 1/2D Attitude, Information Design Journal 10, 3, pp. 171-182, 2001.

Visualizing Graphs as Trees: Plant a Seed and Watch It Grow

Bongshin Lee, Cynthia Sims Parr,
Catherine Plaisant, and Benjamin B. Bederson

Human-Computer Interaction Laboratory,
Computer Science Department, University of Maryland,
College Park, MD 20742, USA
{bongshin, csparr, plaisant, bederson}@cs.umd.edu

Abstract. TreePlus is a graph browsing technique based on a tree-style layout. It shows the missing graph structure using interaction techniques and enables users to start with a specific node and incrementally explore the local structure of graphs. We believe that it supports particularly well tasks that require rapid reading of labels.

1 TreePlus: Interface for Visualizing Graphs as Trees

TreePlus[1] (Fig. 1) transforms graphs into trees by extracting a spanning tree. Users can navigate the tree by clicking on nodes in the main tree browser and preview adjacent nodes of the focus node in the adjacent nodes display (Fig. 2). Animation, zooming, panning, and integrated searching and browsing help users understand the graph.

When users move the cursor over a node, the node gets the focus and TreePlus previews its adjacent nodes on the right. They are left-aligned for readability, as in the main tree browser. Now node color and arrows are relative to the focus node, which is represented by the color green. They are duplicated on the right so that users can focus on one place instead of looking around the whole screen space; they are grouped and shown in gray.

When users select a node, TreePlus moves all of its adjacent nodes (except direct ancestors) to be its children by changing the tree structure. Multi-step animation helps users maintain context. A preview of how fruitful it would be to go down a path is provided by bar graphs showing how many organisms are reachable in each level (Fig. 3). TreePlus provides integrated support for search and lets users choose their own root. By setting a node as a root after search, users can easily gain access to an area of interest.

---

[1] For more information: see http://www.cs.umd.edu/hcil/graphvis

P. Healy and N.S. Nikolov (Eds.): GD 2005, LNCS 3843, pp. 516–518, 2005.

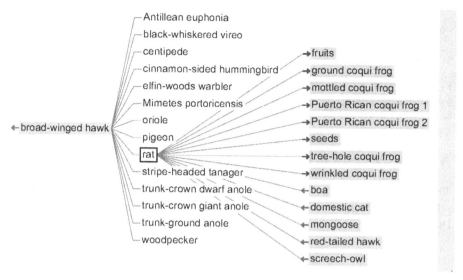

Fig. 1. For the selected node "rat" (distinguished by the black border), TreePlus shows all adjacent nodes as children (on the right), a parent, or an ancestor (both to the left). Node color and arrows help users see the direction of links; red left-pointing arrows mean these animals eat the selected node ("rat"), while blue right-pointing arrows mean the selected node ("rat") eats these animals.

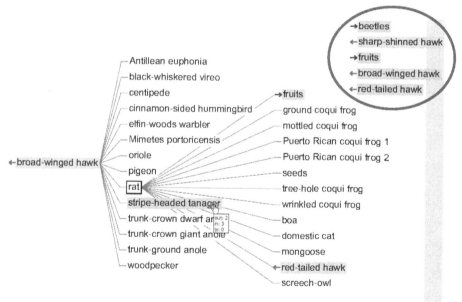

Fig. 2. On mouse over of the focus node, TreePlus lists adjacent nodes, circled here in red. Any adjacent nodes that were already displayed are shown in gray. Here the cursor is hovering over the "stripe-headed tanager," showing a preview of the connected nodes. Given the currently selected node "rat" and the focus node "stripe-headed tanager," the three gray nodes on the right, "fruits," "broad-winged hawk," and "red-tailed hawk" are connected to both nodes. A tool tip shows the number of links in each direction.

518   B. Lee et al.

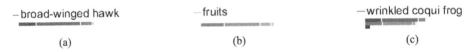

(a)   (b)   (c)

Fig. 3. Bars give a preview of how fruitful it would be to go down a path. "broad-winged hawk" is a start of a chain since it does not have a red bar (nothing eats it) and "fruits" is an end of a chain since it does not have a blue bar (fruits eat nothing).

Acknowledgements

This work is supported by NSF #0219492, Microsoft Research, and ARDA / Booz Allen Hamilton #MDA-904-03-c-0408 Novel Intelligence from Massive Data Program.

On Straightening Low-Diameter Unit Trees*

Sheung-Hung Poon

Department of Mathematics and Computer Science, TU Eindhoven,
5600 MB, Eindhoven, The Netherlands
spoon@win.tue.nl

A *polygonal chain* is a sequence of consecutively joined edges embedded in space. A *k-chain* is a chain of k edges. A *polygonal tree* is a set of edges joined into a tree structure embedded in space. A *unit tree* is a tree with only edges of unit length. A chain or a tree is *simple* if non-adjacent edges do not intersect.

We consider the problem about the reconfiguration of a simple chain or tree through a series of continuous motions such that the lengths of all tree edges are preserved and no edge crossings are allowed. A chain or tree can be *straightened* if all its edges can be aligned along a common straight line such that each edge points "away" from a designed leaf node. Otherwise it is called *locked*. Graph reconfiguration problems have wide applications in contexts including robotics, molecular conformation, rigidity and knot theory. The motivation for us to study unit trees is that for instance, the bonding-lengths in molecules are often similar, as are the segments of robot arms.

A chain in 2D can always be straightened [4, 5]. In 4D or higher, a tree can always be straightened [3]. There exist trees [2] in 2D and 5-chains in 3D that can lock. Alt et al. [1] showed that deciding the reconfigurability for trees in 2D and for chains in 3D is PSPACE-complete. However the problem of deciding straightenability for trees in 2D and for chains in 3D remains open.

It is easy to verify that a tree of diameter at most 3 in 2D or 3D can always be straightened. In this paper, we show that some tree of diameter 4 in 2D or 3D can lock, and a unit tree of diameter 4 in 2D can always be straightened.

In 2D, even a tree with diameter as low as 6 can lock [2] as shown in Figure 1 (*a*). We present a locked tree of diameter 4 in Figure 1 (*b*), which simulates the tree in (*a*). It can be shown locked using the same technique as the proof for (*a*) by assigning the corresponding equilibrium stresses to the tree edges. In 3D, a 5-chain can lock [2]. We present a 3D locked tree of diameter 4, which is shown in Figure 1 (*c*).

We now consider the straightenability of a unit tree T of diameter 4 in 2D. The *center* of tree T, denoted by o, is the middle vertex of any 4-chain in T. We call a path connecting the center to a leaf a *branch* of T. A *direct straightening* of branch $B = ouv$ in T means to rotate v around u until ouv is straightened by passing through the smaller angle. We denote the sweeping region for directly straightening B by $S(B)$. The direct straightening of B is *interfered* by another branch B' if $S(B) \cap B' \neq \emptyset$. There are two kinds of interferences depending on whether B and B' are of the same turn. We say that B' *follows* (resp. *covers*) B

---

* This research was supported in part by the Netherlands' Organisation for Scientific Research (NWO) under project no. 612-065-307.

P. Healy and N.S. Nikolov (Eds.): GD 2005, LNCS 3843, pp. 519–521, 2005.

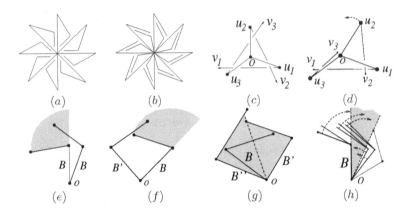

Fig. 1. $(a), (b)$ 2D locked trees. (c) A 3D locked tree of diameter 4, where $d(o, u_i) = 1$ & $d(u_i, v_i) \geq 2$. (d) Straightening unit-tree version of (c). (e) B' follows B. (f) B' covers B. (g) B'' covers B' & B' covers B. (h) Straightening uncovered branch B and all its following branches.

if B is interfered by B' of the same (resp. opposite) turn. See Figure 1 $(e), (f)$ for illustration.

Our algorithm to straighten T relies heavily on the observation of a nice nesting structure on covering relation. Suppose B'' covers B' which in turn covers B. Then B is nested inside the area enclosed by B' and B'', which is the shaded area as shown in Figure 1 (g). Therefore the last branch in a maximal covering sequence is always uncovered. Our algorithm proceeds by successively straightening an uncovered branch and all its following branches. The procedure to straighten an uncovered branch is shown in Figure 1 (h). The whole algorithm can be designed to run in $O(n)$ moves and $O(n \log n)$ time, where n is the number of tree edges.

In 3D, we conjecture that a unit tree of diameter 4 can always be straightened. In particular, it is not hard to see that the unit-tree version of Figure 1 (c) can be straightened. We first rotate v_1 around u_1 until $u_1 v_1$ is very close to ou_3, and then rotate v_3 around u_3 until v_3 is very close to o. Consequently we can rotate u_2 around o to draw $ou_2 v_2$ out. We further conjecture that a unit tree of any diameter in 2D or 3D can always be straightened.

References

1. H. Alt, C. Knauer, G. Rote, and S. Whitesides. The Complexity of (Un)folding. *Proc. 19th ACM Symp. on Comput. Geom. (SOCG)*, 164–170, 2003.
2. T. Biedl, E. Demaine, M. Demaine, S. Lazard, A. Lubiw, J. O'Rourke, S. Robbins, I. Streinu, G. Toussaint, and S. Whitesides. A Note on Reconfiguring Tree Linkages: Trees can Lock. *Disc. Appl. Math.*, 117:1-3, 293–297, 2002.
3. R. Cocan and J. O'Rourke. Polygonal Chains Cannot Lock in 4D. *Comput. Geom.: Theory & Appl.*, 20, 105–129, 2001.

4. R. Connelly, E.D. Demaine, and G. Rote. Straightening Polygonal Arcs and Convexifying Polygonal Cycles. *Disc. & Comput. Geom.*, 30:2, 205–239, 2003.
5. I. Streinu. A combinatorial approach for planar non-colliding robot arm motion planning. *Proc. 41st ACM Symp. on Found. of Comp. Sci. (FOCS)*, 443–453, 2000.

Mixed Upward Planarization - Fast and Robust[*]

Martin Siebenhaller and Michael Kaufmann

Universität Tübingen, WSI, Sand 14, 72076 Tübingen, Germany
{siebenha, mk}@informatik.uni-tuebingen.de

1 Introduction

In a mixed upward drawing of a graph $G = (V, E)$ all directed edges $E_D \subseteq E$ are represented by monotonically increasing curves. Mixed upward drawings arise in applications like UML diagrams where such edges denote a hierarchical structure. The mixed upward planarization is an important subtask for computing such drawings. We outline a fast and simple heuristic approach that provides a good quality and can be applied to larger graphs as before in reasonable time. Unlike other Sugiyama-style [4] approaches, the quality is comparable to the GT based approach [2] even if there are only few directed edges. Furthermore, the new approach is particularly suitable for extensions like clustering and swimlanes.

2 Planarization Approach

We assume that the subgraph induced by the directed edges is acyclic. Our heuristic approach consists of the following three steps:

1. Construction of an upward drawing: We use Sugiyama's approach [4] with a special layering strategy to construct an upward drawing of G including all edges. Common layering strategies are mainly optimized to produce short edges and involve a lot of crossings. Our new layering stems from the vertex ordering obtained in the first phase of the GT-heuristic. To calculate the ordering we use a variant that guarantees a monotonically increasing direction of the directed edges and runs in $O(|V|^2)$ [2]. The layer of a vertex corresponds to its position in this ordering. The crossing reduction is done by a layer-by-layer sweep. We apply a fast variant that keeps the number of dummy vertices small and runs in time $O((|V| + |E|) \log |E|)$ [3]. For the horizontal coordinate assignment we use a linear time approach based on the linear segments model [1].

2. Construction of an upward planar embedding: The embedding of the upward drawing of G can be constructed by detecting all crossings and replace them by dummy vertices. This can be done with a sweep-line approach in time $O((|E| + c) \log |E|)$, where c denotes the number of crossings.

3. Rerouting of undirected edges: Undirected edges are handled too restrictive because they can be routed non-monotonically. Hence we reroute each undirected edge with at least one crossing in randomized order using shortest

---

[*] This work has been supported by DFG-grant Ka812/8-2.

P. Healy and N.S. Nikolov (Eds.): GD 2005, LNCS 3843, pp. 522–523, 2005.

path computations in the extended dual graph. The runtime is $O(((|V| + c)|E|)$. Thus, the overall runtime of our new approach is $O(((|E|+c)\log|E|+(|V|+c)|E|)$.

3   Experiments

For our experiments we used connected directed acyclic random graphs with density 2. In the first experiment (Fig. 3(a)) we compared the number of crossings (using the median heuristic) induced by our new layering to the crossing number induced by a longest path layering, the popular simplex layering and a layering based on a topological sorting of the vertices where each vertex is assigned to exactly one layer. The results indicate that our improvement does not depend solely on the sparse layers. In the second experiment we compared our new approach to the GT based approach described in [2] that runs in $O(|V||E|^2 + (|V| + c)^2|E|)$. We compared random graphs where 1/4 (2/3) of the edges were directed. As the results in Fig. 3(b) show, the quality in terms of crossings is competitive to the GT based approach. Concerning the required time we clearly outperform the GT based approach (Fig. 3(c)). All experiments were performed on a Pentium 4 System, 3 GHz, 1024 MB RAM and Windows XP.

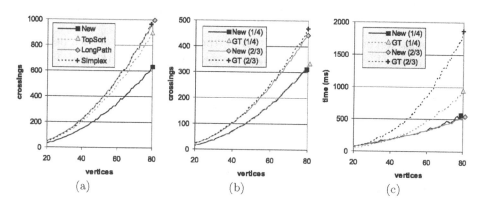

Fig. 1. Results of our experiments (averaged over 100 passes)

References

1. U. Brandes and B. Köpf: *Fast and Simple Horizontal Coordinate Assignment.* Proceedings of Graph Drawing 2001, Springer LNCS 2265, pp. 31–44, 2001.
2. M. Eiglsperger, M. Kaufmann and F. Eppinger: *An Approach for Mixed Upward Planarization.* J. Graph Algorithms Appl., 7(2): pp. 203–220, 2003.
3. M. Eiglsperger, M. Siebenhaller and M. Kaufmann: *An Efficient Implementation of Sugiyama's Algorithm for Layered Graph Drawing.* Proceedings of Graph Drawing 2004, Springer LNCS 3383, pp. 155–166, 2005.
4. K. Sugiyama, S. Tagawa and M. Toda: *Methods for Visual Understanding of Hierarchical System Structures.* IEEE Transactions on Systems, Man and Cybernetics, SMC-11(2): pp. 109–125, 1981.

Network Analysis and Visualisation[*]

Seok-Hee Hong

National ICT Australia, School of Information Technologies,
University of Sydney, Australia
shhong@it.usyd.edu.au

Abstract. A workshop on Network Analysis and Visualisation was held
on September 11, 2005 in Limerick Ireland, in conjunction with 2005
Graph Drawing conference. This report review the background, progress
and results of the Workshop.

1   Motivation

Recent technological advances produce many large and complex network models
in many domains. Examples include web-graphs, biological networks and social
networks.

Visualisation can be an effective tool for the understanding of such networks.
Good visualisation reveals the hidden structure of the networks and amplifies
human understanding, thus leading to new insights, new findings and possible
predictions for the future.

Analysis methods are available for these networks. However, analysis tools
for networks are not useful without visualisation, and visualisation tools are not
useful unless they are linked to analysis.

This workshop aimed to gather researchers interested in the analysis and
visualisation of large and complex networks. More specifically, the workshop
had the objectives identifying new research opportunities in network analysis
and visualisation, and encouraging collaborative solutions in this area.

2   Workshop Overview

The workshop was chaired by Seok-Hee Hong (National ICT Australia and
University of Sydney, Australia), and organised with Dorothea Wagner (Uni-
versity of Karlsruhe, Germany) and Michael Forster (National ICT Australia,
Australia).

It had the following 35 participants from 10 countries: Radoslav Andreev
(University of Limerick, Ireland), Vladimir Batagelj (University of Ljubljana,
Slovenia), Michael Baur (University of Karlsruhe, Germany),Elena Besussi (Uni-

---

[*] This workshop was supported by NICTA (National ICT Australia), funded by the
Australian Government's Backing Australia's Ability initiative, in part through the
Australian Research Council.

P. Healy and N.S. Nikolov (Eds.): GD 2005, LNCS 3843, pp. 524–527, 2005.

versity College London, UK), Ulrik Brandes (University of Konstanz, Germany), Stina Bridgeman (Hobart and William Smith Colleges, US), Ali Civril (Rensselaer Polytechnic Institute, US), Walter Didimo (University of Perugia, Italy), Tim Dwyer (Monash University, Australia), Jean-Daniel Fekete (INRIA Futurs, France), Michael Forster (National ICT Australia, Australia), Marco Gaertler (University of Karlsruhe, Germany), Francesco Giordano (University of Perugia, Italy), Luca Grilli (University of Perugia, Italy), Martin Harrigan (University of Limerick, Ireland), Patrick Healy (University of Limerick, Ireland), Nathalie Henry (INRIA Futurs, France), Joshua Ho (University of Sydney, Australia), Seok-Hee Hong (National ICT Australia and University of Sydney, Australia), Tony Huang (National ICT Australia and University of Sydney, Australia), Dirk Koschtzki (IPK Gatersleben, Germany), Karol Lynch (University of Limerick, Ireland), Anila Mjeda (Waterford Institute of Technology, Ireland), Nikola Nikolov (University of Limerick, Ireland), Andreas Noack (University of Cottbus, Germany), Aaron Quigley (University College Dublin, Ireland), Aimal Rextin (University of Limerick, Ireland), Falk Schreiber (IPK Gatersleben, Germany), Matthew Suderman (McGill University, Canada), Antonios Symvonis (National Technical University of Athens, Greece), Alexandre Tarassov (University of Limerick, Ireland), Ioannis Tollis (University of Crete, Greece), Francesco Trotta (University of Perugia, Italy), Dorothea Wagner (University of Karlsruhe, Germany), Michael Wybrow (Monash University, Australia).

3   Invited Talks

The workshop had two invited talks for one hour each as follows.

- **Vladimir Batagelj: Some Visualization Challenges from Social Network Analysis**
  Network = Graph + Data. The data can be measured or computed/derived from the network. In traditional graph drawing the goal was to produce *the best* layout of given graph. SNA (Social Network Analysis) is a part of data analysis. Its goal is to get insight into the structure and characteristics of given network. There is no single answer - usually we are trying to find interesting facts about the network and present them to the users. The visualization is a tool for network exploration and for presentation of the final results. This process requires combination of analysis and visualization techniques.

  SNA deals also with multi-relational, temporal and often large networks. The standard sheet of paper paradigm is often not appropriate for the amount of information in such networks. It was addressed that one should develop dynamic interactive layouts, introduce new visualization elements to represent typical network substructures, and add some artistic touch to final displays. The slides of the talk are available at:
  http://vlado.fmf.uni-lj.si/pub/networks/doc/mix/GDaSNA.pdf.

- **Ulrik Brandes: Visualization of Dynamic Social Networks**
  Social network visualization is a specific area in the field of information visualization. Models of social structures typically consist of some type of graph data (where variations are with respect to directedness, multiple edges, bipartiteness, etc.) together with an arbitrary set of attributes. While there are many unsolved problems in visualizing social structures, dynamic changes of the graph and/or its attributes introduce an additional dimension of complexity. In this talk, an attempt was made to point out directions relevant to social network analysts and open problems interesting to the graph drawing community. Several examples from recent social network studies were used for illustration.

4   Contributed Talks

The workshop had four contributed talks for 30 minutes each as follows.

- `Falk Schreiber: Visual Network Analysis for Systems Biology`
  Systems biology is a new field in biology that aims at the understanding of complex biological systems, such as a complete cell. It has emerged in the light of the availability of modern high-throughput technologies, which result in huge amounts of molecular data regarding life processes. This data is often related to, or even structured in, the form of biological networks such as metabolic, protein interaction and gene regulatory networks. Visual data exploration methods help scientists to extract information out of the data and thus are very useful for building sophisticated research tools. This presentation gave a brief introduction into systems biology and molecular biological processes. It discussed examples of the analysis of fundamental biological networks and their user-friendly visualisation. These examples range widely from visualising experimental data in the context of the underlying biological networks to structural analysis and subsequent visualisation of biological networks based on motifs, clustering and centralities. Finally new directions and questions in visualising these large and complex networks were considered.
- `Nathalie Henry: Matrices for Visualizing Social Networks`
  Visualizing and interacting with large social networks is a challenging task. Usually, social networks are represented as node-link diagrams. These representations are intuitive and effective for filtered or aggregated social networks. In this talk it was proposed to use matrix-based representations to interact and manipulate social networks prior to node-link diagram visualization. Adjacency matrix representations are easier to interact with in term of navigation and computation such as filtering, clustering or aggregation. These representations may offer the user a good overview of large and dense social networks. The main issue when dealing with matrices is their rows and columns ordering. Ordering is a key to understand a matrix or more generaly a table. An overview of existing methods to reorder tables was presented

with explanation of how automatic methods work in detail and how to apply these methods to adjacency matrices. Finally focus on matrix readability with experimental work was presented.

- **Tim Dwyer**: **New Techniques for Visualisation of Large and Complex Directed Graphs**
  To date, the famous Sugiyama algorithm has been the method of choice for drawing directed graphs. Prime examples are drawings of metabolic pathways and UML diagrams. However, when used to draw large and complex directed graphs - with hundreds or thousands of nodes and a high density of edges - the method does not produce very readable diagrams. This talk argued that a method based on techniques from the field of multi-dimensional scaling coupled with some custom constrained optimisation techniques scales much better to the visualisation of large and complex directed graphs.
- **Michael Wybrow**: **Visualisation of Constraint-Based Relationships in Graphs and Diagrams**
  Constraint-based relationships in diagrams are permanent placement relationships such as alignment, distribution, left-of, right-of, pinning, etc. They are maintained by the diagram editor as objects are added, removed or moved around. It is clear there should be an on-screen representation of such objects, but in practice these tend to quickly clutter the page and make comprehension of the graph and the constraint relationships themselves difficult. This talk demonstrated the problem and offered for discussion some possible solutions using proximity, transparency, time, and knowledge of active/broken constraints.

5   Discussion

The following open problems were suggested and discussed by the participants:

- Ulrik Brandes: Adapt Sugiyama paradigm (i.e. level assignment, crossing reduction, x-coordinates) with partially given y-coordinates
- Vladimir Batagelj:
  - New drawing styles and conventions
  - Interactive network layout
  - Matrix Layout
  - Generalized blockmodeling
  - Visualisation with additional graphical elements
  - Visualisation of multi-relational and dynamic networks
  - Dense directed network layout
- Jean-Daniel Fekete: Evaluation and usability studies of other network visualisation methods (besides node-link diagrams)

Graph-Drawing Contest Report

Christian A. Duncan[1], Stephen G. Kobourov[2], and Dorothea Wagner[3]

[1] University of Miami, Coral Gables, FL 33124, USA
duncan@cs.miami.edu
[2] University of Arizona, Tucson, AZ 85721, USA
kobourov@cs.arizona.edu
[3] Karlsruhe University, 76128 Karlsruhe, Germany
dwagner@ira.uka.de

Abstract. This report describes the Twelfth Annual Graph Drawing Contest, held in conjunction with the 2005 Graph Drawing Symposium in Limerick, Ireland. The purpose of the contest is to monitor and challenge the current state of graph-drawing technology.

1 Introduction

This year's graph drawing contest had three distinct tracks: the graph drawing challenge, the evolving graph contest, and the free-style contest. The graph drawing challenge took place during the conference. The challenge was straight-line crossing minimization of 6 graphs with 20-100 vertices. The contestants were given one hour and were free to use custom designed software or a provided program for manual graph editing, GraphMan. The evolving graph contest asked for visualizations, including animations and static images of the Internet Movie Database graph, emphasizing the evolving nature of the data. The free-style submission offered the opportunity for participants to present their best graph visualizations. Eight teams of one to three participants submitted graphs to the challenge. There were four submissions to the evolving graph contest and five submissions to the free-style contest.

2 Graph Drawing Challenge

Submissions to the challenge were measured by assigning scores between 0 and 6 to each participating team as follows: each of the six submitted graphs G_i was assigned a score: $s_i = \frac{min_i}{cur_i}$, where min_i was the minimal number of crossings found for the i-th graph and cur_i was the number of crossings in G_i. The individual scores for the six graphs were added to obtain the total score for each team. The contest committee awarded one first prize and three honorable mentions.

The first place winner was the team of Markus Chimani, Carsten Gutwenger and Karsten Klein using a custom made tool, `grapla`. The `grapla` tool was written by Andrei Grecu from Vienna University of Technology and was also used by last year's winning team. Three honorable mentions were given to Daniel Stefankovic, Michael Bennett, and the team of Michael Spriggs and Josh Liason. The first two used a combination approach of automated and manual crossing minimization, while the last used only the GraphMan graph editor.

P. Healy and N.S. Nikolov (Eds.): GD 2005, LNCS 3843, pp. 528–531, 2005.

3   Evolving Graph Contest

Four submissions for the evolving graph contest were received. The submissions consisted of papers, animations, SVG-files, and static images. The contest committee awarded first prize to Vlado Batagelj, Andrej Mrvar, Adel Ahmed, Xioyan Fu, Seokhee Hong, and Damian Merrick for their submission "Some Approaches to the Analysis and Visualization of the Internet Movie Database." They used (p, q)-cores, 4-rings, islands and time slices to extract interesting subnetworks of the Internet Movie Database. The densest such cores came from wrestling and adult movies, and some of the islands showed famous series like Charlie Brown, Starkes Team and Doña Macabra. Two visualizations were produced that showed the evolution of collaboration between Hollywood actors, musicians and even presidents, through animated transitions between force-directed layouts. Nodes and edges were assigned sizes and colors according to their relative importance in the graph; see Fig. 1.

Honorable mention was given to Ulrik Brandes, Martin Hoefer, and Christian Pick for their submission "Dynamic Egocentric Layout of Actor Biographies." It consisted of a dynamic visualization of the collaboration and work biography of a single actor, using the collaboration subgraph of that actor. The layouts were generated by implicitly placing the chosen actor in the middle of the drawing, while movies starring the actor were placed on annual rings that moved outwards. Collaborating actors were placed relative to the existing elements with sizes and

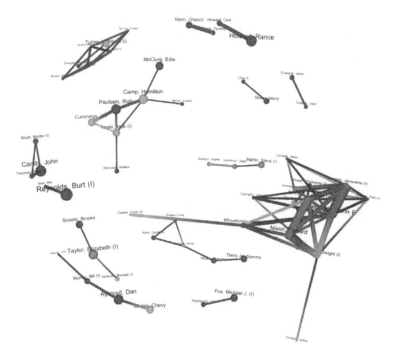

Fig. 1. Snapshot from the winning submission of the evolving graph contest

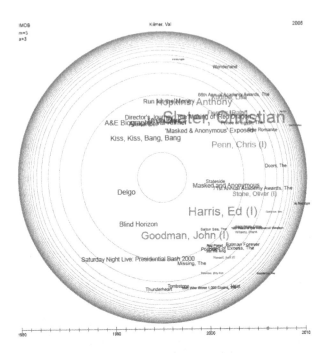

Fig. 2. Snapshot from the submission by Brandes *et al*

positions varying based on importance, age, and structural connectivity; see Fig. 2 for an example using the actor Val Kilmer.

Honorable mention was also given to Michael Baur, Marco Gaertler, and Robert Görke for their submission "Analyzing the Career of Actors (How to Become Famous Fast)." The submission was a Java3D application allowing interactive exploration of the migration of actors through different genres. Combining the genre paths of all actors in the database offers insight into the overall genre migration with an abstracted genre-time graph in space. The nodes and edges of their graph represent different genres at different points in time and migrations of actors between them, respectively. The graph is wrapped around a cylinder, where one axis represents time and the others the different genres; landmarks corresponding to particular movies are also used for ease of navigation; see Fig 3.

4   Free-Style Contest

Five submissions for the free-style category were received. The submissions consisted of individual drawings, SVG files, papers, and movies. The contest committee awarded first prize to Marco Gaertler and Markus Krug for their submission "Flying Through a Graph's Spectrum."

Spectral embeddings offer analytic insight into graph structure, e.g., revealing symmetries. However, traditional representation media restricts spectral embeddings to two or three dimensions. The submission consisted of animations,

Fig. 3. Snapshot from the submission by Baur *et al*

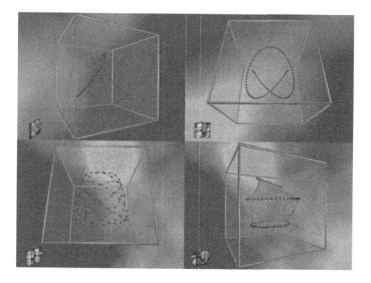

Fig. 4. Snapshot from the winning submission in the free-style contest

where layouts induced by different eigenvectors were morphed into each other. The animation showed many different embeddings in a short time, while offering aesthetically pleasing views. Fig. 4 shows four snapshots of the animation using a path with 100 nodes.

Acknowledgments

We would like to thank Intel and Dell for sponsoring the awards for the contest. We would also like to thank the organizers and all of the contestants for making this another successful contest.

Minimum Cycle Bases and Surface Reconstruction

(Abstract)

Kurt Mehlhorn

Max-Planck-Institut für Informatik, Saarbrücken, Germany

I report on recent work on minimum cycle basis in graphs and their application to surface reconstruction. The talk is based on joint work with C. Gotsmann, R. Hariharan, K. Kaligosi, T. Kavitha, D. Michail, K. Paluch, and E. Pyrga. I refer the reader to [KMMP04, KM05, HKM, GKM$^+$, MM05, Kav05] for details.

References

[GKM$^+$]   C. Gotsman, K. Kaligosi, K. Mehlhorn, D. Michail, and E. Pyrga. Cycle basis and surface reconstruction. in preparation.

[HKM]   R. Hariharan, T. Kavitha, and K. Mehlhorn. A faster deterministic algorithm for minimum cycle basis in directed graphs. `www.mpi-sb.mpg.de/~mehlhorn/ftp/ImprovedDirCycleBasis.ps`.

[Kav05]   T. Kavitha. An $\tilde{O}(m^2 n)$ Randomized Algorithm to compute a Minimum Cycle Basis of a Directed Graph In *ICALP, LNCS*, 2005.

[KM05]   T. Kavitha and K. Mehlhorn. A polynomial time algorithm for minimum cycle basis in directed graphs. In *STACS*, volume 3404 of *LNCS*, pages 654–665, 2005. `www.mpi-sb.mpg.de/~mehlhorn/ftp/DirCycleBasis.ps`.

[KMMP04]   T. Kavitha, K. Mehlhorn, D. Michail, and K. Paluch. A faster algorithm for minimum cycle bases of graphs. In *ICALP*, volume 3142 of *LNCS*, pages 846–857, 2004. `http://www.mpi-sb.mpg.de/~mehlhorn/ftp/MinimumCycleBasis.ps`.

[MM05]   K. Mehlhorn and D. Michail. Implementing minimum cycle basis algorithms. In *Experimental and Efficient Algorithms: 4th International Workshop, WEA 2005*, volume 3503 of *LNCS*, pages 32–XXX, 2005. `www.mpi-sb.mpg.de/~mehlhorn/ftp/CycleBasisImpl.pdf`.

P. Healy and N.S. Nikolov (Eds.): GD 2005, LNCS 3843, p. 532, 2005.

Hierarchy Visualization: From Research to Practice
(Abstract)

George G. Robertson

Microsoft Research, One Microsoft Way, Redmond, WA 98052, USA
ggr@microsoft.com

Hierarchy visualization has been a hot topic in the Information Visualization community for the last 15 years. A number of hierarchy visualization techniques have been invented, with each having advantages for some applications, but limitations or disadvantages for other applications. No technique has succeeded for a wide variety of applications. We continue to struggle with basic problems of high cognitive overhead (e.g., loss of context), poor fit to the data (e.g., problems of scale), and poor fit to the users task at hand (e.g., handling multiple points of focus). At the same time, information access improvements have made available to us much richer sources of information, including multiple hierarchies. In this talk, I will review what we know about hierarchy visualization, then describe our approach to visualization of multiple hierarchies with two techniques (Polyarchy Visualization and Schema Mapping), and conclude with lessons learned for basic hierarchy visualization and suggestions for future work.

P. Healy and N.S. Nikolov (Eds.): GD 2005, LNCS 3843, p. 533, 2005.
© Springer-Verlag Berlin Heidelberg 2005

Author Index

Lecture Notes in Computer Science

For information about Vols. 1–3767

please contact your bookseller or Springer

Vol. 3814: M. Maybury, O. Stock, W. Wahlster (Eds.), Intelligent Technologies for Interactive Entertainment. XV, 342 pages. 2005. (Sublibrary LNAI).

Vol. 3813: R. Molva, G. Tsudik, D. Westhoff (Eds.), Security and Privacy in Ad-hoc and Sensor Networks. VIII, 219 pages. 2005.

Vol. 3810: Y.G. Desmedt, H. Wang, Y. Mu, Y. Li (Eds.), Cryptology and Network Security. XI, 349 pages. 2005.

Vol. 3809: S. Zhang, R. Jarvis (Eds.), AI 2005: Advances in Artificial Intelligence. XXVII, 1344 pages. 2005. (Sublibrary LNAI).

Vol. 3808: C. Bento, A. Cardoso, G. Dias (Eds.), Progress in Artificial Intelligence. XVIII, 704 pages. 2005. (Sublibrary LNAI).

Vol. 3807: M. Dean, Y. Guo, W. Jun, R. Kaschek, S. Krishnaswamy, Z. Pan, Q.Z. Sheng (Eds.), Web Information Systems Engineering – WISE 2005 Workshops. XV, 275 pages. 2005.

Vol. 3806: A.H. H. Ngu, M. Kitsuregawa, E.J. Neuhold, J.-Y. Chung, Q.Z. Sheng (Eds.), Web Information Systems Engineering – WISE 2005. XXI, 771 pages. 2005.

Vol. 3805: G. Subsol (Ed.), Virtual Storytelling. XII, 289 pages. 2005.

Vol. 3804: G. Bebis, R. Boyle, D. Koracin, B. Parvin (Eds.), Advances in Visual Computing. XX, 755 pages. 2005.

Vol. 3803: S. Jajodia, C. Mazumdar (Eds.), Information Systems Security. XI, 342 pages. 2005.

Vol. 3802: Y. Hao, J. Liu, Y.-P. Wang, Y.-m. Cheung, H. Yin, L. Jiao, J. Ma, Y.-C. Jiao (Eds.), Computational Intelligence and Security, Part II. XLII, 1166 pages. 2005. (Sublibrary LNAI).

Vol. 3801: Y. Hao, J. Liu, Y.-P. Wang, Y.-m. Cheung, H. Yin, L. Jiao, J. Ma, Y.-C. Jiao (Eds.), Computational Intelligence and Security, Part I. XLI, 1122 pages. 2005. (Sublibrary LNAI).

Vol. 3799: M. A. Rodríguez, I.F. Cruz, S. Levashkin, M.J. Egenhofer (Eds.), GeoSpatial Semantics. X, 259 pages. 2005.

Vol. 3798: A. Dearle, S. Eisenbach (Eds.), Component Deployment. X, 197 pages. 2005.

Vol. 3797: S. Maitra, C. E. V. Madhavan, R. Venkatesan (Eds.), Progress in Cryptology - INDOCRYPT 2005. XIV, 417 pages. 2005.

Vol. 3796: N.P. Smart (Ed.), Cryptography and Coding. XI, 461 pages. 2005.

Vol. 3795: H. Zhuge, G.C. Fox (Eds.), Grid and Cooperative Computing - GCC 2005. XXI, 1203 pages. 2005.

Vol. 3794: X. Jia, J. Wu, Y. He (Eds.), Mobile Ad-hoc and Sensor Networks. XX, 1136 pages. 2005.

Vol. 3793: T. Conte, N. Navarro, W.-m.W. Hwu, M. Valero, T. Ungerer (Eds.), High Performance Embedded Architectures and Compilers. XIII, 317 pages. 2005.

Vol. 3792: I. Richardson, P. Abrahamsson, R. Messnarz (Eds.), Software Process Improvement. VIII, 215 pages. 2005.

Vol. 3791: A. Adi, S. Stoutenburg, S. Tabet (Eds.), Rules and Rule Markup Languages for the Semantic Web. X, 225 pages. 2005.

Vol. 3790: G. Alonso (Ed.), Middleware 2005. XIII, 443 pages. 2005.

Vol. 3789: A. Gelbukh, Á. de Albornoz, H. Terashima-Marín (Eds.), MICAI 2005: Advances in Artificial Intelligence. XXVI, 1198 pages. 2005. (Sublibrary LNAI).

Vol. 3788: B. Roy (Ed.), Advances in Cryptology - ASIACRYPT 2005. XIV, 703 pages. 2005.

Vol. 3787: D. Kratsch (Ed.), Graph-Theoretic Concepts in Computer Science. XIV, 470 pages. 2005.

Vol. 3785: K.-K. Lau, R. Banach (Eds.), Formal Methods and Software Engineering. XIV, 496 pages. 2005.

Vol. 3784: J. Tao, T. Tan, R.W. Picard (Eds.), Affective Computing and Intelligent Interaction. XIX, 1008 pages. 2005.

Vol. 3783: S. Qing, W. Mao, J. Lopez, G. Wang (Eds.), Information and Communications Security. XIV, 492 pages. 2005.

Vol. 3782: K.-D. Althoff, A. Dengel, R. Bergmann, M. Nick, T.R. Roth-Berghofer (Eds.), Professional Knowledge Management. XXIII, 739 pages. 2005. (Sublibrary LNAI).

Vol. 3781: S.Z. Li, Z. Sun, T. Tan, S. Pankanti, G. Chollet, D. Zhang (Eds.), Advances in Biometric Person Authentication. XI, 250 pages. 2005.

Vol. 3780: K. Yi (Ed.), Programming Languages and Systems. XI, 435 pages. 2005.

Vol. 3779: H. Jin, D. Reed, W. Jiang (Eds.), Network and Parallel Computing. XV, 513 pages. 2005.

Vol. 3778: C. Atkinson, C. Bunse, H.-G. Gross, C. Peper (Eds.), Component-Based Software Development for Embedded Systems. VIII, 345 pages. 2005.

Vol. 3777: O.B. Lupanov, O.M. Kasim-Zade, A.V. Chaskin, K. Steinhöfel (Eds.), Stochastic Algorithms: Foundations and Applications. VIII, 239 pages. 2005.

Vol. 3776: S.K. Pal, S. Bandyopadhyay, S. Biswas (Eds.), Pattern Recognition and Machine Intelligence. XXIV, 808 pages. 2005.

Vol. 3775: J. Schönwälder, J. Serrat (Eds.), Ambient Networks. XIII, 281 pages. 2005.

Vol. 3774: G. Bierman, C. Koch (Eds.), Database Programming Languages. X, 295 pages. 2005.

Vol. 3773: A. Sanfeliu, M.L. Cortés (Eds.), Progress in Pattern Recognition, Image Analysis and Applications. XX, 1094 pages. 2005.

Vol. 3772: M.P. Consens, G. Navarro (Eds.), String Processing and Information Retrieval. XIV, 406 pages. 2005.

Vol. 3771: J.M.T. Romijn, G.P. Smith, J. van de Pol (Eds.), Integrated Formal Methods. XI, 407 pages. 2005.

Vol. 3770: J. Akoka, S.W. Liddle, I.-Y. Song, M. Bertolotto, I. Comyn-Wattiau, W.-J. van den Heuvel, M. Kolp, J. Trujillo, C. Kop, H.C. Mayr (Eds.), Perspectives in Conceptual Modeling. XXII, 476 pages. 2005.

Vol. 3769: D.A. Bader, M. Parashar, V. Sridhar, V.K. Prasanna (Eds.), High Performance Computing – HiPC 2005. XXVIII, 550 pages. 2005.

Vol. 3768: Y.-S. Ho, H.J. Kim (Eds.), Advances in Multimedia Information Processing - PCM 2005, Part II. XXVIII, 1088 pages. 2005.